AF877074

Advances in Experimental Medicine and Biology

Volume 915

Series editors

Irun R. Cohen, Rehovot, Israel
N.S. Abel Lajtha, Orangeburg, USA
Rodolfo Paoletti, Milan, Italy
John D. Lambris, Philadelphia, USA

More information about this series at http://www.springer.com/series/5584

Mark C. Leake
Editor

Biophysics of Infection

Springer

Editor
Mark C. Leake
Biological Physical Sciences Institute
University of York
York
UK

ISSN 0065-2598 ISSN 2214-8019 (electronic)
Advances in Experimental Medicine and Biology
ISBN 978-3-319-32187-5 ISBN 978-3-319-32189-9 (eBook)
DOI 10.1007/978-3-319-32189-9

Library of Congress Control Number: 2016936570

© Springer International Publishing Switzerland 2016
This work is subject to copyright. All rights are reserved by the Publisher, whether the whole or part of the material is concerned, specifically the rights of translation, reprinting, reuse of illustrations, recitation, broadcasting, reproduction on microfilms or in any other physical way, and transmission or information storage and retrieval, electronic adaptation, computer software, or by similar or dissimilar methodology now known or hereafter developed.
The use of general descriptive names, registered names, trademarks, service marks, etc. in this publication does not imply, even in the absence of a specific statement, that such names are exempt from the relevant protective laws and regulations and therefore free for general use.
The publisher, the authors and the editors are safe to assume that the advice and information in this book are believed to be true and accurate at the date of publication. Neither the publisher nor the authors or the editors give a warranty, express or implied, with respect to the material contained herein or for any errors or omissions that may have been made.

Printed on acid-free paper

This Springer imprint is published by Springer Nature
The registered company is Springer International Publishing AG
The registered company address is: Gewerbestrasse 11, 6330 Cham, Switzerland

Preface

Traditional insight into the process of infection has stemmed largely either from conventional bulk ensemble average techniques in vitro or from imaging methods either fixed tissue samples or on living cells but restricted to standard limits of optical resolution. However, recently many state-of-the-art interdisciplinary techniques of modern biophysics have emerged which enable us to understand details of the mechanisms of infection far more clearly than before. Essentially, many of these new methods enhance both the spatial and temporal resolutions of data acquisition. This has enabled us to probe *dynamic* processes of infection directly, and at a precision comparable to the molecular length scale of the key processes involved. These emerging interfacial tools of biophysics include, for example, a range of single-molecule biophysics methods as well as super-resolution microscopy techniques. This volume of The Biophysics of Infection, in the Advances in Experimental Medicine and Biology series includes new protocols, reviews and original research articles for such emerging experimental and theoretical approaches, which have resulted in a substantial improvement to our understanding of the complex processes of infection.

December 2015

Mark C. Leake

Contents

Chapter 1
The Biophysics of Infection

Mark C. Leake

Abstract Our understanding of the processes involved in infection has grown enormously in the past decade due in part to emerging methods of biophysics. This new insight has been enabled through advances in interdisciplinary experimental technologies and theoretical methods at the cutting-edge interface of the life and physical sciences. For example, this has involved several state-of-the-art biophysical tools used in conjunction with molecular and cell biology approaches, which enable investigation of infection in living cells. There are also new, emerging interfacial science tools which enable significant improvements to the resolution of quantitative measurements both in space and time. These include single-molecule biophysics methods and super-resolution microscopy approaches. These new technological tools in particular have underpinned much new understanding of dynamic processes of infection at a molecular length scale. Also, there are many valuable advances made recently in theoretical approaches of biophysics which enable advances in predictive modelling to generate new understanding of infection. Here, I discuss these advances, and take stock on our knowledge of the biophysics of infection and discuss where future advances may lead.

Keywords Single-molecule biophysics · Super-resolution

This volume in the Advances in Experimental Medicine and Biology series consists of a collection of truly cutting-edge research studies, laboratory protocols, experimental and theoretical biophysical techniques and applications in use today by some of the leading international experts in the field of infection research. A key difference in emphasis with this volume compared with other earlier themed collections of infection research is on the emphasis on the utility of *interfacial methods,* which increase the underlying physiological relevance of infection investigation. These developments are manifested through applying methods such as single-molecule

M.C. Leake (✉)
Department of Physics and Biology, Biological Physical Sciences Institute (BPSI), University of York, York YO10 5DD, UK
e-mail: mark.leake@york.ac.uk

© Springer International Publishing Switzerland 2016

M.C. Leake (ed.), *Biophysics of Infection*, Advances in Experimental Medicine and Biology 915, DOI 10.1007/978-3-319-32189-9_1

cellular biophysics which strive to maintain the native physiological context through investigation of living cells (Leake 2013), especially experimental methods using emerging tools of optical microscopy (Wollman et al. 2015), as well as methods which combine in vivo, in vitro and computational approaches to probe biological process such as the interaction of proteins with DNA (Wollman et al. 2015), such as the use of fluorescence microscopy methods to probe functional, living cells, especially so using microbial systems as model organisms (Lenn et al. 2008; Plank et al. 2009; Chiu and Leake 2011; Robson et al. 2013; Bryan et al. 2014; Llorente-Garcia et al. 2014; Reyes-Lamothe et al. 2010; Badrinarayanan et al. 2012; Wollman and Leake 2015; Lenn and Leake 2015; Cordes et al. 2015). The length scale of precision of experimental protocols in this area has improved dramatically over recent years and many cutting-edge methods now utilize state-of-the-art single-molecule approaches, to enable imaging of biomolecule structure to a precision better than the standard optical resolution limit (Miller et al. 2015), as well as emerging biophysics tools which use single-molecule force spectroscopy (Leake et al. 2003, 2004, 2006; Linke and Leake 2004; Bullard et al. 2006). This volume also includes more complex representative methods to investigate infection through the use of advanced mathematical analysis and computation.

It is clear is that combining pioneering molecular biology, biochemistry, structural biology and genetics methods with emerging, exciting tools from the *younger* areas of biophysics, bioengineering, computer science and biomathematics, that our understanding of the processes of infection are being transformed. Improvements in all of these fields are likely to add yet more insight over the next years in the near future into the complex interactions between multiple key molecular players involved in infection.

Acknowledgements MCL was assisted by a Royal Society URF and research funds from the Biological Physical Sciences Institute (BPSI) of the University of York, UK.

References

Badrinarayanan A, Reyes-Lamothe R, Uphoff S et al (2012) In vivo architecture and action of bacterial structural maintenance of chromosome proteins. Science 338:528–531

Bryan SJ, Burroughs NJ, Shevela D et al. (2014) Localisation and interactions of the Vipp1 protein in cyanobacteria. Mol Microbiol

Bullard B, Benes V, Tzintzuni G, Leake MC, Linke WA, Oberhauser AF (2006) The molecular elasticity of the insect flight muscle proteins projectin and kettin. Proc. Natl. Acad. Sci. U S A. 103:4451–4456

Chiu S-W, Leake MC (2011) Functioning nanomachines seen in real-time in living bacteria using single-molecule and super-resolution fluorescence imaging. Int J Mol Sci 12:2518–2542

Cordes T, Moerner W, Orrit M, Sekatskii S, Faez S, Borri P, Prabal Goswami H, Clark A, El-Khoury P, Mayr S, Mika J, Lyu G, Cross D, Balzarotti F, Langbein W, Sandoghdar V, Michaelis J, Chowdhury A, Meixner AJ, van Hulst N, Lounis B, Stefani F, Cichos F, Dahan M, Novotny L, Leake M, Yang Frsc H. (2015) Plasmonics, tracking and manipulating, and living cells: general discussion. Faraday Discuss. In Press. doi:10.1039/C5FD90093J

Leake MC (2013) The physics of life: one molecule at a time. Philos Trans R Soc Lond B Biol Sci 368(1611):20120248

Leake MC, Wilson D, Bullard B, Simmons RM (2003) The elasticity of single kettin molecules using a two-bead laser-tweezers assay. FEBS Lett 535:55–60

Leake MC, Wilson D, Gautel M, Simmons RM (2004) The elasticity of single titin molecules using a two-bead optical tweezers assay. Biophys J 87:1112–1135

Leake MC, Grutzner A, Kruger M, Linke WA (2006) Mechanical properties of cardiac titin's N2B-region by single-molecule atomic force spectroscopy. J Struct Biol 155:263–272

Lenn T, Leake MC (2015) Single-molecule studies of the dynamics and interactions of bacterial OXPHOS complexes. Biochim Biophys Acta. pii: S0005-2728(15)00215-7. doi:10.1016/j.bbabio.2015.10.008

Lenn T, Leake MC, Mullineaux CW (2008) Are Escherichia coli OXPHOS complexes concentrated in specialized zones within the plasma membrane? Biochem Soc Trans 36:1032–1036

Linke WA, Leake MC (2004) Multiple sources of passive stress relaxation in muscle fibres. Phys Med Biol 49:3613–3627

Llorente-Garcia I, Lenn T, Erhardt H et al (2014) Single-molecule in vivo imaging of bacterial respiratory complexes indicates delocalized oxidative phosphorylation. Biochim Biophys Acta 1837:1811–1824

Miller H, Zhaokun Z, Wollman AJM et al. (2015) Superresolution imaging of single DNA molecules using stochastic photoblinking of minor groove and intercalating dyes. Methods

Plank M, Wadhams GH, Leake MC (2009) Millisecond timescale slimfield imaging and automated quantification of single fluorescent protein molecules for use in probing complex biological processes. Integr Biol A Quant Biosci Nano Macro 1:602–612

Reyes-Lamothe R, Sherratt DJ, Leake MC (2010) Stoichiometry and architecture of active DNA replication machinery in Escherichia coli. Science 328:498–501

Robson A, Burrage K, Leake MC (2013) Inferring diffusion in single live cells at the single-molecule level. Philos Trans R Soc Lond B Biol Sci 368:20120029

Wollman A, Leake MC (2015) Single Molecule Microscopy: Millisecond single-molecule localization microscopy combined with convolution analysis and automated image segmentation to determine protein concentrations in complexly structured, functional cells, one cell at a time. Faraday Discuss 184:401–424

Wollman AJM, Nudd R, Hedlund EG et al (2015a) From Animaculum to single molecules: 300 years of the light microscope. Open Biol 5:150019–150019

Wollman AJM, Miller H, Zhou Z et al (2015b) Probing DNA interactions with proteins using a single-molecule toolbox: inside the cell, in a test tube and in a computer. Biochem Soc Trans 43:139–145

Chapter 2
Single-Molecule Observation of DNA Replication Repair Pathways in *E. coli*

Adam J.M. Wollman, Aisha H. Syeda, Peter McGlynn and Mark C. Leake

Abstract The method of action of many antibiotics is to interfere with DNA replication—quinolones trap DNA gyrase and topoisomerase proteins onto DNA while metronidazole causes single- and double-stranded breaks in DNA. To understand how bacteria respond to these drugs, it is important to understand the repair processes utilised when DNA replication is blocked. We have used tandem *lac* operators inserted into the chromosome bound by fluorescently labelled *lac* repressors as a model protein block to replication in *E. coli*. We have used dual-colour, alternating-laser, single-molecule narrowfield microscopy to quantify the amount of operator at the block and simultaneously image fluorescently labelled DNA polymerase. We anticipate use of this system as a quantitative platform to study replication stalling and repair proteins.

Keywords Single-molecule · Super-resolution · Fluorescent protein · In vivo imaging · DNA repair

2.1 Introduction

2.1.1 Antibiotics Interfere with DNA Replication

Different types of antibiotics kill bacteria by interfering with DNA replication. In bacteria, a sophisticated complex of protein machinery, called the replisome, replicates DNA by unwrapping its double helix and using the two exposed single strands as templates for DNA synthesis creating a structure called the replication fork (Reyes-Lamothe et al. 2010). Failure to copy DNA completely or accurately results in potentially disastrous consequences for the cell. The antibiotic family of Quinolones bind to two bacterial complexes associated with DNA replication, DNA

A.J.M. Wollman (✉) · A.H. Syeda · P. McGlynn · M.C. Leake
Department of Physics and Biology, Biological Physical Sciences Institute, University of York, York YO10 5DD, UK
e-mail: adam.wollman@york.ac.uk

© Springer International Publishing Switzerland 2016

M.C. Leake (ed.), *Biophysics of Infection*, Advances in Experimental Medicine and Biology 915, DOI 10.1007/978-3-319-32189-9_2

gyrase and topoisomerase IV (Mustaev et al. 2014). These complexes remove positive DNA supercoils generated by the replisome and also disentangle intertwined sister chromosomes as replication proceeds. These reactions occur by binding of gyrase or topoisomerase IV to one DNA duplex, cleavage of that duplex and passage of another region of the chromosome through the break prior to sealing the break to reform an intact chromosome. The outcome of this complex reaction is the release of torsional strain or chromosome disentanglement (Drlica and Zhao 1997). Quinolones trap these topoisomerases on the DNA by stabilising a covalent protein-DNA complex that is a normal part of the reaction cycle for both gyrase and topoisomerase IV, generating a protein block to replication and disrupting the DNA architecture (Mustaev et al. 2014). Another antibiotic, metronidazole, disrupts replication by inducing single-strand and double-strand breaks on DNA in anaerobic pathogens (Edwards 1977). Metronidazole is readily reduced, creating DNA damaging compounds in anaerobes but is easily re-oxidised in aerobes. There is evidence of increased DNA repair in *Helicobacter pylori* when exposed to metronidazole (Goodwin et al. 1998).

2.1.2 Replisome Response to Blocks

DNA damage occurs naturally in *E. coli*, due to reactive oxygen species, chemicals and radiation causing double- and single- stranded breaks on DNA. There are also natural protein blocks to replication. Transcription occurs concurrently with DNA replication and as RNA polymerases are an order of magnitude slower than replisomes in bacteria, collisions can occur (McGlynn et al. 2012). RNA polymerases can also become stalled on the DNA by template damage, leading to the build-up of many polymerases (McGlynn et al. 2012). Thus, the replisome encounters many blocks to replication during the normal cell cycle and has been shown to pause frequently (Gupta et al. 2013). Many of these stalled replisomes can continue if the block is removed, which is advantageous, as reloading the replisome can lead to genome rearrangements (Syeda et al. 2014). However, replisomes lose functionality over time (Yeeles and Marians 2011) and so replisome reloading mechanisms are required, for when replisome barriers are not cleared sufficiently rapidly, prior to the blocked replisome losing activity (Duch et al. 2013).

DNA replication is initiated from *oriC* in a sequence-specific manner on the genome. However, the replisome can stall anywhere and so different reloading and re-initiation mechanisms are required for stalled replication forks that are DNA structure-rather than DNA sequence-specific. These mechanisms are not fully understood. Two proteins, PriA and PriC, can both reload the replicative helicase DnaB back onto replication fork structures. DnaB plays a central role in the replisome, unwinding the two DNA template strands and also acting as an organising hub for the entire replisome complex. PriA and PriC recognise different forked DNA structures that together represent all possible types of fork structure on the chromosome (Yeeles et al. 2013). *priA* and *priC* can be separately deleted from the genome but a knock-out mutant of both is not viable, thus these repair pathways are essential for cell survival. The in vivo dynamics of these proteins are unknown and there is evidence that DNA at forks needs processing by other proteins to allow repair or bypassing the block (Lecointe et al. 2007; Atkinson and McGlynn 2009). There is also an accessory helicase, Rep, which promotes the movement of replisomes through protein blockages on DNA (Guy et al. 2009; Boubakri et al. 2010).

It is therefore important to study blocks to replication not only to understand the effect of antibiotics but also to understand how DNA replication is successfully completed in the face of the many natural blocks to replisomes inside cells. To study stalled replication, we have used a model protein block to replication by inserting tandem binding sites (34 copies of *lacO*) for the *lacI* transcription factor into the *E. coli* genome and over expressing the LacI protein. The Lac repressor-operator complex mimics naturally occurring protein-DNA complexes and inhibits fork movement with an affinity typically encountered during genome duplication. Since the majority of forks continue through a single complex unhindered, multiple complexes are required to give detectable inhibition of fork movement (Payne et al. 2006; Guy et al. 2009). Studying replisomes stalled at these blocks is an ideal problem for single-molecule microscopy, as it requires observation of individual replication machineries at blocks in the natural cell environment and also the associated repair proteins.

2.1.3 Single-Molecule Fluorescence Microscopy

Fluorescent protein fusions can act as reporters to provide significant insight into a wide range of biological processes and molecular machines. They can be used to gain insight into stoichiometry and architecture as well as details of molecular mobility inside living, functional cells with their native physiological context intact (Lenn et al. 2008; Plank et al. 2009; Chiu and Leake 2011; Robson et al. 2013; Llorente-Garcia et al. 2014; Bryan et al. 2014; Corbes et al. 2015). These fusion proteins can be used in conjunction with single-molecule narrowfield microscopy, and its similar counterpart Slimfield microscopy, as a versatile tool to investigate a diverse range of protein dynamics in live cells to generate enormous insight into biological processes at the single-molecule level. It has been used in *E. coli* to investigate DNA replication by determining the stoichiometry of the components of the bacterial replisome (Reyes-Lamothe et al. 2010) and the proteins involved in the structural maintenance of chromosomes (Badrinarayanan et al. 2012).

In narrowfield microscopy, the normal fluorescence excitation field is reduced to encompass only a single cell and produce a Gaussian excitation field (~20 μm^2) with 100–1000 times the laser excitation intensity of standard epifluorescence microscopy. Using such intense illumination causes fluorophores to emit many more photons, greatly increasing the signal to noise. This allows millisecond timescale imaging of individual fluorescently labelled proteins in their native

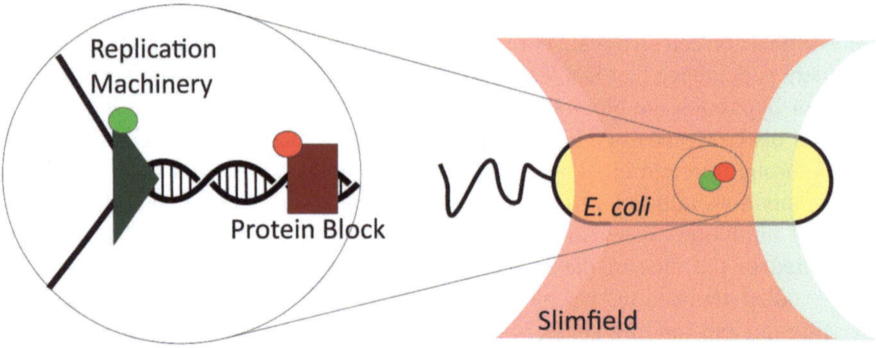

Fig. 2.1 Schematic of slimfield observation of fluorescently labelled replisome components encountering a fluorescently labelled protein block in *E. coli*

cellular environment. This time scale is fast enough to observe the diffusional motion of proteins and the dynamics which may occur around the replication fork.

We have labelled the *lac* operator replication block with a fluorescent *lac* repressor-mCherry fusion protein together with the dnaQ replisome component fused to the monomeric green fluorescent protein (GFP) allowing simultaneous imaging of the replisome and block. Using a bespoke narrowfield microscope, we have observed complexes of these proteins in live cells (see schematic in Fig. 2.1). To reduce the impact of autofluorescence caused by the blue GFP-excitation light, we have used high speed alternating-laser excitation (ALEX) to alternately excite each fluorophore at high speed. This enables the relatively dim mCherry protein to be observed without autofluorescence contamination and co-localised with GFP at high speed. Using custom software (Miller et al. 2015; Wollman et al. 2015a), we can quantify the number of fluorescently labelled proteins present in molecular complexes. Here, we demonstrate quantification of a replisome component and model protein replication block and show simultaneous imaging of both in the same live cell.

2.2 Methods

2.2.1 Generating Fluorescent Strains

2.2.1.1 Construction of Chromosomal *dnaQ-mGFP* Fusion

To create a *dnaQ-mGFP* C-terminal fusion, a PCR fragment containing mGFP and a downstream kanamycin resistance cassette amplified from pDHL580 (Landgraf et al. 2012) using primers oAS77 and oAS78 was recombineered as described (Datsenko and Wanner 2000) immediately downstream of *dnaQ* into PM300, a derivative of MG1655. The recombinants were selected for kanamycin resistance and successful integration was confirmed by PCR and subsequent sequencing. The verified strain was called AS217.

2.2.1.2 Construction of *LacI-mCherry* Fusion

A synthetic *lacI-mCherry* C-terminal fusion from plasmid pAS13 (Eurofins MWG operon synthesis) was subcloned into pBAD24 between *Nco*I and *Xba*I sites that placed it under the control of the arabinose inducible promoter. The presence and orientation of the construct was confirmed by restriction digestion, sequencing and phenotypic testing in a reporter strain. The verified plasmid was called pAS17.

2.2.1.3 Construction of Strain with Lac Repressor Array, *dnaQ-mGFP* and *LacI-mCherry*

dnaQ-mGFP was moved from AS217 to a laboratory stock strain AS249 carrying $lacO_{34}$ (Payne et al. 2006) by phage P1-mediated transduction. The transductants were selected for kanamycin resistance and presence of the *dnaQ-mGFP* allele was confirmed by PCR. The resulting $lacO_{34}$ *dnaQ-mGFP* strain (AS271) was transformed with pAS17 to create a dual-labelled strain with an inducible roadblock to replication.

2.2.2 Growing Strains and Inducing the Lac Repressor

Single colonies from transformation of AS271 with pAS17 were grown in 5 ml Luria-Bertani (LB) ampicillin and Isopropyl-β-D-thiogalactopyranoside (IPTG) in 15 ml culture tubes overnight. 1 ml of the overnight culture was washed twice with 1X 56 salts and inoculated into 10 ml 1X 56 salts together with ampicillin, glucose for growth and arabinose for Lac repressor induction and grown to an A_{650} of 0.4–0.6 (mid log phase). Concentrations of ampicillin, glucose, arabinose and IPTG were 100 μg/ml, 0.1 %, 0.02 % and 1 mM respectively. Cells from 1 ml of culture were resuspended in 100 μl of fresh 1X 56 salts medium for visualisation.

2.2.3 Fluorescence Microscopy

2.2.3.1 The Microscope

Our bespoke inverted fluorescence microscope was constructed from a Zeiss microscope body using a 100x TIRF 1.49 NA Olympus oil immersion objective lens and a *xyz* nano positioning stage (Nanodrive, Mad City Labs). Fluorescence excitation used 50 mW Obis 488 and 561 nm lasers, modulated using TTL pulses sent from National Instruments digital modulation USB module. A dual pass GFP/mCherry dichroic with 25 nm transmission windows centred on 525 and 625 nm was used underneath the objective lens turret. The beam was reduced 0.5x, to generate an excitation field of intensity ~6 Wcm^{-2}. The beam intensity profile was measured directly by raster scanning in the focal plane while imaging a sample of fluorescent beads. A high speed camera (iXon DV860-BI, Andor Technology, UK) was used to image at 5 ms/frame with the magnification set at ~50 nm per pixel. Laser emission was modulated such that each laser was on for 5 ms in alternating frames to give a 10 ms sampling time with 5 ms exposure time. The camera CCD was split between a GFP and mCherry channel using a bespoke colour splitter consisting of a dichroic centred at pass wavelength 560 nm and emission filters with 25 nm bandwidths centred at 525 and 594 nm. The microscope was controlled using our in-house bespoke LabVIEW (National Instruments) software.

2.2.3.2 Preparing Samples and Obtaining Fluorescence Data

E. coli cells were imaged on agarose pads suffused with media (Reyes-Lamothe et al. 2010). In brief, gene frames (Life Technologies) were stuck to a glass microscope slide to form a well and 500 μl 56 salts media plus 1 % agarose was pipetted into the well. The pad was left to dry at room temperature before 5 μl *E. coli* culture was pipetted in 6–10 droplets onto the pad. This was covered with a plasma-cleaned glass coverslip and imaged immediately. For each sample 10–30 cells were imaged in fluorescence and brightfield.

2.2.4 Analysing the Data

Single fluorescent proteins or complexes of proteins can be considered point sources of light and so appear as spatially extended spots in a fluorescence image due to diffraction by the microscope optics (Wollman et al. 2015b). Narrowfield fluorescence microscopy data consists of a time-series of images of spots which require *in silico* analysis to track each spot. We used custom Matlab™ tracking software to automatically identify spots, quantify them and link them into trajectories (Miller et al. 2015; Wollman et al. 2015a). The software identifies candidate bright spots using a combination of pixel intensity thresholding and image transformation. The threshold is set using the pixel intensity histogram as the full width half maximum of the peak in the histogram which corresponds to background pixels. A series of morphological transformations including erosion and dilation is applied to the thresholded image to remove individual bright pixels due to noise and leave a single pixel at each candidate spot co-ordinate. The intensity centroid of candidate spots is found using iterative Gaussian masking (Thompson et al. 2002) and the characteristic intensity is defined as the sum of the pixel intensities inside a 5 pixel radius region of interest around the spot minus the local background (Xue and Leake 2009) and corrected for non-uniformity in the excitation field. If this spot is above a pre-set signal to noise ratio, defined as the characteristic intensity divided by the standard deviation of the local background, it is accepted. Trajectories are formed by linking together spots in adjacent frames based on their proximity and intensity.

The number of fluorophores present in a molecular complex is determined by dividing its intensity by the intensity of a single fluorophore. The characteristic intensity of a single fluorophore can either be determined from in vitro measurements of purified fluorophore or from the in vivo data itself using the intensity of spots found after bleaching the cell.

2.3 Results and Discussion

2.3.1 Quantifying dnaQ

The single labelled dnaQ-GFP strain was imaged using narrowfield microscopy. An example cell is shown in Fig. 2.2. The brightfield image of the cell is shown in Fig. 2.2a and the fluorescence image of dnaQ-GFP shown in Fig. 2.2b. Two spots of dnaQ can be seen in the fluorescence image corresponding to the two copies of the replisome, consistent with previous observations (Reyes-Lamothe et al. 2010). Spots found by software over all frames are shown as green and blue circles in Fig. 2.2a with their intensity values plotted against time in Fig. 2.2c in units of characteristic GFP intensity. The spots have a stoichiometry of 3 dnaQ-GFP per replisome, consistent with previous observations (Reyes-Lamothe et al. 2010).

The distribution of dnaQ replisome stoichiometries was obtained from a kernel density estimation and is shown in Fig. 2.3. The stoichiometry peaks at 2 and ranges up to 6 dnaQ per replisome. This agrees well with previous observations of 2–3 per replication fork. These forks appear to be overlapping when replication is initiated from the origin leading to the observation of double stoichiometries. These results combined with recent measures of the total copy number of dnaQ (Wollman and Leake 2015) are in good agreement with a previous study which labelled dnaQ with the Ypet fluorophore. It has been suggested that the fluorophore used in a fusion protein can effect the stoichiometry of native complexes (Landgraf et al. 2012), but here we observe no difference between Ypet and monomeric GFP fusions.

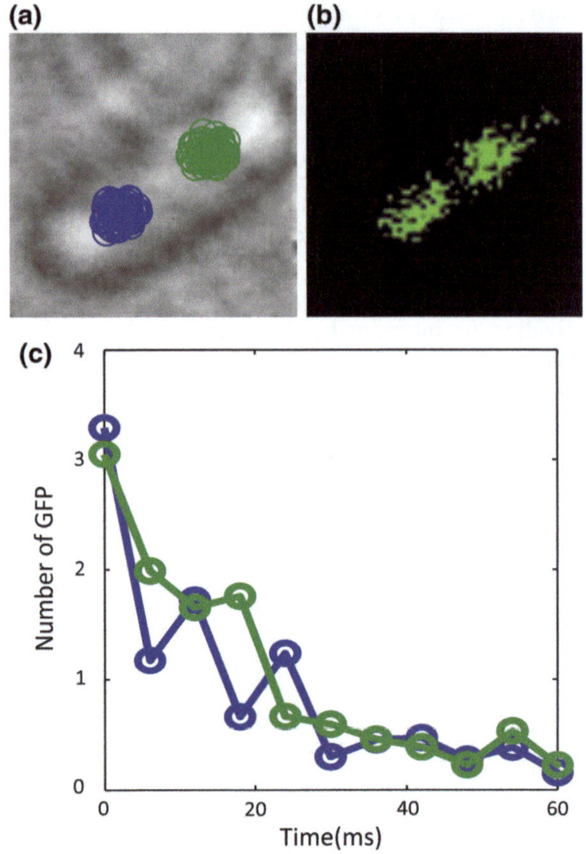

Fig. 2.2 **a** Brightfield image of an *E. coli* cell with tracked dnaQ-GFP overlaid, **b** fluorescence micrograph of dnaQ-GFP, **c** intensity of each spot over time in GFPs

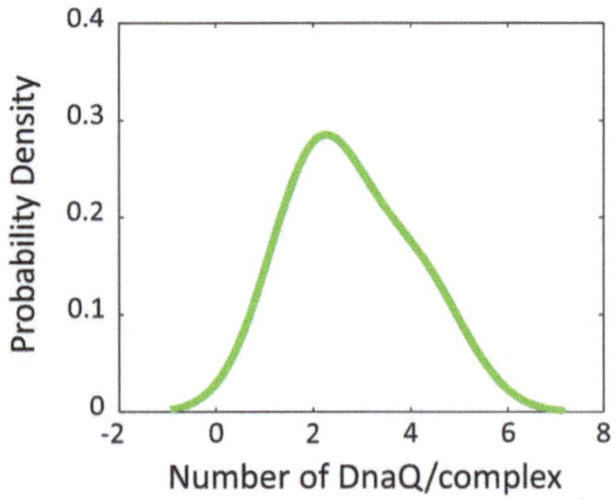

Fig. 2.3 Kernal density estimation of the number of dnaQ-GFP per spot

2.3.2 Quantifying LacI

We then imaged the *lac* operator blocks. This required optimisation of the growth conditions and expression levels as the fluorescently labelled *lac* protein is not endogenously expressed in this strain. Cells were grown in minimal media so that growth is slowed and there is, on average, only one replisome per cell. This not only eliminates the noise caused by LB autofluorescence but also the signal from multiple replisomes. Thus, having a single replisome greatly simplifies its tracking on the chromosome when it encounters the block and also makes downstream analysis easier by eliminating complexities due to multiple factors coming into play.

The results are shown in Fig. 2.4, with a brightfield image in Fig. 2.4a and mCherry fluorescence image in Fig. 2.4b. Two mCherry spots are seen in the fluorescence image, consistent with the *lac* operator sites having been replicated. All spots found over time are marked as red circles in Fig. 2.4a and their intensity plotted over time in Fig. 2.4c in units of mCherry intensities. The stoichiometry of the complexes is much lower than the 34 possible sites on the DNA and is closer to 5–10. This is unlikely to be caused by low expression levels as there is a significant diffuse background in the cell from unbound LacI-mCherry molecules. These results imply that the *lac* operators are not saturated with repressor. Further study is needed to understand the basis of this lack of saturation. The expression level could be varied and the number of potential binding sites on the DNA changed.

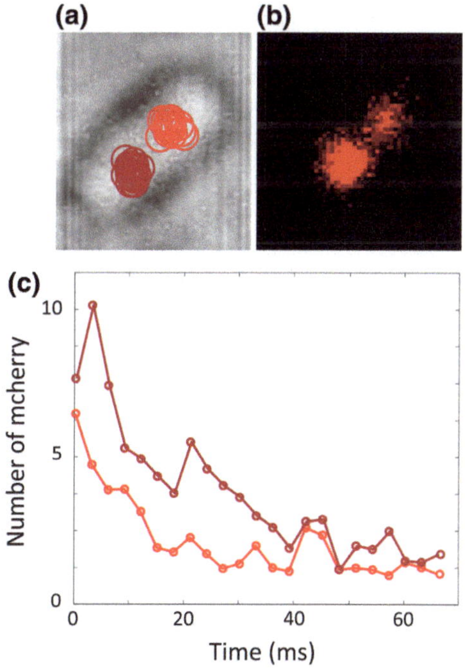

Fig. 2.4 **a** Brightfield image of an *E. coli* cell with tracked LacI-mCherry overlaid, **b** fluorescence micrograph of LacI-mcherry, **c** intensity of each spot over time in mcherrys

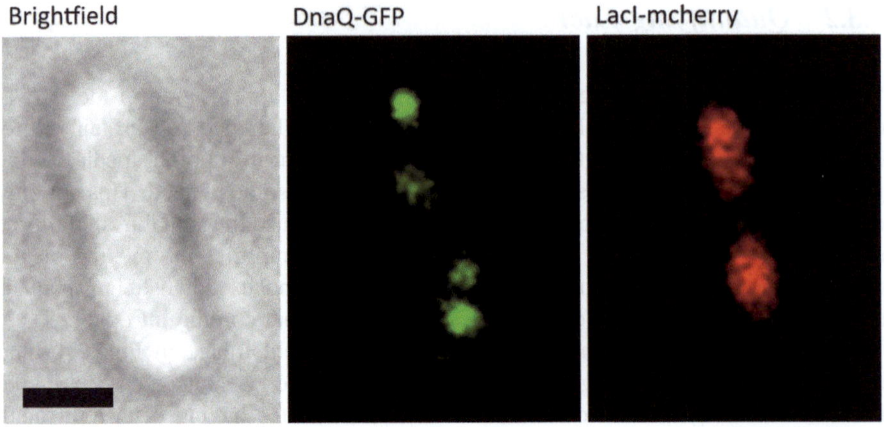

Fig. 2.5 *Left* brightfield image of an *E. coli* cell, *middle* fluorescence micrograph of dnaQ-GFP, *right* fluorescence micrograph of LacI-mCherry in the same cell

2.3.3 Dual-Colour Experiments

The *lac* operator block plus LacI-mCherry has been incorporated into the dnaQ-GFP strain and preliminary data obtained. Figure 2.5 shows brightfield and fluorescence micrographs of the dual-labelled strain. Our intention is to use this strain as a platform to study stalled replication by observing the behaviour of the replisome as it encounters different blocks with varying numbers of *lac* operators. This system could also be used to study repair proteins and could be combined with three colour microscopy, labelling the dnaQ with CyPet, the repair protein with Ypet and retaining the mCherry labelled *lac* operator array.

2.4 Summary

We have used tandem *lac* operators inserted into the chromosome bound by fluorescently labelled *lac* repressors as a model protein block to replication in *E. coli*. This block is a model for the action of some antibiotics such as quinolones which trap gyrases and topoisomerases on DNA. We have used dual-colour, alternating-laser, single-molecule narrowfield microscopy to quantify the amount of operator at the block and simultaneously image fluorescently labelled DNA polymerase. This quantitative platform for studying replication stalling will underpin future investigations.

References

Atkinson J, McGlynn P (2009) Replication fork reversal and the maintenance of genome stability. Nucleic Acids Res 37:3475–3492. doi:10.1093/nar/gkp244

Badrinarayanan A, Reyes-Lamothe R, Uphoff S et al (2012) In vivo architecture and action of bacterial structural maintenance of chromosome proteins. Science 338(80):528–31. doi: 10.1126/science.1227126

Boubakri H, de Septenville AL, Viguera E, Michel B (2010) The helicases DinG, Rep and UvrD cooperate to promote replication across transcription units in vivo. EMBO J 29:145–157. doi:10.1038/emboj.2009.308

Bryan SJ, Burroughs NJ, Shevela D et al (2014) Localisation and interactions of the Vipp1 protein in cyanobacteria. Mol Microbiol. doi:10.1111/mmi.12826

Chiu S-W, Leake MC (2011) Functioning nanomachines seen in real-time in living bacteria using single-molecule and super-resolution fluorescence imaging. Int J Mol Sci 12:2518–2542. doi:10.3390/ijms12042518

Cordes T, Moerner W, Orrit M, Sekatskii S, Faez S, Borri P, Prabal Goswami H, Clark A, El-Khoury P, Mayr S, Mika J, Lyu G, Cross D, Balzarotti F, Langbein W, Sandoghdar V, Michaelis J, Chowdhury A, Meixner AJ, van Hulst N, Lounis B, Stefani F, Cichos F, Dahan M, Novotny L, Leake M, Yang Frsc H (2015) Plasmonics, tracking and manipulating, and living cells: general discussion. Faraday Discuss 184:451–473. doi:10.1039/c5fd90093j

Datsenko KA, Wanner BL (2000) One-step inactivation of chromosomal genes in Escherichia coli K-12 using PCR products. Proc Natl Acad Sci USA 97:6640–6645. doi:10.1073/pnas.120163297

Drlica K, Zhao X (1997) DNA gyrase, topoisomerase IV, and the 4-quinolones. Microbiol Mol Biol Rev 61:377–392

Duch A, Felipe-Abrio I, Barroso S et al (2013) Coordinated control of replication and transcription by a SAPK protects genomic integrity. Nature 493:116–119. doi:10.1038/nature11675

Edwards DI (1977) The action of metronidazole on DNA. J Antimicrob Chemother 3:43–48. doi:10.1093/jac/3.1.43

Goodwin A, Kersulyte D, Sisson G et al (1998) Metronidazole resistance in Helicobacter pylori is due to null mutations in a gene (rdxA) that encodes an oxygen-insensitive NADPH nitroreductase. Mol Microbiol 28:383–393. doi:10.1046/j.1365-2958.1998.00806.x

Gupta MK, Guy CP, Yeeles JTP et al (2013) Protein-DNA complexes are the primary sources of replication fork pausing in Escherichia coli. Proc Natl Acad Sci USA 110:7252–7257. doi:10.1073/pnas.1303890110

Guy CP, Atkinson J, Gupta MK et al (2009) Rep provides a second motor at the replisome to promote duplication of protein-bound DNA. Mol Cell 36:654–666. doi:10.1016/j.molcel.2009.11.009

Landgraf D, Okumus B, Chien P et al (2012) Segregation of molecules at cell division reveals native protein localization. Nat Methods 9:480–482. doi:10.1038/nmeth.1955

Lecointe F, Sérèna C, Velten M et al (2007) Anticipating chromosomal replication fork arrest: SSB targets repair DNA helicases to active forks. EMBO J 26:4239–4251. doi:10.1038/sj.emboj.7601848

Lenn T, Leake MC, Mullineaux CW (2008) Are Escherichia coli OXPHOS complexes concentrated in specialized zones within the plasma membrane? Biochem Soc Trans 36:1032–1036. doi:10.1042/BST0361032

Llorente-Garcia I, Lenn T, Erhardt H et al (2014) Single-molecule in vivo imaging of bacterial respiratory complexes indicates delocalized oxidative phosphorylation. Biochim Biophys Acta 1837:811–824. doi:10.1016/j.bbabio.2014.01.020

McGlynn P, Savery NJ, Dillingham MS (2012) The conflict between DNA replication and transcription. Mol Microbiol 85:12–20. doi:10.1111/j.1365-2958.2012.08102.x

Miller H, Zhaokun Z, Wollman AJM, Leake MC (2015) Superresolution imaging of single DNA molecules using stochastic photoblinking of minor groove and intercalating dyes. Methods. doi:10.1016/j.ymeth.2015.01.010

Mustaev A, Malik M, Zhao X et al (2014) Fluoroquinolone-gyrase-DNA complexes two modes of drug binding. J Biol Chem 289:12300–12312. doi:10.1074/jbc.M113.529164

Payne BTI, van Knippenberg IC, Bell H et al (2006) Replication fork blockage by transcription factor-DNA complexes in Escherichia coli. Nucleic Acids Res 34:5194–5202. doi:10.1093/nar/gkl682

Plank M, Wadhams GH, Leake MC (2009) Millisecond timescale slimfield imaging and automated quantification of single fluorescent protein molecules for use in probing complex biological processes. Integr Biol (Camb) 1:602–612. doi:10.1039/b907837a

Reyes-Lamothe R, Sherratt DJ, Leake MC (2010) Stoichiometry and architecture of active DNA replication machinery in Escherichia coli. Science 328:498–501. doi:10.1126/science.1185757

Robson A, Burrage K, Leake MC (2013) Inferring diffusion in single live cells at the single-molecule level. Philos Trans R Soc Lond B Biol Sci 368:20120029. doi:10.1098/rstb.2012.0029

Syeda AH, Hawkins M, McGlynn P (2014) Recombination and replication. Cold Spring Harb Perspect Biol 6:a016550. doi:10.1101/cshperspect.a016550

Thompson RE, Larson DR, Webb WW (2002) Precise nanometer localization analysis for individual fluorescent probes. Biophys J 82:2775–2783. doi:10.1016/S0006-3495(02)75618-X

Wollman A, Leake MC (2015) FD2015-Single Molecule Microscopy: Millisecond single-molecule localization microscopy combined with convolution analysis and automated image segmentation to determine protein concentrations in complexly structured, functional cells, one cell at a time. Faraday Discuss 184:401–424. doi:10.1039/C5FD00077G

Wollman AJM, Miller H, Zhou Z, Leake MC (2015a) Probing DNA interactions with proteins using a single-molecule toolbox: inside the cell, in a test tube and in a computer. Biochem Soc Trans 43:139–145. doi:10.1042/BST20140253

Wollman AJM, Nudd R, Hedlund EG, Leake MC (2015b) From Animaculum to single molecules: 300 years of the light microscope. Open Biol 5:150019–150019. doi:10.1098/rsob.150019

Xue Q, Leake MC (2009) A novel multiple particle tracking algorithm for noisy in vivo data by minimal path optimization within the spatio-temporal volume. In: Proceedings—2009 IEEE International symposium on biomedical imaging: from nano to macro, ISBI 2009, pp 1158–1161

Yeeles JTP, Marians KJ (2011) The Escherichia coli replisome is inherently DNA damage tolerant. Science 334:235–238. doi:10.1126/science.1209111

Yeeles JTP, Poli J, Marians KJ, Pasero P (2013) Rescuing stalled or damaged replication forks. Cold Spring Harb Perspect Biol 5:a012815. doi:10.1101/cshperspect.a012815

Chapter 3
Investigating the Swimming of Microbial Pathogens Using Digital Holography

K.L. Thornton, R.C. Findlay, P.B. Walrad and L.G. Wilson

Abstract To understand much of the behaviour of microbial pathogens, it is necessary to image living cells, their interactions with each other and with host cells. Species such as *Escherichia coli* are difficult subjects to image: they are typically microscopic, colourless and transparent. Traditional cell visualisation techniques such as fluorescent tagging or phase-contrast microscopy give excellent information on cell behaviour in two dimensions, but no information about cells moving in three dimensions. We review the use of digital holographic microscopy for three-dimensional imaging at high speeds, and demonstrate its use for capturing the shape and swimming behaviour of three important model pathogens: *E. coli*, *Plasmodium* spp. and *Leishmania* spp.

Keywords Optical microscopy · Holography · Image analysis · Leishmania · Plasmodium

3.1 Introduction

Microbial pathogens are responsible for the majority of annual mortality and morbidity. A high-profile example is malaria, which infects around 200 million people, primarily in the developing world (World Health Organization 2014).

K.L. Thornton · R.C. Findlay · L.G. Wilson (✉)
Department of Physics, University of York, Heslington, York YO10 5DD, England
e-mail: laurence.wilson@york.ac.uk

K.L. Thornton
e-mail: klt531@york.ac.uk

R.C. Findlay
e-mail: rcf511@york.ac.uk

R.C. Findlay · P.B. Walrad
Centre for Immunology and Infection, Department of Biology, University of York, Heslington, York YO10 5DD, England
e-mail: pegine.walrad@york.ac.uk

© Springer International Publishing Switzerland 2016
M.C. Leake (ed.), *Biophysics of Infection*, Advances in Experimental Medicine and Biology 915, DOI 10.1007/978-3-319-32189-9_3

Pathogens responsible for malaria and other diseases have diverse evolutionary histories found in both the bacterial and eukaryotic domains.[1] For many pathogens, cell motility is an important means of facilitating infection. The ability to move allows extension of territory, search for nutrients and avoidance of host immune response.

Studies of pathogenic cell motility have long been dictated by the experimental tools available. Optical microscopes have been central in this effort, and have undergone extensive refinement. Microscope objective lenses are now highly optimised, and allow the operator to image objects separated by as little as half a wavelength—a limitation set by the fundamental physics of waves and known as the diffraction limit. Advances in contrast generation techniques such as Zernike phase contrast (Born and Wolf 2005) or differential interference contrast (Allen 1985) have improved image fidelity, and biochemistry has played a role in developing new variant markers and stains to facilitate the visualisation of microscopic subjects under fluorescent illumination. Recently, super-resolution techniques have provided a way to beat the diffraction limit, imaging structures in the 10 nm size range, through the combined advancement of laser illumination with digital image sensors and computing technology.

In fact, digital sensors and high-performance microscopes are nearly ubiquitous in modern microbiology labs. This equipment is often used for routine sample screening and documentation, but provides other opportunities for quantitative measurements. Video tracking of particles or cells (Crocker and Grier 1996) gives a single-cell resolution picture of motility, while scattering-based methods can assay motility in a larger population (Wilson et al. 2011; Colin et al. 2014). Another example is digital inline holographic microscopy (DIHM). Although this method was first described some 20 years ago (Schnars and Jüptner 1994), computing power has recently developed to the stage that DIHM can be routinely implemented in desktop machines. There are two distinct implementations of this method: a quantitative phase contrast method to assess the axial thickness of microscopic subjects (Cuche et al. 1999; Park et al. 2009; Edwards et al. 2014); and a method for three-dimensional localisation and imaging of objects (Xu et al. 2001; Fung et al. 2011; Su et al. 2012; Merola et al. 2013). This paper concentrates on the latter, and its application to microbial pathogens.

Different forms of cellular motility are observed among microbial cells, with the exact mode dictated to some extent by the cell's environment. Within one species, several different methods for movement may exist, triggered by distinct environmental context. *E. coli*, implicated in over half of urinary tract infections in a recent study based in Canada (Foxman 2010), demonstrates a strong connection between motility and ability to infect patients. *E. coli* cells are capable of swimming freely in a liquid, propelled by a helical bundle of thin flagellar filaments. When present on surfaces in sufficient numbers, the cells grow longer and produce more flagella, becoming hyper-flagellated swarmer cells that swim at increased speeds within

[1]Although, intriguingly, not yet among the archaea (Cavicchioli et al. 2003).

surface fluid films. These modes of motion—certainly swimming, and possibly swarming too—can be modulated by a chemical-sensing response (chemotaxis). This allows cells to navigate to more favourable environments. A variety of motility strategies convey competitive advantages. Studies have revealed that efficient swimming behaviours and chemotactic responses are highly beneficial in the early colonisation of the urinary tract, bladder and kidney, and may provide an increase in fitness essential for competition with other pathogens (Wright et al. 2005). Beyond swimming and swarming, bacterial motions include 'twitching' in *Pseudomonas aeruginosa* and 'gliding' in *Flavobacterium* and *Mycoplasma* (Shrivastava et al. 2015; Nakane and Miyata 2012).

Eukaryotic pathogens also display a diverse range of motilities. Some of the most rapid swimming motion exhibited by eukaryotes is due to the action of eukaryotic flagella, or motile cilia.[2] These whip-like organelles are widely conserved across the eukaryotes, and are found in plants, animals, fungi and protozoa. Eukaryotic flagella are distinct from their bacterial and archaeal counterparts in terms of their more sophisticated internal machinery (the bacterial and archaeal appendages are simply rigid propellers), and consequently, their size. The underlying structure of eukaryotic flagella, known as the axoneme, is a 200 nm-diameter cylinder of microtubule doublets bridged by dynein molecules. These dynein 'molecular motors' cause the doublets to shift longitudinally with respect to each other, causing bending waves that run along the whole structure. This outer cylinder sometimes surrounds microtubule singlets, which can occur individually, in pairs or in triplets. The exact role of these additional microtubules remains obscure, although they have been implicated in beat regulation and may help to define an anisotropic bending rigidity. Other internal machinery along the axoneme is responsible for transporting protein cargo to the flagellum's proximal and distal ends, although this is not present in all species.

The molecular motors that allow flagellar motion act in concert to deform the flagellum, creating a beating stroke that pumps external fluid. Observing flagella in action, and understanding the physics and biology of their operating mechanism is a key challenge in modern biophysics. The complex life cycles of some of the most medically-relevant eukaryotic parasites requires cells to move in hostile environments. This motion is enabled in several key cases by the adaptation and action of eukaryotic flagella. Parasites such as *Plasmodium* spp. and *Leishmania* spp. (responsible for malaria and leishmaniasis, respectively) encounter a range of physical and chemical challenges in their journeys through insect vectors and vertebrate hosts. In doing so, the cells radically remodel their physiology to adapt to external conditions. A striking example is the sexually reproductive stage of *Plasmodium*'s life cycle. Here, the male gametocyte of the intracellular parasite synthesizes eukaryotic flagella when taken up into the insect vector's midgut. These flagellated *Plasmodium* parasites, also known as microgametes, must leave the red blood cell

---

[2]Eukaryotic flagella and motile cilia are structurally very similar; their names appear to be used somewhat interchangeably within the motility literature.

in which they were synthesised and convey nuclear material through a dense blood meal to find their female counterparts. This is the only motile form of the parasite's life cycle, and its motility is known to be critical to the parasite's survival (Talman et al. 2014). To study the motility of *Plasmodium* in this sexual reproduction stage, we have used the rodent-infecting species *Plasmodium berghei* (Wass et al. 2012).

Leishmania are protozoan parasites able to infect human leukocytes and cause a range of clinical symptoms, from minor skin lesions to fatal visceral disease (Kaye and Scott 2011). Although our current understanding of host-pathogen interaction is incomplete, the life cycle of *Leishmania* has been well characterised. There are two basic life cycle forms of *Leishmania*: the amastigote form is an obligate intracellular, non-motile form which multiplies by binary fission within mammalian host cells, the promastigote form is a flagellated, extracellular form which infects the sandfly vector (Forestier et al. 2011). In the sandfly, the amastigotes transform into weakly motile procyclic promastigotes which have a flagellum the same size as the cell body localised at their anterior end of the cell. The assembly of a flagellum in the promastigotes allows for the motility and attachment in the sandfly's gut and is critical for the parasite's transmission through the vector (Bates and Rogers 2004) and into the mammalian hosts. These procyclic parasites replicate and migrate towards the anterior midgut and foregut of the fly. Most individuals transform into non-replicating metacyclic promastigotes, which have increased motility in culture possibly due to an elongated flagellum twice the length of the cell body, at the anterior end (Mosser and Brittingham 1997). A gel secreted by the promastigotes (PSG) creates a plug forcing the sand fly's stomodeal valve to open and extend into the pharynx. The fly then regurgitates and expels the PSG along with metacyclic promastigotes into the host it has bitten (Rogers et al. 2009).

The molecular aspects of this infection process are relatively well known and several host cell receptors that recognise the parasite's surface molecules have been identified (Kaye and Scott 2011; Rogers et al. 2010). However, much of the parasite cell biology is poorly understood, specifically the involvement of motility in the processes of parasite attachment and engulfment by host macrophages (Rogers 2012). A recent study highlighted the importance of the role of promastigote polarity and motility during parasite entry to host cells (Forestier et al. 2011). Using high spatiotemporal resolution microscopy to image an infection assay it was shown the motile promastigotes enter the macrophages in a polarized manner through their flagellar tip. The persistent intracellular flagellar activity is thought to lead to reorientation of the parasite flagellum towards the host cell periphery resulting in oscillatory parasite movement. The *Leishmania* infection process is assumed to rely on the phagocytic capacity of the host cell (Forestier et al. 2011). Thus, understanding how morphology and motility of the promastigote parasites affect macrophage colonisation may provide important insights into *Leishmania* spp. infection.

Three key species described above—*E. coli*, *P. berghei* and *L. mexicana*—are shown in Fig. 3.1, alongside some relevant motility characteristics. We have used DIHM to study all three; we present some results below, with a brief discussion of the physical principles of holography and practical experimental considerations.

Organism	Normalised Holographic Image	Physical Features
Escherichia coli 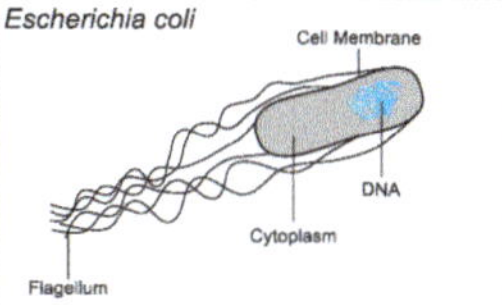	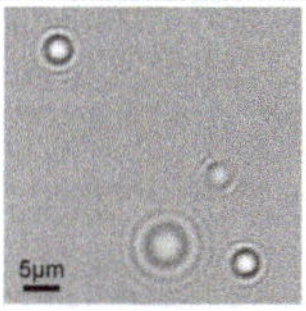	• Body size ~ 2 x 1 μm • Flagellar filaments measure ~ 6-10 μm in length, ~ 20 nm in diameter • Cells peritrichously flagellated, 6-10 flagella per cell • Flagellar rotation rate ~ 100 Hz • Swimming speed ~ 20 μm/s
Plasmodium berghei 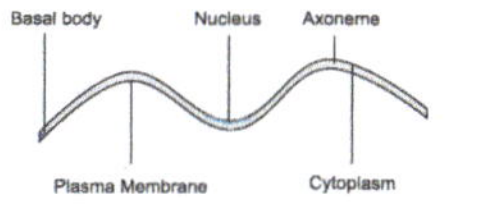	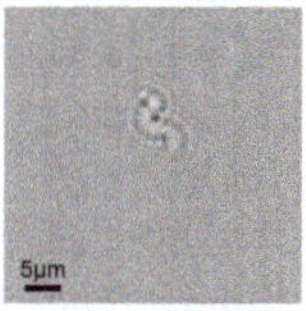	• Free swimming eukaryotic flagellum • Length ~ 8-12 μm, diameter ~ 250 nm • Nuclear material inside flagellum • Flagellar beat frequency ~ 10 Hz • Swimming speed ~ 5 μm/s
Leishmania mexicana 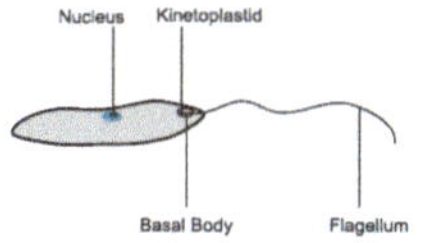	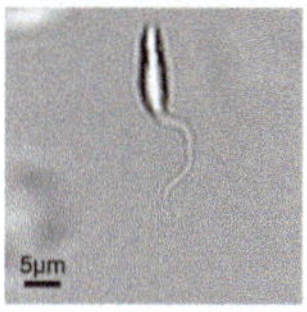	• Body size ~ 10 x 2 μm (procyclic promastigote) • Eukaryotic flagellum ~ 15 μm in length, 250 nm diameter • One flagellum per cell • Flagellar beat frequency ~ 30 Hz • Swimming speed ~ 35 μm/s

Fig. 3.1 Table showing three pathogens studied using DIHM. *Top row: E. coli* bacteria, showing an example holographic image and characteristic physical parameters dimensions. *Middle row: P. berghei* microgamete. These cells are essentially an isolated eukaryotic flagellum with nuclear material distributed along the length of the axoneme, inside the cell membrane. *Bottom row: Leishmania mexicana* procyclic promastigote. The relatively large cell body can be clearly seen in the raw data (*centre* panel)

3.1.1 Optics Background

The study of optics is one of the oldest branches of physics, certainly dating back to antiquity (Born and Wolf 2005). The description of light as oscillations in an electromagnetic field is nineteenth century physics, and our understanding of how light is detected, both by our eyes and by electronic sensors, is now well established. Oscillations in an electric field are conveniently described as sinusoidal waves, with well-defined amplitude and phase (position in the wave cycle). Almost all light-recording material—the retina, photographic film, digital imaging sensors —are sensitive to the *intensity* of incident light (given by the amplitude squared), but not the *phase*. As a result, phase information is lost in the process of recording an optical field by conventional means. This is not generally a serious issue; human perception is adept at interpreting data from the intensity of optical fields, and so photographs and digital images are perfectly intelligible.

There are ways of capturing the phase information, however, and holography is one example. Holography was originally developed by Dennis Gabor (Gabor 1948) as a method of correcting aberrations in an electron microscope (in which the electrons can be thought of as waves), but has subsequently been applied to optical and acoustic imaging as well (Goodman 2005). The word 'holography' comes from the Greek *'holos'*, meaning whole. This is appropriate because the phase and amplitude information that a hologram contains is sufficient to reconstruct the optical field at any other point along its propagation direction. There are many different methods for achieving reconstruction, even in the field of microscopy, but we will focus on DIHM, as it is one of the simplest and most robust microscopic holographic imaging methods.

Interference between two waves is the central concept in holography: a light wave scattered by an object of interest (the 'object wave') is superimposed on to a wave with a known phase (the 'reference wave'). This reference wave is often a plane wave, in which the phase is constant across any plane perpendicular to the direction of propagation. When the electric fields of the object and reference waves combine, they produce an interference pattern of light and dark fringes, corresponding to constructive or destructive interference. This interference pattern reveals the phase of the object wave, as compared the quasi-constant reference, and encodes phase differences within the recorded amplitude. Examples of holographic images can be seen in the central column of Fig. 3.1, in which the cells' rough outlines can be discerned, surrounded by light and dark fringes. These images were obtained using a DIHM setup with LED illumination, as shown in Fig. 3.2a, which we describe in more detail in the next section.

3.1.2 Optical Setup

Figure 3.2 shows two alternative geometries for an inline holographic microscope. Both of the geometries are based on a standard commercial inverted microscope; the only deviation from standard bright field microscopy is in the illumination assembly. Figure 3.2a shows a setup in which an LED is used as a light source, and the nearly-closed condenser aperture is used to create a point-like source in the back focal plane of the condenser lens. The lens turns the point-source emission into plane waves at the sample, as indicated. The alternative setup in Fig. 3.2b uses a laser for illumination, coupled to a single-mode optical fibre. The fibre is held at a distance of 2–5 cm above the sample, meaning that the optical field is well approximated as a plane wave when it is incident on the sample. Note that the condenser lens is unnecessary in this second setup, as laser sources can have high intensity, and do not require collimation, unlike the LED source. When light passes through either system, some of it is scattered by objects in the sample. This scattered light (the object wave) interferes with the remaining unscattered light (the reference wave) at the image plane. The fact that the object and reference wave both take substantially the same path through the imaging system is the reason why these geometries are described as 'inline' holographic microscopes.

The two configurations shown in Fig. 3.2 take advantage of the different levels of optical coherence offered by each source. Optical coherence is a nuanced topic beyond the scope of this article (Mandel and Wolf 1995), but manifests here as the deviation that the object wave can take from the reference while still producing interference fringes. An LED source has low coherence, so the object and reference

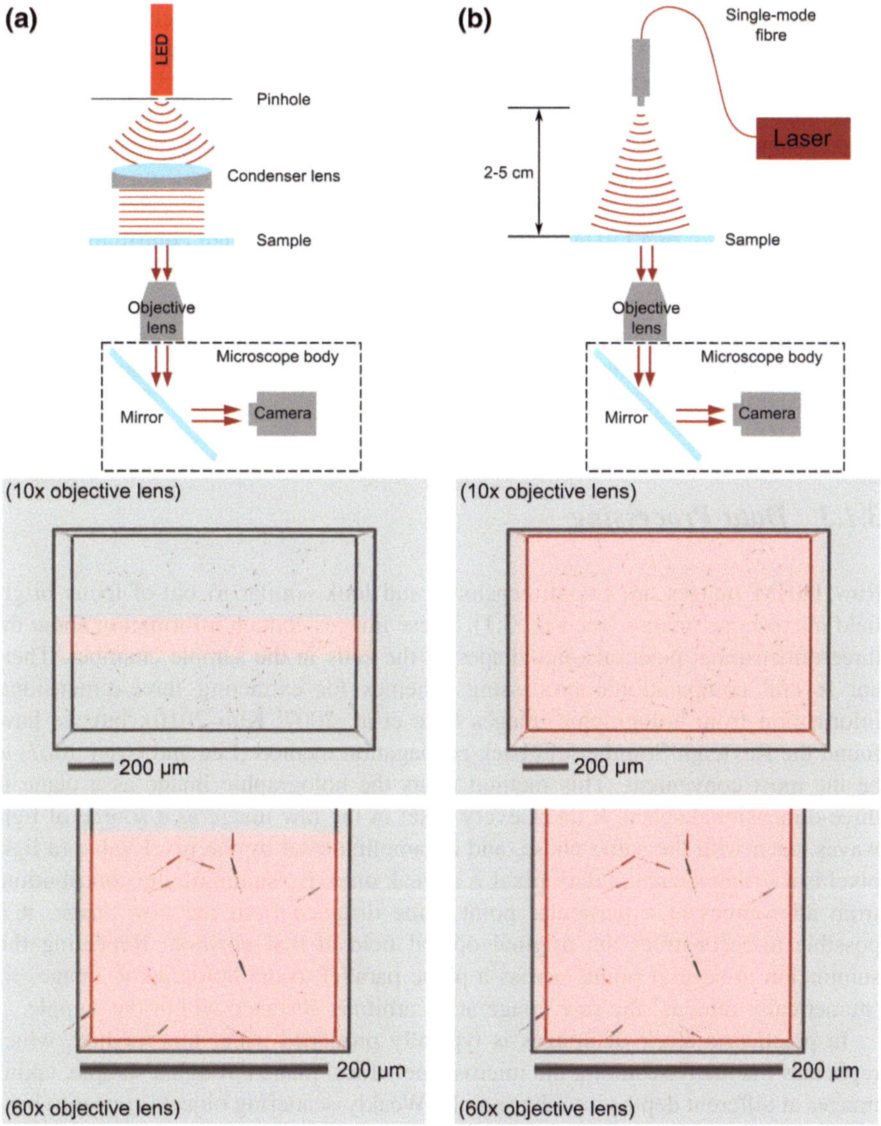

Fig. 3.2 Two layouts for digital inline holographic microscopy (DIHM). At the *bottom* of each panel are two 'cartoon' image of the sample, highlighting the volume from which data is recorded, using typical 10× and 60× objective lenses. These lower four panels contain 'cartoon' cells, drawn to scale, to represent *Leishmania mexicana* and *E. coli*. The latter are barely visible in each case, due to their small size. The system in panel **a** uses an LED and a pinhole (nearly-closed condenser aperture) for quasi-coherent illumination. Panel **b** uses a diode laser coupled to a single mode fibre, which eliminates the need for a pinhole

wave must be nearly colinear in order to produce fringes. In practice, this means that only a small thickness of the sample volume can be imaged at one time, as indicated in the red highlighted regions in the lower panels of Fig. 3.2 (*L. mexicana* cells drawn to scale). A laser is highly coherent and so can be used to image much thicker sample volumes, at both high and low magnification. At first sight, it would seem that a laser is therefore preferable as a source for producing holographic images, but this is not always the case. Use of a laser increases the size of the 'sensitive volume' in a DIHM setup, but this includes unwanted contributions from dust particles that cross the optical path during experiments, irregularities on the surface of the sample chamber and other sources that can produce a confounding signal. For the study of small, very weakly scattering subjects, the smaller volume accessed by an LED is more selective and less susceptible to noise from other parts of the experiment.

3.1.3 Data Processing

Raw DIHM images are two-dimensional and look similar to out-of-focus bright field microscope images (see Fig. 3.1). These images contain information about the three-dimensional positions and shapes of the cells in the sample chamber. There are several computational processing schemes for extracting three-dimensional information from holographic images (Lee et al. 2007; Kim 2010), but we have found the Rayleigh-Sommerfeld back-propagation method (Lee and Grier 2007) to be the most convenient. This method treats the holographic image as a plane in three-dimensional space. It takes every pixel in the raw image as a source of light waves, each with the same phase, and an amplitude set by the pixel value (a light pixel is a strong source, a dark pixel is a weak one). By summing the contributions from all sources at a particular point some distance from the raw image, it is possible to reconstruct the original optical field at that position. Repeating this summation at several points across a plane parallel to the holographic image, we 'numerically refocus' the raw image at an arbitrary distance within the sample.

In practice, a stack of images is typically produced using this method, which replicates manually scanning the microscope's focal plane through a sample, taking images at different depths into the sample. Weakly-scattering objects appear to have bright or dark centres, depending upon which side of the focal plane they lie on. An object's appearance changes (from light to dark or vice versa) when it passes through the focal plane (Wilson and Zhang 2012; Giuliano et al. 2014). Regions where this light-dark transition occurs in a short distance in the axial (z) direction are therefore associated with scattering regions and are used to localise objects in the three-dimensional volume. Each small volume of the object hit by the laser light will scatter some light, and the total scattered field can then be measured as the sum of the individual contributions. This approach is an approximation of the light scattering physics known as the Rayleigh-Gans-Debye (RGD) approximation. RGD scattering is a viable approximation if the scatterer's refractive index relative to the suspending medium ($m = \frac{n_s}{n_m}$) and characteristic dimension, d, adhere to the following conditions:

$$|m - 1| \ll 1, \quad \text{and} \quad kd|m - 1| \ll 1, \tag{3.1}$$

where the wavenumber $k = 2\pi n_m/\lambda$, and λ is the illumination wavelength. Under this assumption each small portion of the subject behaves as an independent scattering centre, scattering light as if isolated from the rest of the sample. This independence means that each notional scattering centre appears in the refocused image stack as a localised region in which there is a dark-to-light transition as the image is numerically refocused. These dark-to light transitions are isolated in an image stack using a three-dimensional image processing filter based on the two-dimensional Sobel-Feldman operator (Wilson and Zhang 2012). The use of this operator transforms a stack of numerically refocused images into one in which axial intensity gradients are picked out as bright objects on a dark background. The objects are simple to isolate by performing a thresholding operation on the image stack. Multiple scattering centres within the same extended object can be associated with each other, or single scattering centres tracked over time, as seen in the next section.

3.1.4 *Example Results*

The holographic recording and reconstruction process described in the previous section results in data that can be analysed at a range of different length scales. Previous studies have broadly taken one of two approaches: examine the shape of single cells at high resolution, or establish a 'centroid' position within the cell and track this over time. We will give examples of both in the following sections, but note that the intermediate length scale (larger than a cell, but smaller than a sample chamber) could be a fruitful area for future investigation.

3.1.4.1 Long Range Tracking

The description in previous sections, and references therein, concern how an object can be localised within a reconstructed optical field. At low magnification, single cells appear point-like. At low cell concentrations and high enough video frame rates, individual points can be tracked across multiple frames, to provide information on the swimming patterns of cells (Jikeli et al. 2015). This type of information is shown in Fig. 3.3a–d. The data show the track of a single *E. coli* cell as it swims in a sample chamber filled with motility buffer. The cell covers a distance of around 300 μm in a period of 65 s, and displays the characteristic 'run and tumble' behaviour of a peritrichously flagellated cell first observed by Berg and Brown (Berg and Brown 1972). This can be seen as relatively straight trajectories, interspersed with angular turns.

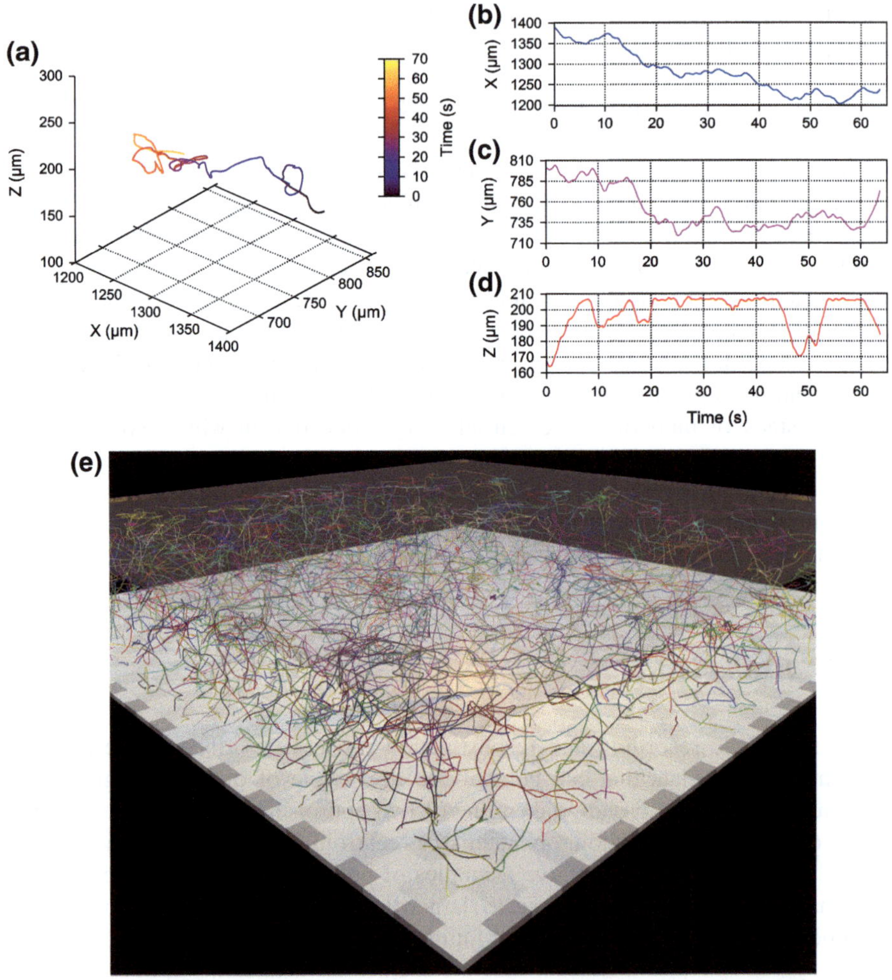

Fig. 3.3 **a** Reconstructed track of a single swimming *E. coli* bacterium, over a period of around 65 s (time indicated by the *colour* of the track). **b–d** Single-axis data from the track shown in panel **a**. The height of the sample chamber was around 210 μm, hence the large amount of data close to that position in panel **d**; the cell was hydrodynamically/sterically trapped at this location for some time (see text). **e** Computer rendering of bacterial tracks from over 1000 cells, captured over a period of 80 s. The squares on the floor of the rendering represent a distance of 50 μm

Figure 3.3a shows an isometric projection of the three-dimensional track, while Figs. 3.3b–d show the cell motion in three orthogonal directions, *X*, *Y* and *Z*. The *Z* (axial) direction is of particular interest, as the cell can be seen colliding with the chamber wall located at $Z \approx 210$ μm. It has been known for some time that swimming cells such as *E. coli* and *Caulobacter crescentus* are influenced by the presence of a wall (Li and Tang 2009), and often become confined to surfaces as a result of hydrodynamic (Lauga et al. 2006) or steric (geometrical) constraints. Moreover, recent results using holographic microscopy (Molaei et al. 2014) have demonstrated that aside of physical constraints, the cell biophysics is also affected by the presence of a nearby wall, which suppresses tumbling events in *E. coli*. This result would not have been accessible without the rapid three-dimensional tracking afforded by DIHM.

At relatively low cell concentrations, many cells can be tracked simultaneously. Figure 3.3e shows over 1000 tracks, all from an 80 s video acquired at 50 frames per second. The video resolution was 1024 × 1024 pixels, and using our 10× magnification objective lens, this corresponds to a field of view of 1440 × 1440 μm. The computer rendering in the figure is drawn to scale; the grey/white squares on the ground represent 50 μm, showing the extent of the bacterial motion over this time. The tracks are not all the same length, as some cells swim faster than others, and cells may swim into and out of the field of view (horizontally) during recording. Nevertheless, data on the position of each bacterium as a function of time makes it easy to obtain parameters such as swimming speed, both on a single-cell level and across a population. This information offers insight into how cells spread into new territory, and interact with each other and their environment.

3.1.4.2 High Resolution Shape Studies

At higher magnification, the shape of individual cells can be examined in three dimensions. Eukaryotic flagella are an ideal subject for this type of analysis, as they are weakly scattering, ensuring that the RGD scattering limit is valid. We have used DIHM on *P. berghei* microgametes (Wilson et al. 2013) and cells of *L. mexicana*, to examine the shape and dynamics of the flagellum. Example data from both species can be seen in Fig. 3.4. Figure 3.4a shows the DIHM reconstruction procedure, from raw data (on the left) through to the fitted contour that represents the shape of the microgamete (on the right-hand side). Assuming that the microtubules are bound together at one end (the 'no basal sliding' approximation), the relative displacement of microtubules within the microgamete can be extracted from the flagellar shape, as seen in Fig. 3.4b. The colour map shows the sliding of microtubule doublets in the axoneme relative to the gamete centre line, from the relatively passive 'tail' end of the microgamete (position $s = 0$ μm on the horizontal axis) to the more active head end ($s = 9$ μm). The vertical axis maps position around the circumference of the microgamete, with an angle as indicated in the left-hand panel of Fig. 3.4b. A bright yellow band extends from bottom left to top right of the sliding map; this helical wave of microtubule sliding gives rise to the helical waveform seen in the rendering on the left.

DIHM on the *L. mexicana* cells is more challenging due the presence of a cell body that scatters light more strongly than the flagellum. Some example data are shown in Fig. 3.4c, where the raw data are on the left, and the reconstructed shape can be seen on the right. The lower object in the reconstruction is the raw data; note the rather noisy data surrounding the position of the cell body. Fortunately, the cell

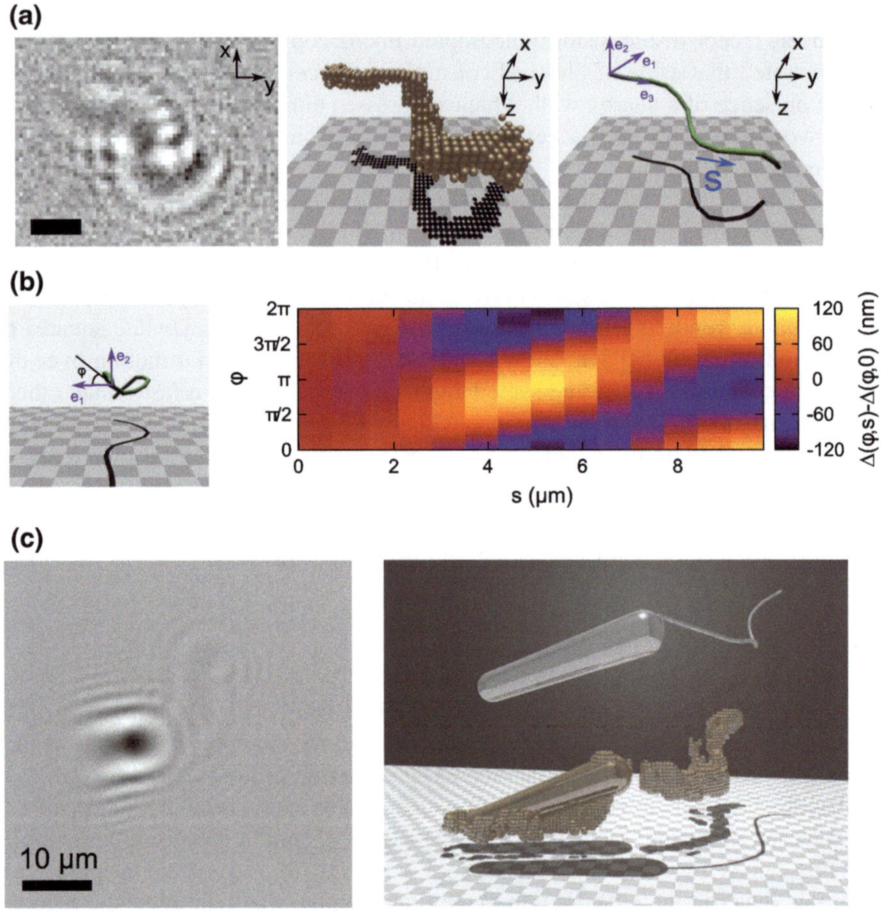

Fig. 3.4 **a** Three stages in reconstructing the shape of a *P. berghei* microgamete. The *left-hand* image is raw holographic data, scale bar 3 μm; the *central* panel shows the reconstructed voxels (volume pixels) that indicate the volume in which the microgamete is most likely located; the *right-hand* panel shows the fitted configuration of the microgamete, subject to known constraints of length and width. **b** Geometric analysis of one particular frame of data from a *P. berghei* microgamete. The *left-hand* panel shows the reconstruction of the raw holographic data, and the *right-hand* panel shows the best estimate of inter-tubule sliding that has to occur in order to achieve this waveform geometry. **c** Raw holographic data from a procyclic cell of *L. mexicana*. The cell body scatters light much more strongly than the flagellum, and is surrounded by higher-contrast fringes. The panel on the *right* shows the best estimate of the cell body position, and the shape of the flagellum obtained a modified version of the routine used to fit the malaria microgametes. The squares on the floor of the reconstructed image represent 1 μm. *[Panels a and b reproduced from Ref. (Wilson et al.* 2013*)]*

body shape of any one individual is constant on the time scale necessary for observing flagellar beating, simplifying analysis somewhat. The ability to study swimming behaviour on a single-cell level in these parasitic species offers new insight into the way that cells move during a critical phase of their life cycles (Hill 2003). High-quality data on the motion of these cells allows us to test existing models of how eukaryotic flagella work (Riedel-Kruse et al. 2007; Lindemann 2011), which in turn could lead to new pathways for treatment.

3.1.5 Discussion and Outlook

We have outlined a few ways in which DIHM can be used to study motile microbial pathogens. The advantages of DIHM are three-fold: (i) the low cost of the apparatus required, (ii) the availability of free software algorithms for post-processing, and (iii) the unique information offered by high-speed, three-dimensional imaging.

The first of these advantages is clear from the optical layout schemes shown in Fig. 3.2. An existing optical microscope may be altered in order to perform DIHM, by the introduction of an alternative illumination source. Any CMOS or CCD camera can be used for imaging, as shown in recent studies that have used cellphone cameras to perform DIHM in a field-ready instrument (Mudanyali et al. 2010). The second advantage is one that is becoming increasingly important in the adoption of image processing standards. Programmes such as ImageJ (Schneider et al. 2012) can be freely downloaded, and custom-written add-ons that provide unique functionality find a broader user base if the source code is made available. The last advantage is probably the most compelling, however. Other competing methods for three-dimensional imaging, such as rapid scanning of the microscope focal plane (Corkidi et al. 2008) are technically demanding and only accessible by dedicated experimental apparatus. Laser scanning confocal microscopy is another approach for three-dimensional imaging of microscopic subjects; this scheme allows high—or super-resolution imaging, but typically requires that the subject is fluorescently labelled. Moreover, confocal schemes require the sample to be mechanically translated between axial sections, restricting the speed at which three-dimensional information can be acquired. In the case of multiple freely-swimming cells, DIHM has the distinct advantage because of its high data acquisition rate; this is limited only by the frame rate of the camera used for imaging, and modern cameras with frame rates in excess of 1 kHz at megapixel resolution can be obtained for small fraction of the cost of a confocal scanning system.

DIHM is certainly not without shortcomings, however. The use of lasers invokes the usual safety concerns, especially where optical instruments are involved: even at relatively low beam powers, a microscope can concentrate laser light to levels that are hazardous to the eyes if appropriate safety interlocks are not in place. The resolution of digital holographic microscopy is similar to that of standard bright field microscopy (Giuliano et al. 2014), although typically with slightly poorer resolution in the axial direction, and improvements in the resolution are bound by the classical diffraction limit.

Looking towards the future, the biggest single step would be coupling DIHM and fluorescence methods. This would take advantage of increasingly fine biochemical and genetic control over microbiological subjects, allowing experimentalists to label and track structures in three dimensions, and potentially at high speeds. Holography performed with the light emitted by fluorescent subjects has been demonstrated on test samples (Rosen and Brooker 2007), but will require a significant investment of research effort before it is suitable for use with weakly-fluorescent biological samples, due to serious inefficiencies in the light-gathering optics. Nevertheless, the idea that fluorescently labelled cellular components could be tracked in three dimensions is a compelling one, and would surely enable new insights into the structure and function of microbial pathogens.

References

Allen RD (1985) New observations on cell architecture and dynamics by video-enhanced contrast optical microscopy. Ann Rev Biophys Biophys Chem 14:265–290

Bates PA, Rogers PA (2004) New insights into the developmental biology and transmission mechanisms of *leishmania*. Curr Mol Med 4:601–609

Berg HC, Brown DA (1972) Chemotaxis in *Escherichia coli* analysed by three-dimensional tracking. Nature 239:500

Born M, Wolf E (2005) Principles of optics, 7th edn. Cambridge University Press

Cavicchioli R, Curmi PMG, Saunders N, Thomas T (2003) Pathogenic archaea: do they exist? BioEssays 25(11):1119–1128

Colin R, Zhang R, Wilson LG (2014) Fast, high-throughput measurement of collective behaviour in a bacterial population. J R Soc Interface 11:20140486

Corkidi G, Taboada B, Wood CD, Guerrero A, Darszon A (2008) Tracking sperm in three-dimensions. Biochem Biophys Res Comm 373:125–129

Crocker JC, Grier DG (1996) Methods of digital video microscopy for colloidal studies. J Colloid Interface Sci 179:298–310

Cuche E, Bevilacqua F, Depeursinge C (1999) Digital holography for quantitative phase-contrast imaging. Opt Lett 24:291–293

Edwards C, Zhoui R, Hwang S-H, McKeown SJ, Wang K, Bhaduri B, Ganti R, Yunker PJ, Yodh AG, Rogers JA, Goddard LL, Popescu G (2014) Diffraction phase microscopy: monitoring nanoscale dynamics in materials science. Appl Optics 53(27):G33–G43

Forestier C-L, Machu C, Loussert C, Pescher P, Späth GF (2011) Imaging host cell-leishmania interaction dynamics implicates parasite motility, lysosome recruitment, and host cell wounding in the infection process. Cell Host Microbe 9(4):319–330

Foxman B (2010) The epidemiology of urinary tract infection. Nat Rev Urol 7:653–660

Fung J, Martin KE, Perry RW, Katz DM, McGorty R, Manoharan VN (2011) Measuring translational, rotational, and vibrational dynamics in colloids with digital holographic microscopy. Opt Express 19(9):8051–8065

Gabor D (1948) A new microscopic principle. Nature 161(161):18275–18282

Giuliano CB, Zhang R, Wilson LG (2014) Digital inline holographic microscopy (dihm) of weakly-scattering subjects. J Vis Exp 84:e50488

Goodman JW (2005) Introduction to fourier optics, 3rd edn. Roberts and Company

Hill KL (2003) Biology and mechanism of trypanosome cell motility. Eukaryot Cell 2(2):200–208

Jikeli JF, Alvarez L, Friedrich BM, Wilson LG, Pascal R, Colin R, Pichlo M, Rennhack A, Brenker C, Kaupp UB (2015) Sperm navigation along helical paths in 3d chemoattractant landscapes. Nat Commun 6:7985

Kaye P, Scott P (2011) Leishmaniasis: complexity at the host-pathogen interface. Nat Rev Micriobiol 9:604–615

Kim MK (2010) Principles and techniques of digital holographic microscopy. SPIE Rev 1:018005
Lauga E, DiLuzio WR, Whitesides GM, Stone HA (2006) Swimming in circles: motion of bacteria near solid boundaries. Biophys J 90:400–412
Lee S-H, Grier DG (2007) Holographic microscopy of holographically trapped three-dimensional structures. Opt Express 15(4):1505–1512
Lee SH, Roichman Y, Yi G-R, Kim S-H, Yang S-M, van Blaaderen A, van Oostrum P, Grier DG (2007) Characterizing and tracking single colloidal particles with video holographic microscopy. Opt Express 15(26):18275–18282
Li G, Tang JX (2009) Accumulation of microswimmers near a surface mediated by collision and rotational brownian motion. Phys Rev Lett 103:078101
Lindemann CB (2011) Experimental evidence for the geometric clutch hypothesis. Curr Top Dev Biol 95:1–31
Mandel L, Wolf E (1995) Optical coherence and quantum optics. Cambridge University Press
Merola F, Miccio L, Memmolo P, Di Caprio G, Galli A, Puglisi R, Balduzzi D, Coppola G, Netti P, Ferraro P (2013) Digital holography as a method for 3d imaging and estimating the biovolume of motile cells. Lab Chip 13:4512–4516
Molaei M, Barry M, Stocker R, Sheng J (2014) Failed escape: solid surfaces prevent tumbling of *Escherichia coli*. Phys Rev Lett 113:068103
Mosser DM, Brittingham A (1997) *Leishmania*, macrophages and complement: a tale of subversion and exploitation. Parasitology 115:S9–S23
Mudanyali O, Tseng D, Oh C, Isikman SO, Sencan I, Bishara W, Oztoprak C, Seo S, Khademhosseini B, Ozcan A (2010) Compact, light-weight and cost-effective microscope based on lensless incoherent holography for telemedicine applications. Lab Chip 10:1417–1428
Nakane D, Miyata M (2012) *Mycoplasma mobile* cells elongated by detergent and their pivoting movements in gliding. J Bacteriol 194(1):122–130
Park YK, Choi W, Yaqoob Z, Dasari R, Badizadegan K, Feld MS (2009) Speckle-field digital holographic microscopy. Opt Express 17(15):12285–12292
Riedel-Kruse IH, Hilfinger A, Howard J, Jülicher F (2007) How molecular motors shape the flagellar beat. HFSP J 1(3):192–208
Rogers ME (2012) The role of *Leishmania* proteophosphoglycans in sand fly transmission and infection of the mammalian host. Front. Microbiol 3:223
Rogers M, Kropf P, Choi B-S, Dillon R, Podinovskaia M, Bates P, Müller I (2009) Proteophosophoglycans regurgitated by *leishmania*-infected sand flies target the l-arginine metabolism of host macrophages to promote parasite survival. Microbes Infect 5:e1000555
Rogers ME, Corware K, Müller I, Bates PA (2010) *Leishmania infantum* proteophosphoglycans regurgitated by the bite of its natural sand fly vector, *Lutzomyia longipalpis*, promote parasite establishment in mouse skin and skin-distant tissues. Microbes Infect 12:875–879
Rosen J, Brooker G (2007) Fluorescence incoherent color holography. Opt Express 15:2244–2250
Schnars U, Jüptner W (1994) Direct recording of holograms by a ccd target and numerical reconstruction. Appl Opt 33:179–181
Schneider CA, Rasband WS, Eliceiri KW (2012) Nih image to imagej: 25 years of image analysis. Nat Meth 9(7):671–675
Shrivastava A, Lele PP, Berg HC (2015) A rotary motor drives *Flavobacterium* gliding. Curr Biol 25:338–341
Su TW, Xue L, Ozcan A (2012) High-throughput lensfree 3d tracking of human sperms reveals rare statistics of helical trajectories. Proc Natl Acad Sci 109:16018–16022
Talman AM, Prieto JH, Marques S, Ubaida-Mohien C, Lawniczak M, Wass MN, Xu T, Frank R, Ecker A, Stanway RS, Krishna S, Sternberg MEJ, Christophides GK, Graham DR, Dinglasan RR, Yates JR III, Sinden RE (2014) Proteomic analysis of the *Plasmodium* male gamete reveals the key role for glycolysis in flagellar motility. Malaria J 13:315
Wass MN, Stanway R, Blagborough AM, Lal K, Prieto JH, Raine D, Sternberg MJE, Talman AM, Tomley F, Yates J III, Sinden RE (2012) Proteomic analysis of *Plasmodium* in the mosquito: progress and pitfalls. Parasitology 139:1131–1145

Wilson Laurence, Zhang Rongjing (2012) 3d localization of weak scatterers in digital holographic microscopy using Rayleigh-Sommerfeld back-propagation. Opt Express 20(15):16735–16744
Wilson LG, Martinez VA, Schwarz-Linek J, Tailleur J, Bryant G, Pusey PN, Poon WCK (2011) Differential dynamic microscopy of bacterial motility. Phys Rev Lett 106(1):018101
Wilson LG, Carter LM, Reece SE (2013) High-speed holographic microscopy of malaria parasites reveals ambidextrous flagellar waveforms. Proc Natl Acad Sci USA 110(47):18769–18774
World Health Organization (2014) World malaria report. WHO Press
Wright KJ, Seed PC, Hultgren SJ (2005) Uropathogenic *Escherichia coli* flagella aid in efficient urinary tract colonization. Infect Immun 73(11):7657–7668
Xu W, Jericho MH, Meinertzhagen IA, Kreuzer HJ (2001) Digital in-line holography for biological applications. Proc Natl Acad Sci 98(20):11301–11305

Chapter 4
What Is the 'Minimum Inhibitory Concentration' (MIC) of Pexiganan Acting on *Escherichia coli*?—A Cautionary Case Study

Alys K. Jepson, Jana Schwarz-Linek, Lloyd Ryan, Maxim G. Ryadnov and Wilson C.K. Poon

Abstract We measured the minimum inhibitory concentration (MIC) of the antimicrobial peptide pexiganan acting on *Escherichia coli*, and found an intrinsic variability in such measurements. These results led to a detailed study of the effect of pexiganan on the growth curve of *E. coli*, using a plate reader and manual plating (i.e. time-kill curves). The measured growth curves, together with single-cell observations and peptide depletion assays, suggested that addition of a sub-MIC concentration of pexiganan to a population of this bacterium killed a fraction of the cells, reducing peptide activity during the process, while leaving the remaining cells unaffected. This pharmacodynamic hypothesis suggests a considerable inoculum effect, which we quantified. Our results cast doubt on the use of the MIC as 'a measure of the concentration needed for peptide action' and show how 'coarse-grained' studies at the population level give vital information for the correct planning and interpretation of MIC measurements.

Keywords Antimicrobial peptide · Pexiganan · *Escherichia coli* · Minimum inhibitory concentration · Killing curves · Inoculum effect

4.1 Introduction

The discovery of the β-lactam antibiotic penicillin by Fleming in 1928 was a milestone of twentieth-century medicine. Today, the rampant spread of antimicrobial resistance (AMR) constitutes a grand challenge facing medical science in the new century

A.K. Jepson · J. Schwarz-Linek · W.C.K. Poon (✉)
SUPA and School of Physics and Astronomy, The University of Edinburgh, Peter Guthrie Tait Road, Edinburgh EH9 3FD, Scotland, UK
e-mail: w.poon@ed.ac.uk

L. Ryan · M.G. Ryadnov
National Physical Laboratory, Hampton Road, Teddington TW11 0LW, UK

© Springer International Publishing Switzerland 2016

M.C. Leake (ed.), *Biophysics of Infection*, Advances in Experimental Medicine and Biology 915, DOI 10.1007/978-3-319-32189-9_4

(Aminov 2010), which, if not met, will turn a 'strep throat' back into a life threatening illness. To confront AMR, more effective ways of using of existing agents, including dosing regimes less prone to generating AMR, are urgently needed, as are new agents. In either case, standardised measures of effectiveness allowing science-based comparison between different agents are clearly required.

In the global effort to confront AMR, antimicrobial peptides (AMPs) have attracted considerable attention (Fjell et al. 2012). These short peptides ($\approx$10–50 residues) are widely distributed among metazoans, with diverse sequence and structure. Certain motifs recur, e.g. high net positive charge under physiological conditions, and/or α-helical conformation in solution or upon binding to membranes; but these motifs are not universal. AMPs are effective against a wide spectrum of bacteria, viruses and fungi *in natura*. The hope is that some natural or synthetic AMPs may be suitable therapeutic antimicrobial agents, especially against AMR strains.

The 23-residue AMP magainin-2 secreted by the African clawed frog *Xenopus laevis* and its 'relatives' have attracted particular attention. Pexiganan (or MSI-78), a synthetic 22-residue magainin analogue (Gottler and Ramamoorthy 2009), was trialed for topical treatment of diabetic foot ulcers, but was denied approval in 1999 because it seemed no more effective than antibiotics already in use for such ulcers (Moore 2003); however, future clinical approval remains a possibility (Gottler and Ramamoorthy 2009). Partly due to on-going efforts to secure such approval, pexiganan has been well studied.

A large biophysical literature exists on AMPs in general, and pexiganan in particular, focussing on the molecular *modus operandi*. Partly as a result of substantial research into how pexiganan and similar AMPs interact with lipid bilayers in unilamellar vesicles (Gottler and Ramamoorthy 2009), it is widely believed that this and other α-helical AMPs lyse bacteria by membrane poration.

We consider the other end of the length scale spectrum, and report a study of the *modus operandi* of pexiganan on *E. coli* at the population level. Such 'coarse-grained' studies using rather classical methods (albeit in updated, high-throughput forms) are seldom performed today. Our results show how such work is needed to complement molecular-level studies, preventing the misinterpretation of pharmacodynamic measurements performed to judge the concentration required for antimicrobial action against live bacteria.

We start by measuring the minimum inhibitory concentration (MIC), which is the most important 'one-number characterisation' of the effectiveness of an antimicrobial agent against a target organism. Loosely, it is the minimum concentration of an antimicrobial agent necessary to cause stasis (no growth). We redetermine the MIC of pexiganan on *E. coli*, but using more repeat experiments than has ever been reported before. A critique of using MIC to characterise potency based on our measurements leads us to study the effect of pexiganan on the growth curve of *E. coli*, which turns out to be strikingly different from the way many classical antibiotics change the growth curve of the same bacterium.

Our growth curves, along with single-cell observations, peptide depletion assays and time-kill curves, suggest that adding sub-MIC concentrations of this AMP to a population of *E. coli* rapidly kills a fraction of the cells, leaving the rest to grow unaffected while at the same time removing active AMP molecules from the medium. As Udekwu et al. (2009) have previously suggested, the depletion or deactivation of an antibiotic causes an 'inoculum effect' (dependence of MIC on initial inoculum concentration), which we quantify for pexiganan. We end by discussing the implications of our findings on the interpretation of the MIC in mechanistic and pharmacodynamic contexts.

4.2 Materials and Methods

4.2.1 Bacteria Culture

We worked with *E. coli* K-12 derived strain MG1655 (Blattner et al. 1997). Five colonies, grown on an Mueller Hinton Broth (MHB) agar plate, were touched with a sterile loop and introduced to a 5 ml MHB liquid culture, which were grown at 37 °C and 200 rpm to OD = 0.5 (600 nm). We also grew bacteria in the filtered supernatant of cells lysed by sonication. Ten 30 s pulses of sonication applied to 10 ml of *E. coli* culture resting on ice achieved a 99.98 % reduction in viability. The supernatant was filtered (0.22 μm) to remove the surviving cells.

4.2.2 Pexiganan

Pexiganan (GIGKFLKKAKKFGKAFVKILKK-NH_2) was synthesised on a Liberty microwave peptide synthesiser (CEM Corporation) using standard solid phase Fmoc protocols on Rink amide-MBHA resins with HCTU/DIPEA as coupling reagents. Peptides were purified by semi-preparative RP-HPLC on a JASCO HPLC system (model PU-980, Tokyo, Japan) and confirmed by MALDI-ToF mass spectrometry (Bruker Daltonics Ltd, UK), with α-cyano-4-hydroxycinnamic acid as the matrix.

We prepared stock solutions at 2 mM in sterile, distilled water, which were stored at −20 °C and defrosted immediately before using and refreezing.

4.2.3 Growth Curves and MIC

We followed a published protocol for MIC determination using microdilution assays in microtiter plates (Wiegand et al. 2008) that is consistent with the guidelines of the Clinical and Laboratory Standards Institute and the European Committee on Antimicrobial Susceptibility testing. MIC assays were prepared in 96-well polystyrene microtiter plates (Greiner) with 200 μl cylindrical wells. Initial inoculum sizes of $n_0 = 5 \times 10^5$ cell/ml were incubated in and their optical density read (at 600 nm) by a FLUOstar Optima (BMG Labtech) plate reader with lid-covered plates that allowed air flow. We checked that our MIC results were the same using either polystyrene or polypropylene plates. Note that, we report pexiganan concentrations in μM to facilitate the comparison with cell concentrations in cells/ml, where 0.4 μM = 1 μg/ml (using the molecular weight of 2477 g mol^{-1}). Literature values in μg/ml have been converted to and quoted in μM.

4.2.4 Single-Cell Imaging

To determine times to first division, we took time-lapsed phase-contrast images of cells using a Nikon TE300 Eclipse inverted microscope with a 100× PH3 oil-immersion objective and a CoolSNAP HG^2 CCD camera (Photometrics). MHB agar pads were set into adhesive Gene Frames (Thermo Scientific). Pexiganan at 3 μM was added to *E. coli* at OD = 0.5 and the solution was left to incubate for 3 min before pipetting 1 μl onto the agar pad. The *E. coli* were spread by tipping the microscope slide and within ~4.5 min all liquid had, by eye, disappeared. We then mounted a glass coverslip to the gene frame in contact with the agar pad. The sample was immediately transferred to the microscope and preheated to 37 °C in a temperature-controlled box for observation.

4.2.5 Time-Kill Curves

We grew *E. coli* MG1655 following the same protocol as for MIC assays and diluted to 5×10^5 cell/ml. We worked with 3 ml of suspension in two tubes (Greiner, polystyrene, 50 ml). Pexiganan was added to one of these at time $t = 0$ min and both incubated at 37 °C and shaken at 200 rpm. A 100 ml sample was removed from both tubes 1 min after peptide addition, a range of tenfold dilutions were spread onto MHB agar plates in triplicate. Not more than five samples were removed, resulting in a 17 % volume reduction, which is somewhat above the recommended maximum reduction in standard protocols. The agar plates were incubated at 30 °C for 16 h before the colonies were counted manually and density of cells was calculated in cell-forming units (CFU) per ml.

4.2.6 Pharmacodynamic Studies

We followed literature procedures (Udekwu et al. 2009) to determine the effect of residual AMP. Two tubes containing 5 ml of MHB and 40 μM of pexiganan were prepared, one of which was inoculated with 5×10^6 cells/ml. After 18 h incubation, both suspensions were filtered and the supernatant was used to set up two MIC assays in a 96-well plate each, alongside two replicate control MIC assays using the standard protocol. Further, two assays were set up with supernatant from the tube which had contained both cells and peptide, with 20 μM of freshly added pexiganan.

4.3 Minimum Inhibitory Concentration

Figure 4.1a shows the representative raw data for our determination of the MIC of pexiganan for *E. coli* MG1655 using microdilution assays in microtiter plates, with growth after 24 h detected by a plate reader.[1] In our 96-well plate, each of the 12 columns constitutes a separate MIC determination. Thus, for example, in column 4 of the plate shown, the minimum concentration showing no growth at 24 h is 5 μM, which therefore by definition is the MIC from this particular dilution series. In these 12 replicates, five return a MIC of 2.5 μM (columns 1, 2, 8, 9 and 11), four return a MIC of 5 μM (columns 4, 6, 7 and 12) and three columns (3, 5 and 10) show a seemingly 'impossible' pattern of 're-entrant growth': after being inhibited at 2.5 μM, growth apparently restarted at 5 μM.[2]

What, then, is *the* MIC of pexiganan acting on *E. coli* MG1655? Our data, Fig. 4.1a, do not allow us to assign a unique value, but if such a value exists, then it lies in the region of 2.5–5 μM. Previous studies have returned values in the range of

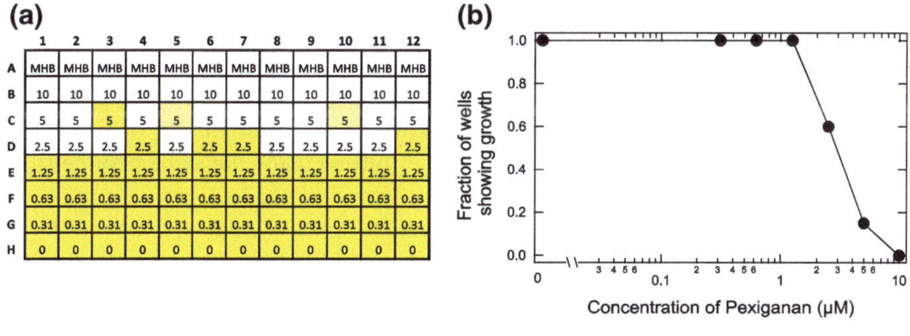

Fig. 4.1 **a** Data from one dilution experiment in a 96-well plate. Numbers are pexiganan concentrations in μM in the 200 μl wells, each inoculated with $n_0 = 5 \times 10^5$ cells/ml. MHB = buffer with no cells. Each column is one dilution series. *Yellow* wells with an OD ≈ 10× the OD of MHB wells after 24. **b** Fraction of wells that showed apparent bacterial growth for 20 replicates as a function of pexiganan concentration, with the zero pexiganan point plotted on the *left* end of the logarithmic *horizontal axis*

[1]See http://datashare.is.ed.ac.uk/handle/10283/1885 to access relevant data on which this article is based.

[2]Continuing the experiment to 72 did not change the observed growth/no-growth pattern.

3.2–12.8 μM against various isolates (Fuchs et al. 1998; Ge et al. 1999), placing our range of 2.5–5 μM at the low end of the spectrum.[3]

However, to compare our results and previous work in terms of range of values masks a fundamental qualitative difference between our work and almost all previous MIC measurements. We, like the majority of the literature, follow best practice guidelines (Wiegand et al. 2008; CLSI 2012). The difference is that, we perform multiple (12) replicates of the same MIC assay on a single plate. This is almost never done in the literature. Instead, commercial 'MIC plates' are used with wells preloaded to give single-dilution series of multiple antimicrobial agents. Alternatively, the same dilution series of a single agent is used to test multiple organisms, again with one dilution series per organism.[4] The range of the literature values quoted above for pexiganan acting on *E. coli* therefore arises from single-dilution series measurements on many isolates, whereas our own range of values arises from variability between multiple replicates on a single strain, Fig. 4.1a.

When a single-dilution series is used, as is the case in the majority literature, the issue of disagreement between replicates does not arise. On the rare occasion, where two replicates are prepared and they disagree, such as our columns 7 and 8 (Fig. 4.1), the disagreement is typically ascribed to 'dilution error'. Literature protocols do occasionally mention 're-entrant growth', but would ascribe this to accidental 'skip' or to 'single well contamination'. In particular, if we follow Hendriksen (2010), we should identify our wells D3, D5, D10 as 'skips', and in these cases take the 'true MIC' to be 10 μM.[5] Repeated dilution error, skip or contamination at the same point of our multiple dilution series seem highly improbable. Moreover, repeated measurements showed similar patterns of variability seen in Fig. 4.1a, including 'reentrant growth'. Thus, the variability revealed by our results is intrinsic, and reporting a single MIC is misleading. A better way to summarise our findings is to plot the fraction of wells showing growth as a function of pexiganan concentration. Figure 4.1b shows such a plot for the 12 replicates shown in Fig. 4.1a and another 8 replicates performed using the same peptide stock and inoculum.

Intrinsic variability is consistent with a previous meta-study (Annis and Craig 2005), which ascribed half of the variability uncovered in a survey of literature values of the MICs of various antibiotics against *E. coli* and *Staphylococcus aureus* to laboratory-to-laboratory differences. Presumably, then, it is possible that at least part of the other half of the variability is attributable to intrinsic causes, although Annis and Craig attributed this 'commonly shown threefold dilution range' entirely to environmental factors such as temperature, inoculum size and incubation time.

---

[3]This is perhaps unsurprising given that MG1655 is a laboratory strain that has been described as 'deceitful delinquents growing old disgracefully' (Hobman et al. 2007).

[4]E.g., no replicates are suggested in the guidelines for loading a 96-well plate for testing a range of agents on *E. coli* and *Salmonella* isolates in a WHO project (Hendriksen 2010).

[5]See Fig. 2 in Hendriksen (2010). If, alternatively, we identify our wells C3, C5 and C10, Fig. 4.1, as contaminated, then the 'true MIC' would be taken to be 2.5 M.

To begin to elucidate the source of the MIC variability, we turn to consider the full growth curves that were collected and used to generate the MIC data.

4.4 Sub-MIC Growth Curves

Figure 4.2a shows a typical set of growth curves of *E. coli* MG1655 at sub-MIC concentrations of pexiganan (background subtracted in Fig. 4.2b, c), from the MIC assays displayed in Fig. 4.1a. While such curves are collected in every microtiter assay, they are seldom, if ever, presented or interpreted. Exceptions concern the response of *E. coli* to tetracycline and amoxicillin (Schuurmans et al. 2009) and to cefotaxime (Baraban et al. 2011). At sub-MIC concentrations, tetracycline and cefotaxime reduce the population growth rate and stationary level whilst amoxicillin does not significantly affect the growth curves until the MIC is reached.

Pexiganan does not influence the maximum growth rate (measured to be $\alpha = 0.034 \pm 0.01\ \text{min}^{-1}$) or the stationary level of the population (a background

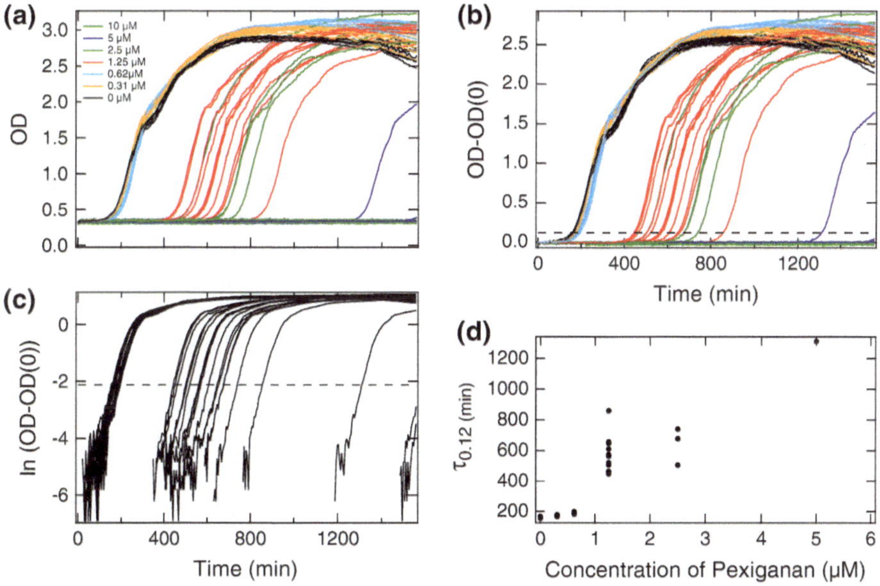

Fig. 4.2 Sub-MIC growth curves for *E. coli* with an initial inoculum size of $n_0 = 5 \times 10^5$ cells/ml at the same pexiganan concentrations used for the 96-well plate experiment in Fig. 4.1 and stated in the legend. **a** Raw OD data against time. **b** The same data with OD at $t = 0$ subtracted. **c** Log-linear plot of the data in (**b**). **d** The time to reach OD = 0.12, $\tau_{0.12}$, plotted against concentration of pexiganan. In parts (**b**) and (**c**), OD = 0.12 is shown as *dashed lines*

subtracted OD of $\approx$2.5), but lengthens the time until growth is detected[6] in a concentration-dependent manner. Figure 4.2c plots the time taken to measure an OD of 0.12 above the OD of MHB ($\tau_{0.12}$), which increases with pexiganan concentration. Note that without pexiganan, the population takes 160 $\pm$ 10 min to grow to OD = 0.12 = 1 $\pm$ 0.4 $\times$ 10^8 cell/ml. An initial inoculum of 5 $\times$ 10^5 cells/ml will reach 1 $\times$ 10^8 cell/ml in $\alpha^{-1}\ln(2 \times 10^2) = 156$ min with our measured α. The population lag time is therefore immeasurably small.

The variation in $\tau_{0.12}$ between wells increases with pexiganan concentration. In particular, the large differences in $\tau_{0.12}$ between replicates at 2.5 μM correlate well with the observation, Fig. 4.1, that some wells do not show growth within 24 h at this concentration. No replicate experiments were reported in previous growth curve studies. In the case of cefotaxime (Baraban et al. 2011), the smooth variation of growth curves at closely spaced antibiotic concentrations allows us to conclude that there should be little variation between replicates. We therefore infer an intriguing difference vis-à-vis stochasticity between pexiganan and cefotaxime. We cannot draw a similar conclusion from data for tetracycline and amoxicillin (Schuurmans et al. 2009) because only a few concentrations were studied.

4.5 Single-Cell Observations

Next, we probe the cause of the observed effect of pexiganan concentration on $\tau_{0.12}$, the growth curve detection time. There are two generic ways in which the peptide could affect cells at early times to increase $\tau_{0.12}$. Either some of the initial inoculum dies, and the remaining cells take longer to grow to a given density even if their growth rate remains unaltered, or growth of all the cells is retarded, giving the same macroscopic observed effect on $\tau_{0.12}$. To observe directly the early-time effect of sub-MIC concentrations of pexiganan, we imaged cells spread on agar and recorded the time to first division (TTFD) of each cell, thought to be the sum of the lag time and first generation time (Rasch et al. 2007; Métris et al. 2005). We chose conditions (caption, Fig. 4.3) that resulted in sufficient cell death to demonstrate the effect of pexiganan, whilst leaving enough live cells to collect meaningful data.

The TTFD distribution for the pexiganan-free control, Fig. 4.3 (red), is spread over ~50 min due to heterogeneities in growth stages, single-cell lag times[7] and generation times. On exposure to 3 μM of pexiganan for 3 min before being placed

[6]We detected growth in a well at 2 $\times$ 10^7 cells/ml, similar to what was reported previously for multi-well plate readers (Pin and Baranyi 2006; Métris et al. 2006). Lengthened detection times due to peptide action recalls the 'virtual colony count' (VCC) approach developed to measure defensin activity (Ericksen et al. 2005). However, unlike in VCC, bacteria in our case were exposed to peptides in their growth medium rather than grown in a peptide-free medium after exposure.

[7]Cells with the shortest single cell lag times dominate the population lag (Baranyi 1998), which can be crudely calculated to be ~10 min after the transfer procedure from liquid culture to agar.

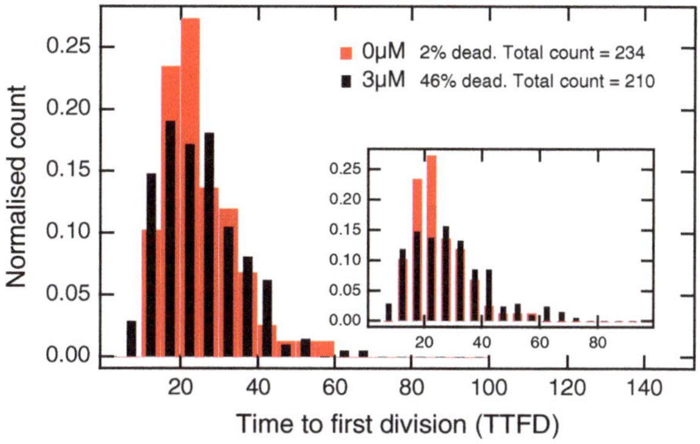

Fig. 4.3 Normalised histogram of times to first division for *E. coli* on agar, unexposed to pexiganan (*red*) and exposed to 3 μM pexiganan for 3 min (*black*) *Inset* shows the same plot if the time to second division is counted for all first divisions resulting in the death of a daughter cell

on the agar, a fraction of the cells never go on to divide. Some are clearly dead (showing less contrast and do not grow) by the time the first image was taken, but others grow initially and then stop at later times. Some cells lyse suddenly whilst other fade over time. Interestingly, 29 % of the cells which grow and divide give rise to a daughter cell that subsequently dies. Importantly, the TTFD distribution of cells that survive and divide to form colonies does not seem to differ greatly to that of the control sample, Fig. 4.3 (black and red, respectively). Including cells whose daughters subsequently die shifts the distribution only slightly (inset).

These data suggest that exposure to pexiganan results either in death or leaves the cells unaffected to go on to grow and divide, at least for cells inoculated onto agar. A number of caveats are, however, in order. First, the concentration of peptide the cells are exposed to is uncontrolled while the inoculum drop is evaporating after being placed on the agar surface. Thereafter, the peptides are free to diffuse into the agar away from the cells. It is therefore possible that under more prolonged exposure than is possible under our experimental conditions, e.g. in liquid medium, the growth of surviving cells would be retarded. Nevertheless, our experiments certainly show that the main effect of pexiganan is to kill a proportion of the cells rapidly.

4.6 Depletion of Active Peptide

It is important to know whether this rapid initial killing of cells significantly affects the concentration of free peptides. Towards this end, we first determine if the peptide is depleted with time in our assays; we performed the MIC experiments detailed in Sect. 4.2.6. MIC assays using preincubated peptide showed that the activity of the peptide degrades by a factor of 2–4 times over 24 h in MHB at 37 °C, perhaps due to peptide aggregation, common for magainins, or adhesion to components of the medium. However, this is a small contribution to what was found when bacteria had also been in the solution. An assay containing 40 μM of pexiganan and 5×10^6 cell/ml, after being incubated and then filtered to remove any survivors, no longer had any measurable affect on the growth of a new inoculum, so that (from Sect. 4.3) we know that the pexiganan concentration must be below 2.5 μM. In other words, the activity of the peptide was reduced more than 16-fold. When an additional 20 μM of peptide was added to this solution and serially diluted with MHB, it resulted in a MIC 4× greater than that of a control. Not only had the original peptide been depleted but the additional peptide showed less activity in this solution than in MHB. Similar experiments conducted without filtering gave the same results, so that the filtering was not responsible for removing active peptide molecules or complexes.

Following these findings, we used a suspension containing the filtered remains of sonicated cells grown in MHB to prepare a MIC assay instead of MHB alone. Even at the highest concentration of 20 μM pexiganan, there was no measurable effect on the inoculum of 5×10^5 cell/ml, suggesting a reduction in the antimicrobial action of the peptide by 10–20 fold. It appears that the presence of lysed bacteria reduces peptide activity. A plausible hypothesis is that positively charged peptides are depleted by adhesion to negatively charged DNA and proteins, and/or by the action of proteolytic enzymes from lysed cells.

4.7 Sub-MIC Time-Kill Curves

If we are right that pexiganan kills a fraction of cells at short times upon addition, our growth curves ought to show an initial decrease in numbers. This is not borne out by our plate reader data, Fig. 4.2. We now demonstrate that this is simply because of lack of detection sensitivity. By using plate counts to measure the number of viable *E. coli* as a function of time after the addition of pexiganan to their medium, i.e. by determining time-kill curves (Sect. 4.2.5), it is possible to detect much lower cell densities than is possible using a plate reader. We added pexiganan at $t = 0$ min to concentrations of 1, 2.5, 5 or 10 μM and collected plate count data for the first 200 min, and then once the following day.

Figure 4.4a shows time-kill curves for an inoculum size of 5×10^5 cell/ml, the same as that used for MIC assays in the plate reader. As can be expected from the results shown in Sect. 4.3, no viable cells were recorded at any time point at 10 μM. At lower concentrations the density of viable cells in the suspension drops at early times; the higher the peptide concentration, the larger the drop in initial viability. This is as expected from the only previous publication of pexiganan time-kill curves (Ge et al. 1999). These measurements however were performed at above-MIC concentrations, and only monitored for 120 min.

We measured at lower concentrations and for longer and therefore observed the viability of *E. coli* increasing again at later times. The higher the peptide concentration, the longer before recovery starts; however, after 1 day, all samples have recovered to the same level.

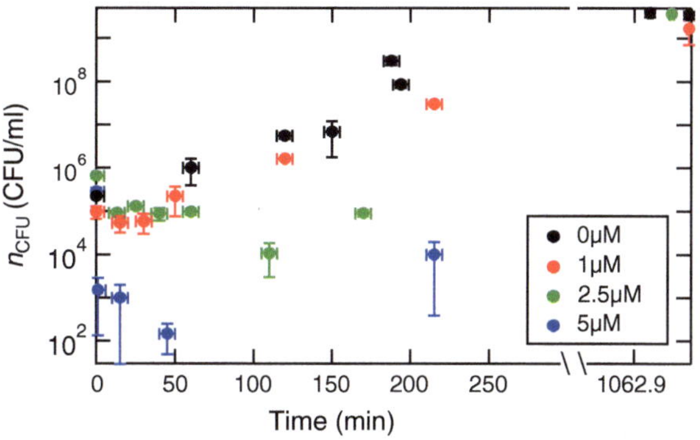

Fig. 4.4 Plate counted time-kill curves in MHB for pexiganan concentrations shown in the legend for the inoculum size $n_0 = 5 \times 10^5$ cells/ml

Regrowth after an initial drop in viability has been reported in the time-kill curves of other AMPs (Matsuzaki et al. 1998; McGrath et al. 2013; Spindler et al. 2011). It has been attributed to 'resistance', with little evidence given in support. To verify that no resistance had developed in our case, we exposed *E. coli* that had regrown in our experiments to 2.5 and 5 μM pexiganan, and found them as susceptible as cells from the parent population.

Our results confirm that the lack of an initial phase of negative growth in our plate reader data, Fig. 4.2, is simply due to the lack of detection sensitivity, and strengthens the case for our suggestion that pexiganan kills a fraction of the cells at early times, leaving the rest to grow unaffected. The literature has only recently suggested that such non-monotonic time-kill curves can be due to a dynamic state of balanced division and death rather than the historically assumed subpopulation of non-dividing persister cells (Wakamoto et al. 2013).

The killing of a fraction of cells in a clonal population leaving the rest of the cells unaffected is striking. Given the large number of peptides per bacterium in the system ($\sim 10^9$, much more than is needed to cover each cell in a monolayer however the peptide is oriented) (Wimley 2010), the observed heterogeneity is unlikely to be due to fluctuations in the number of peptides strongly interacting with each cell. Instead, the heterogeneity most likely arises from phenotypic variations in a property or properties of single cells. Our evidence shows that the survivors are not a subpopulation of non-dividing cells. Perhaps, then, cell surface heterogeneities, e.g. in the structure of the lipopolysaccharides (Lerouge and Vanderleyden 2002) or in the expression of fimbrae (Abraham et al. 1985), are responsible for our observations.

Note that at peptide concentrations approaching the MIC, e.g. $\gtrsim$2.5 μM for the inoculum size relevant to Fig. 4.4, the number of viable cells drops to rather low levels (the minima at 2.5 and 5 μM in Fig. 4.4 correspond to 2000 and 20 cells per well, respectively) before net growth begins. Such low numbers lead to large fluctuations; in particular, some replicates could easily have all cells eradicated at these peptide concentrations. This is the source of the observed fluctuations in Fig. 4.1, which make it difficult to determine 'the MIC' based on a single measurement.

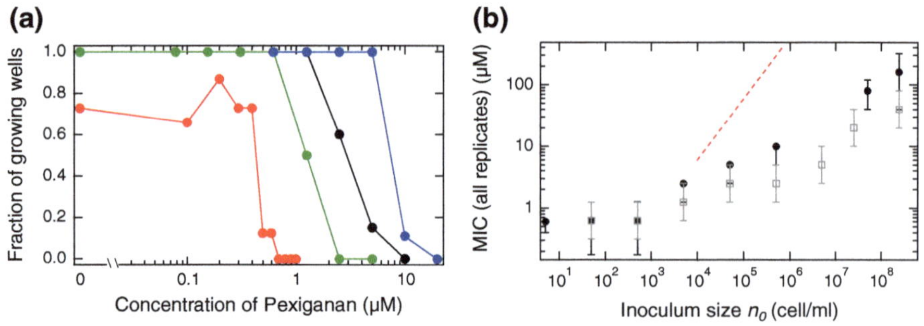

Fig. 4.5 **a** Fraction of growing wells as a function of pexiganan concentration, for inocula of $n_0 = 2.5 \times 10^7$ (*blue*), $n_0 = 5 \times 10^5$ (*black*), $n_0 = 5 \times 10^3$ (*green*) and $n_0 =$ 5 cells/ml ($\equiv$ 1 cell/well) (*red*). **b** The MIC, defined as the concentration of pexiganan at which no replicate grows, plotted against inoculum size, n_0, for one set of experiments using the same pexiganan stock (*grey*) and data collated from many tests (*black*). The *dashed line* has unit slope

4.8 The Inoculum Effect

If pexiganan kills a certain fraction of cells outright, depletes the number of active peptides, and leaves the remaining cells to grow, then there should be a strong inoculum effect (IE): the MIC should be higher for higher inoculum densities. The IE is important because the in vitro population density of pathogens could be significantly higher than typical inocula cell densities in MIC assays. The magnitude of the IE for a small number of inocula concentrations is reported in the antimicrobial literature (Udekwu et al. 2009), including for some AMPs (Levison et al. 1993; Jones et al. 1994); but possible causative mechanisms have seldom been discussed, partly because data for MICs over a large range of inocula densities are, with very few exceptions (Udekwu et al. 2009), not reported.

We report such data for pexiganan acting on *E. coli*. Figure 4.5a shows the fraction of growing wells against pexiganan concentration for four different inoculum sizes, $n_0 = 2.5 \times 10^7$, 5×10^5 (our standard inoculum size), 5×10^3 and 5 cell/ml.[8] Qualitatively, these data are similar to those shown in Fig. 4.2. Quantitatively, the pexiganan concentration at which the fraction of growing wells drops to zero, which we take in this context to be the MIC, increases with inoculum size, n_0. Figure 4.5b plots this MIC as a function of n_0 from two sets of results: one which collates data from multiple assays done over a long period of time, and a second set where the data were collected in one experiment using the same peptide stock for each inoculum size. In both cases, the MIC approaches a constant value at low n_0, and increases sublinearly at higher n_0 (the dashed line in Fig. 4.5b has unit slope).

[8]Note that 5 cells/ml ≡ 1 cell/well, so that the fraction of growing wells is <1 at zero pexiganan.

Perhaps the simplest way to account for an inoculum effect, given our observation of depletion of active peptides by cell lysis, is that each lysed cell inactivates a fixed number of antimicrobial molecules. However, in its simplest form, this mechanism should lead to a linear dependence of the MIC on n_0 at high enough n_0. It is possible that our observed sublinear n_0 dependence is *en route* to such a linear dependence, but we do not reach high enough inoculum concentrations to observe it. On the other hand, a speculative explanation for a sublinear dependence at all n_0 is that cells aggregate at high densities and high peptide concentrations,[9] which may (Moiset et al. 2013) allow peptide molecules or complexes to affect two cells simultaneously.

4.9 Discussion: What Really Is the MIC?

The overwhelming majority of existing MIC measurements rely on a single series of dilution assays performed at one inoculum size (typically 5×10^5 cells/ml) to determine a one-off concentration of antimicrobial agent at which no growth is observed after 24 h. This 'one-off MIC' is frequently assumed to represent, in the words of Melo et al. (2009), 'the macroscopically observable threshold for the onset of [antimicrobial] activity'. This implies that sub-MIC concentrations of antimicrobial agents are sublethal towards the target cells. In the light of our findings, both the practice of determining a 'one-off MIC' value and the quoted interpretation of the significance of this value are problematic, at least for pexiganan acting on *E. coli*.

First, we have found, using multiple replicates, that the 'one-off MIC' shows large intrinsic variability, Fig. 4.1a. This is because at these peptide concentrations, cell numbers drop very low before net regrowth starts, Fig. 4.4, and such low numbers give rise to significant stochasticity. Instead of a 'one-off MIC' from a single-dilution series, findings from multiple replicates, reported as the fraction of samples in which no growth is observed after 24 h as a function of the peptide concentration, Fig. 4.1b, are necessary to give adequate information.

Secondly, we have shown that exposure to sub-MIC levels of pexiganan kills a fraction of *E. coli* in a population quickly, leaving the remaining cells to grow at a rate that is indistinguishable from that of cells never exposed to the AMP, Fig. 4.3. Indeed, the MIC for a standard inoculum size (5×10^5 cell/ml) is $\sim 10\times$ above its single-cell action threshold: compare the black and red data sets in Fig. 4.5a. Thus, it is *not* true that the MIC of pexiganan is 'the macroscopically observable threshold for the onset of [antimicrobial] activity'. Instead, if a single value is desired to quantify the 'single-cell action threshold' of pexiganan on *E. coli* MG1655, a reasonable candidate is the concentration for the onset of lengthening detection times, $\tau_{0.12}$: compare Fig. 4.2d with the red curve in Fig. 4.5a.

[9]We have seen such aggregation in optical microscopy (data not shown).

The interpretation that the MIC itself is a 'single-cell action threshold' clearly lies behind a substantial body of the literature. Thus, for example, Ramamoorthy et al. (2006) found a MIC of 4 μM for pexiganan acting on *E. coli* using an inoculum of 10^7 cell/ml, but observed outer membrane perturbation (according to an ANS uptake assay) at the significantly lower concentrations of 0.67 μM after only 5 min of incubation. Ramamoorthy et al. (2006) infer that factors other than membrane disruption must be involved in the bacterial killing process. A later study came to a similar conclusion on similar grounds (Pius et al. 2012). Our findings mean that this inference is neither necessary nor likely correct: a fraction of cells should be killed (presumably by mechanisms involving membrane perturbation) at 0.67 μM $\lesssim 0.2 \times$ MIC. In general, understanding what MIC values mean is crucial if pharmacodynamic results are to be correctly compared with mechanistic studies.

Thirdly, there is strong evidence that peptides causing cell death are sequestered and unavailable for further bactericidal action. This finding should influence the way in which comparative data between different AMPs are interpreted. Thus, for example, Fluorogainin-1 is designed to be a 'more stable' analogue of pexiganan. It displays a lower MIC against *S. aureus* (Gottler et al. 2008). Our findings suggest that this could be because Fluorogainin-1 is not depleted as rapidly. Separately, one could postulate that mutant strains more resistant to pexiganan (Perron et al. 2006) could have evolved more efficient mechanisms for peptide inactivation in lysed cells, thus raising the measured MIC. These scenarios are speculative, but illustrate once more the need to consider carefully the meaning of MIC measurements.

Finally, we have quantified the inoculum effect (IE) of pexiganan acting on *E. coli*, Fig. 4.5b. There is no IE for $n_0 \lesssim 10^3$ cells/ml; thereafter, MIC $\propto n_0^{\alpha}$ with $\alpha \approx 0.4$. This finding has significant practical and theoretical implications.

Practically, while the IE is well known, it is seldom quantified. In lieu of quantification, it is presumably natural to assume that the MIC is essentially proportional to n_0, especially if the antimicrobial agent is known to be depleted by the process of killing bacteria. Making this assumption in the case of pexiganan acting on *E. coli* MG1655 would lead to very substantial errors, with knock-on implication for dosage regimes arising from pharmacokinetic studies based on such an assumption. Theoretically, we note that our power law with $\alpha \approx 0.4$ also fits the data for 5 out of the 6 antibiotics for which the IE has been quantified (Udekwu et al. 2009) (we find α between 0.3 and 0.5 fitting their data to power laws). Intriguingly, Udekwu et al. (2009) found no depletion of the antibiotic for 3 of these 5 antibiotics. It is a challenge for the future theoretical modelling to decide whether a deeper generic mechanism underly such similarity, or whether two or more mechanisms fortuitously lead to similar quantitative IEs.

Overall, we suggest that data like those presented in Fig. 4.5a should form the minimum basis for interpreting MIC measurements in the context of mechanistic studies, pharmacodynamic modelling and clinical decision-making for all antimicrobial agents. More specifically, given that many bactericidal AMPs are structurally similar to pexiganan, we surmise that our findings concerning the *modus operandi* of pexiganan, e.g. that its action is likely crucially linked to phenotypic heterogeneity of the target organism, may generalise to other compounds. For example, preliminary data on the MIC of amhelin, a pore-forming peptide designed by Rakowska et al. (2013), show the same growth curve (Fig. 4.2) and IE (Fig. 4.5b) as pexiganan. Future research should explore the validity of this surmise for a wider range of AMPs, especially those that, like pexiganan, are bactericidal.

Acknowledgements AKJ was funded by an EPSRC studentship. JSL was funded by the National Physical Laboratory and SUPA. LR and MGR were funded by the UK Department of Business, Innovation and Skills. WCKP was funded by EPSRC Programme Grant EP/J007404/1. We thank Simon Titmuss for illuminating discussions, Angela Dawson for assistance in biological lab work and Vincent Martinez for assistance with data analysis.

References

Abraham JM, Freitag CS, Clements JR, Eisenstein BI (1985) An invertible element of DNA controls phase variation of type 1 fimbriae of *Escherichia coli*. Proc Natl Acad Sci USA 82:5724–5727

Aminov RI (2010) A brief history of the antibiotic era: lessons learned and challenges for the future. Front Microbiol 1:art. 134

Annis DH, Craig BA (2005) The effect of interlaboratory variability on antimicrobial susceptibility determination. Diagn Microbiol Infect Dis 53:61–64

Baraban L, Bertholle F, Salverda MLM, Bremond N, Panizza P, Baudry J, de Visser JAGM, Bibette J (2011) Millifluidic droplet analyser for microbiology. Lab Chip 11:4057–4062

Baranyi J (1998) Comparison of stochastic and deterministic concepts of bacterial lag. J Theor Biol 192(3):403–408

Blattner F, Plunkett G, Bloch C, Perna N, Burland V, Riley M, ColladoVides J, Glasner J, Rode C, Mayhew G, Gregor J, Davis N, Kirkpatrick H, Goeden M, Rose D, Mau B, Shao Y (1997) The complete genome sequence of *Escherichia coli* K-12. Science 277:1453–1462

CLSI (2012) Methods for Dilution Antimicrobial Susceptibility Tests for Bacteria That Grow Aerobically; Approved Standard, 9th edition. CLSI document M07-A9. Clinical and Laboratory Standards Institute, Wayne, PA

Fjell CD, AH J, Hancock REW, Schneider G (2012) Designing antimicrobial peptides: form follows function. Nat Rev Drug Discov 11:37–51

Ericksen B, Wu Z, Lu W, Lehrer RI (2005) Antibacterial activity and specificity of the six human α-defensins. Antimicrob. Agents Chemother. 49:269–275

Fuchs PC, Barry AL, Brown SD (1998) In vitro antimicrobial activity of MSI-78, a magainin analog. Antimicrob Agents Chemother 42:1213–1216

Ge Y, MacDonald DL, Holroyd KJ, Thornsberry C, Wexler H, Zasloff M (1999) In vitro antibacterial properties of pexiganan, an analog of magainin. Antimicrob Agents Chemother 43:782–788

Gottler LM, Ramamoorthy A (2009) Structure, membrane orientation, mechanism, and function of pexiganan: a highly potent antimicrobial peptide designed from magainin. Biochim Biophys Acta 1788:1680–1686

Gottler LM, Lee HY, Shelburne CE, Ramamoorthy A, Marsh ENG (2008) Using fluorous amino acids to modulate the biological activity of an antimicrobial peptide. ChemBioChem 9:370–373

Hendriksen RS (ed) (2010) Laboratory Protocols: Level 1 Training Course: MIC determination by broth dilution using Sensititre. No publisher given, available online at http://goo.gl/57Pu7F. Accessed Sat Mar 19 15:45:46 2016

Hobman JL, Penn CW, Pallen MJ (2007) Laboratory strains of *Escherichia coli*: model citizens or deceitful delinquents growing old disgracefully? Mol Microbiol 64:881–885

Jones E, Smart A, Bloomberg G, Burgess L, Millar M (1994) Lactoferricin, a new antimicrobial peptide. J Appl Bact 77:208–214

Lerouge I, Vanderleyden J (2002) O-antigen structural variation: mechanisms and possible roles in animal/plantâmicrobe interactions. FEMS Microbiol Rev 26:17–47

Levison ME, Pitsakis PG, May PL, Johnson CC (1993) The bactericidal activity of magainins against *Pseudomonas aeruginosa* and *Enterococcus faecium*. J Antimicrob Chemother 32:577–585

Matsuzaki K, Mitani Y, Akada KY, Murase O, Yoneyama S, Zasloff M, Miyajima K (1998) Mechanism of synergism between antimicrobial peptides magainin 2 and PGLa. Biochemistry 37:15,144–53

McGrath DM, Barbu EM, Driessen WHP, Lasco TM, Tarrand JJ, Okhuysen PC, Kontoyiannis DP, Sidman RL, Pasqualini R, Arap W (2013) Mechanism of action and initial evaluation of a membrane active all-D-enantiomer antimicrobial peptidomimetic. Proc Natl Acad Sci USA 110:3477–3482

Melo MN, Ferre R, Castanho MARB (2009) Antimicrobial peptides: linking partition, activity and high membrane-bound concentrations. Nat Rev Microbiol 7(3):245–250

Métris A, Le Marc Y, Elfwing A, Ballagi A, Baranyi J (2005) Modelling the variability of lag times and the first generation times of single cells of *E. coli*. Int J Food Microbiol 100(1):13–19

Métris A, George SM, Baranyi J (2006) Use of optical density detection times to assess the effect of acetic acid on single-cell kinetics. Appl Environ Microbiol 72(10):6674–6679

Moiset G, Cirac AD, Stuart MCA, Marrink SJ, Sengupta D, Poolman B (2013) Dual action of BPC194: a membrane active peptide killing bacterial cells. PLoS ONE 8:e61541

Moore A (2003) The big and small of drug discovery. EMBO Rep 4:114–117

Perron GG, Zasloff M, Bell G (2006) Experimental evolution of resistance to an antimicrobial peptide. Proc Biol Sci/Royal Soc 273:251–256

Pin C, Baranyi J (2006) Kinetics of single cells: observation and modeling of a stochastic process. Appl Environ Microbiol 72:2163–2169

Pius J, Morrow MR, Booth V (2012) H solid-state nuclear magnetic resonance investigation of whole Escherichia coli interacting with antimicrobial peptide MSI-78. Biochemistry 51:118–125

Rakowska PD, Jiang H, Ray S, Pyne A, Lamarre B, Carr M, Judge PJ, Ravi J, Gerling UIM, Koksch B, Martyna GJ, Hoogenboom BW, Watts A, Crain J, Grovenor CRM, Ryadnov MG (2013) Nanoscale imaging reveals laterally expanding antimicrobial pores in lipid bilayers. Proc Natl Acad Sci USA 110:8918–8923

Ramamoorthy A, Thennarasu S, Lee DK, Tan A, Maloy L (2006) Solid-state NMR investigation of the membrane-disrupting mechanism of antimicrobial peptides MSI-78 and MSI-594 derived from magainin 2 and melittin. Biophys J 91:206–216

Rasch M, Métris A, Baranyi J, Bjørn Budde B (2007) The effect of reuterin on the lag time of single cells of Listeria innocua grown on a solid agar surface at different pH and NaCl concentrations. Int J Food Microbiol 113(1):35–40

Schuurmans JM, Nuri Hayali AS, Koenders BB, ter Kuile BH (2009) Variations in MIC value caused by differences in experimental protocol. J Microbiol Meth 79:44–47

Spindler EC, Hale JDF, Giddings TH, Hancock REW, Gill RT (2011) Deciphering the mode of action of the synthetic antimicrobial peptide Bac8c. Antimicrob Agents Chemother 55:1706–1716

Udekwu KI, Parrish N, Ankomah P, Baquero F, Levin BR (2009) Functional relationship between bacterial cell density and the efficacy of antibiotics. J Antimicrob Chemother 63:745–757

Wakamoto Y, Dhar N, Chait R, Schneider K, Signorino-Gelo F, Leibler S, McKinney JD (2013) Dynamic persistence of antibiotic-stressed mycobacteria. Science (New York, NY) 339:91–95

Wiegand I, Hilpert K, Hancock REW (2008) Agar and broth dilution methods to determine the minimal inhibitory concentration (MIC) of antimicrobial substances. Nat Protoc 3:163–175

Wimley WC (2010) Describing the mechanism of antimicrobial peptide action with the interfacial activity model. ACS Chem Biol 5:905–917

Chapter 5
Evolution of Drug Resistance in Bacteria

B. Waclaw

Abstract Resistance to antibiotics is an important and timely problem of contemporary medicine. Rapid evolution of resistant bacteria calls for new preventive measures to slow down this process, and a longer-term progress cannot be achieved without a good understanding of the mechanisms through which drug resistance is acquired and spreads in microbial populations. Here, we discuss recent experimental and theoretical advances in our knowledge how the dynamics of microbial populations affects the evolution of antibiotic resistance. We focus on the role of spatial and temporal drug gradients and show that in certain situations bacteria can evolve de novo resistance within hours. We identify factors that lead to such rapid onset of resistance and discuss their relevance for bacterial infections.

Keywords Antibiotic resistance · Modelling infection · Biological evolution

5.1 Introduction

Since their inception in the beginning of the twentieth century, antibiotics have been an indispensable tool used to treat and prevent bacterial infections and, together with other health care improvements, have revolutionized medicine. However, penicillin-resistant bacteria emerged in just a few years after the introduction of penicillin (Finch 2012), and it is now widely acknowledged that resistance to antibiotics is becoming a global thread (Walsh 2013). Different counter-measures have been proposed (CDC 2013; GOV-UK 2013; WHO 2001), including reducing the usage of antibiotics, better surveillance methods or incentives for the pharmaceutical industry to develop new antibiotics. However, the battle against antimicrobial

B. Waclaw (✉)
SUPA, School of Physics and Astronomy, University of Edinburgh,
Peter Guthrie Tait Road, Edinburgh EH9 3FD, UK
e-mail: bwaclaw@staffmail.ed.ac.uk

© Springer International Publishing Switzerland 2016
M.C. Leake (ed.), *Biophysics of Infection*, Advances in Experimental Medicine
and Biology 915, DOI 10.1007/978-3-319-32189-9_5

resistance (AMR) cannot be won unless we fully understand the molecular basis of resistance, how it spreads among microorganisms, and how it is selected for.

This chapter reviews theoretical and experimental progress in our understanding of the evolution of AMR. We focus on the role of population dynamics and spatial and temporal heterogeneities—the areas which have strongly benefited from a physics-like approach. We briefly discuss relevant experiments and show how their outcomes can be understood using simple mathematical models. In the last section, we also speculate on the significance of in vitro experiments and theoretical models to bacterial infections in vivo.

We do not discuss molecular mechanisms behind AMR (Benveniste and Davies 1973; Davies and Davies 2010) except where necessary. This gives our discussion a more general flavour. In fact, many analogies exist between AMR and resistance to cancer chemotherapy (Lambert et al. 2011), and theoretical models similar to discussed in this chapter have been applied to cancerous tumours (Foo et al. 2015; Fu et al. 2014) and viral infections (Moreno-Gamez et al. 2014).

5.2 Quantifying Resistance

Drug resistance refers to the ability of a pathogen to grow in the presence of a drug. Many different measures have been proposed to quantify resistance. One of them is the minimal inhibitory concentration (MIC), above which no growth can be detected in a standardized experimental protocol (Andrews 2001). Different bacteria and antibiotics have very different MICs (Andrews 2001). MIC can be as low as ~ 10 ng/ml (fluoroquinolones, susceptible bacteria) and as high as ~ 10 mg/ml (beta-lactams, beta-lactamase carrying bacteria) depending on what enzyme the antibiotic targets, the permeability of the cell membrane, and other factors. Another popular way of quantifying resistance is IC_{50}—the concentration of the drug at which the growth rate decreases by 50 % in the exponential phase of growth. A yet more complete characterization of the response to an antibiotic can be obtained by measuring the growth rate at different drug concentrations (Greulich et al. 2015; Regoes et al. 2004). These "kill curves" are non-universal and depend on the antibiotic, bacterial strain and growth conditions (Greulich et al. 2015), but they often follow a sigmoidal shape (Fig. 5.1, left).

High levels of resistance are typically mediated by genes that code enzymes degrading or pumping out the drug (Finch 2012). Such genes are often acquired from other microbial species in the process of gene exchange. However, low-to-moderate resistance can also evolve de novo, sometimes following only a brief exposure to the drug. A single-nucleotide changing mutation that alters the binding site of the antibiotic may be enough to increase the MIC tens of times or more (Brandis et al. 2012; Heisig and Tschorny 1994; Marcusson et al. 2009), and a combination of a few such mutations may be enough to increase it a few thousand times (Marcusson et al. 2009). Such mutations occur with probabilities 10^{-9}–10^{-7}

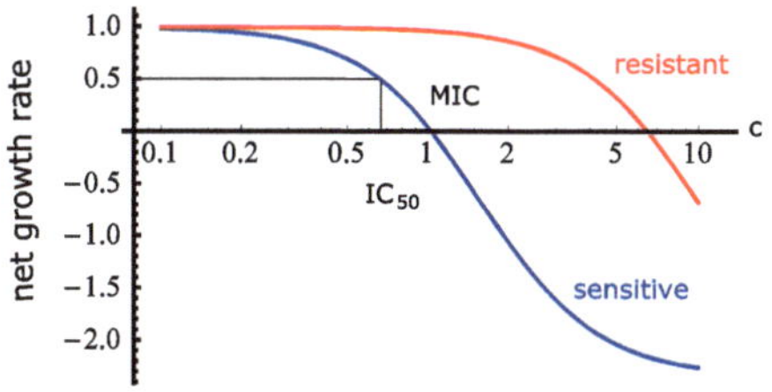

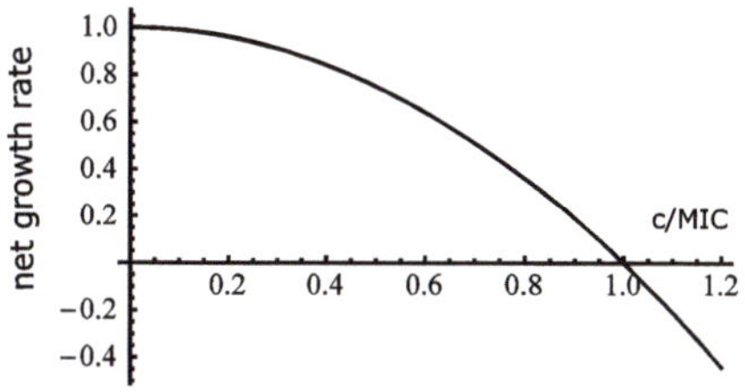

Fig. 5.1 *Left* A typical kill curve (*blue*) versus antibiotic concentration *c*. Net growth = replication minus death. IC_{50} and MIC are defined as *c* at which growth is 50 % of the maximal rate and zero, respectively. The net growth can become negative above MIC if the drug kills the cells or the cells are removed by another mechanism. The curve for the resistant strain (*red*) is shifted to larger *c* and hence it has higher MIC and IC_{50}. *Right* growth curve $g = 1 - (c/\mathrm{MIC})^2$ from Eq. (5.2) used in mathematical models in this chapter

per nucleotide per replication (Gillespie et al. 2005; Lee et al. 2012; Oliver et al. 2000; Sniegowski et al. 1997). In what follows we shall consider only de novo evolution of AMR.

5.3 Evolution of Resistance in Homogeneous Environments

In many laboratory experiments on AMR (Gullberg et al. 2011; Kim et al. 2014; Maisnier-Patin et al. 2002; Perron et al. 2012), bacteria grow in a shaken flask, a microtiter plate, or a chemostat. Shaking or stirring mixes the content so that all bacteria experience identical conditions, including the same concentration of the antibiotic. After a few tens of generations, bacteria often evolve resistance in these experiments when exposed to sub-lethal concentrations of the drug.

To gain insights into what determines the rate of AMR evolution in such experiments, we shall study a simple mathematical model of a population of N bacteria of different genotypes m with different resistance levels (different MICs). Let $m = 0$ be the most sensitive (least resistant) genotype and let the MIC increase monotonously with m (Fig. 5.2, top). Genotype m reproduces at rate g_m which increases with m. Reproduction is assumed to be balanced by death so that the population size remains approximately constant.[1] Genotype m can mutate into genotypes $m \pm 1$ with probability μ per cell per replication.

We want to calculate the rate μ at which a new, more resistant mutant reproducing at a higher rate $g_{m+1} > g_m$ will emerge and take over the population. If the population size is much smaller than the inverse of the mutation rate, the

[1]This can be achieved experimentally by serially diluting the culture when it reaches saturation, or by continuously diluting it in a chemostat.

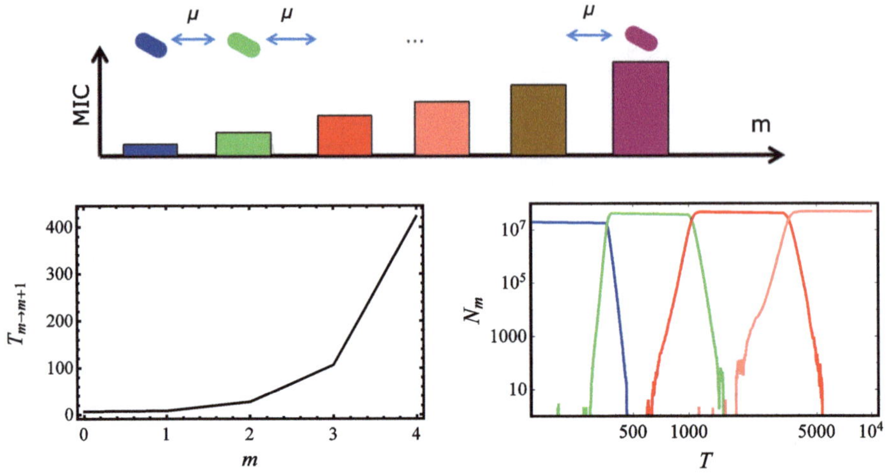

Fig. 5.2 *Top* allowed mutations between genotypes $m = 0, 1, 2, \ldots$. *Bottom left* plot of the times $T_{m\to m+1}$ to the next mutant $m+1$, for $c = 0.9$ and $\mathrm{MIC}_m = 2^m$. *Bottom right* a single simulation run for a stochastic model with carrying capacity $K = 10^8$, death rate $d = 0.5$, mutation probability $\mu = 10^{-9}$, and $c = 0.6$. Colours *blue*, *green*, *red*, and *pink* (as in the *top* panel) correspond to $m = 0, 1, 2, 3$. We can see consecutive fixation of the *green*, *red* and *pink* mutants which replace the previous genotype, as well as failed fixation attempts ("spikes" with $N_m < 10$). This process will continue until there are no mutations that could further increase the MIC, or the selective advantage F of the new mutant becomes comparable to $1/K$. All further mutations behave as neutral mutations and their fixation is extremely rare (probability $F = 1/K$)

rate-limiting step is the process of generating resistant mutants. The rate u is then the product of the rate $\mu g_m N$ with which the mutants are created, times the probability F that a mutant fixes (becomes the dominant organism) in the population. The fixation probability F can be obtained using the standard result of population genetics (Nowak 2006) and is approximately equal to the selective advantage of mutant $m + 1$ over mutant m: $F = (g_{m+1} - g_m)/g_{m+1}$.[2] The rate u is then $\mu g_m NF$ and the average time to resistance can be estimated as

$$T_{m\to m+1} = 1/u = \frac{g_{m+1}}{\mu N g_m (g_{m+1} - g_m)}. \tag{5.1}$$

Each of the growth rates $\{g_m\}$ is a function of the antibiotic concentration c. For the sake of simplicity, we shall assume that

$$g_m = g_{\max}\left(1 - (c/\mathrm{MIC}_m)^2\right), \tag{5.2}$$

[2] This is true as long as $F \gg 1/N$.

so that all functions g_m look the same when plotted against the "normalized" concentration c/MIC_m (Fig. 5.1, right). Equation (5.2) is of course only an approximation to the real kill curves, but for $c < \mathrm{MIC}_m$, it qualitatively reproduces the curves for many antibiotics (cf. Ref. (Regoes et al. 2004)).

5.3.1 Time-Independent Drug Concentration

Let us now apply Eq. (5.1) to the case of fixed concentration c and the initial population consisting of the sensitive strain $m = 0$. If $c > 0$, the antibiotic imposes a selection pressure on the sensitive strain. If the antibiotic concentration is below MIC_0 and the sensitive strain can grow ($g_0 > 0$), a faster-growing mutant $m = 1$ will eventually emerge. After some time $T_{0\to1}$ the population will consist mainly of genotype $m = 1$, which will then produce mutant $m = 2$ etc. Importantly, however, the time $T_{1\to2}$ will typically be larger than $T_{0\to1}$ because the difference $g_2 - g_1$ is smaller than $g_1 - g_0$. To see this more clearly, let us consider a particular, if slightly artificial, example of g_m given by Eq. (5.2) with $\mathrm{MIC}_m = 2^m$. The MICs of the genotypes $m = 0, 1, 2, 3, \ldots$ in this example are $1, 2, 4, 8, \ldots$ (arbitrary units) and the growth rate decreases quadratically with increasing c. This is our "standard model" that we shall use in this chapter. From (5.1) we obtain that

$$T_{m\to m+1} = \frac{4^m(4^{m+1} - c^2)}{g_{\max} 3\mu N c^2 (4^m - c^2)}. \tag{5.3}$$

Figure 5.2, bottom left, shows how quickly $T_{m\to m+1}$ increases with m. This is mainly caused by the fixation probability F decreasing fast with m. Intuitively, this is to be expected—each new mutation increases the MIC but does not increase the growth rate as significantly as the previous mutation, if drug concentration is kept constant. This means that although evolving a moderately resistant strain requiring only a single mutation ($m = 1$) does not take very long, evolving a strain with multiple mutations $m \gg 1$ is a very slow process because the total time $T_{0\to1} + T_{1\to2} + \cdots + T_{m-1\to m}$ increases quickly with the number m of mutational steps.

This is confirmed in a stochastic simulation in Fig. 5.2 bottom right. We assume that cells die with rate d and replicate with rates $g_m = (1 - (c/2^m)^2)(1 - N/K)$. Here N is the total population size, and K is the carrying capacity. The term $1 - N/K$ decreases the growth rate as the carrying capacity is approached, limiting the total population size. Figure 5.2, bottom right shows that it takes 10× less time to evolve mutant $m = 1$ than to evolve the 4×-more resistant mutant $m = 3$. Note that Eq. (5.3) agrees only quantitatively with this result because it uses maximal rather than actual growth rates. When the actual rates are taken into account, the agreement becomes quantitative (Greulich et al. 2012).

5.3.2 Evolution in Time-Varying Drug Concentration

We shall consider two processes in which the concentration $c(t)$ of the antibiotic changes in time:

(1) $c(t)$ varies according to a fixed protocol, independent of the actual state of the population. This is the case of a typical antimicrobial therapy in which a patient takes the medicine at regular intervals, regardless of the present size or genetic composition of the microbial population. In an in vitro experiment, this amounts to, e.g. adding progressively more and more antibiotic to a batch bacterial culture, or gradually increasing the antibiotic influx in a chemostat.
(2) $c(t)$ is continuously adjusted to maximize selection while avoiding killing the population. The size of the population, monitored, e.g. by measuring the turbidity of the sample is used in a feedback loop to control the influx of the drug. An example of a device with such a feedback is the morbidostat (Toprak et al. 2012).

Process (1) can lead to a rapid evolution of resistance only if $c(t)$ increases fast enough to be always close to the MIC of the most resistant genotype in the population, but not exceeding it—otherwise, the population dies out. For a low removal rate d, the critical time T_{crit} over which the concentration increases from $c \ll \text{MIC}$ to $c \approx \text{MIC}$ must be larger than the time to produce at least one resistant mutant. In the case of subsequent fixation of beneficial mutants and for small populations ($\mu N \ll 1$), this critical time is

$$T_{\text{crit}} \approx 1/(\mu g_m N), \tag{5.4}$$

whereas in a large populations in which new mutants occur frequently, the rate-limiting step is the growth from one to enough many cells to produce a mutant and hence

$$T_{\text{crit}} \approx (1/g_m)\ln(1/\mu). \tag{5.5}$$

The rate at which the drug concentration increases must be such that the above equations are fulfilled for all mutants; too slow and the evolution of resistance takes very long, while too fast kills the population. This is hard to achieve in practice unless we know the growth curves, mutation rates, etc. parameters very precisely which is almost never the case. This is illustrated in a computer simulation in Fig. 5.3, top—a small change in the increase rate of the antibiotic concentration over time completely changes the fate of the population (rapid evolution of resistance versus extinction).

Process (2) does not require any knowledge of the MICs, except that the initial concentration of the antibiotic must be below the MIC of the sensitive strain. By systematically increasing c, the time to evolve $m + 1$ from m can be kept very small (cf. Eq. (5.3)). Figure 5.3 bottom shows a simulated evolution of our standard

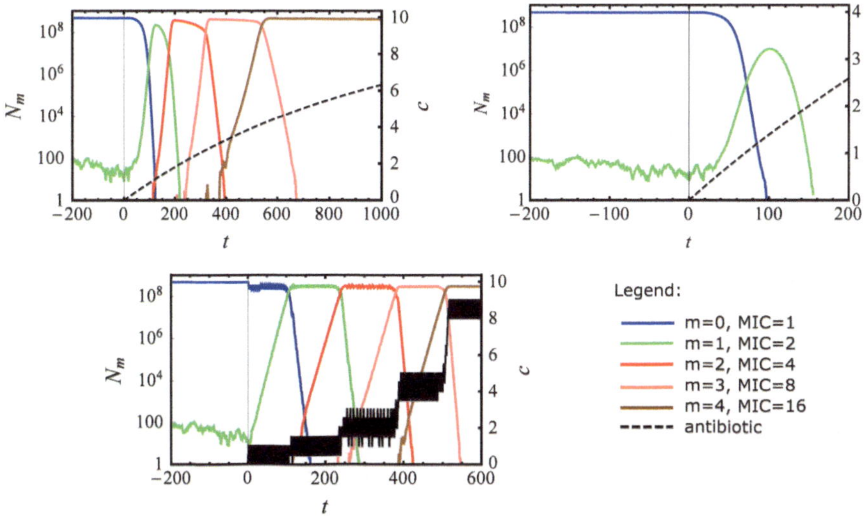

Fig. 5.3 *Top* examples of simulation runs in which antibiotic concentration increases in a pre-determined way as $c(t) = 10(1 - e^{-\alpha t})$ for $t > 0$ (*dashed curves*). α is the increase rate of the drug concentration. Such $c(t)$ would be typical for a chemostat experiment with constant influx of the drug for $t > 0$. *Left* $\alpha = 0.001$, *right* $\alpha = 0.0015$. Different coloured curves correspond to different mutants from Fig. 5.2 (*top*), see also the legend. A 50 % change in the increase rate α causes a switch from fast evolution (*left*) to no evolution and extinction (*right*). Bottom: simulated "morbidostat" in which $c(t)$ is adjusted in response to population's growth to exert strong selection pressure at all times. Parameters: $K = 10^9$, the rest as in Sect. 5.3.1

model in the case when c is increased by 0.5 every hour if the total bacterial count is above 30 % of the carrying capacity K, and decreased by the same amount if the bacterial count is below that value. As expected, evolution proceeds very quickly and soon the most resistant mutant takes over. This rapid evolution agrees with what has been experimentally observed in Ref. (Toprak et al. 2012).

5.4 Evolution of Resistance in Heterogeneous Environments

A uniform distribution of the drug is easy to achieve in the lab but is very unlikely in real infections, where imperfect tissue permeability, drug degradation, binding to proteins, etc. cause the concentration to vary in space. The effect of a non-uniform drug distribution was first studied experimentally in a 2 × 2 cm microfluidic "death galaxy" (Bos et al. 2015; Zhang et al. 2011), in which a gradient of the antibiotic ciprofloxacin was created by pumping two different growth media (with and without the drug) along the opposite sides of the device. Diffusion and partial mixing created a stable gradient in the centre of the device. Surprisingly, bacteria bacterium can increase or decrease its MIC by mutating along a one-dimensional mutational pathway as in our standard model from Fig. 5.2, top.

A kymograph of a typical simulation run for the model from Ref. (Greulich et al. 2012) is plotted in Fig. 5.4, bottom. As in the experiments (Zhang et al. 2011), resistance evolves faster and higher MICs are obtained in the presence of an antibiotic gradient than if the antibiotic is uniformly distributed. In the gradient case, the population advances in waves, each corresponding to a new, better adapted mutant. Although the mathematical theory explaining the difference between the two cases involves some algebra, a qualitative understanding can be achieved using a much simpler and intuitive reasoning described below.

5.4.2 Intuitive Explanation

It turns out that a system with a spatially non-uniform drug distribution behaves somewhat similar to a homogeneous system in which drug concentration increases in time in a way that maximizes selection pressure (Sect. 5.3.2). We can understand this using the well-mixed formula (5.1) applied to the spatial model near the population's edge. Albeit the effective population size N can be quite small as replication is possible only in a narrow region at the edge, this is offset by a large increase in the selective advantage new mutants have at the edge. Let us look at what happens to a population with the same MICs and growth curves as in Sect. 5.3.1. The population of genotype m will move forward until the antibiotic concentration $c \approx \mathrm{MIC}_m = 2^m$ becomes high enough so that the net growth rate (replication minus death) becomes zero. However, if a mutant $m + 1$ occurs near the front where $g_m = \epsilon$ is non-zero but small, its growth rate will be $g_{m+1} \approx (1/2)g_{\max}$ and hence

$$T_{m \to m+1} \approx \frac{1}{\mu N \epsilon}. \tag{5.6}$$

This is independent of m and hence the total time to evolve the m-th mutant increases only linearly with m and not exponentially as in the uniform case (5.3). Drug gradients thus provide optimal conditions for the evolution of resistance: strong selection for more resistant mutants, and free resources (space and nutrients) to explore by successful genotypes.

Whether the absolute time for any given m will be larger or smaller in the gradient than in the uniform case depends on the population size N at the front from which new mutants emerge. In the case of steep gradients, N can be very small and gradients may actually slow down evolution compared to the uniform distribution of the drug. In addition, our reasoning breaks down if a mutation $m \to m+1$ lowers the resistance instead of increasing it. In this case, the genotype $m + 1$ almost never fixes; the probability of fixation is exponentially small in N (Nowak 2006). The only way for the population to move forward is to produce a double (or higher) mutant

$m \to m+2$ which has a positive fitness advantage over genotype m. The rate at which this happens is proportional to $\mu^2 N$ and, as μ^2 is very small, the adaptation time is determined mainly by this process and reads $1/(\mu^2 N)$. Since the effective population size N is smaller in the non-uniform case, drug gradients are likely to slow down evolution if a fitness valley is present in the mutational pathway.

5.5 Real Infections

We shall now discuss how theoretical considerations presented above apply to de novo evolution of resistance in real infections. We shall consider two generic scenarios: (i) a single mutation substantially increases resistance, (ii) multiple mutations are necessary to increase the MIC above the drug concentration achieved during a typical therapy. In (i), resistance will evolve as long as the population is large enough at the onset of treatment. In a large population, resistant mutants are always created even in the absence of the drug but they do not have any selective advantage over the sensitive strain and can grow to high abundance only after treatment has begun. We shall see that this may be the case for chronic infections in cystic fibrosis (Sect. 5.5.1), but this scenario may actually be quite common since bacterial densities often reach 10^8/ml in soft tissue infections (König et al. 1998). Such high densities diminish drug activity, sometimes by 100×, which lowers the required level of resistance and makes AMR easier to evolve (Meredith et al. 2015).

Case (ii) requires a sufficiently slow increase of drug concentration at the site of infection so that multiple mutations have enough time to fix (Sects. 5.3.2 and 5.4.2). The critical time T_{crit} to evolve a more resistant mutant must be much smaller than the time T_{MIC} it takes the antibiotic to reach the MIC of the parent strain. This amounts to the following condition:

$$T_{\text{crit}} \ll T_{\text{MIC}}, \tag{5.7}$$

where T_{crit} is given by Eqs. (5.4), (5.5). If the short-time evolution of the drug concentration is affected mostly by diffusion from a source such as a small blood vessel (Fig. 5.5), this can be rewritten as

$$T_{\text{crit}} \ll \frac{x^2}{D \ln(c_0/\text{MIC})}, \tag{5.8}$$

where x is the distance to the infection site, D is the diffusion constant and c_0 is the source (blood) concentration of the antibiotic. According to this estimate, in vivo evolution of drug resistance mediated by more than a single mutation is likely if: (i) the distance x is large, (ii) the drug diffuses slowly into the tissue (small D),

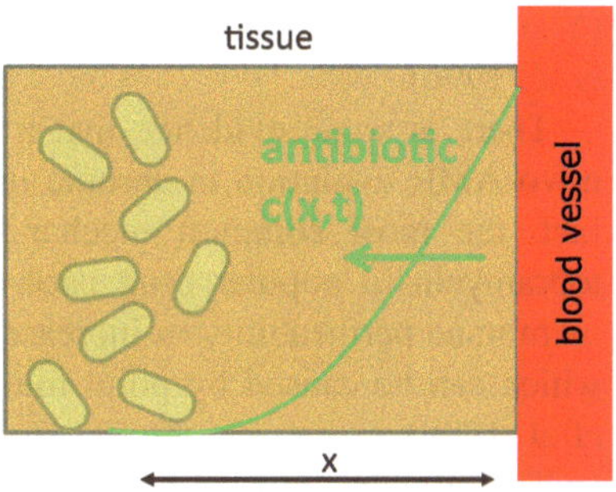

Fig. 5.5 A cartoon view of an infected tissue. The antibiotic diffuses with diffusion constant D from the blood where its concentration is c_0

(iii) bacteria replicate fast (large g_m), (iv) mutation rate μ is large. The last two conditions enter Eq. (5.8) via Eqs. (5.4), (5.5).

Are these requirements fulfilled in a normal tissue? In a healthy and well vascularized tissue, the inter-capillary distance is about 100–200 μm (Casley-Smith et al. 1976), and so the infection site cannot be further than $x \approx 100$ μm from the nearest capillary (drug source). The estimates on the diffusion constants of antibiotics in tissues vary (Meulemans 1992; Meulemans et al. 1989), but the lower limit is $D = 10^{-7}\ \text{cm}^2/\text{s} = 10\ \mu\text{m}^2/\text{s}$.[4] The mutation probability μ ranges from 10^{-9} to 10^{-7}, depending on the antibiotic and bacterial strain (Lee et al. 2012; Oliver et al. 2000). Fastest-reproducing bacteria have $g \sim 10^{-3}\text{s}^{-1}$ (Schlegel and Zaborosch 1993). Using Eq. (5.5) (relevant for sequential fixation) and inserting all these estimates into Eq. (5.8), we obtain that the condition is not fulfilled by a large margin. It is thus very unlikely to evolve drug resistance through the above mechanism in normal tissues. However, not all sites of microbial infections are well vascularised and we shall see an example of that in Sect. 5.5.2.

5.5.1 *Quasi-Uniform Drug Distribution*—P. aeruginosa *Infection in Cystic Fibrosis Patients*

Cystic fibrosis (CF) is a genetic disease characterized by thicker-than-normal secretions such as digestive fluids or mucus. CF affects mostly (but not exclusively) the lungs. The clinical picture is characterised by the presence of thick mucus plugging small and medium-size (0.3–5 mm) bronchioles in the lungs—the tubes that carry air into pulmonary alveoli where gas exchange takes place. CF patients often suffer from lung infections caused by many pathogens, of which *P. aeruginosa* is the most significant one, accounting for 80 % of chronic infections in adult CF patients (Gibson et al. 2003). Bacteria live in the mucus, where they form micro-colonies and biofilms (Singh et al. 2000), with doubling times 100–200 min

[4]Note that this value is much less than D measured in water or agar-based gels often used to study the diffusion of biomolecules (Meulemans et al. 1989).

(Yang et al. 2008) and densities up to 10^7–10^8 bacteria per ml of mucus (Barclay et al. 1996).

There is some evidence that a prolonged course of antibiotics can lead to de novo AMR evolution in chronic infections in CF patients (Gibson et al. 2003). In fact, the most common mechanism of resistance to aminoglycosides such as tobramycin (a popular drug used to treat *P. aeruginosa* infections) is decreased membrane permeability or increased activity of efflux pumps (Poole 2005), both of which can be caused by point mutations (Vogne et al. 2004; Westbrock-Wadman et al. 1999).

To estimate the likelihood of de novo evolution of AMR in a CF patient, we carried out computer simulations of a typical treatment against *P. aeruginosa* using the same model as in Sect. 5.3.1. We modelled a chronic infection with bacteria replicating with max. rate 0.34 h^{-1} before treatment and being removed with rate $d = 0.27$. Treatment begun at $t = 0$ and consisted of two inhalations of 300 mg tobramycin per inhalation per day (Gibson et al. 2003). We assumed that immediately after application, the concentration of tobramycin was 20 μg/ml[5] and its decay followed the first-order kinetics with half-life time 2 h (Barclay et al. 1996). We also took the mutation probability to be $\mu = 10^{-9}$, the rate of point mutations in the PAO1 strain of *P. aeruginosa*. Following the experimental evidence from Ref. (Vogne et al. 2004) we assumed that only one such mutation was required to increase the MIC from ~ 1 to ~ 10 μg/ml.

The population was first allowed to grow to size N starting from a single cell and at the onset of treatment it consisted almost exclusively of sensitive cells (Fig. 5.6). A small sub-population of resistant cells was typically present at $t = 0$ but as these cells did not have any selective advantage, their number was low. However, during therapy ($t > 0$), after the sensitive cells had been killed, resistant cells were able proliferate and re-establish the infection after some time (Fig. 5.6, top). Figure 5.6, bottom, shows that the regrowth probability was very small if the initial size $N < 10^8$, but it approached 100 % for $N > 10^{10}$ cells—a number expected in about 100 ml of mucus.

In reality, the situation may be worse (for the patient) than in our simulations. First, higher mutation rates (100× the basal rate) have been reported in *P. aeruginosa* isolates from CF patients (Oliver et al. 2000). Second, part of the population may be screened from the drug by forming a biofilm (Gibson et al. 2003; Poole 2005; Stewart and William Costerton 2001). Third, bacteria can develop adaptive, non-genetic resistance to antibiotics (Barclay et al. 1996). This type of resistance caused by a temporal reduction in drug uptake (Poole 2005) is unstable and bacteria revert to the sensitive phenotype when the antibiotic is withdrawn, but

[5]We assumed that the concentration was 5× higher than the experimentally determined 3–5 μg/ml, following a single dose of 80 mg (Barclay et al. 1996). The actual concentration may be lower—drug binding to serum proteins lowers the effective concentration of the drug, see e.g. (Bergan 1981; Rowland 1995). Reference (Nickel et al. 1985) claims 8 μg/ml in the plasma, and sputum concentrations can be even lower (Honeybourne 1994). All this can promote the evolution of resistance.

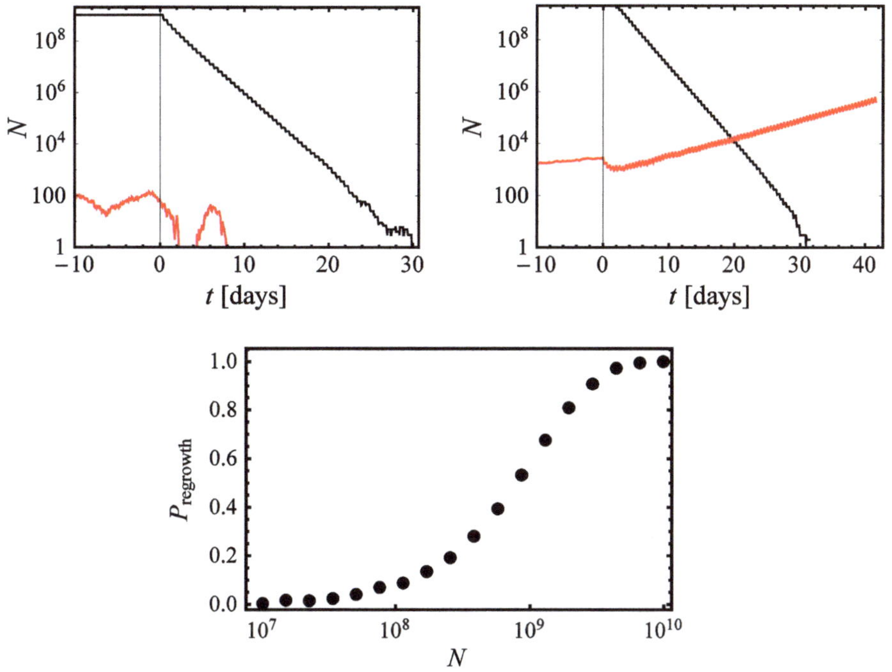

Fig. 5.6 Size and composition of a simulated microbial population following a course of tobramycin. *Left* population size $N = 10^9$ at the onset of treatment. *Right* $N = 10^{10}$. *Bottom* regrowth probability, equal to the fixation probability of the resistant mutant, for different N

it can enable them to evolve genetic resistance by providing a temporary "safe niche" (Levin and Rozen 2006).

Finally, there may be spatial inhomogeneities in the distribution of the drug in the lung. Although difficult to model computationally in a realistic way, such heterogeneities will certainly play a role if a biofilm is formed, or when drugs with low diffusion rates are used. This may be particularly relevant to new drug classes such as antimicrobial peptides (Zasloff 2002) whose molecules are bigger than many antibiotics and hence diffuse slower.

5.5.2 *Drug Gradients—Infections of Avascular Tissues*

We shall now consider a situation in which the distance x from the infected site to the source of antibiotic is much larger than 100 μm. As explained above, the distance between capillaries which deliver blood (and drug) to the tissue is ~100 μm and hence long-lasting antibiotic gradients do not occur in well perfused, normal tissues. However, if blood circulation is impaired, large, cm-size avascular necrotic areas can develop in which dead cells decay and become a rich source of

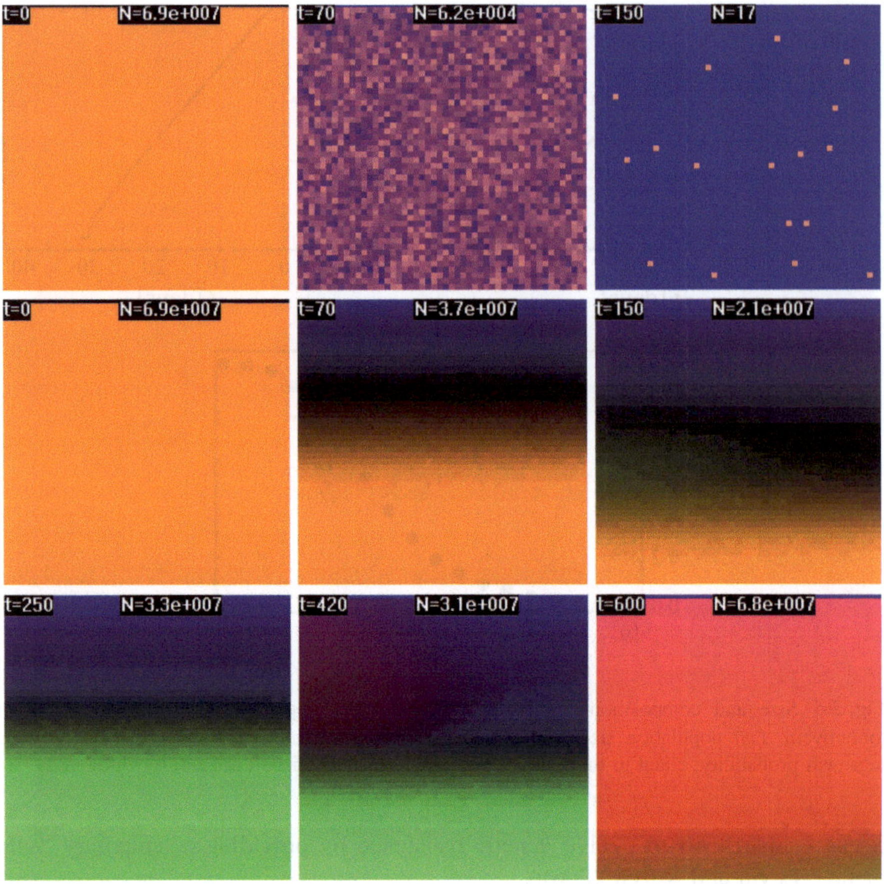

Fig. 5.7 Snapshots of a simulated infected tissue. Time is in hours, *N* is the total bacterial count. The concentration of the antibiotic/bacterial density is shown as different shades of *blue/orange*, respectively. *Top* the tissue is well perfused and the antibiotic diffuses quickly. Bacteria are eradicated after approx. 70 h. *Bottom* avascular tissue, diffusion constant $D = 10^{-7}$ cm^2/s. After the initial decline, the bacterial population grows back due to the evolution of resistant mutants (*green* and *red*)

nutrients fuelling microbial growth. This is common in chronic wounds, especially in the elderly who often develop pressure sores or ulcers due to poor mobility or underlying conditions such as diabetes (Mustoe 2004; Nicolle and Strausbaugh 1996). Bacterial infections of such wounds are common and are treated with both topical (antibiotic applied directly to the wound) and systemic therapy (antibiotic taken orally/intravenously) (Lipsky and Hoey 2009).

Although de novo AMR evolution is thought to be rare (Nicolle et al. 1996), the following numerical simulation demonstrates that it could easily occur in areas with deep necrosis or scarring. We model a 1 cm^3 of ill-perfused tissue inhabited by 10^5

bacteria per mm^3 (Mustoe 2004) at the onset of treatment. We use the same model as in the previous section[6] but we subdivide the tissue into ~200 μm-size regions, each treated as a well-mixed system but with restricted migration between the regions (migration rate per bacterium = 1 h^{-1}). This leads to a saturated colonization in about 5 days. We take *E. coli* as the pathogen and ciprofloxacin as the drug—though neither of them is the most common pathogen/therapeutic drug in wound treatment, ciprofloxacin has been used to treat soft tissue and wound infections (Eron et al. 1985; Smith et al. 1989) and *E. coli* can be isolated from wounds (Bowler et al. 2001). Our choice is motivated by data availability: the nature of mutations and their effect on resistance have been well characterised for this particular combination (Heisig and Tschorny 1994; Marcusson et al. 2009). We assume the sensitive strain has MIC = 30 ng/ml, and two subsequent mutations increase the MIC to 300 and 3000 ng/ml, respectively.

Treatment begins when the total bacterial count is about 10^8. Ciprofloxacin diffuses into the model system from the top side where its concentration is 1 μg/ml —a typical concentration found in tissues and necrotic areas during treatment (Adam et al. 2001; Crump et al. 1983). Figure 5.7 shows that resistance evolves if the antibiotic cannot penetrate the tissue fast enough (diffusion constant $D = 10^{-7}\ \mathrm{cm}^2/\mathrm{s}$) but not if the tissue is well vascularized and the antibiotic quickly reaches a uniform and high concentration.

5.6 Summary

We have reviewed here the theory and recent experiments on the evolution of AMR. We have focused on the population dynamics and the role of spatial heterogeneities in drug distribution. Many models of bacterial infections and treatment were proposed in the past (see e.g. Austin et al. 1999; Bonhoeffer et al. 1997; Lipsitch and Levin 1997) but they usually treated all resistant genotypes as a single species. Here we have discussed the role of the spectrum of mutants on AMR evolution. We have shown that if a sequence of mutations progressively increases the resistance, a highly-resistant strain will often evolve faster if the drug is non-uniformly distributed. We hypothesize that this may be the case in some soft tissue infections.

Although we have not explored this avenue here, it is also likely that *de novo* evolution takes place in biofilms (Donlan and Costerton 2002), where drug gradients can be stronger than in tissues due to decreased permeability and active degradation of the drug. Although resistance in biofilms often relies on non-genetic alterations (Stewart and William Costerton 2001) such as decreased permeability or

[6]We take the maximal growth rate g = 0.34 h^{-1} and a lower removal rate $d = 0.1$ than before because the tissue is not perfused and bacteria are less likely to be removed by the immune system.

switching to a different phenotype, these alterations may actually promote genetic evolution through the "save niche" mechanism mentioned in Sect. 5.5.1.

Further theoretical and experimental research may address AMR evolution in the presence of interactions between different drugs (Bollenbach et al. 2009; Chait et al. 2007), and the role of social interactions between the microbes such as reported in a recent work (Koch et al. 2014) on de novo evolution of vancomycin-intermediate resistant *S. aureus.*

Acknowledgements I thank Rosalind Allen (Edinburgh) and Philip Greulich (Cambridge) for discussions, and Johannes Brennecke (Edinburgh) for reading the manuscript and many helpful suggestions. Financial support of the Royal Society of Edinburgh is acknowledged.

References

Adam U. Herms S, Werner U, Strubelt H, Makowiec F, Hopt UT, Drewelow B (2001) The penetration of ciprofloxacin into human pancreatic and peripancreatic necroses in acute necrotizing pancreatitis. Infection 29(6):326–331. doi:10.1007/s15010-001-1158-y. http://link.springer.com/article/10.1007/s15010-001-1158-y

Andrews JM (2001) Determination of minimum inhibitory concentrations. J Antimicrob Chemother 48(suppl 1):5–16. doi:10.1093/jac/48.suppl_1.5. http://jac.oxfordjournals.org/content/48/suppl_1/5

Antibiotic resistance threats in the United States (2013). http://www.cdc.gov/drugresistance/pdf/ar-threats-2013-508.pdf

Austin D, Kristinsson K, Anderson R (1999) The relationship between the volume of antimicrobial consumption in human communities and the frequency of resistance. Proc Natl Acad Sci 96 (3):1152

Barclay ML, Begg EJ, Chambers ST, Thornley PE, Pattemore PK, Grimwood K (1996) Adaptive resistance to tobramycin in Pseudomonas aeruginosa lung infection in cystic fibrosis. J Antimicrob Chemother 37(6):1155–1164. http://jac.oxfordjournals.org/content/37/6/1155.short

Benveniste R, Davies J (1973) Mechanisms of antibiotic resistance in bacteria. Annu Rev Biochem 42(1):471–506. doi: 10.1146/annurev.bi.42.070173.002351. http://www.annualreviews.org/doi/abs/10.1146/annurev.bi.42.070173.002351

Bergan T (1981) Pharmacokinetics of tissue penetration of antibiotics. Rev Infect Dis 3(1):45–66. http://cid.oxfordjournals.org/content/3/1/45.short

Bollenbach T, Quan S, Chait R, Kishony R (2009) Nonoptimal microbial response to antibiotics underlies suppressive drug interactions. Cell 139(4):707–718. doi:10.1016/j.cell.2009.10.025. http://linkinghub.elsevier.com/retrieve/pii/S0092867409013154

Bonhoeffer S, Lipsitch M, Levin B (1997) Evaluating treatment protocols to prevent antibiotic resistance. Proc National Acad Sci 94(22):12106–12111

Bos J, Zhang Q, Vyawahare S, Rogers E, Rosenberg SM, Austin RH (2015) Emergence of antibiotic resistance from multinucleated bacterial filaments. Proc National Acad Sci 112 (1):178–183. doi:10.1073/pnas.1420702111. http://www.pnas.org/content/112/1/178

Bowler PG, Duerden BI, Armstrong DG (2001) Wound microbiology and associated approaches to wound management. Clin Microbiol Rev 14(2):244–269. doi:10.1128/CMR.14.2.244-269.2001. http://cmr.asm.org/content/14/2/244

Brandis G, Wrande M, Liljas L, Hughes D (2012) Fitness-compensatory mutations in rifampicin-resistant RNA polymerase: fitness compensation of RifR. Mol Microbiol 85

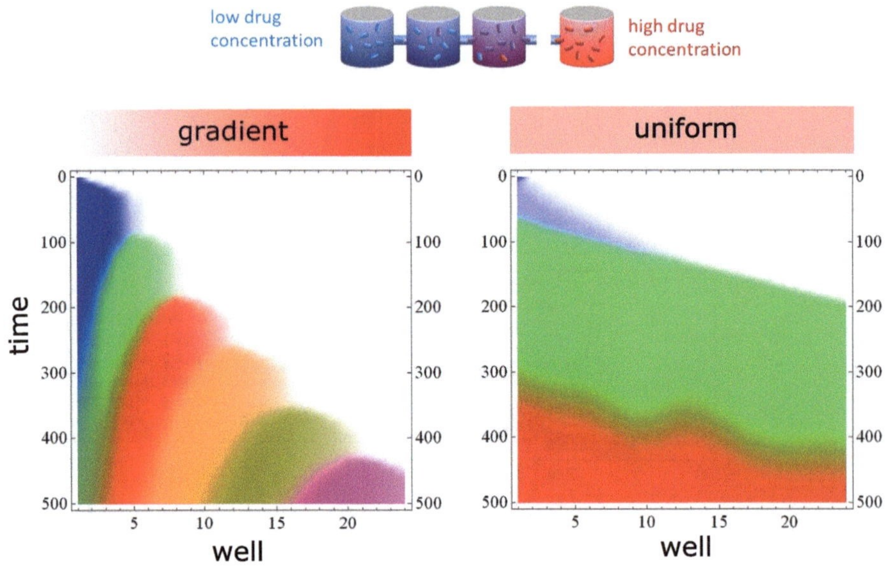

Fig. 5.4 *Top* spatial structure of the model with non-uniform drug distribution. Antibiotic concentration $c(x)$ increases from *left* to *right*. *Left/right* kymographs of two simulation runs for the uniform $c(x) = 0.9$ and non-uniform $c(x) = \exp(3x/24) - 1$ drug distribution. *Horizontal* axis represents space (wells), *vertical* axis represents time. Colours represent different mutants as in Figs. 5.2 and 5.3. The *purple* mutant (MIC = 32) evolves in the gradient case at $t \approx 450$, whereas only a moderately resistant *red* mutant (MIC = 4) can be seen at the same time in the uniform case. In each case, $L = 24$ wells, $K = 10^5$ per well, mutation probability $\mu = 10^{-5}$, migration rate (rate at which bacteria jump between the compartments) $m = 0.02$; other parameters are the same as in Sect. 5.3.1 and Fig. 5.3

with MIC more than 200× that of the parental strain emerged in as little as 10 h in this experiment, whereas no resistant bacteria were detected in a well-mixed control experiment. Resistance was caused by point mutations in four different genes (Zhang et al. 2011).

5.4.1 Computer Model

Two computer models have been proposed (Greulich et al. 2012; Hermsen et al. 2012)[3] to explain these results. Although slightly different in details, both models assume the same spatial and genetic structure. Bacteria inhabit a series of weakly-connected compartments containing increasing (from left to right) amounts of the antibiotic (Fig. 5.4, top). Bacteria migrate between the compartments with rate m and replicate if the antibiotic concentration is below their MIC. Each

[3]See also earlier work (Kepler and Perelson 1998) on HIV infections.

(1):142–151. doi:10.1111/j.1365-2958.2012.08099.x. http://doi.wiley.com/10.1111/j.1365-2958.2012.08099.x

Casley-Smith JR, Sims MA, Harris JL (1976) Capillary lengths and areas, and intercapillary distances in tissue near the human knee. Experientia 32(1):64–66. doi:10.1007/BF01932625. http://link.springer.com/article/10.1007/BF01932625

Chait R, Craney A, Kishony R (2007) Antibiotic interactions that select against resistance. Nature 446(7136):668–671. doi:10.1038/nature05685. http://www.nature.com/doifinder/10.1038/nature05685

Crump B, Wise R, Dent J (1983) Pharmacokinetics and tissue penetration of ciprofloxacin. Antimicrob Agents Chemother 24(5):784–786. doi:10.1128/AAC.24.5.784. http://aac.asm.org/content/24/5/784

Davies J, Davies D (2010) Origins and evolution of antibiotic resistance. Microbiol Mol Biol Rev 74(3):417–433. doi:10.1128/MMBR.00016-10. http://mmbr.asm.org/cgi/doi/10.1128/MMBR.00016-10

Donlan RM, Costerton JW (2002) Biofilms: survival mechanisms of clinically relevant microorganisms. Clin Microbiol Rev 15(2):167–193. doi:10.1128/CMR.15.2.167-193.2002. http://cmr.asm.org/content/15/2/167

Eron LJ, Harvey L, Hixon DL, Poretz DM (1985) Ciprofloxacin therapy of infections caused by Pseudomonas aeruginosa and other resistant bacteria. Antimicrob Agents Chemother 28 (2):308–310. doi:10.1128/AAC.28.2.308. http://aac.asm.org/content/28/2/308

Finch R et al (2012) Antimicrobial chemotherapy, 6 edn. Oxford University Press

Foo J, Choi N, Leder K, Mumenthaler S, Pao W, Michor F, Mallick P, Heise N (2015) The Impact of microenvironmental heterogeneity on the evolution of drug resistance in cancer cells. Cancer Inf 19. doi:10.4137/CIN.S19338. http://www.la-press.com/the-impact-of-microenvironmental-heterogeneity-on-the-evolution-of-dru-article-a4945

Fu F, Nowak MA, Bonhoeffer S (2014) Spatial heterogeneity in drug concentrations can facilitate the emergence of resistance to cancer therapy. arXiv preprint arXiv:1411.6684. http://arxiv.org/abs/1411.6684

Gibson RL, Burns JL, Ramsey BW (2003) Pathophysiology and management of pulmonary infections in cystic fibrosis. Am J Respir Crit Care Med 168(8):918–951. doi:10.1164/rccm.200304-505SO. http://www.atsjournals.org/doi/abs/10.1164/rccm.200304-505SO

Gillespie SH, Basu S, Dickens AL, O'Sullivan DM, McHugh TD (2005) Effect of subinhibitory concentrations of ciprofloxacin on Mycobacterium fortuitum mutation rates. J Antimicrob Chemother 56(2):344–348. doi:10.1093/jac/dki191. http://jac.oxfordjournals.org/content/56/2/344

Greulich P, Waclaw B, Allen RJ (2012) Mutational pathway determines whether drug gradients accelerate evolution of drug-resistant cells. Phys Rev Lett 109(8). doi:10.1103/PhysRevLett.109.088101. http://link.aps.org/doi/10.1103/PhysRevLett.109.088101

Greulich P, Scott M, Evans MR, Allen RJ (2015) Growth-dependent bacterial susceptibility to ribosome-targeting antibiotics. Mol Syst Biol 11(3):796–796. doi:10.15252/msb.20145949. http://msb.embopress.org/cgi/doi/10.15252/msb.20145949

Gullberg E, Cao S, Berg OG, Ilbäck C, Sandegren L, Hughes D, Andersson DI (2011) Selection of resistant bacteria at very low antibiotic concentrations. PLoS Pathog 7(7), e1002,158. doi:10.1371/journal.ppat.1002158. http://dx.plos.org/10.1371/journal.ppat.1002158

Heisig P, Tschorny R (1994) Characterization of fluoroquinolone-resistant mutants of escherichia coli selected in vitro. Antimicrob Agents Chemother 38(6):1284–1291. http://aac.asm.org/content/38/6/1284.short

Hermsen R, Deris JB, Hwa T (2012) On the rapidity of antibiotic resistance evolution facilitated by a concentration gradient. Proc Natl Acad Sci 109(27):10775–10780

Honeybourne D (1994) Antibiotic penetration into lung tissues. Thorax 49(2):104–106. http://thorax.bmj.com/content/49/2/104.full.pdf

Kepler T, Perelson A (1998) Drug concentration heterogeneity facilitates the evolution of drug resistance. Proc National Acad Sci 95(20):11,514

Kim S, Lieberman TD, Kishony R(2014) Alternating antibiotic treatments constrain evolutionary paths to multidrug resistance. Proc National Acad Sci 201409800. doi:10.1073/pnas.1409800111. http://www.pnas.org/content/early/2014/09/16/1409800111

Koch G, Yepes A, Förstner KU, Wermser C, Stengel ST, Modamio J, Ohlsen K, Foster KR, Lopez D (2014) Evolution of resistance to a last-resort antibiotic in staphylococcus aureus via bacterial competition. Cell 158(5):1060–1071. doi:10.1016/j.cell.2014.06.046. http://www.sciencedirect.com/science/article/pii/S0092867414008733

König C, Simmen HP, Blaser J (1998) Bacterial concentrations in pus and infected peritoneal fluid–implications for bactericidal activity of antibiotics. J Antimicrob Chemother 42(2):227–232. doi:10.1093/jac/42.2.227. http://jac.oxfordjournals.org/content/42/2/227

Lambert G, Estévez-Salmeron L, Oh S, Liao D, Emerson B, Tlsty T, Austin R (2011) An analogy between the evolution of drug resistance in bacterial communities and malignant tissues. Nat Rev Cancer 11(5):375–382

Lee H, Popodi E, Tang H, Foster PL (2012) Rate and molecular spectrum of spontaneous mutations in the bacterium Escherichia coli as determined by whole-genome sequencing. Proc National Acad Sci U.S.A. 109(41):E2774–E2783. doi:10.1073/pnas.1210309109. http://www.ncbi.nlm.nih.gov/pmc/articles/PMC3478608/

Levin BR, Rozen DE(2006) Non-inherited antibiotic resistance. Nature Reviews Microbiology 4 (7):556–562. http://www.nature.com/articles/nrmicro1445

Lipsitch M, Levin B (1997) The population dynamics of antimicrobial chemotherapy. Antimicrob Agents Chemother 41(2):363

Lipsky BA, Hoey C (2009) Topical antimicrobial therapy for treating chronic wounds. Clin Infect Dis 49(10):1541–1549. doi:10.1086/644732. http://cid.oxfordjournals.org/content/49/10/1541

Maisnier-Patin S, Berg OG, Liljas L, Andersson DI (2002) Compensatory adaptation to the deleterious effect of antibiotic resistance in Salmonella typhimurium. Mol Microbiol 46 (2):355–366. doi:10.1046/j.1365-2958.2002.03173.x. http://onlinelibrary.wiley.com/doi/10.1046/j.1365-2958.2002.03173.x/abstract

Marcusson LL, Frimodt-Møller N, Hughes D (2009) Interplay in the selection of fluoroquinolone resistance and bacterial fitness. PLoS Pathog 5(8):e1000,541. doi:10.1371/journal.ppat.1000541. http://dx.plos.org/10.1371/journal.ppat.1000541

Meredith HR, Srimani JK, Lee AJ, Lopatkin AJ, You L (2015) Collective antibiotic tolerance: mechanisms, dynamics and intervention. Nat Chem Biol 11(3):182–188. doi:10.1038/nchembio.1754. http://www.nature.com/nchembio/journal/v11/n3/full/nchembio.1754.html

Meulemans A(1992) A model of cefoperazone tissue penetration: diffusion coefficient and protein binding. Antimicrob Agents Chemother 36(2):295–298. http://aac.asm.org/content/36/2/295.short

Meulemans A, Paycha F, Hannoun P, Vulpillat M (1989) Measurement and clinical and pharmacokinetic implications of diffusion coefficients of antibiotics in tissues. Antimicrob Agents Chemother **33**(8):1286–1290. http://aac.asm.org/content/33/8/1286.short

Moreno-Gamez S, Hill AL, Rosenbloom DI, Petrov DA, Nowak MA, Pennings P (2014) Imperfect drug penetration leads to spatial monotherapy and rapid evolution of multi-drug resistance. bioRxiv 013003. http://biorxiv.org/content/early/2014/12/19/013003.abstract

Mustoe T (2004) Understanding chronic wounds: a unifying hypothesis on their pathogenesis and implications for therapy. Am J Surg 187(5, Supplement 1), S65–S70. doi:10.1016/S0002-9610(03)00306-4. http://www.sciencedirect.com/science/article/pii/S0002961003003064

Nickel JC, Ruseska I, Wright JB, Costerton JW (1985) Tobramycin resistance of Pseudomonas aeruginosa cells growing as a biofilm on urinary catheter material. Antimicrob Agents Chemother 27(4):619–624. doi:10.1128/AAC.27.4.619. http://aac.asm.org/content/27/4/619

Nicolle LE, Strausbaugh LJ, Garibaldi RA (1996) Infections and antibiotic resistance in nursing homes. Clin Microbiol Rev 9(1):1–17. http://cmr.asm.org/content/9/1/1

Nowak MA (2006) Evolutionary dynamics. Harvard University Press

Oliver A, Cantón R, Campo P, Baquero F, Blázquez J (2000) High frequency of hypermutable pseudomonas aeruginosa in cystic fibrosis lung infection. science 288(5469):1251–1253. doi:10.1126/science.288.5469.1251. http://www.sciencemag.org/content/288/5469/1251

Organization WH, others (2001) WHO global strategy for containment of antimicrobial resistance. http://apps.who.int/iris/handle/10665/66860

Perron GG, Kryazhimskiy S, Rice DP, Buckling A (2012) Multidrug therapy and evolution of antibiotic resistance: when order matters. Appl Environ Microbiol 78(17):6137–6142. doi:10.1128/AEM.01078-12. http://aem.asm.org/content/78/17/6137

Poole K (2005) Aminoglycoside resistance in pseudomonas aeruginosa. Antimicrob Agents Chemother 49(2):479–487. doi:10.1128/AAC.49.2.479-487.2005. http://aac.asm.org/content/49/2/479

Regoes R, Wiuff C, Zappala R, Garner K, Baquero F, Levin B (2004) Pharmacodynamic functions: a multiparameter approach to the design of antibiotic treatment regimens. Antimicrob Agents Chemother 48(10):3670–3676

Rowland M, Tozer TN (1995) Clinical pharmacokinetics: concepts and applications

Schlegel HG, Zaborosch C, Kogut M (1993) General microbiology. Cambridge University Press

Singh PK, Schaefer AL, Parsek MR, Moninger TO, Welsh MJ, Greenberg EP (2000) Quorum-sensing signals indicate that cystic fibrosis lungs are infected with bacterial biofilms. Nature 407(6805):762–764. doi:10.1038/35037627. http://www.nature.com/nature/journal/v407/n6805/abs/407762a0.html

Smith SM, Eng RH, Tecson-Tumang F (1989) Ciprofloxacin therapy for methicillin-resistant Staphylococcus aureus infections or colonizations. Antimicrob Agents Chemother 33(2):181–184. doi:10.1128/AAC.33.2.181. http://aac.asm.org/content/33/2/181

Sniegowski PD, Gerrish PJ, Lenski RE (1997) Evolution of high mutation rates in experimental populations of E. coli. Nature 387(6634):703–705. doi:10.1038/42701. http://www.ncbi.nlm.nih.gov/pubmed/9192894

Stewart PS, William Costerton J (2001) Antibiotic resistance of bacteria in biofilms. The Lancet 358(9276):135–138. http://www.sciencedirect.com/science/article/pii/S0140673601053211

Toprak E, Veres A, Michel JB, Chait R, Hartl DL, Kishony R (2012) Evolutionary paths to antibiotic resistance under dynamically sustained drug selection. Nat Genet 44(1):101–105. doi:10.1038/ng.1034. http://dx.doi.org/10.1038/ng.1034

UK 5 year antimicrobial resistance strategy 2013–2018. https://www.gov.uk/government/publications/uk-5-year-antimicrobial-resistance-strategy-2013-to-2018

Vogne C, Aires JR, Bailly C, Hocquet D, Plésiat P (2004) Role of the multidrug efflux system Mexxy in the emergence of moderate resistance to aminoglycosides among Pseudomonas aeruginosa isolates from patients with cystic fibrosis. Antimicrob Agents Chemother 48(5):1676–1680. doi:10.1128/AAC.48.5.1676-1680.2004. http://aac.asm.org/content/48/5/1676

Walsh F(2013) Antibiotics resistance 'as big a risk as terrorism'—medical chief. http://www.bbc.co.uk/news/health-21737844

Westbrock-Wadman S, Sherman DR, Hickey MJ, Coulter SN, Zhu YQ, Warrener P, Nguyen LY, Shawar RM, Folger KR, Stover CK (1999) Characterization of a Pseudomonas aeruginosa efflux pump contributing to aminoglycoside impermeability. Antimicrob Agents Chemother 43(12):2975–2983. http://aac.asm.org/content/43/12/2975

Yang L, Haagensen JAJ, Jelsbak L, Johansen HK, Sternberg C, Hoiby, N., Molin, S (2008) In situ growth rates and biofilm development of Pseudomonas aeruginosa populations in chronic lung infections. J Bacteriol 190(8):2767–2776. doi:10.1128/JB.01581-07. http://www.ncbi.nlm.nih.gov/pmc/articles/PMC2293235/

Zasloff M (2002) Antimicrobial peptides of multicellular organisms. Nature 415(6870):389–395. doi:10.1038/415389a. http://www.nature.com/nature/journal/v415/n6870/full/415389a.html

Zhang Q, Lambert G, Liao D, Kim H, Robin K, Tung CK, Pourmand N, Austin RH (2011) Acceleration of emergence of bacterial antibiotic resistance in connected microenvironments. Science 333(6050):1764–1767. doi:10.1126/science.1208747. http://www.sciencemag.org/cgi/doi/10.1126/science.1208747

Chapter 6
Using Biophysics to Monitor the Essential Protonmotive Force in Bacteria

Mei-Ting Chen and Chien-Jung Lo

Abstract Protonmotive force is an essential biological energy format in all levels of cells. Protonmotive force comprises electrical and chemical potential difference across biological membrane. In bacteria, protonmotive force couples to metabolism and ATP production. Moreover, protonmotive force directly provides driving energy of bacterial flagellar motor that is critical for bacterial motility and infection. Due to the small size of bacterial cells, there were limited experimental tools to measure protonmotive force in bacteria. Recent developments of optical membrane potential and intracellular pH indicators provide valuable information on bacterial studies. These new biophysical techniques allow us to monitor the protonmotive force even in single bacterial cell level that shed the light of next generation single-cell physiological experiments towards the understanding of bacterial infection process.

Keywords Prontonmotive force · Fluorescent indicator · pHluorin · Membrane potential · Single-cell measurement

6.1 Introduction

Protonmotive force (Δp) is the sum of electrical and chemical potential across cellular membrane (Mitchell 1966; Kashket 1985). It is important to understand the energy flow in bacteria from metabolism to maintaining protonmotive force and the following coupling to ATP production and membrane-associated reactions. For example, bacteria use protonmotive force for several import cellular functions such

M.-T. Chen · C.-J. Lo
Department of Physics, National Central University, Jhongli, Taiwan 32001, Republic of China

C.-J. Lo (✉)
Graduate Institute of Biophysics,
National Central University, Jhongli, Taiwan 32001, Republic of China
e-mail: cjlo@phy.ncu.edu.tw

© Springer International Publishing Switzerland 2016

M.C. Leake (ed.), *Biophysics of Infection*, Advances in Experimental Medicine and Biology 915, DOI 10.1007/978-3-319-32189-9_6

as cell flagellar motility (Sowa 2008; Baker 2009; Lo 2013), ATP synthesis (Okuno 2011), and active membrane transport (Shultis 2006). Flagellar motility and protonmotive force energetic are critical for infection (Ottemann 2002; Stingl 2002) and biofilm formation (Blair 2008).

The Protonmotive force is expressed as

$$PMF = V_m - Z\Delta pH \tag{6.1}$$

where $Z = 2.303 \times RT/F = 2.303 \times k_BT/e$, R the gas constant, T the absolute temperature (K), F the Faradays constant, k_B the Boltzmann's constant, and e the electrical charge.

The protonmotive force is not physical force (in the unit of *Newton*) but a protonic potential difference (in the unit of *volt*) across cellular membranes (from outside to inside). It consists of V_m, the electrical potential difference across membrane, and ΔpH, the chemical potential difference (intracellular pH—extracellular pH).

Measuring Protonmotive force of bacteria is not a simple task because of cell size and membrane properties. V_m measurement probing the electrical properties across the membrane evolves all of the charges, while pH measurement only involves with specific ion species, proton. Direct measurement of both V_m and pH in such a small compartment is limited by the direct physical tools such as electrodes and glass micropipette. Therefore, most of the protonmotive force measurements use the property changes of designed molecules in response to different protonmotive forces. Because of the limitation of signal detection from these designed molecules, it was only possible for bulk measurement of protonmotive force (Kashket 1985). Recent development of various fluorescent indicators makes single-cell measurement possible (Lo 2006, 2007; Martines II 2012). It is also important to monitor individual cells towards the single-cell physiology understanding.

This chapter is divided into three parts. In the first part, my aim is to present a general assessment of understanding the physical origin of protonmotive force and the concepts behind the measurement methods. In the second part, I will introduce modern single-cell protonmotive force measurements using optical membrane voltage indicator and pH indicator. The final part introduces the use of bacterial flagellar motor as a fast protonmotive force measurement tool. The second and third parts are focused on the protonmotive force measurement in *Escherichia coli (E. coli)*. *E. coli* is a pathogen but also these proof-of-concept techniques could then be applied to much more dangerous pathogens if required.

6.2 Protonmotive Force

6.2.1 Membrane Voltage Measurements

Membrane potential (V_m) is generated by ion transportation across membrane. Although most of the membrane potential indicators are developed for the eukaryotic excitable cells, these indicators are potentially applicable to prokaryotic cells with extra cares of the loading mechanisms. There are two different types of techniques to measure membrane potential: (1) Nernst potential indicator and (2) Local electrical field indicator. Typically Nernst potential indicators are slow (in the order of mins) and local electrical field indicators are fast (in the order of μs-ms) in response to the membrane potential change. But the signal changes of Nernst potential indicator could be as high as 100 times/100 mV comparing with 2–30 %/ 100 mV for local electrical field indicator. Therefore, Nernst potential indicators are suitable for the steady state measurement, while local electrical field indicators are suitable for the dynamical change of membrane potential.

Extra care must be taken into the loading process because the membrane structure is different from cell to cell. For bacterial cells, high permeability of Nernst potential indicator is the key for successful membrane potential measurement.

6.2.1.1 Nernst Potential Indicator

For a cationic and membrane-permeable molecule, it can be used to measure the membrane potential because the distribution of the indicator molecules in equilibrium would follow Boltzmann's Law:

$$C_{in}/C_{ex} = exp(-qV_m/k_BT) \tag{6.2}$$

where q the charge of the indicator, and C_{in} and C_{ex} the intracellular and extracellular concentrations, respectively. For typical negative membrane potential and positive charged indication, the indicator molecules would accumulate inside the cell. Assuming that the external indicator concentration is given, the membrane potential can be calculated by the Nernst equation

$$ln(C_{in}/C_{ex}) = -q(V_m)/k_BT \tag{6.3}$$

Since lnN ≈ 2.303logN, membrane potential can be rewritten in a general form for simpler calculation

$$V_m = 2.303\frac{k_BT}{q}\log\frac{C_{in}}{C_{ex}} \tag{6.4}$$

By measuring the C_{in} with the given C_{ex}, the membrane potential can be found. There are various techniques to measure the C_{in} by the indictor's properties such as radioactivity and fluorescence. The indicators must have high membrane permeability for rapid equilibrium and low toxicity to avoid damaging cells. Further requirements including low membrane binding and low cellular compartment binding would ensure the accuracy of the membrane potential measurements, Fig. 6.1a.

For room temperature T = 25 °C = 298 K, $2.303k_BT/q \approx 59$ mV. A typical bacterial membrane potential is around −120 mV, corresponding to $C_{in}/C_{ex} \approx 110$.

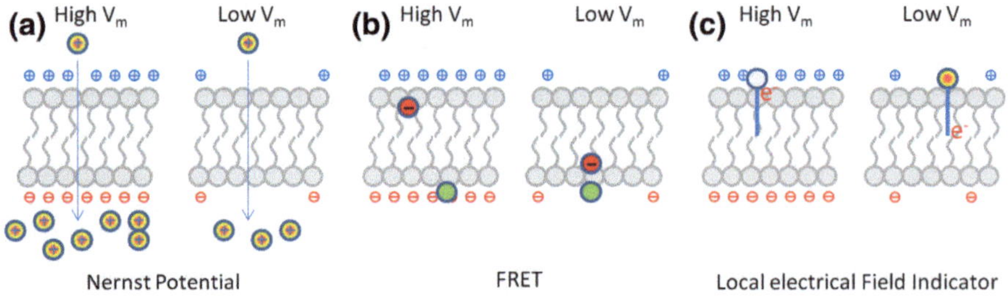

Fig. 6.1 Mechanisms of optical membrane potential indicators. **a** Nernst potential indicators are charged fluorescent membrane-permeable molecules that the distribution inside and outside of cells will obey Nernst equation. **b** FRET-pair membrane potential indicator has two counterparts where one is immobilized fluorophore (*green*) and one is mobile charged fluorophore (*red*) to detect membrane potential difference. **c** Local electrical field indicators are peptide-like molecules inserted into membrane that the local electric field affects the fluorophore function

This can lead to aggregation and quenching of fluorescence. Therefore, careful indicator loading test and low indicator concentration are recommended.

Optical measurement has the great advantage to monitor the C_{in} in many respects. First, it is a fast measurement and can be monitored by fluorescent-based laboratory equipments such as fluorescent microscope, fluorescent plate reader, flow cytometry and fluorescent spectroscopy. Second, due to the high signal-to-noise ratio of fluorescent indicators and small population, even single-cell membrane potential measurement is possible.

There are carbocyanine (cationic) dyes, oxonol (anionic) dyes, and rhodamine (cationic) dyes as commercial available Nernst potential indicators. It has been demonstrated that rhodamin serious dye, TMRM/TMRE, is useful for the membrane potential measurement in *E. coli* and other bacterium in the single-cell level (Lo 2006; Kurre 2013).

6.2.1.2 Local Electrical Field Indicator

To measure the local electrical field properties of cell membrane, the probing molecules are inserted into the membrane. The plasma membrane of a cell is typically in the ~ -120 mV. Unlike excitable eukaryote cells, prokaryote cells maintain their membrane potential stably. Typical optical probing molecules can be cataloged into two types: charge transfer type and FRET-pair type.

Charge transfer type membrane potential indicator can be electrochromic indicator or photo-induced electron transfer (PeT) indicators (Woodford 2015), Fig. 6.1c. Typical electrochromic indicator has small membrane potential dependent fluorescence changes (2–10 % per 100 mV). Newly developed PeT indicator shows higher response up to 30 % per 100 mV and 100 μs response time. PeT indicator shows very promising results in neuron cells (Woodford 2015). However, structure modification may be required to apply on bacterial membrane.

In order to enhance the response signal of membrane potential measurement with sufficient response speed, Forster resonance energy transfer (FRET) type membrane potential indicators have been developed. FRET-pair voltage indicator uses lipophilic negatively charges dye which positions in the membrane depending on the overall membrane voltage, to quench its immobilized FRET partner, Fig. 6.1b. The FRET-pairs could be nongenetic dyes or genetic fluorescent proteins (Bradley 2009). However, limited development of application on bacterial cells can be found.

6.2.2 *Intracellular pH Measurements*

6.2.2.1 Nongenetic pH Indicator

There are several dyes that change their color in response to a pH change such as phenol red. On the other hand, fluorescent dyes provide fluorescent intensity which change in response to pH change. There are two important factors regarding measuring intracellular pH in living cells. First, the binding affinity of the experimental system that limits the measurable pH range. Second, mostly the nongenetic pH indicators are changing the fluorescent intensity in response to different pH. Careful calibration is required for the accurate pH measurements by using these indicators.

6.2.2.2 Genetic pH Indicator

Genetically encoded pH indicators are fluorescent proteins whose fluorescent functions are sensitive to the local environmental pH. There are several advantages of using genetically encoded pH indicator in the living bacteria. First, the loading process is replaced by the protein expression that is typically uniform among cells. Second, their spectral responses to the pH allow ratiometric measurements that provide accurate intracellular pH measurement.

6.3 Protonmotive Force Measurements in Bacteria

In this section, single cell optical protonmotive measurements are presented.

6.3.1 *Nernst Potential Indicator*

The Nernstian fluorescent dye, tetramethyl rhodamine methyl ester (TMRM) has been applied to measure intracellular membrane potential in different bacteria

(Lo 2007; Kurre 2013). A short treatment (10 min) of 10 mM EDTA buffer can increase permeability to TMRM.

For *E. coli* dye loading, cells at a density of ~10^8 cells/ml were suspended in PBS with 0.1 μM TMRM for 40 min. Then the cells can be transferred and immobilized on coverglass for the epifluorescence microscopy observation. The excitation and emission wavelength peak are at 552 nm and 575 nm, respectively. Standard TRITC fluorescent microscopy cube and excitation light source can be used to observe TMRM signal.

To calculate the membrane potential from Eq. (6.4), one needs to measure the C_{in}/C_{ex} precisely. Bacterial cells are about 1 μm in size that is close to the diffraction limit (about half of the wavelength ~0.25 μm). Therefore, the blurring effect of fluorescent images would seriously affect the calculation of accurate membrane voltage from the fluorescent ratio (F_{in}/F_{ex}). A careful estimation of the correction factor is listed below.

6.3.1.1 Point Spread Function (PSF)

The PSF, $P(x_i, y_i, z_i)$, of your microscopy system can be measured by a z-stack images of 20 nm fluorescent beads stuck on surface. After background subtraction and normalization, the averaged PSF can be obtained from multiple beads data, Fig. 6.2b.

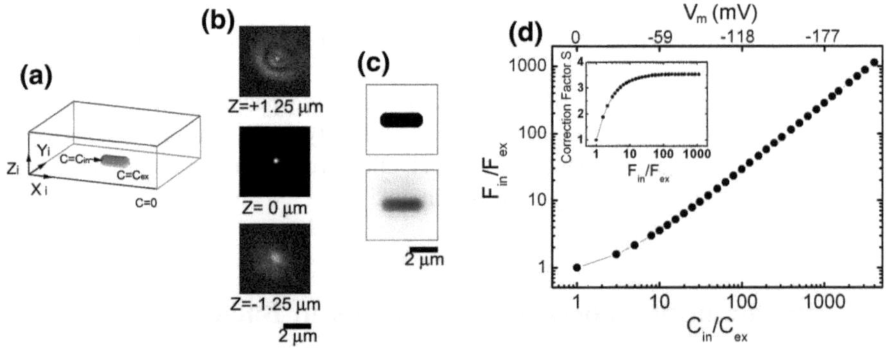

Fig. 6.2 Calculation of the correction factor for convolution of the dye distribution with the microscope PSF. **a** Model of the dye distribution for a bacterium attached to a coverglass. **b** Cross sections of the measured PFS at different values of z. **c** Dye distribution at the mid-cell height (*upper*) and the corresponding image after convolution (*lower*). **d** the calculated fluorescence intensity ratio, F_{in}/F_{ex}, versus the concentration ratio C_{in}/C_{ex}. The correction factor, $S(F_{in}/F_{ex}) = (C_{in}/C_{ex})/(F_{in}/F_{ex})$ is shown in the *inset* (Lo 2007)

6.3.1.2 Optical Convolution Model

One can create a 3D model of a chamber with a bacterium with their shape stuck to the surface, Fig. 6.2a. The dye concentrations in the model are

$$C_i(x_i, y_i, z_i) = \begin{Bmatrix} 0 & : & z \leq 0,\ glass \\ C_{ex} & : & medium \\ C_{in} & : & cell \end{Bmatrix} \tag{6.5}$$

In this demonstration, the cell is an *E. coli* cell that is a cylinder (2 μm long, 0.9 μm diameter) capped by two hemispheres (0.9 μm diameter) and stuck on the surface. The remaining volume of the chamber is filled with medium. The blurred image, with the focal plane at the middle of the cell, was calculated as the convolution of the dye distribution and the PSF:

$$I_m(x_j, y_j, z_0) = \Sigma\left[C_i(x_i, y_i, z_i)P(x_j - x_i, y_j - y_i, z_0 - z_i)\right] \tag{6.6}$$

where $I_m(x_j, y_j, z_0)$ was the modeled intensity of image pixel j, Fig. 6.2c lower. By comparing the images before and after convolution, Fig. 6.2c, the correction factor $S(F_{in}/F_{ex}) = (C_{in}/C_{ex})/(F_{in}/F_{ex})$ can be found, Fig. 6.2d. Please note that the width of bacteria is an important factor for the correction factor while bacterial length is less important. One should generate their correction factor according to their bacterial shape.

6.3.1.3 Calculation of Membrane Potential

The V_m can be calculated using the Nernst equation, Eq. (6.4), and the corrected concentration ratio

$$C_{in}/C_{ex} = (F_{in}/F_{ex})S(F_{in}/F_{ex}) \tag{6.7}$$

where F_{in} and F_{ex} were measured experimentally and the correction factor $S(F_{in}/F_{ex})$ was calculated using the optical convolution model. Take *E. coli* for example, the membrane potential varies linearly (−80 to −140 mV) in response to external pH change (pH5–pH7), Fig. 6.3.

6.3.2 pHluorin

pHluorin, a GFP-derivative protein, is a ratiometric pH indicator with two excitation peaks at 410 and 470 nm and single emission peak at 508 nm (Miesenböck et al. 1998). It has been applied to measure intracellular pH in different bacteria (Morimoto 2011; Kurre 2013; Martinez II 2012). The different responses in two

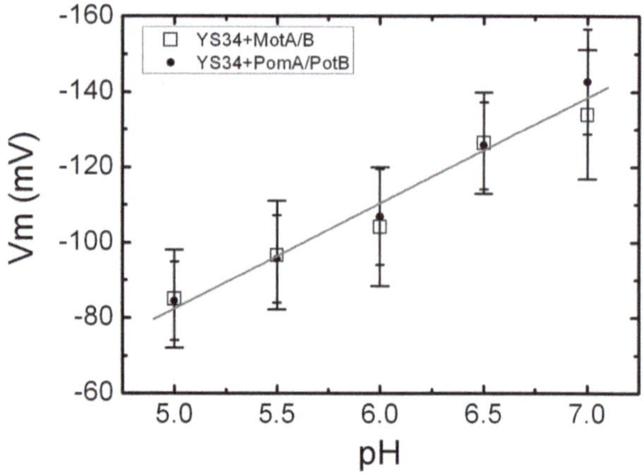

Fig. 6.3 V_m versus pH_{ex} in *E. coli* cells with proton-driven (*open squares*) and sodium-driven flagellar motors (*solid dots*). Mean ± SD of the measurements from 50 cells are shown (Lo 2007)

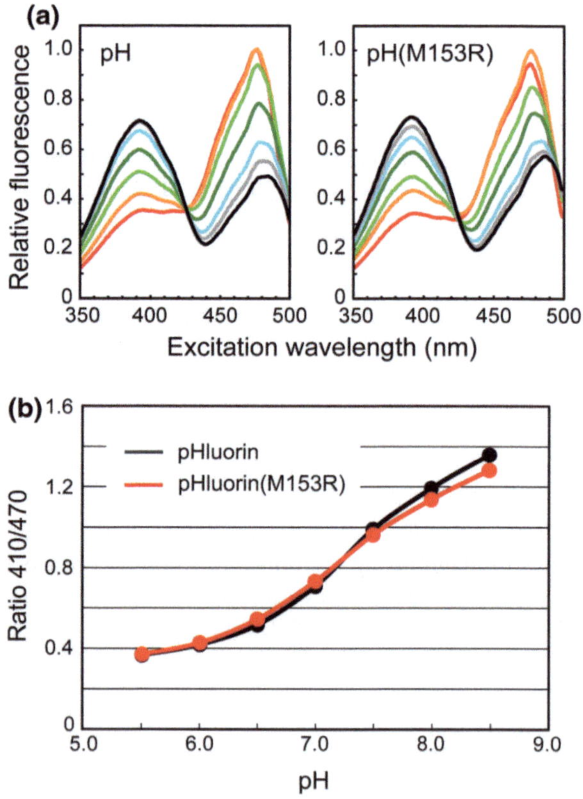

Fig. 6.4 pH dependence of fluorescence excitation spectra and the $Ratio_{410/470}$ of pHluorin and pHluorin (M153R). **a** Fluorescence excitation spectra measured in different pH: *red*, pH 5.5; *orange*, pH 6.0; *light green*, pH 6.5; *green*, pH 7.0; *cyan*, pH 7.5; *grey*, pH 8.0; *black*, pH 8.5. **b** The calibration curve of pHluorin fluorescence $Ratio_{410/470}$ in different pH (Morimoto 2011)

excitation peaks to the environmental pH provide a useful feature for the ratiometric measurement, Fig. 6.4. The fluorescent intensity ratio of 410 and 470 nm bands is monotonically increasing with pH value. This is an important feature for reliable pH measurement and calibration. In addition to the pHluorin, the pHluorin M153R mutant is stable with its fusion protein that is useful for the local pH measurement of specific intracellular target (Morimoto 2011).

Genetic pH indicator, pHluorin, can be expressed from plasmid or chromosome. The dual excitation wavelength measurements can be achieved using high-speed wavelength switcher or motorized fluorescent cube turret. High NA 100X objectives and sensitive cameras are recommended for single bacterial cell observation.

In vivo intracellular pH, calibration is necessary for single-cell measurement. The calibration curve can be obtained by adding 40 mM potassium benzoate and 40 mM methylamine hydrochloride in different pH to collapse the pH difference across the membrane, (Martinez II 2012). For long time and continuous measurement, pHluorin shows little photobleaching speed difference in 410 and 470 nm excitation wavelengthes. Photobleaching effect can be calibrated by the photobleaching curve fitting.

6.4 Bacterial Flagellar Motor

The bacterial flagellar motor (BFM) is an electrical rotary molecular machine existing in many species of bacteria. The flagellar motor is using either proton or sodium ions driven by the ion motive force. The function of this protein complex is a physical motion, rotation, from several hertz at high load to several hundred hertz at low load (Reid 2006; Lo 2013). The BFM has two parts, rotor and stator. MotA/B proteins form stator units and interact with FliG on the rotor to rotate. There are up to 11 stator-units in a BFM and each of them can function independently. The BFM rotation speed is proportional to PMF (Gabel and Berg 2003; Lo 2013). Therefore, the measurement of BFM rotation speed can be a fast relative PMF indicator. But extra care must be taken while using this method because there are other intracellular controls of BFM function. For example, the biofilm formation associated protein EpsE could disable BFM rotation (Blair 2008). A c-di-GMP binding protein YcgR also inhibits cell motility (Paul 2010).

6.4.1 Fast Protonmotive Force Measurement Using Bacterial Flagellar Motor

For *E. coli*, there are two simple BFM rotation assays, tethered cell and beads assay (Sowa 2008; Baker 2009). In tethered cell assay, flagellum was attached to the coverglass surface that the whole cell body is rotating, typically at few hertz,

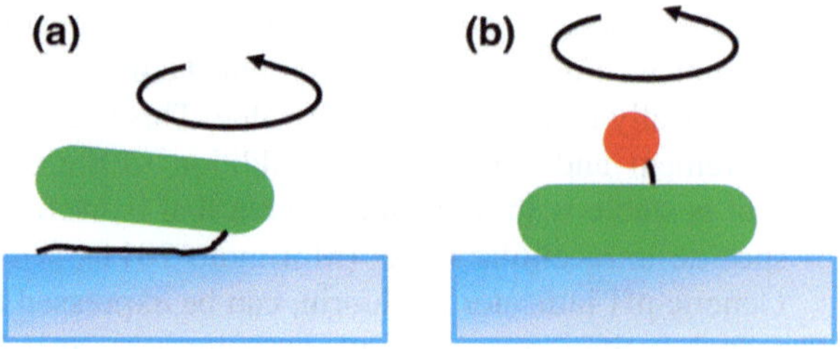

Fig. 6.5 Schematic of two bacterial flagellar motor rotational speed measurement assay. **a** Tethered cell assay. **b** Beads assay

Fig. 6.5a. The rotation can be easily recorded by video rate camera at 30 Hz. However, because the flagella are located randomly on *E. coli* membrane, the rotational load is different from cell to cell. To compare the protonmotive force from different cells, speed normalization is necessary. Tethered cell assay has also been applied to other bacteria such as *Vibrio alginolyticus* polar and lateral BFM (Kojima et al. 2007).

In bead assay, *E. coli* flagella are truncated by shearing force (Reid 2006) and cells are immobilized on coverglass, Fig. 6.5b. Micron-size polystyrene beads can be attached to the truncated flagella by either anti-flagellin antibody interaction or by hydrophobic interaction to the "sticky flagella" (Reid 2006). For 1 μm bead attached to a fully energized BFM, it can rotate at ~60 Hz (Reid 2006). The fast rotation can be recorded by fast camera mounted on a biological microscope with bright field set-up. Phase contrast or dark field set-up can dramatically enhance the contrast for these high-speed recordings (Terasawa 2011). Beads assay has been applied to other bacteria such as *Vibrio alginolyticus* polar flagella (Sowa et al. 2003). By using high-speed camera and beads assay, fast multiple cell relative protonmotive force tracking can be achieved.

6.5 Conclusions

Protonmotive force is an important energetic component in bacteria for motility, survival, and infection. Single-cell physiology is a new direction towards the understanding of cells in response to the changing world. Cell variation is a critical factor for the survival. Appling new optical protonmotive force measurements in the single bacterial cell measurement would provide new insight into the mechanism of single-cell physiology.

Acknowledgments I am grateful to Dr. Teuta Pilizota and Dr. Bai Fan for their assistance of single-cell protonmotive force measurement. The work in my lab was supported by the Ministry of Science and Technology of the Republic of China under contract No. MOST-103-2112-M-008-013-MY3.

References

Baker M, Berry RM (2009) An introduction to the physics of the bacterial flagellar motor: a nanoscale rotary electric motor. Contemp Phys 50:617–632

Blair KM et al (2008) A molecular clutch disables flagella in the Bacillus subtilis biofilm. Science 320:1636

Bradley J et al (2009) Submillisecond optical reporting of membrane potential *in situ* using a neuronal tracer dye. J Neuronsci 29:9197–9209

Gabel C, Berg HC (2003) The speed of the flagellar rotary motor of Escherichia coli varies linearly with protonmotive force. Proc Natl Acad Sci USA 100:8748–8751

Kashket ER (1985) The proton motive force in bacteria: a critical assessment of methods. Ann Rev Microbiol 39:219–42

Kojima M et al. (2007) The bidirectional polar and unidirectional lateral flagellar motors of Vibrio alginolyticus are controlled by a single CheY species. Mol Microbiol 64:57–67

Kurre R et al. (2013) Speed switching of Gonococcal surface motility correlates with proton motive force. PLoS ONE 8(6):e67718

Lo CJ, Leake MC, Berry RM (2006) Fluorescence measurement of intracellular sodium concentration in single *Escherichia coli* cells. Biophys J 90:357–365

Lo CJ, Leake MC, Pilizota T, Berry RM (2007) Non-equivalence of membrane voltage and ion-gradient as driving forces for the bacterial flagellar motor at low load. Biophys J 93:294–302

Lo CJ, Sowa Y, Pilizota T, Berry RM (2013) The mechanism and kinetics of a sodium-driven bacterial flagellar motor. Proc Natl Acad Sci USA 110:2544–2551

Martines K II et al (2012) Cytoplasmic pH response to acid stress in individual cells of Escherichia coli and Bacillus subtilis observed by fluorescence ratio imaging microscopy. App Env Microbiol 78:3706

Miesenböck G et al. (1998) Visualizing secretion and synaptic transmission with pH-sensitive green fluorescent proteins. Nature 394:192–195

Mitchell P (1966) Chemiosmotic coupling in oxidative and photosynthetic phosphorylation. Bio Rev 41:445–501; reprinted (2011) BBA-Bioenerg 1807:1507–1538

Morimoto Y, Kojima S, Namba K and Minamino T (2011) M153R mutation in a pH-sensitive green fluorescent protein stabilizes its fusion proteins. PLoS ONE 6(5): e19598

Okuno D, Iino R, Noji H (2011) Rotation and structure of FoF1-ATP synthase. J Biochem 149 (6):655–664

Ottemann KM, Lowenthal AC (2002) *Helicobacter pylori* uses motility for initial colonization and to attain robust infection. Infect Immun 70:1984–1990

Paul K et al (2010) The c-di-GMP binding protein YcgR controls flagellar motor direction and speed to affect chemotaxis by a "backstop brake" mechanism. Mol Cell 38:128–139

Reid S, Leake MC, Chandler JH, Lo CJ, Armitage JP, Berry RM (2006) The maximum number of torque-generating units in the flagellar motor of *Escherichia coli* is at least 11. Proc Natl Acad Sci USA 103:8066–8071

Shultis DD, Purdy MD, Banchs CN, Wiener MC (2006) Outer membrane active transport: structure of the BtuB:TonB complex. Science 312(5778):1396–1399

Sowa Y et al. (2003) Torque–speed relationship of the Na-driven Flagellar motor of *Vibrio alginolyticus.* J Mol Biol 327:1043–1051

Sowa Y, Berry RM (2008) Bacterial flagellar motor. Q Rev Biophys 41(2):103–132

Stingl K et al. (2002) Energetics of *Helicobacter pylori* and its implications for the mechanism of urease-dependent acid tolerance at pH 1. J Bacteriol 184:3053–3060

Terasawa S et al (2011) Coordinated reversal of flagellar motors on a single *Escherichia coli* cell. Biophys J 100:2193

Woodford CR et al (2015) Improved PeT molecules for optical sensing voltage in neurons. JACS 137:1817–1824

Chapter 7
The Type I Restriction Enzymes as Barriers to Horizontal Gene Transfer: Determination of the DNA Target Sequences Recognised by Livestock-Associated Methicillin-Resistant *Staphylococcus aureus* Clonal Complexes 133/ST771 and 398

Kai Chen, Augoustinos S. Stephanou, Gareth A. Roberts, John H. White, Laurie P. Cooper, Patrick J. Houston, Jodi A. Lindsay and David T.F. Dryden

Abstract The Type I DNA restriction–modification (RM) systems of *Staphylococcus aureus* are known to act as a significant barrier to horizontal gene transfer between *S. aureus* strains belonging to different clonal complexes. The livestock-associated clonal

---

The original version of this chapter was revised: The mention in the publication details was modified in this chapter. The erratum PDF available at 10.1007/978-3-319-32189-9_22

Kai Chen, Augoustinos S. Stephanou and Gareth A. Roberts. These authors contributed equally.

K. Chen · A.S. Stephanou · G.A. Roberts · J.H. White · L.P. Cooper · D.T.F. Dryden (✉)
EaStCHEM School of Chemistry, University of Edinburgh the King's Buildings,
Edinburgh EH9 3JJ, UK
e-mail: david.dryden@ed.ac.uk

P.J. Houston · J.A. Lindsay (✉)
Institute of Infection and Immunity, St George's, University of London,
Cranmer Terrace, London SW17 0RE, UK
e-mail: jlindsay@sgul.ac.uk

Present Address:
K. Chen
Shenyang Research Institute of Chemical Industry, 8 Shenliao Dong Road,
Shenyang Liaoning, People's Republic of China

Present Address:
P.J. Houston
The Pirbright Institute, Ash Road, Pirbright, Woking GU24 0NF, UK

© Springer International Publishing Switzerland 2016
M.C. Leake (ed.), *Biophysics of Infection*, Advances in Experimental Medicine and Biology 915, DOI 10.1007/978-3-319-32189-9_7

complexes CC133/771 and CC398 contain Type I RM systems not found in human MRSA strains as yet but at some point transfer will occur. When this does take place, horizontal gene transfer of resistance will happen more easily between these strains. The reservoir of antibiotic resistance, virulence and host-adaptation genes present in livestock-associated MRSA will then potentially contribute to the development of newly evolving MRSA clones. The target sites recognised by the Type I RM systems of CC133/771 and CC398 were identified as C$\underline{A}$G(N)$_5$R$\underline{T}$GA and $\underline{A}$CC(N)$_5$R$\underline{T}$GA, respectively. Assuming that these enzymes recognise the methylation state of adenine, the underlined A and T bases indicate the unique positions of methylation. Target methylation points for enzymes from CC1 were also identified. The methylation points for CC1-1 are CC$\underline{A}$Y(N)$_5$T$\underline{T}$AA and those for CC1-2 are CC$\underline{A}$Y(N)$_6$$\underline{T}$GT with the underline indicating the adenine methylation site as confirmed recently by Monk et al. (2015), which finally clears up the ambiguity noted previously (Roberts et al. 2013, Nucleic Acids Res 41:7472–7484) for the half sites containing two adenine bases.

Keywords Restriction enzyme · Target recognition domain · *Staphylococcus aureus*

Abbreviations

CC	Clonal complex
ST	Sequence type
MRSA	Methicillin-resistant *Staphylococcus aureus*
HA	Healthcare associated
CA	Community associated
LA	Livestock associated
MGE	Mobile genetic element
RM	Restriction–modification
hsd	Host specificity for DNA
REase	Restriction endonuclease
MTase	Methyltransferase
TRD	Target recognition domain
EGFP	Enhanced green fluorescent protein
PCR	Polymerase chain reaction

7.1 Introduction

Staphylococcus aureus, a member of the Micrococcaceae family, is the causative agent of a broad range of infectious diseases, such as postoperative wound infections and necrotising pneumonia, in humans and various animal species (Lowy 1998). The widespread, and sometimes indiscriminate, use of antibiotics combined with the adaptive genetic capacity of *S. aureus* has led to the emergence of lineages with multiple drug resistance (Lindsay 2010). Multi-locus sequence typing, microarray analysis and whole genome sequencing divide lineages of *S. aureus* and close relatives into clonal complexes (CC) (Feil et al. 2003; Sung et al. 2008; Méric et al. 2015), each of which carries a different range of mobile genetic elements and antibiotic resistance genes on the genome (McCarthy and Lindsay 2010, 2012; McCarthy et al. 2012a). Each CC is further subdivided into sequence types (ST). Although horizontal gene transfer (HGT) is considered the commonest mechanism through which bacteria acquire drug resistance, HGT between different CC is extremely inefficient in *S. aureus* (Waldron and Lindsay 2006; Lindsay 2010, 2014; Monk et al. 2012).

The rapid dissemination of antibiotic resistance among *S. aureus* strains within the same CC is a major worldwide health issue (Hiramatsu et al. 2001). For example, methicillin was launched in 1959, but clinical isolates of methicillin-resistant *S. aureus* (MRSA) were subsequently reported in early 1961 (Jevons, 1961). Since then, MRSA has gradually spread, and began causing serious hospital infections in the 1970s throughout the world. MRSA is encountered most frequently in the hospital setting and is a major source of nosocomial infections resulting in increased morbidity and mortality (Klevens et al. 2007). Healthcare-associated MRSA (HA-MRSA) is often associated with exposure to intravenous catheters, surgical procedures, and contact with devices found in hospitals. However, the epidemiology of MRSA appears to be changing because infections are no longer limited to hospitalised patients or individuals with predisposing risk factors (Gorak et al. 1999). Over the past 20 years, there has been a significant increase in the occurrence of infections unrelated to the hospital environment caused by so-called community-associated MRSA (CA-MRSA) (Conly and Johnston 2003) that have evolved independently (Lindsay 2010).

In addition to humans, *S. aureus* can infect a variety of other animal species, including farmed animals. For example, *S. aureus* infections in animals can cause mastitis in dairy-producing animals, including cattle and goats, and 'bumblefoot' in chickens (McNamee and Smyth 2000). These animal-associated *S. aureus* strains are distinct from those infecting humans, suggesting the presence of host-specific lineages that only rarely cross the species barrier (Shepheard et al. 2013). Nonetheless, workers involved in animal husbandry are known to be at risk from livestock-associated MRSA (LA-MRSA). The widespread use of antibiotics on farms, such as feed additives for growth promotion in industrial livestock and poultry, for prevention of disease within herds, or for treatment of an existing disease outbreak, has led to the emergence of drug-resistant *S. aureus* strains. Antibiotics commonly used in an agricultural setting include many classes that are relevant for human health, including tetracyclines, sulphonamides, penicillins, fluoroquinolones and macrolides. As such, animal staphylococcal strains represent a significant reservoir of antibiotic resistance determinants. For example, MRSA ST398 colonisation of pigs and calves is widespread, and the overall prevalence rates for pigs are reported to be as high as 49 % (de Neeling et al. 2007). Moreover, cases of cross-species transfer of MRSA ST398 from swine to human have been reported (Schijffelen et al. 2010; Uhlemann et al. 2012). Similarly, the pathogenic lineage

CC133 is now established in small ruminants such as sheep and goats (Guinane et al. 2010). The spread of mobile genetic elements carrying virulence and antibiotic resistance markers from this pool of livestock-associated *S. aureus* (LA-MRSA) may be an important factor in the future emergence of new MRSA clones. Furthermore, the spread of human-adapted mobile genetic elements (MGE) to LA-MRSA may lead to increased zoonotic infection (McCarthy et al. 2012b).

Restriction and modification (RM) systems of Types I, II and III are found in the majority of the sequenced bacterial genomes (REBASE; Roberts et al. 2015) and represent a significant barrier to HGT (Loenen et al. 2014a). Type I RM systems comprise three host specificity for DNA (*hsd*) genes namely, *hsdR* encoding the restriction endonuclease (REase) subunit, *hsdM* encoding the modification methyltransferase (MTase) subunit and *hsdS* encoding the sequence specificity subunit (Tock and Dryden 2005; Loenen et al. 2014b). The complex $R_2M_2S_1$ forms an enzyme capable of switching between restriction of DNA containing unmodified target sequences and modification of the hemimethylated target sequences produced after each round of host DNA replication (Kennaway et al. 2009, 2012) Modification can also be performed by an M_2S_1 complex.

The Type I RM systems found in *S. aureus* are of particular interest as they define the boundaries between the CC groups into which *S. aureus* is divided (Waldron and Lindsay 2006). DNA cannot easily transfer between CC because it lacks the appropriate methylation pattern to circumvent the Type I REase activity. *S. aureus* Type I RM systems are unusual in that they often have two *hsdM* genes, two *hsdS* genes but only one *hsdR* gene on the genome. The single *hsdR* gene encodes the subunit that has to associate with each M_2S_1 core (Kennaway et al. 2009) to form two variants of the Type I RM system in each CC or ST. The two *hsdS* genes encode two discrete S subunits with different sequence specificities. The S subunits recognise the DNA target sequence and this is typically a bipartite sequence, such as $CCAY(N)_5TTAA$ and $CCAY(N)_6TGT$, for the two Type I RM systems characteristic of *S. aureus* CC1 (Roberts et al. 2013). Each part of the target sequence is recognised by a target recognition domain (TRD) in the S subunit, which contains two different TRDs. The livestock-acquired MRSA Type I RM *hsd* genes appear to be present in only a single copy in CC133/ST771 strain 32320 (Sung et al. 2008) and CC(ST)398 strain S0385 (Schijffelen et al. 2010). Strain ED133 in CC133 contains the same *hsdS* gene as strain 32320 but also contains a second, different *hsdS* (Guinane et al. 2010).

Determining the target sequences recognised by Type I RM systems is unfortunately rather difficult as the cleavage of DNA occurs at random locations remote from the target sequence (Tock and Dryden 2005; Loenen et al. 2014a, b). Consequently, sequencing around the cleavage site as done for other restriction enzymes is not applicable for Type I RM enzymes. While single-molecule real-time (SMRT) sequencing shows considerable promise for locating the target sequences by their methylation pattern (Fang et al. 2012; Murray et al. 2012; Lluch-Senar et al. 2013), this technique is not commonly available as yet. Hence, the conventional method of examining the restriction of libraries of plasmids and using a computer programme, RMsearch (Ellrot et al. 2002; Kasarjian et al. 2003), to identify sequences unique to plasmids that are restricted but absent in unrestricted plasmids is still valuable. We have recently used this methodology to determine the target sequences for the major hospital acquired MRSA CC1, CC8 and CC5 (Roberts et al. 2013).

In this study, we have determined the target sequences recognised by the LA-MRSA Type I RM systems by investigation of strain 32320, an *S. aureus* from bovine milk and belonging to ST771, a sequence type of the lineage CC133 (Sung et al. 2008) and of strain S0385 of the CC398 lineage (Schijffelen et al. 2010). In addition, the particular adenines targeted for methylation have been determined using an ATPase assay.

7.2 Materials and Methods

7.2.1 Preparation of Enzymes

The *hsd* genes for the Sau1 MTases, Table 7.1, were amplified from genomic DNA and used to replace the genes for the EcoKI MTase in the expression plasmid pJFMSEGFP (Chen et al. 2010). The Sau1 MTase operons lack BamHI restriction sites in their open reading frames thus allowing a common cloning strategy employing only one vector. Following the success of producing EcoKI MTase with the S subunit tagged with the enhanced green fluorescent protein (EGFP) and a hexa-histidine tag from plasmid pJFMSEGFP (Chen et al. 2010), we engineered an expression construct for the two MTases. Through the use of the polymerase chain reaction (PCR), employing oligonucleotides pJFMSEGFPhisBS (5′-GAGTG AATCCCCGGGGATCCGTCGACC-3′) and pJFMSEGFPhisTS (5′-AGTCAGTC AGGGATCCATGGTGAGCAAGGGCGAGGAGCTG-3′) with pJFMSEGFP as template, we obtained a linear PCR derivative of the vector. The resulting PCR product allows a coding sequence to be introduced downstream of the tac promoter and fused in frame with EGFP-His-tag following digestion of the ends of the coding sequence with BamHI. The *hsdM–hsdS* operon was amplified from the appropriate

Table 7.1 The *S. aureus* strains examined in this work showing their clonal complex identification, their assigned names, genome coordinates and animal host

S. aureus strain	*S. aureus* clonal complex	Official REBASE names	Genbank number for HsdS	Our name	Host
S0385	CC398	SauSTORF213P M.SauSTORF499P S.SauSTORF499P	AM990992	CC398-1	Pig
32320	CC133/ST771	Sau133OR140P M. Sau133ORF451P S.Sau133ORF451P	ADI96988.1	CC133/771-1	Goat

The REBASE names and Genbank number for CC133/ST771 are those for strain ED133 described in Guinane et al. (2010) as they are identical in sequence

S. aureus genomic DNA by PCR, employing a universal Sau1 M oligonucleotide for the start of the *hsdM* sequence, (5′-AGTCAGTCAGGGATCCAAGAAGG AGATATACATATGTCTATTACTGAAAAACAACG-3′) in every reaction, in combination with a locus-specific oligonucleotide, homologous to the end of the appropriate *hsdS* as follows:

CC398 (5′-GATCGAATTCCGGATCCAATAAACATCTTTTGAAGTAATGAC-3′)
CC133/771 (5′-GATCGAATTCCGGATCCAATAAACATCTTTTGAAGTAATG CC-3′).

PCR products were purified, cut with BamHI and ligated into the BamHI interval of the vector PCR product. The universal Sau1 M oligonucleotide assumes that the M subunit commences with the sequence MSITEKQRQQQ and ignores unconserved sequences upstream of the conserved ATG start codon for methionine. The plasmid constructs containing CC398 and CC133/771 were named pCC398 and pCC133/771, respectively.

The MTases and the R subunit were purified separately. Plasmid encoding MTase (pCC398 or pCC133/771) or the R subunit of the Sau enzymes were used to transform *E. coli* BL21(DE3). Heterologous gene expression was induced by adding isopropyl β-D-1-thiogalactopyranoside (final concentration 1 mM) to cultures growing at 37 °C once the optical density had reached ~0.6. Induction was for 3–4 h at 25 °C or 30 °C. The cells were disrupted by intermittent sonication for approximately 20 min with cooling on ice using a Soniprep 150 sonicator (Sanyo, Tokyo, Japan) fitted with a 9 mm diameter probe and then centrifuged at 20000 g for 90 min at 4 °C. Proteins were purified by HisTrap chromatography, size exclusion chromatography, DEAE anion exchange chromatography and, if necessary, heparin HiTrap chromatography (GE Healthcare, Uppsala, Sweden) and were >98 % pure as judged by Coomassie Blue staining of SDS-polyacrylamide gels. The NaCl concentration was at 0.2 M in the size exclusion chromatography step for CC133/771 MTase and 0.5 M for the CC398 MTase. The purified proteins were stored at−20 °C after addition of glycerol to 50 % (v/v). NaCl was also added to 0.5 M for the CC398 MTase preparation to maintain protein solubility. Extinction coefficients were calculated at 280 nm (Gill and von Hippel 1989) assuming an M_2S_1 stoichiometry for the MTases (and including the EGFP and hexa-His tag on the S subunit) and a monomeric R subunit (Dryden et al. 1997; Kennaway et al. 2012). The R subunit was prepared as described previously (Roberts et al. 2013).

7.2.2 Target Recognition Site Determination

Endonuclease cleavage assays were performed by incubating a library of plasmids, based upon insertion of known DNA sequences ligated into the EcoRI-BamHI interval of pUC19, with MTase and R subunit for 15 min at 37 °C. The reaction products were analysed by agarose gel electrophoresis. Cleavage sites are distant from the target site for these enzymes, so a computer programme was used to search for target sequences present in plasmids cut by the enzyme and not present in uncut plasmids (Ellrot et al. 2002; Kasarjian et al. 2003). The main sets of plasmids were based upon the DNA sequence of phage PhiED1 (a kind gift from Dr Garry Blakely, School of Biological Sciences, University of Edinburgh) and inserts from phage lambda (a kind gift from Dr Iain Murray of New England Biolabs, Ipswich, MA) as described previously (Roberts et al. 2013). Typically, about 40 plasmids were analysed for each enzyme. Cleavage sites were then confirmed by inserting a short, defined oligonucleotide sequence containing the putative target into pUC19. Reaction digests had a total volume of 50 μl and a typical digest was performed in 1xNEB4 buffer (50 mM potassium acetate, 20 mM Tris-acetate, 10 mM magnesium acetate, 1 mM dithiothreitol, pH 7.9) containing 2 mM ATP, 3.0 nM plasmid DNA, 0.5 mM *S*-adenosyl-L-methionine, 0.1 mg/ml bovine serum albumin and 1 μl of the enzyme stock (3.0 μM). The enzyme stock was prepared in a volume of 50 μl in 1xNEB4 buffer with final concentrations of 7.5 μM R subunit and 3.0 μM MTase thus ensuring an excess of R over the MTase to give formation of the $R_2M_2S_1$ RM enzyme.

We used the programme "RMsearch" (Ellrot et al. 2002; Kasarjian et al. 2003) to find the common DNA sequence in all cleavable plasmids and absent from all uncleaveable plasmids and thereby determine the DNA target sequences recognised by the *S. aureus* Type I RM enzymes. In both cases, a single target recognition site was identified from analysing the pattern of digestion.

7.2.3 Determination of Methylation Sites for CC1-1 and CC1-2 Type I RM Enzymes

Oligonucleotides (≥47mer) containing a centrally located target recognition sequence for the CC1-1 or CC1-2 RM enzymes described previously (Roberts et al. 2013) were designed in which one or other of the potential target adenine bases was methylated. The sequence of the unmethylated forward strand for CC1-1 was

5-AGATGATGGAATCAATGC*CCAT*GTCCA*TTAA*GCCCTATACGATATA A-3′

and the complementary strand was

5′-TTATATCGTATAGGGC*TTAA*TGGAC*ATGG*GCATTGATTCCATCATCT-3′ in which one or other of the underlined adenine bases was N6 methylated (site highlighted in italics). The sequence of the unmethylated forward strand for CC1-2 was

5′-AGATGATGGAATCAATGC*CCAT*GTACCA*TGT*AGCCCTATACGATAT AA-3′

and the complementary strand was

5′-TTATATCGTATAGGGCT*ACA*TGGTAC*ATGG*GCATTGATTCCATCAT CT-3′ where one or other of the underlined adenine bases was N6 methylated (site highlighted in italics). Each non-methylated oligonucleotide was separately annealed to an equimolar amount of one of two methylated complementary oligonucleotides to generate double-stranded DNA duplexes containing the two possible hemimethylated target recognition sites. As a negative control, a DNA duplex of similar length that lacks the target recognition sequence was also generated (sequence not shown). The various DNA duplexes were then used in a coupled enzyme assay (Keatch et al. 2004) to determine the rate of ATPase activity in the presence of reconstituted nuclease CC1-1 or CC1-2 (final concentration of 100 nM in the assay mix). An extinction coefficient of 6220 $M^{-1}cm^{-1}$ for NADH absorption at 340 nm was used.

7.3 Results

7.3.1 Protein Purification

The MTase enzymes for both CC133/771 and CC398 were overexpressed and purified in the same manner as described previously for other Type I RM enzymes from *S. aureus* (Roberts et al. 2013). The purified proteins were analysed by SDS-PAGE and estimated to be >98 % pure upon staining with Coomassie Blue, Fig. 7.1. To form an active restriction enzyme, the MTase preparations were mixed with an excess of the R subunit (Roberts et al. 2013).

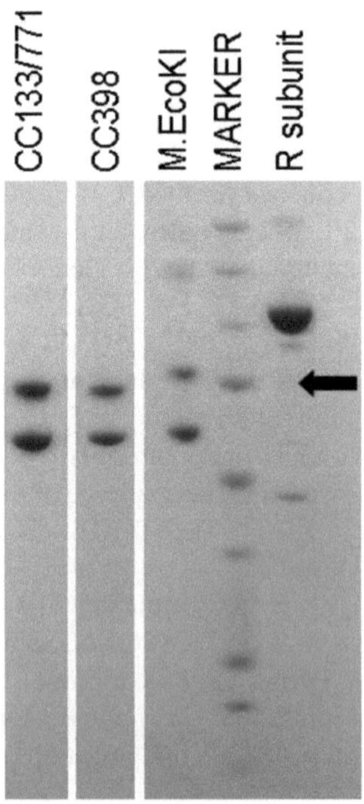

Fig. 7.1 SDS-PAGE analysis of purified proteins. The upper band in the MTase preparations is the EGFP-His-tagged S subunit and the lower band is the M subunit. The EcoKI MTase with the EGFP-His-tagged S subunit is shown for comparison. The purified R subunit is also shown. The markers have molecular masses of 250, 150, 100, 75, 50, 37, 25 and 20 kDa (Biorad precision plus protein standards). The *arrow* indicates the 75 kDa size marker

7.3.2 *Naming of the SauI Systems*

As the putative open reading frames for the SauI systems from CC133/771 and CC398 have been overexpressed and shown to be active, they can be assigned formal names according to the convention described previously (Roberts et al. 2003). These names are given in Table 7.1. While the formal names should be used for descriptions of the individual enzymes, for clarity we use names based upon the CC groups when comparing groups of these *S. aureus* enzymes. Table 7.1 also gives our suggested names based upon CC groups, which we will use in the remainder of this paper, i.e. CC133/771-1 and CC398-1 for the enzymes from strain 32320 and strain S0385, respectively. Although only one Type I RM system appears to be present in each strain, the −1 is useful in case further Type I enzymes are found in other strains of these CC. For instance, Guinane et al. (2010) located a second Type I RM system in strain ED133 (M.Sau133ORF1794P and S.Sau133ORF1794P in REBASE).

7.3.3 *DNA Cleavage by the Enzymes*

Mixing the MTase preparations with the purified R subunit produced an active restriction endonuclease capable of cleaving circular plasmids containing either a single copy or multiple copies of the target sequence. Figure 7.2 shows the cleavage patterns obtained using a range of plasmids of known sequence. Table 7.2 shows the complete set of cutting data for the complete set of plasmids described previously (Roberts et al. 2013). As observed previously with other Type I RM enzymes from *S. aureus* (Roberts et al. 2013), a nicking activity was present in our samples. This was not assessed as being evidence for the presence of a target site as it was previously shown to be ATP-independent and therefore not due to the Type I RM enzyme. Only full linearisation of the plasmid and absence of any remaining supercoiled plasmid was taken as evidence of a target site on the plasmid. For these two enzymes, no evidence of smearing of the DNA was observed indicating that the plasmids contain at most a single site for each enzyme.

The data for cleaved and uncleaved plasmids were analysed by the RMsearch programme to identify sequences found in cleaved plasmids but not in uncleaved plasmids. It is known that the second TRDs for the SauI enzymes from CC133/771-1 and CC398-1 are of identical amino acid sequence and thus recognise the same DNA half-site, Fig. 7.3. This observation determines the direction in which the potential target recognition sequence must be read. These TRD sequences are also different from those found previously in our analysis of CC1 and CC5 (Roberts et al. 2013). Our analysis resulted in unique sites of CAG(N)$_5$RTGA and ACC(N)$_5$RTGA for CC133/771-1 and CC398-1, respectively. Assuming that these enzymes recognise the methylation state of adenine, the underlined positions, either on the A shown or on the A on the complementary strand, indicate the unique

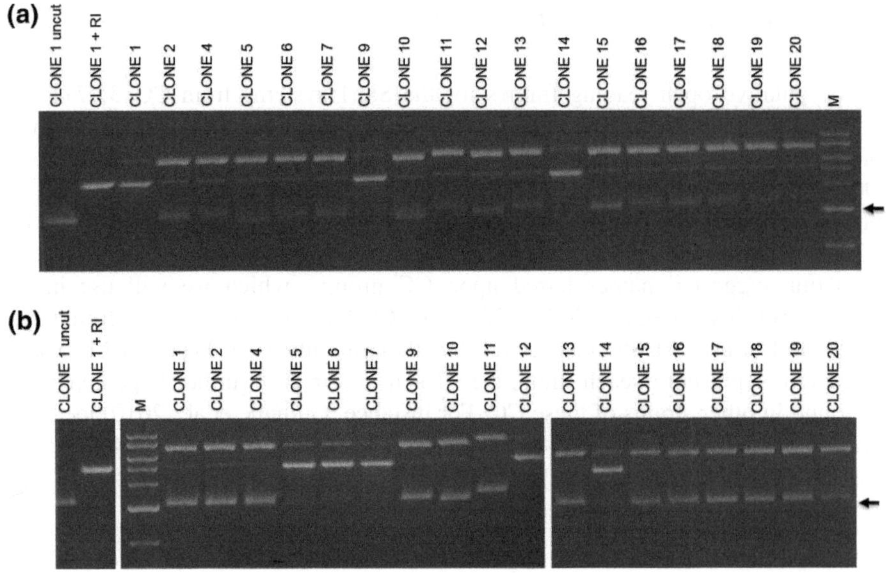

Fig. 7.2 Agarose gel analysis of DNA cleavage activity of the prepared Sau1 restriction enzymes against a series of plasmids based upon pUC19 containing different 2.5 kb inserts (Roberts et al. 2013). **a** CC133/771-1. **b** CC398-1. M = 1 kb markers with the *arrow* indicating the 3 kb size (New England Biolabs), uncut indicates the supercoiled plasmid, +RI indicates the plasmid linearised by EcoRI

location of methylation Thus, in both instances, the methylation sites can be unambiguously determined because only a single location in each part of the target recognition sequence is possible.

7.3.4 Identification of Adenines Targeted for Methylation When There Is Ambiguity in the Target Site

Although the orientation of the site and the respective adenine bases that are subject to methylation is clearly determined for CC133/771-1 and CC398-1, this is not always the case using the outlined approach. For example, in our previous paper we assigned the orientation of CC5-1 (SauN315I) as ATC(N)$_5$CCT (Roberts et al. 2013), but subsequent work using Type I RM systems with S subunits sharing sequence identity to one of the TRDs in CC5-1 clarified that the recognition site of CC5-1 (SauN315I) should be defined in the opposite orientation as AGG(N)$_5$GAT (unpublished results). Moreover, in our previous work on CC1-1 and CC1-2 we were unable to assign the adenine base in one half of the site that is subject to methylation because there were two possible locations within each target recognition sequence. We have since determined the methylation site using an ATPase

Table 7.2 Plasmid cleavage data showing positive ("yes") and negative ("no") results for cleavage of the complete set of plasmids containing 2 kb, 200 bp and lambda phage DNA inserts in pUC19 described previously (Roberts et al. 2013)

2 kb library summary			200 bp library summary		
Plasmid	CC133-771-1	CC398-1	Plasmid	CC133-771-1	CC398-1
pUC19	No	No	5E	No	Yes
1E	Yes	No	P5I	–	No
2E	No	No	P5II	–	No
4E	No	No	P5IV	–	Yes
5E	No	Yes	P5 V	–	No
6E	No	Yes	P5VI	–	No
7E	No	Yes	P5VII	–	No
9E	Yes	No	P5VIII	–	No
10E	No	No	P5IX	–	No
11E	No	No	P5X	–	No
12E	No	Yes	14E	Yes	Yes
13E	No	No	P14I	No	No
14E	Yes	Yes	P14II	No	No
15E	No	No	P14III	No	No
16E	No	No	P14IV	Yes	No
17E	No	No	P14 V	No	No
18E	No	No	P14VI	No	No
19E	No	No	P14VII	No	No
20E	No	No	P14VIII	No	No
4λ	?	No	P14IX	No	No
33λ	?	No	P14X	No	Yes
107λ	Yes	No			
124λ	Yes	No			
225λ	No	Yes			
378λ	No	No			
401λ	No	No			
493λ	No	No			
524λ	Yes	No			
627λ	No	No			
667λ	No	Yes			

assay. Specifically, hemimethylated DNA duplexes containing the target recognition sites for either CC1-1 or CC1-2 in which one or other of the possible methylation sites was methylated were designed and generated. A DNA duplex of a similar length lacking the target recognition site was used as a negative control. An ATP-coupled assay was then employed to assess which of the alternative DNA duplexes promote ATPase activity after addition of the reconstituted nuclease for CC1-1 or CC1-2, Fig. 7.4. A duplex containing the unmethylated recognition site will promote the ATPase activity of the complete nuclease. Using this approach we

```
CC5-1           KM SNTQKKNVPELRFPGFEGEWEEKQLGDLTDRVIR--KNKNLESKKP-LTISGQLGLID 57
CC133/771-1     KM SNTQTKNVPELRFPGFEGEWEEKKLGDLGLFQKSYSFSRAKEGNGKTKHIHYGDIHSK 60
CC398-1         KM SNTQKKNVPELRFPGFEGEWEEKKLGEF--------AGKVTQKNVDKKYIETLTNSAE 52
CC1-2           KM SNTQTKNVPELRFPGFEGEWEEKKLGNLTTKIGS----GKTPKGGSENYTNKGIPFLR 56
CC5-2           KM SNTQTKNVPELRFPGFEGEWEEKKLGNLTTKIGS----GKTPKGGSENYTNKGIPFLR 56
CC1-1           KM SNTQKKNVPELRFPGFEGEWEEKKLGDLTTKIGS----GKTPKGGSENYTNKGIPFLR 56
                ** ****.******************::*::

CC5-1           QTEYFSK------SVSSKNLENYTLIKNGEFAYNKSYSNGYPLGAIKRLTRYDSGVLSSL 111
CC133/771-1     FKTVLDS---DGNIPNIIEKAVFELIQKGDIVFADASEDYSDLGKAVMIDFKPNSLISGL 117
CC398-1         LGIISQKDYFDKEISNIDNIKKYYVVEENDFVYNPRMSNYAPFGPVNRNKLGKKGVMSPL 112
CC1-2           SQNIRNGKLNLNDLVYISKDIDDEMKNSRTYYGDVLLNITGASIGRTAINSIVETHANLN 116
CC5-2           SQNIRNGKLNLNDLVYISKDIDDEMKNSRTYYGDVLLNITGASIGRTAINSIVETHANLN 116
CC1-1           SQNIRNGKLNLNDLVYISKDIDDEMKNSRTYYGDVLLNITGASIGRTAINSIVEIHANLN 116
                                                                         .

CC5-1           YICFSIKS--EMSKDFMEAYFDSTHWYREVSGIAVEGARNHGLLNVSVNDFFTILIKYPS 169
CC133/771-1     HTH-LFRPLNNAISNFLIFYTKTLSYKKFIRQQGTG---ISVLGISKKSLLNLNVLIPRS 173
CC398-1         YT--VFK-IQNIDLNFIEFYFKSSKWYRFMALNGDSGARADRFSIKDRTFMEMPLHIP-C 168
CC1-2           QHVCIIRLKKEYYYNFFGQYLLSRKGKRKIFLAQSGGSREGLNFKEIAN---LKIFTPTI 173
CC5-2           QHVCIIRLKKEYYYNFFGQYLLSRKGKRKIFLAQSGGSREGLNFKEIAN---LKIFTPTI 173
CC1-1           QHVCIIRLKKEYYYNFFGQYLLSRKGKRKIFLAQSGGSREGLNFKEIAN---LKIFTPTI 173
                        :                      :                         .

CC5-1           LEEQQKIGKFFSKLDRQIELEEQKLELLQQQKKGYMQKIFSQELRFKDENGEDYPDWENS 229
CC133/771-1     ELEQQKVGKFFSKLDRQIELEEQKIELLQQQKKGYIQKIFSQELRFKDENGDDYPEWEET 233
CC398-1         MDEQIKIGQFFSKLDRQIELEEQKLELLQQQKKGYMQKIFSQELRFKDENGKDYPEWEET 228
CC1-2           FEEQQKIGKFFSKLDRQIELEEQKLELLQQQKKGYMQKIFTQELRFKDENGEEYPEWENK 233
CC5-2           FEEQQKIGQFFSKLDQQIELEEQKLELLQQQKKCYIQKIFSQELRFKDEEGNYYKGWNKK 233
CC1-1           FEEQQKIGEFISKLDRQIELEEQKLELLQQQKKGYMQKIFSQELRFKDEEGKDYPDWKSK 233
                  ** *:*.*: ***:********:******** *:****:******:*:.. *  *.

CC5-1           K----IEKYLKERN--------ERSDKG--QMLSVTIN----SGIIKFSELDRK---DNS 268
CC133/771-1     TIKEIAQINTGKKD---------TKDAITNGSYDFYVR---SPIVYKINTFSYEG----- 276
CC398-1         TIKEIAQINTGKKD---------TKDAITNGSYDFYVR---SPIVYKINTFSYEG----- 271
CC1-2           FIKDIFIFEN----NRRK-----PITSSLREKGLYP--YYGATGIIDYVKDYLFN----- 277
CC5-2           QLKDVLEFSN----KRTINENEYPVLTSSRQGLILQSDYYKDRKTFAESNIGYFI----- 284
CC1-1           SIQEIFENKGGTALETEFNFDGNYKVISIGSYSINSTYNDQNIRVNKNKKTEKYI----- 288

CC5-1           SKDKSNYKVVRKNDIAYNSMRMWQGASGKSNYNGIVSPAYTVLYPTQNTSSLFIGYKFKT 328
CC133/771-1     ------------EAILTVGDGVGVGKVFHYVNGKFDYHQRVYKISDFKNYYGLLLFYYFS 324
CC398-1         ------------EAILTVGDGVGVGKVFHYVNGKFDYHQRVYKISDFKNYYGLLLFYYFS 319
CC1-2           --NEER--LLIGEDGAKWG-QFETSSFIANGQYWVN--NHAHVVKSNDHNLFFMNYYLNF 330
CC5-2           --LPKN--HITYRSRSDDG-IFKFNLNLMIDVGIIS--KYYPVFKGIDANQYYLTLHLNY 337
CC1-1           --LSKGDLAMVLNDKTKDGKIIGRSIFIDKDNQYIYNQRTERLIPFAENDNKFLWFLMNT 346
                             .                                         :

CC5-1           HRMIHKFKINSQGLTSDTWNLKYKQLKNINIDIP-VLEEQEKIGDFFKKMDILISKQKMK 387
CC133/771-1     QNFLKETKKYSAKTSVD--SVRKDMVANMKVPRP-IYIEQEKIGQFIKKVDNKIKIQKQV 381
CC398-1         QNFLKETKKYSAKTSVD--SVRKDMIANMKVPRP-IYIEQKKIGQFIKRVDNKTKIQKQV 376
CC1-2           ----KELRAFVTGNAPA--KLTHANLCNINLKIP-CLTEQDKVSALLKSIDNKMNNQMNR 383
CC5-2           QLK-KEYIKYATGTSQL--VLSQKDLQNIKTKLP-SYEEQQKIGDFFSEIDRLVEKQSSK 393
CC1-1           DLIRNKIKGMMQGATQV--YINYSSIKLISIQLP-LLEEQQKIRGFLEVLSGITTKQLHK 403
                    :.               :    .       *    **.::  ::  :.       :

CC5-1           IEILEKEKQSFLQKMFL 404
CC133/771-1     IELLKQRKKALLQKMFI 398
CC398-1         IELLKQRKKSLLQKMFI 393
CC1-2           IELLKERKKELLQKMFI 400
CC5-2           VGRLKVRKKELLQKMFV 410
CC1-1           IDQLKERKKAFLQKMFI 420
                :  *:  *: :*****:
```

Fig. 7.3 Amino sequence comparison of the S subunits from different CC obtained using CLUSTAL (Sievers et al. 2011). Sequences from CC1 and CC5 strains have been previously described (Roberts et al. 2013). Target recognition domains are shown either coloured or highlighted in colour). TRDs with the same target have the same colour. All of our S subunits assume that the initial KM is not present in the translated protein

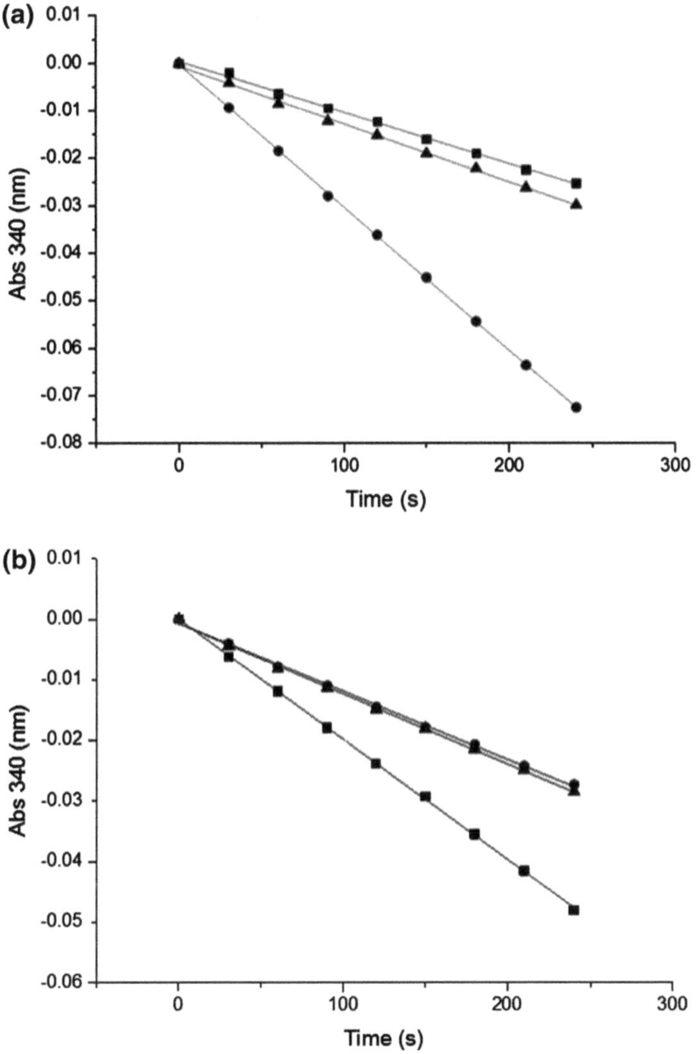

Fig. 7.4 ATPase assay to determine the methylation sites for CC1-1 and CC1-2. Rate of ATPase consumption was followed by analysing the NADH absorption at 340 nm in the coupled assay. In each case the rate of reaction was corrected for the negative control. *Panel a* A 47mer DNA duplex containing the recognition sequence for CC1-1 (CCAY-5-TTAA) in which the A on the opposite strand of the first (*circle*) or second (*triangle*) T in the 3′ defined region of the bipartite recognition site was methylated. A duplex of the same length lacking the recognition site was used as a negative control (*square*). Apparent rate of reaction for the duplex containing the sequence CCAY-5-TTAA (where the underline represents the site of methylation on the opposite strand) was 31.0 nmoles/s/L, while that for the duplex containing the sequence CCAY-5-TTAA was 2.2 nmoles/s/L. *Panel b* A 48mer DNA duplex containing the recognition sequence for CC1-2 (CCAY-6-TGT) in which the A on the opposite strand of the first (*triangle*) or second (*square*) T in the 3′ defined region of the bipartite recognition site was methylated. A duplex of the same length lacking the recognition site was used as a negative control (*circle*). Apparent rate of reaction for the duplex containing the sequence CCAY-6-TGT (where the underline represents the site of methylation on the opposite strand) was 0.6 nmoles/s/L, while that for the duplex containing the sequence CCAY-6-TGT was 13.8 nmoles/s/L

were able to verify that the methylation sites for CC1-1 are CCAY(N)$_5$TTAA and those for CC1-2 are CCAY(N)$_6$TGT (underlined bases indicate methylation site either on the A shown or on the A on the complementary strand). These methylation sites have also been confirmed recently using single molecule real-time sequencing methods (Monk et al. 2015).

7.4 Discussion

The target sites recognised by CC133/771-1 and CC398-1 were identified as CAG (N)$_5$RTGA and ACC(N)$_5$RTGA, respectively. Assuming that these enzymes recognise the methylation state of adenine, the underlined A and T bases indicate the unique positions of methylation. Target methylation points for enzymes from CC1 were identified. The methylation points for CC1-1 are CCAY(N)$_5$TTAA and those for CC1-2 are CCAY(N)$_6$TGT with the underline indicating the adenine methylation site thus clearing up the ambiguity noted by Roberts et al. (2013) for the half sites containing two adenine bases.

The Type I RM system is known to act as a significant barrier to horizontal gene transfer between *S. aureus* strains belonging to different CC. Thus, horizontal gene transfer between *S. aureus* lineages of different CC is severely limited. The CC133/771-1 and CC398-1 RM systems are not found in HA-MRSA or CA-MRSA as yet but at some point TRD swaps will occur. When this TRD swap does take place, horizontal gene transfer will happen more easily between these strains. The reservoir of antibiotic resistance, virulence and host-adaptation genes present in livestock-associated MRSA will then potentially contribute to the development of newly evolving MRSA clones. Analysis of the distribution of these newly identified DNA recognition sites in a range of conjugative plasmids found in *S. aureus* strains that carry multiple antibiotic resistance markers (McCarthy and Lindsay 2012) may help to predict the likely generation of new MRSA lineages with the potential to cause widespread infections in humans. The dissemination of antibiotic resistance markers between these strains may also have a significant effect on the agricultural sector. However, knowledge gained from analysing the distribution of the relevant recognition sites on mobile genetic elements may allow the likely impact of these newly evolved MRSA clones to be anticipated.

In addition to the potential impact of our research findings in the clinical and agricultural setting, determination of the recognition sites for these Type I RM systems has implications in terms of fundamental research. Specifically, the transformation of *S. aureus* strains in the laboratory is often prevented due to the presence of a RM barrier (Waldron and Lindsay 2006). However, such problems can be circumvented by choosing vectors that lack the specific Type I RM sites (Roberts et al. 2013) or by modifying cloned DNA carrying target sites using methylation enzymes in cloning strains prior to gene transfer (Jones et al. 2015; Monk et al. 2015). Knowledge of the Type I RM recognition sequence for CC133/771-1 and CC398-1 will allow a rationale approach to be developed in order to evade restriction when working with these strains. As such, knowledge of the Type I RM sites will assist in fundamental research where transformation of these *S. aureus* strains has proved to be problematic.

Acknowledgements We gratefully acknowledge financial support from the Wellcome Trust (GR080463MA to D.T.F.D and 090288/Z/09/ZA to D.T.F.D. and J.A.L). We thank Drs Iain Murray and Rich Roberts of New England Biolabs for plasmids and the RMsearch programme.

References

Chen K, Roberts GA, Stephanou AS, Cooper LP, White JH, Dryden DTF (2010) Fusion of GFP to the M.EcoKI DNA methyltransferase produces a new probe of Type I DNA restriction and modification enzymes. Biochem Biophys Res Commun 398:254–259

Conly JM, Johnston BL (2003) The emergence of methicillin-resistant Staphylococcus aureus as a community-acquired pathogen in Canada. Can J Infect Dis 14:249–251

de Neeling AJ, van den Broek MJM, Spalburg EC, van Santen-Verheuvel MG, Dam-Deisz WDC, Boshuizen HC, van de Giessen AW, van Duijkeren E, Huijsdens XW (2007) High prevalence of methicillin resistant Staphylococcus aureus in pigs. Vet Microbiol 122:366–372

Dryden DTF, Cooper LP, Thorpe PH, Byron O (1997) The in vitro assembly of the EcoKI Type I DNA restriction/modification enzyme and its in vivo implications. Biochemistry 36:1065–1076

Ellrott KP, Kasarjian JK, Jiang T, Ryu J (2002) Restriction enzyme recognition sequence search program. Biotechniques 33:1322–1326

Fang G, Munera D, Friedman DI, Mandlik A, Chao MC, Banerjee O, Feng Z, Losic B, Mahajan MC, Jabado OJ, Deikus G, Clark TA, Luong K, Murray IA, Davis BM, Keren-Paz A, Chess A, Roberts RJ, Korlach J, Turner SW, Kumar V, Waldor MK, Schadt EE (2012) Genome-wide mapping of methylated adenine residues in pathogenic Escherichia coli using single-molecule real-time sequencing. Nat Biotechnol 30:1232–1239

Feil EJ, Cooper JE, Grundmann H, Robinson DA, Enright MC, Berendt T, Peacock SJ, Smith JM, Murphy M, Spratt BG, Moore CE, Day NP (2003) How clonal is Staphylococcus aureus? J Bacteriol 185:3307–16

Gill SC, von Hippel PH (1989) Calculation of protein extinction coefficients from amino acid sequence data. Anal Biochem 182:319–326

Gorak E, Yamada S, Brown J (1999) Community-acquired methicillin-resistant Staphylococcus aureus in hospitalized adults and children without known risk factors. Clin Infect Dis 29:797–800

Guinane CM, Ben Zakour NL, Tormo-Mas MA, Weinert LA, Lowder BV, Cartwright RA, Smyth DS, Smyth CJ, Lindsay JA, Gould KA, Witney A, Hinds J, Bollback JP, Rambaut A, Penadés JR, Fitzgerald JR (2010) Evolutionary genomics of Staphylococcus aureus reveals insights into the origin and molecular basis of ruminant host adaptation. Genome Biol Evol 12:454–66

Hiramatsu K, Cui L, Kuroda M, Ito T (2001) The emergence and evolution of methicillin-resistant Staphylococcus aureus. Trends Microbiol 9:486–493

Jevons MP (1961) Celbenin-resistant staphylococci. Br Med J 1:124–125

Jones MJ, Donegan NP, Mikheyeva IV, Cheung AL (2015) Improving transformation of Staphylococcus aureus belonging to the CC1, CC5 and CC8 clonal complexes. PLoS ONE 10: e0119487. doi:10.1371/journal.pone.0119487

Kasarjian JK, Iida M, Ryu J (2003) New restriction enzymes discovered from Escherichia coli clinical strains using a plasmid transformation method. Nucleic Acids Res 31:e22

Keatch SA, Su TJ, Dryden DTF (2004) Alleviation of restriction by DNA condensation and non-specific DNA binding ligands. Nucleic Acids Res 32:5841–5850

Kennaway CK, Obarska-Kosinska A, White JH, Tuszynska I, Cooper LP, Bujnicki JM, Trinick J, Dryden DTF (2009) The structure of M.EcoKI Type I DNA methyltransferase with a DNA mimic antirestriction protein. Nucleic Acids Res 37:762–770

Kennaway CK, Taylor JE, Song CF, Potrzebowski W, Nicholson W, White JH, Swiderska A, Obarska-Kosinska A, Callow P, Cooper LP, Roberts GA, Artero JB, Bujnicki JM, Trinick J, Kneale GG, Dryden DTF (2012) Structure and operation of the DNA-translocating Type I DNA restriction enzymes. Genes Dev 26:92–104

Klevens RM, Morrison MA, Nadle J, Petit S, Gershman K, Ray S, Harrison LH, Lynfield R, Dumyati G, Townes JM, Craig AS, Zell ER, Fosheim GE, McDougal LK, Carey RB, Fridkin SK; Active Bacterial Core surveillance (ABCs) MRSA Investigators. (2007) Invasive methicillin-resistant Staphylococcus aureus infections in the United States. J Am Med Ass 298:1763–1771

Lindsay JA (2010) Genomic variation and evolution of Staphylococcus aureus. Intl J Med Microbiol 300:98–103

Lindsay JA (2014) Staphylococcus aureus genomics and the impact of horizontal gene transfer. Intl J Med Microbiol 304:103–109

Lluch-Senar M, Luong K, Lloréns-Rico V, Delgado J, Fang G, Spittle K, Clark TA, Schadt E, Turner SW, Korlach J, Serrano L (2013) Comprehensive methylome characterization of Mycoplasma genitalium and Mycoplasma pneumoniae at single-base resolution. PLoS Genet 9:e1003191. doi:10.1371/journal.pgen.1003191

Loenen WAM, Dryden DTF, Raleigh EA, Wilson GG, Murray NE (2014a) Highlights of the DNA cutters: a short history of the restriction enzymes. Nucleic Acids Res 42:3–19

Loenen WAM, Dryden DTF, Raleigh EA, Wilson GG (2014b) Type I Restriction Enzymes and Their Relatives. Nucleic Acids Res 42:20–44

Lowy FD (1998) Staphylococcus aureus infections. N Engl J Med 339:520–532

McCarthy AJ, Lindsay JA (2010) Genetic variation in Staphylococcus aureus surface and immune evasion genes is lineage associated: implications for vaccine design and host-pathogen interactions. BMC Microbiol 10:173–187

McCarthy AJ, Lindsay JA (2012) The distribution of plasmids that carry virulence and resistance genes in Staphylococcus aureus is lineage associated. BMC Microbiol 12:104–111

McCarthy AJ, Witney AA, Lindsay JA (2012a) Staphylococcus aureus temperate bacteriophage: carriage and horizontal gene transfer (HGT) is lineage associated. Front Cell Infect Microbiol 2:6. doi:10.3389/fcimb.2012.00006

McCarthy AJ, van Wamel W, Vandendriessche S, Larsen J, Denis O, Garcia-Graells C, Uhlemann AC, Lowy FD, Skov R, Lindsay JA (2012b) Staphylococcus aureus CC398 clade associated with human-to-human transmission. Appl Environ Microbiol 78:8845–8

McNamee PT, Smyth JA (2000) Bacterial chondronecrosis with osteomyelitis ('femoral head necrosis') of broiler chickens: a review. Avian Pathol 29:253–270

Méric G, Miragaia M, de Been M, Yahara K, Pascoe B, Mageiros L, Mikhail J, Harris LG, Wilkinson TS, Rolo J, Lamble S, Bray JE, Jolley KA, Hanage WP, Bowden R, Maiden MC, Mack D, de Lencastre H, Feil EJ, Corander J, Sheppard SK (2015) Ecological overlap and horizontal gene transfer in Staphylococcus aureus and Staphylococcus epidermidis. Genome Biol Evol 7:1313–28

Monk IR, Shah IM, Xu M, Tan MW, Foster TJ. (2012) Transforming the untransformable: application of direct transformation to manipulate genetically Staphylococcus aureus and Staphylococcus epidermidis. MBio 3:pii:e00277–11. doi:10.1128/mBio.00277-11

Monk IR, Tree JJ, Howden BP, Stinear TP, Foster TJ (2015) Complete Bypass of Restriction Systems for Major Staphylococcus aureus Lineages. MBio 6:e00308–15. doi:10.1128/mBio.00308-15

Murray IA, Clark TA, Morgan RD, Boitano M, Anton BP, Luong K, Fomenkov A, Turner SW, Korlach J, Roberts RJ (2012) The methylomes of six bacteria. Nucleic Acids Res 40:11450–11462

Roberts RJ, Belfort M, Bestor T, Bhagwat AS, Bickle TA, Bitinaite J, Blumenthal RM, Degtyarev SKh, Dryden DTF, Dybvig K, Firman K, Gromova ES, Gumport RI, Halford SE, Hattman S, Heitman J, Hornby DP, Janulaitis A, Jeltsch A, Josephsen J, Kiss A, Klaenhammer TR, Kobayashi I, Kong H, Krüger DH, Lacks S, Marinus MG, Miyahara M, Morgan RD, Murray NE, Nagaraja V, Piekarowicz A, Pingoud A, Raleigh E, Rao DN, Reich N, Repin VE, Selker EU, Shaw PC, Stein DC, Stoddard BL, Szybalski W, Trautner TA, Van Etten JL, Vitor JM, Wilson GG, Xu SY (2003) A nomenclature for restriction enzymes, DNA methyltransferases, homing endonucleases and their genes. Nucleic Acids Res 31:1805–1812

Roberts GA, Houston PJ, White JH, Chen K, Stephanou AS, Cooper LP, Dryden DTF, Lindsay JA (2013) Impact of target site distribution for Type I restriction enzymes on the evolution of methicillin-resistant Staphylococcus aureus (MRSA) populations. Nucleic Acids Res 41:7472–7484

Roberts RJ, Vincze T, Posfai J, Macelis D (2015) REBASE—a database for DNA restriction and modification: enzymes, genes and genomes. Nucleic Acids Res 43:D298–D299

Schijffelen MJ, Boel CH, van Strijp JA, Fluit AC (2010) Whole genome analysis of a livestock-associated methicillin-resistant Staphylococcus aureus ST398 isolate from a case of human endocarditis. BMC Genom 11:376–386

Shepheard MA, Fleming VM, Connor TR, Corander J, Feil EJ, Fraser C, Hanage WP (2013) Historical zoonoses and other changes in host tropism of Staphylococcus aureus, identified by phylogenetic analysis of a population dataset. PLoS ONE 8:e62369. doi:10.1371/journal.pone.0062369

Sievers F, Wilm A, Dineen DG, Gibson TJ, Karplus K, Li W, Lopez R, McWilliam H, Remmert M, Söding J, Thompson JD, Higgins D (2011) Fast, scalable generation of high-quality protein multiple sequence alignments using Clustal Omega. Mol Syst Biol 7:539. doi:10.1038/msb.2011.75

Sung JM, Lloyd DH, Lindsay JA (2008) Staphylococcus aureus host specificity: comparative genomics of human versus animal isolates by multi-strain microarray. Microbiol 154:1949–1959

Tock MR, Dryden DTF (2005) The biology of restriction and anti-restriction. Curr Op Microbiol 8:466–472

Uhlemann AC, Porcella SF, Trivedi S, Sullivan SB, Hafer C, Kennedy AD, Barbian KD, McCarthy AJ, Street C, Hirschberg DL, Lipkin WI, Lindsay JA, DeLeo FR, Lowy FD (2012) Identification of a highly transmissible animal-independent Staphylococcus aureus ST398 clone with distinct genomic and cell adhesion properties. MBio 3:pii:e00027–12. doi:10.1128/mBio.00027-12

Waldron DE, Lindsay JA (2006) Sau1: a novel lineage-specific Type I Restriction-Modification system that blocks horizontal gene transfer into Staphylococcus aureus, and between S. aureus isolates of different lineages. J Bacteriol 188:5578–5585

Chapter 8
Biomechanical Analysis of Infectious Biofilms

David Head

Abstract The removal of infectious biofilms from tissues or implanted devices and their transmission through fluid transport systems depends in part of the mechanical properties of their polymeric matrix. Linking the various physical and chemical microscopic interactions to macroscopic deformation and failure modes promises to unveil design principles for novel therapeutic strategies targeting biofilm eradication, and provide a predictive capability to accelerate the development of devices, water lines, etc, that minimise microbial dispersal. Here, our current understanding of biofilm mechanics is appraised from the perspective of biophysics, with an emphasis on constitutive modelling that has been highly successful in soft matter. Fitting rheometric data to viscoelastic models has quantified linear and nonlinear stress relaxation mechanisms, how they vary between species and environments, and how candidate chemical treatments alter the mechanical response. The rich interplay between growth, mechanics and hydrodynamics is just becoming amenable to computational modelling and promises to provide unprecedented characterisation of infectious biofilms in their native state.

Keywords Biomechanics · Biofilm viscoelasticity · Constitutive modelling

8.1 Introduction

The majority of chronic infections are due to sessile, surface-associated microbial communities known as biofilms (Costerton 2007; Hall-Stoodley et al. 2004). This protected mode of existence resists external challenges including many standard clinical treatments via a range of physiological and physical mechanisms, and this realisation is hastening the development of novel therapeutic strategies targeting biofilm-specific properties (Balaban 2008; Wilson and Devine 2003). Infectious biofilms may grow directly on tissue, but are also to be found attached to the

D. Head (✉)
School of Computing, University of Leeds, Leeds, UK
e-mail: d.head@leeds.ac.uk

© Springer International Publishing Switzerland 2016

M.C. Leake (ed.), *Biophysics of Infection*, Advances in Experimental Medicine and Biology 915, DOI 10.1007/978-3-319-32189-9_8

surfaces of implanted devices where they are responsible for the majority of hospital-related infections (Shirtliff and Leid 2009). Additionally, the transmission of pathogens between individuals is facilitated by transport through artificial water lines and airways where biofilms lay dormant, detaching cells under a combination of genetic control and flow-imposed mechanical stresses to colonise new biofilms downstream (Busscher and van der Mei 2006; Gilkeson et al. 2013); see Fig. 8.1.

In many applications the expansion, detachment and ultimately dispersal of potentially pathogenic microbes depends in part on the mechanical properties of their biofilm phenotype. This is perhaps most familiar in our daily attempts to reduce the bacterial load of the most accessible and well-studied biofilm of relevance to human health, dental plaque, by regular brushing (Marsh and Martin 2009; Verkaik et al. 2010). Other shear-based removal devices include microbubble jets for oral hygiene (Howlin et al. 2015), and automated wipers (Flemming et al. 2009) and wall-scraping 'pigs' (Szomolay and Cogan 2015) in industry. An integrated removal strategy first exposes biofilms to sublethal concentrations of antibiotics or other chemical agents, which can lead to reduced stiffness and enhanced removal, although crosslinkers and multivalent ions can increase stiffness (Lieleg et al. 2011; Zrelli et al. 2013). In addition, modifying the topographic, chemical or elastic properties of abiotic surfaces can reduce cell attachment or promote weak bonding that can be more easily removed (Brady and Singer 2000; Dexter 1979; Flemming et al. 2009; Fu et al. 2011; Katsikogianni and Missirlis 2010).

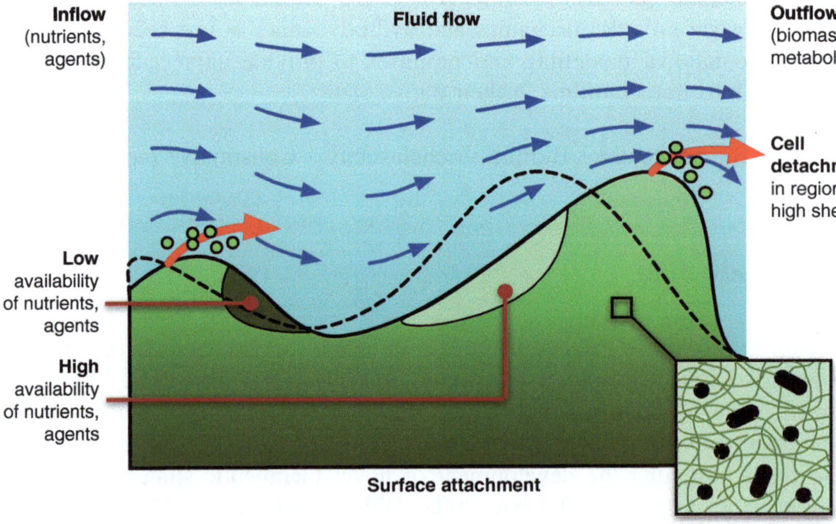

Fig. 8.1 Schematic representation of some issues relevant to biofilm mechanics. Fluid flow generates shear stresses that deform the biofilm (*solid line*) from its nominal unstressed profile (*dashed line*). Cellular detachment due to high shear stress is one mechanism for dispersal (*large arrows*). The direction of flow is modulated by the free surface, resulting in a non-uniform advection of dispersed phases such as nutrients, chemical agents, and metabolic products. Their availability therefore varies across the deformed biofilm profile. At a microscopic level, the mechanical properties ultimately derive from microbial cells and the polymeric matrix they produce (*inset*)

Biofilm mechanics is not just important under physical brushing or scraping, however. Hydrodynamic flow is known to affect biofilm formation and propagation in dental unit water lines (Walter et al. 2004), catheters (both urinary tract and intravascular) and hemodialysis machines (Fux et al. 2004; Kim et al. 2014; Shirtliff and Leid 2009; Wilson and Devine 2003), bile drains, stents, and voice prostheses (Busscher and van der Mei 2006). In addition, the response to surface airflow and known airborne dispersal modes should be considered in the design of nosocomial ventilation systems (Gilkeson et al. 2013; Shirtliff and Leid 2009). Flow is also a key consideration for industrial biofilms as overgrowth can cause clogging in biofilm reactors (Knutson et al. 2005; Thullner and Baveye 2008), although here the goal is to maximise metabolic efficacy rather than microbial eradication.

The aim of this chapter is to summarise our current understanding of biofilm mechanics from the perspective of biophysics, in the hope of informing future investigations targeting the control or elimination of infectious biofilms under mechanical challenge. The primary distinction from a recent review (Guélon et al. 2011) is the focus here on constitutive modelling following an approach common to soft matter systems, i.e. solutions of macromolecules such as colloids, flexible and semi-flexible polymers, which have been recognised as an abiotic counterpart of biofilms with regards mechanical response (Wilking et al. 2011). A pragmatic benefit of deriving and validating quantitative models is their predictive capability, which can be used to guide the design of novel technologies targeting reduced biofilm-related infections, for instance water lines that minimise microbial spread. However, the advantages of constitutive modelling go beyond characterisation. Identifying the relevant microscopic mechanisms responsible for observed bulk macroscopic behaviour generates fundamental insights with potentially far-reaching consequences, for instance in suggesting targets for novel therapeutics that would not otherwise have been considered, or allowing extrapolative predictions to situations not yet assayed experimentally. A wide range of theoretical and experimental methods have been developed over decades of soft matter physics research, and the potential benefits of porting these to the study of biofilms is substantial.

Biofilm mechanics is inherently a multi-disciplinary topic and as such there is a need to summarise a number of key concepts, some of which will undoubtedly already be known to readers. These are outlined in Sect. 8.2 with due apologies for any overfamiliarity of the material. It should be clearly stated from the outset what will *not* be considered here. The interactions between the microbes and the surface influence initial biofilm formation and generate adhesive complexes that need to be overcome or dismantled prior to large-scale motion or detachment (Song et al. 2015; Tuson and Weibel 2013), but this important topic is simply too extensive to cover here. In addition, self-propelled bacterial swimmers are most commonly associated with the planktonic phenotype but are known to arise in some biofilms, both early stage and mature (Houry et al. 2012; Rogers et al. 2008), and could potentially be tractable to the analytical tools of active matter currently undergoing vigorous investigation (Marchetti et al. 2013), but since their relationship with biofilm mechanics has not been explored they will not be discussed further. The coupling between mechanics and community dynamics is likely to be of key importance in the future and is discussed in Sect. 8.4.

8.2 Background

In most situations, biofilms are subjected to deformations or stresses that vary over length scales far exceeding that of the constituent molecules, making it impractical to characterise the mechanical response taking microscopic quantities as the degrees of freedom. It is often more expedient to adopt a coarse-grained continuum approximation defined by spatio-temporally averaged fields, much as the equations for hydrodynamics involve fluid velocity and density rather than the atomistic quantities from which they derive. In this section, some relevant aspects of continuum mechanics are briefly overviewed, followed by a survey of the primary experimental tools employed to quantify them in vitro. It is nonetheless important to remain mindful of the microscopic basis for these macroscopic quantities, and these will be discussed first.

8.2.1 Molecular Origins of Biofilm Mechanics

From a physical perspective, biofilms can be crudely described as two-component systems comprising cells of various well-defined forms (spheres, rods, etc.) embedded in a hydrated polymeric mesh (Stoodley et al. 1999). If the volume fraction ϕ occupied by the cells is high, greater than roughly 0.2–0.5, the mechanical response will be dominated by steric hinderance acting between the stiff cell walls, and an analogy with colloidal systems is possible (Wilking et al. 2011). More commonly, however, $\phi \ll 1$ and the response is determined by the inter-dispersed matrix which reacts via purely physico-chemical mechanisms on times scales shorter than the cells' metabolic response. It is on these time scales that the analogy with abiotic systems is closest.

The multi-functional biofilm matrix consists of a variety of polysaccharides, extracellular DNA, proteins (including amyloid fibrils, e.g. curli), biosurfactants, and other macromolecules, with the composition depending on the cell strains and their environment (Blanco et al. 2012; Flemming and Wingender 2010). Its cohesive properties derive from chemical and strong physical inter-molecular bonds (e.g. covalent, ionic) that are permanent on relevant time scales, and weak physical bonds (e.g. hydrogen bonds, van der Waals) with energies comparable to or less than the thermal scale $k_{\mathrm{B}}T$ and are thus transient (Rubinstein and Colby 2003). This description is strongly reminiscent of polymer solutions and gels, whose mechanical properties are well understood after decades of study in an industrial context (The Theory of Polymer Dynamics, M. Doi, S. Edwards (Oxford 1986); Rubinstein and Colby 2003). Following the established procedure outlined in Fig. 8.2, the bulk

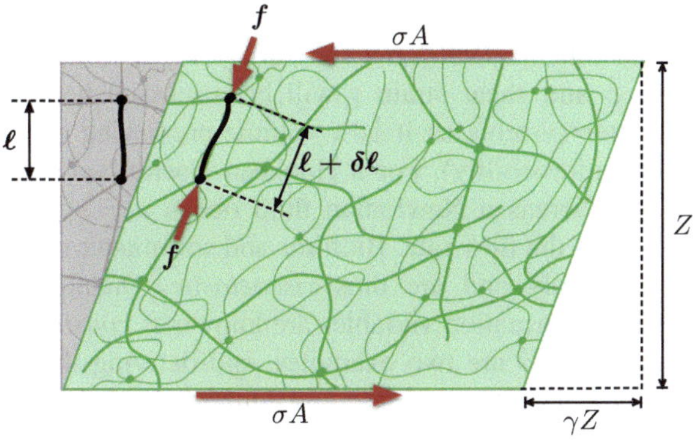

Fig. 8.2 A conceptual representation of a small volume of biofilm matrix under an imposed shear strain γ. The lines and nodes in the background (pre-strain) and foreground (post-strain) images correspond to polymers and inter-molecular interactions, respectively. The highlighted polymer segment extends from a length ℓ to $\ell + \delta\ell$ under the strain, resulting in equal-and-opposite tensile forces f acting on the nodes at each of its ends. Integrating over all such microscopic forces results in bulk forces σA acting across the body, with σ the shear stress and A the area of the opposing surfaces

mechanical response can be theoretically derived by summing the force-extension curves for individual polymers over a hypothesised molecular composition. These force-extension relations derive from thermodynamic relations if thermal fluctuations are thought to dominate, or by treating polymers as slender elastic bodies if only enthalpic contributions (e.g. backbone stretching) are relevant.

Compared to flexible synthetic polymers, biopolymers can be thick, of the order of 1–10 nm or more (Alberts et al. 2008), and this imbues them with a natural resistance to bending which significantly modifies their dynamics (Morse 1998). A highly successful force-extension relation deriving from the worm-like chain (Kratky-Porod) model has been shown to describe well a range of biomolecules, including DNA (Smith et al. 1996), collagen (Chang and Buehler 2014), intermediate filaments (Lin et al. 2010) and actin (Broedersz and MacKintosh 2014; Pritchard et al. 2014). Incorporating increases in backbone length produces the extensible worm-like chain model applicable for large (i.e. nonlinear) deformations (Storm et al. 2005). Note that closed-form analytical expressions require multiple assumptions to be made, including the *affine* assumption that the local deformation field is a scaled-down version of the bulk strain (Basu et al. 2011; Broedersz and MacKintosh 2014). Spatially extended models can relax such assumptions but typically require numerical solution.

8.2.2 *Continuum Viscoelasticity*

There are two central quantities in continuum mechanics; the stress, which is the force normalised to area and has units of pressure, and the strain or relative deformation which is dimensionless (Barnes et al. 1989). Both are properly rank-2 tensors (Landau and Lifshitz 1986), but for clarity only a scalar treatment in terms of the shear stress σ and shear strain γ will be discussed here (shear is usually chosen for soft matter systems as it is a volume-preserving mode that does not invoke solvent incompressibility). For a Hookean elastic body, $\sigma = G_0\gamma$ with G_0 the shear modulus, whereas a Newtonian fluid obeys $\sigma = \eta\dot{\gamma}$ with η the (shear) viscosity and $\dot{\gamma} \equiv d\gamma/dt$ the flow rate. Biofilms and soft matter systems exhibit both behaviours simultaneously and are thus *viscoelastic*. Quantitative relationships between stresses and deformation variables are known as *constitutive equations*, of which $\sigma = G_0\gamma$ and $\sigma = \eta\dot{\gamma}$ are two examples, and deriving these relationships is the goal of the analytical coarse-graining procedure discussed in Sect. 8.2.1. Thus, fitting validated constitutive equations to experimental rheometry data generates estimates of microscopic parameters in what can be thought of as a 'rheological microscope'.

A key issue to identify early on is whether or not the material response is linear, i.e. if all stresses, strains and their time derivatives obey proportionality relations (Barnes et al. 1989). If linearity holds, the stress at time t can be written as the superposition of stresses due to infinitesimal strain increments $d\gamma = \dot{\gamma}(t)\mathrm{d}t$ at all previous times,

$$\sigma(t) = \int_{-\infty}^{t} G(t-s)\dot{\gamma}(s)\mathrm{d}s, \tag{8.1}$$

where $G(t)$ is the linear step-strain response (Fielding et al. 2000; Rubinstein and Colby 2003). Alternatively, the strain at time t can be written as an integral over previous stress increments in terms of the compliance $J(t)$, where $J(t)$ gives the strain due to a step shear stress at $t = 0$ as measured in creep compliance tests (Barnes et al. 1989). As long as linearity holds, these two representations are equivalent (Rubinstein and Colby 2003). Note that (8.1) assumes time translational invariance which would not hold if the material properties varied over the measurement time, when at least one additional time variable would be required (Fielding et al. 2000).

To quantify the linear response at a particular frequency, it is common to apply an oscillatory shear $\gamma(t) = \gamma_0\cos(\omega t)$ and measure components of the stress response that are in phase ($\propto \cos\omega t$) or out of phase ($\propto \sin\omega t$) with this driving. As shown in Fig. 8.3, this produces two moduli, the in-phase storage modulus $G'(\omega)$ and the out-of-phase loss modulus $G''(\omega)$. To avoid a proliferation of trigonometric functions it is expedient to adopt the complex representation $\gamma(t) = \mathrm{Re}\{\gamma_0 e^{i\omega t}\}$, where 'Re' takes the real part. Inserting this into (8.1) gives

$$\sigma(t) = \mathrm{Re}\{\gamma_0 G^*(\omega)e^{i\omega t}\}, \quad G^*(\omega) = i\omega\int_0^{\infty} e^{-i\omega t}G(t)\mathrm{d}t, \tag{8.2}$$

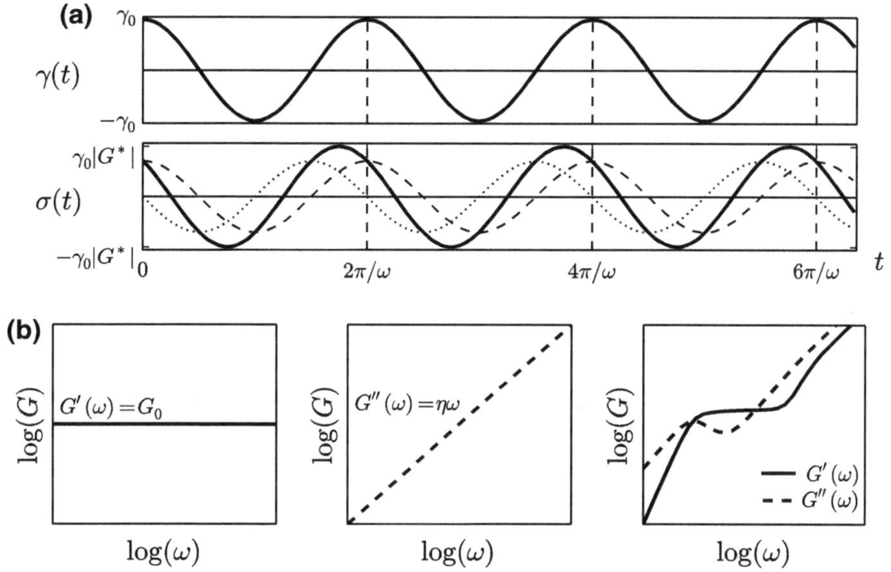

Fig. 8.3 **a** Example of an applied oscillatory shear strain $\gamma(t) = \gamma_0 \cos\omega t$ and the induced shear stress $\sigma(t)$, which can be decomposed into the in-phase contribution $\gamma_0 G'(\omega)\cos\omega t$ (*dashed line*) and the out-of-phase contribution $-\gamma_0 G''(\omega)\sin\omega t$ (*dotted line*). **b** Examples of viscoelastic spectra for a Hookean solid (*left*), a Newtonian fluid (*middle*), and an entangled solution of semi-flexible polymers (Morse 1998) (*right*). *Solid lines* are $G'(\omega)$ and *dashed lines* are $G''(\omega)$

where $G^*(\omega) = G'(\omega) + iG''(\omega)$. The entirety of a material's linear response (for scalar shear) is contained in the two viscoelastic spectra $G'(\omega)$ and $G''(\omega)$. Examples are given in Fig. 8.3b, including an elastic solid with $G'(\omega) = G_0$ and $G''(\omega) = 0$, and a viscous fluid obeying $G'(\omega) = 0$ and $G''(\omega) = \eta\omega$ (alternatively, $G^*(\omega) = G_0$ and $G^*(\omega) = i\eta\omega$ respectively). Just as the Hookean solid and Newtonian fluid represent idealised elastic and viscous bodies, schematic models have been devised for idealised viscoelastic materials, including the Maxwell model which relaxes to a viscous liquid over a single relaxation time, and the Kelvin–Voigt model which relaxes to an elastic solid (Barnes et al. 1989). Such schematic models are known as spring-dashpot models as they can be represented as combinations of (elastic) springs and (viscous) dashpots in serial, parallel or both.

When linearity does not hold, superposition (8.1) is not valid, the various definitions of stress and strain tensors start to deviate (Ogden 1984), and both quantifying and modelling the material response becomes more challenging. It is common to consider continuous shear $\gamma(t) = \dot{\gamma}t$ when investigating the nonlinear response, although small-amplitude oscillatory shear about a fixed prestrain has also been employed for viscoelastic solids (Storm et al. 2005). The nonlinear regime is of most practical interest as stress-based detachment and failure are both nonlinear phenomena, and for this reason the majority of biofilm rheology studies have focussed on this regime (see Sect. 8.3). Without supporting modelling, however, it is difficult to gauge the generality of reported findings with regards varying strains or environmental conditions. Successful predictive modelling for strongly deformed polymeric systems typically follows validation of linear models for the same material (The Theory of Polymer Dynamics, M. Doi, S. Edwards (Oxford 1986; Storm et al. 2005), suggesting that pursuing a similar progression for biofilms would lead to representative constitutive equations for both the linear and nonlinear response.

8.2.3 Experimental Biofilm Rheology

The nature of biofilms renders the application of standard rheology measurement protocols problematic. Biofilms are thin, surface-associated, living colonies that cannot be removed from their native environment without risk of altering their properties. They are also strongly heterogeneous, typically exhibiting a complex three-dimensional morphology with structured cell-rich and cell-deficient domains, the latter permitting flow (Stewart 2012; Vogt et al. 2013). Measuring their mechanical properties has required modification of existing methods, or the development of novel biofilm-specific ones. Some commonly used methods are summarised below; for a more extensive discussion see (Guélon et al. 2011).

Macrorheometry: Standard rheometers constrain the sample between two surfaces of various geometries (parallel plate, cone and plate, Taylor-Couette cell etc.), with the motion of one controlled to impose a predefined strain or stress schedule. Biofilms can be grown ex situ and transplanted to the rheometer, either intact on plates (He et al. 2013; Ochoa et al. 2007) or by a destructive process (Lieleg et al. 2011; Wloka et al. 2004), or the rheometer modified to permit growth in situ (Pavlovsky et al. 2015; Pavlovsky et al. 2012). Macrorheometry can extract bulk properties (Paramonova et al. 2007; Shaw et al. 2004) but is not well suited to studying heterogeneities.

Macro and micro-indentation: The free surfaces of biofilms in their native state can be compressed globally (He et al. 2013; Peterson et al. 2013), or locally such as performed by atomic force microscopy (Brindle et al. 2011; Lau et al. 2009), and force-displacement relations and relaxation curves measured. Heterogeneity can be probed by multi-site sampling (Hwang et al. 2015).

Microrheology: This methodology can be active or passive, and one or two particle (Mizuno et al. 2008). For one particle, the motions of micron-scale particles undergoing passive Brownian motion are converted to $G^*(\omega)$ or $J^*(\omega)$ using thermodynamic relations (Evans et al. 2009; Mason 2000; Mason and Weitz 1995), or they are actively driven via external forces and the moduli extracted directly (Galy et al. 2012; Zrelli et al. 2013). These particles are either added, or are endogenous such as the cells themselves (Rogers et al. 2008). Two-particle microrheology tracks correlated motion to reduce artefacts at the particle surfaces. Microrheology is well suited to studying heterogeneities and has been widely used in soft matter.

Flow cells: Imaging biofilms grown in situ in continuous flow cells is of direct relevance to many applications (Busscher and van der Mei 2006). Nonlinear streamers often form in rapid flow (Stoodley et al. 1999), whose visually determined deformation fields can be fitted to those of simple elastic bodies to estimate static moduli (Aravas and Laspidou 2008; Stoodley et al. 1999). More recently, microfluidic devices have been employed to provide better environmental control and ensure laminar flow (Knutson et al. 2005; Rusconi et al. 2011; Song et al. 2014; Weaver et al. 2012), and fitting of visually identified streamers (Kim et al. 2014) or partitioning walls (Hohne et al. 2009) permits parameter estimation.

Biofilm reactors: Although industrially motivated, laboratory-scale attached growth reactors can be used to quantify the relationship between detachment and flow for generic biofilms, generating basic insight that may also apply to clinical situations. Reactor geometries are designed to expose high surface areas of the biofilm to flow, such as porous media (Chang et al. 1991; Vogt et al. 2013) or fibres (Huang et al. 2013; Tierra et al. 2015), and biomass effluent and other quantities measured.

8.3 Biofilm Mechanics

The complex physical chemistry of biofilms, and in particular the multi-component biofilm matrix, makes it difficult to interpret rheology data in terms of identifiable microscopic processes. Even restricting attention to just the linear response regime still leaves a range of inter-molecular bonds or junctions that need to be considered to reproduce the full viscoelastic spectra $G'(\omega)$ and $G''(\omega)$. A common approach has been to fit rheometric data to a small number of viscoelastic constitutive equations, tracking parameter changes as microbial species, environmental conditions, etc, are varied, but the proposed mechanisms underlying each fit remain unproven without independent measurements to confirm model validity. These problems are exacerbated in the nonlinear regime, where additional processes such as cellular rearrangement must also be considered. A summary of our current understanding of biofilm rheology is presented here, starting with the bulk response before turning to consider more applied protocols.

8.3.1 Bulk Rheology

$G'(\omega)$ and $G''(\omega)$ have been measured for biofilms of the well-characterised bacterium *P. aeruginosa* (Ehret and Böl 2012; Lieleg et al. 2011) with similar results emerging from both studies, i.e. a weak frequency dependence over the range 10^{-3}–10 Hz, with a slight decay at low frequencies. Although $G'(\omega)$ was roughly an order of magnitude larger than $G''(\omega)$, indicating elastic-dominated behaviour, the two spectra became comparable on lowering the concentration of Ca^{2+}, suggesting a reduction in strong ionic matrix bonds and increased fluidity, consistent with fits to a superposition of worm-like chains (Ehret and Böl 2012). Broadly similar spectra have been observed in a range of polymeric gels, including peptide fibrils (Roberts et al. 2012), block copolymers (Rombouts et al. 2013) and intermediate filaments (Lin et al. 2010). F pilus producing *E. coli* biofilms, but not curli producing ones, have also been compared to actin protein gels based on active microrheology experiments (Galy et al. 2012). *P. aeruginosa* biofilms rapidly stiffen for frequencies exceeding 10 Hz (Lieleg et al. 2011), an effect not seen in the other systems mentioned.

Fitting stress relaxation in compressed biofilms to a generalised Maxwell model has been used to infer the relevance of extracellular DNA to the viscoelasticity of single species *S. mutans*, *P. aeruginosa*, *S. aureus* and *S. epidermidis* systems, where principal component analysis was employed to determine the number of modes (Peterson et al. 2013). Biomass rearrangement and flow modes were inferred from similar analysis of two oral species (*S. oralis* and *A. naeslundii*) and intact dental plaque (He et al. 2013). Despite the intuitive appeal of these findings, it should be noted that spring-dashpot models are idealised and admit only a small number of relaxation times, in contrast to *P. aeruginosa* matrix extracts which have been shown to relax over a broad distribution of time scales (Wloka et al. 2004). Model-free data presentation in the form of $G'(\omega)$ and $G''(\omega)$ is preferable as it permits the development of theoretical tools subsequent to the experimental study (Ehret and Böl 2012). Spring-dashpot models are also linear, so their application to nonlinear phenomena should only be attempted after careful consideration.

Quantifying changes in fit parameters as environmental factors are altered can identify controllable modifications leading to reduced stiffness and enhanced clearance. Parallel plate rheometry of *S. epidermidis* biofilms demonstrated a reduction in stiffness with increasing temperature, presumably by reducing the dissociation time for weak inter-molecular matrix bonds, with potential applications to medical devices (Pavlovsky et al. 2015). Urea was found to weaken similar biofilms, possibly by interfering with hydrogen bonding (Brindle et al. 2011). Cellular growth under continuous flow resulted in stiffer biofilms for the same species, which may be a physiological response to sustained mechanical stress, although the failure strain did not significantly vary (Aggarwal and Hozalski 2012). Antibiotics of sufficient concentration to destroy the majority of cells had no significant effect on the stiffness of the *E. coli* matrix, but it was drastically weakened by the protease trypsin, suggesting a dominant role for proteins in the matrix mechanics of this species (Zrelli et al. 2013). By contrast, a broad range of chemical agents had no significant affect of the mechanics of *P. aeruginosa* biofilms, but could influence the rate of recovery from nonlinear strains (Lieleg et al. 2011), possibly due to chemical modification of inter-molecular associations between matrix components.

These results highlight that different microbial species and strains produce different matrix polymers with broadly varying mechanical properties (Aggarwal et al. 2010; Lau et al. 2009), although commonality in the longest relaxation time has been claimed from linear creep fits in a parallel plate rheometer (Shaw et al. 2004). The nonlinear response is also variable, with both shear thinning (Lieleg et al. 2011) and shear thickening (Stoodley et al. 1999) reported, although the empirical Cox–Merz rule relating linear oscillatory viscoelasticity to nonlinear flow has been verified for *P. aeruginosa* matrix extracts (Wloka et al. 2004) which, if found to be generally true, would aid the characterisation of biofilms in flow. An additional problem is the significant run-to-run variation, consistent with a broad, log-normal distribution (Aggarwal and Hozalski 2012), and the variation in stiffness, decreasing near the surface and increasing with age (Galy et al. 2012; Lau et al. 2009; Rogers et al. 2008). Such spatiotemporal variation can in principle be incorporated into suitably parameterised constitutive models, at the expense of an increased number of fit parameters and greater validation challenge.

8.3.2 Fluid-Structure Coupling

As shown in Fig. 8.1, the relationship between the deformation and growth of immersed biofilms and the shear stresses generated by the surrounding fluid flow is complex and bidirectional. Biofilm morphology influences local flow patterns and hence the fluxes of dispersed phases such as nutrients, metabolites, and quorum sensing molecules (Song et al. 2014; Stewart 2012; Weaver et al. 2012), and flow can drive biofilm morphogenesis to rippled beds (Hödl et al. 2014; Stoodley et al. 1999), streamers (Rusconi et al. 2011; Stoodley et al. 1999a, b; Weaver et al. 2012), and rolling clusters (Rupp et al. 2005). The relevance of such process to the formation, growth, and dispersal of both infectious and non-infectious biofilms demands quantitative modelling with a predictive capability, but this situation is far from being realised. Even if a validated constitutive model for the biofilm were available, its integration into these complex, dynamic geometries is challenging even with computational solution.

Partial insight can be attained by considering limited scenarios before tackling the full complexity. Streamers are elongated oscillating structures aligned with the flow direction that can form in turbulent conditions (Stoodley et al. 1999a, b) or in rapid laminar flow in microfluidic devices (Rusconi et al. 2011; Weaver et al. 2012), and can be visualised with optical or confocal microscopy and the profiles fitted to predictions for slender elastic filaments (Aravas and Laspidou 2008). Note that fitting to elastic bodies only allows static quantities to be extracted, i.e. $G'(0) \equiv G_0$ and non-shear quantities such as the Poisson ratio (Landau and Lifshitz 1986) ($G''(0) \equiv 0$ by symmetry; Barnes et al. 1989). Modelling the two-way coupling between streamer elasticity and fluid flow presents numerically difficulties, but qualitative agreement has been reached in 2D, identifying vortices as the source of the oscillations (Taherzadeh et al. 2012). An alternative reduced geometry is the lateral expansion of circular biofilms by osmosis (i.e. uptake of fluid into the matrix), for which good quantitative agreement between experiments and continuum modelling is possible (Seminara et al. 2012). Enhanced models have demonstrated that elastic instabilities can generate wrinkles, even without confinement (Ben Amar and Wu 2014; Espeso et al. 2015).

A complete description of a growing 3D biofilm in flow is only now becoming possible, but some modelling insight has already been achieved. Lattice models with cellular automata-like rules for biofilm growth have predicted a wide range of biofilm morphologies (Rodriguez et al. 2012), and also graphically demonstrated the advection of substrates around the fluid-biofilm interface (Picioreanu et al. 2001). Continuum representations of the biofilm as constitutively linear elastic (Coroneo et al. 2014; Duddu et al. 2009; Eberl and Sudarsan 2008; Taherzadeh et al. 2012), non-linear elastic (Böl et al. 2009; Dupin et al. 2001), viscous (Cogan 2011), or viscoelastic (Towler et al. 2007) bodies is challenging as the interface must be tracked using stress and displacement matching; simplifications such as a one-way coupling or reduced dimensionality are sometimes employed. Interface tracking is not required for phase field models, which have been employed to argue that low matrix elasticity is required for streamers to form (Tierra et al. 2015) and to investigate the role of cohesion on interface stabilisation (Klapper and Dockery 2006). The non-overlapping range of feature sets and problems considered makes it difficult to discern a coherent picture from these modelling studies, but it is hoped that the advent of full-featured simulations (e.g. Coroneo et al. 2014) will herald reproducible investigations of an incrementally expanded parameter space generating consistent predictions for experiments.

8.3.3 Mechanically Induced Detachment

Dispersal is part of the biofilm life cycle and as such under a degree of genetic control (McDougald et al. 2012), but mechanics also plays a role. The inter-polymeric bonds that provide the matrix with its mechanical resilience and structural integrity must either be separated from cell surface proteins, enzymatically lysed, or overcome by tensile forces deriving from the external fluid shear (Hall-Stoodley et al. 2004), before cells can escape the biofilm envelope. Erosion of single cells, multi-cellular clusters (which maintain their resistance to antibiotics Fux et al. 2004) and sudden large-scale sloughing events also result from fluid shear (Davies 2011), limiting biofilm thickness (Chang et al. 1991).

Detachment is a key issue for industrial biofilm reactors and has been widely studied in this context. Reactors maximise the area of contact between microbes and fluid by growing the biofilms on highly porous supports, but this entails the risk of overgrowth leading to reduced flow rates and eventual blockage. Quantifying biomass effluent from complex biofilms grown in a glass-bead reactor in terms of a spatially averaged model has revealed design principles that could be exploited to control growth and detachment (Chang et al. 1991). Spatially extended models reveal a complex biomass distribution on the scale of reactor pores, with reduced nutrient availability and biofilm growth downstream from clogged pores (Knutson et al. 2005; Thullner and Baveye 2008). A suite of techniques has been employed to measure velocity profiles and metabolic fluxes in terms of biofilm age and microbial composition (Huang et al. 2013), generating a range of data that could be used to validate highly sophisticated and predictive models.

Early models for shear-based detachment imposed height-dependent rates rather than solve for the explicit hydrodynamic flow field (Alpkvist et al. 2006), but still demonstrated that combining multiple detachment mechanisms can generate a variety of morphologies (Chambless and Stewart 2007). To consider the effects of flow, some continuum fluid-structure coupling models have been extended to include detachment as reduced interfacial growth (Coroneo et al. 2014; Duddu et al. 2009) or a critical stress threshold (Böl et al. 2009; Picioreanu et al. 2001), and suggest erosion smoothens while sloughing roughens the biofilm surface. Note that not all of these include a two-way coupling between biofilm mechanics and fluid shear stress. Single-cell detachment events are best modelled using particle-based methods, but these do not typically include elasticity. One that does is the immersed boundary method in which the biofilm is represented as a spring-node assembly (Alpkvist and Klapper 2007; Vo and Heys 2011), but this requires extension to include growth and dispersed-phase advection. Another is an off-lattice method from soft matter physics known as dissipative particle dynamics that has been applied to 2D biofilms (Xu et al. 2011), and includes thermal fluctuations of cells within the matrix which has been experimentally measured (Rogers et al. 2008) but is often neglected in cell-scale models.

8.4 Outlook

Constitutive modelling of biofilm viscoelasticity, and in particular that of the biofilm matrix provides quantitative insight into the mechanisms by which microscopic modifications cascade up through the length scales to influence the bulk properties relevant to the dispersal and removal of infectious colonies. Linear analysis is the most amenable to mathematical modelling and has already been employed as a tool to extract the importance of different inter-molecular interactions, as well as representing a first step towards fully nonlinear models relevant to dispersal. Validated models can be employed to accelerate the design of treatments that beneficially modulate biofilm mechanics, and optimise the geometries of fluid transport systems to minimise microbial spread, to give just two examples. While it is clear that the multi-component matrix represents a greater analytical challenge than classical soft matter systems, the difference is one of quantity rather than quality: Each inter-molecular interaction can be modelled in isolation; the goal is to combine them into a comprehensive picture. This is most immediately possible in the linear regime where superposition applies.

While there has already been an array of investigations into specific pathogens, geometries and applications, progress towards a synthesis embracing biofilm mechanics and environmental conditions in general has been comparatively subdued. Whereas this may reflect the scope and complexity of the problem, the significant benefits of uncovering commonalities, such as the transferal of insight and analytical tools between problems, suggests such higher-level investigations deserve more attention. Overlapping studies would increase the chances of discovering such commonalities while helping to identify hidden relevant parameters that need to be controlled for reproducibility and hence predictability. This emphasis on universal properties is perhaps indicative of a physics viewpoint and will therefore only provide partial insight into biological problems, but as part of an integrated biophysical approach promises to reveal fundamental, general principles that will both elucidate biofilm mechanics and guide the development of future, robust treatments.

One aspect of natural biofilms not addressed here is that they are properly regarded as ecosystems, i.e. communities of different species and strains interacting with each other and their environment, including the host (Costerton 2007). This is established for oral biofilms, where the ecological plaque hypothesis asserts that dental caries (tooth decay) results from excessive dietary carbohydrates promoting the overgrowth of acid-producing bacteria, and the resulting drop in pH promotes enamel demineralisation (Marsh 1994). A similar picture has been argued for periodontal (gum) disease (Marsh et al. 2011) (note that, according to Koch's postulates (Hogg 2005), these are *not* infections as the causative agents are present in healthy plaque, albeit in small numbers). Multi-species biofilms present different mechanical properties than their single-species counterparts (Hwang et al. 2015), and therapies targeting mixed biofilms must ensure they do not preferentially remove commensal species, as this could conceivably lead to regrowth with an increased pathogenic fraction. Given the potential for mechanically altering the composition of infectious biofilms to reduce the numbers of pathogenic microbes, this fascinating and potentially rich coupling surely deserves greater attention.

Acknowledgment DAH is funded by the Biomedical Health Research Centre, University of Leeds, UK.

References

Alberts B, Johnson A, Lewis J, Raff M, Roberts K, Walter P (2008) Molecular biology of the cell, 5th edn. Garland
Alpkvist E, Klapper I (2007) Water Sci Tech 55:265
Alpkvist E, Picioreanu C, van Loosdrecht MCM, Heyden A (2006) Biotech Bioeng 94:961
Aggarwal S, Hozalski RM (2012) Langmuir 28:2812
Aggarwal S, Poppele EH, Hozalski RM (2010) Biotech Bioeng 105:924
Aravas N, Laspidou CS (2008) Biotech Bioeng 101:196
Balaban N (ed) (2008) Control of biofilm infections by signal manipulation. Springer, Berlin
Barnes HA, Hutton JF, Walters K (1989) An introduction to rheology. Elsevier
Basu A, Wen Q, Mao X, Lubensky TC, Janmey PA, Yodh AG (2011) Macromolecules 44:1671
Ben Amar M, Wu M (2014) Europhys Lett 108:38003
Blanco LP, Evans ML, Smith DR, Badtke MP, Chapman MR (2012) Trends Microbiol 20:66
Böl M, Möhle RB, Heasner M, Neu TR, Horn H, Krull R (2009) Biotech Bioeng 103:177
Brady RF, Singer IL (2000) Biofouling 15:73–81
Brindle ER, Miller DA, Stewart PS (2011) Biotech Bioeng 108:2968
Broedersz CP, MacKintosh FC (2014) Rev Mod Phys 86:995

Busscher HJ, van der Mei HC (2006) Clin Microbiol Rev 19:127–141
Chambless JD, Stewart PS (2007) Biotech Bioeng 97:1573
Chang HT, Rittmann BE, Amar D, Heim R, Ehlinger O, Lesty Y (1991) Biotech Bioeng 38:499
Chang S-W, Buehler MJ (2014) Mater Today 17:70
Cogan NG (2011) Int J Numer Meth Biomed Eng 27:1982
Coroneo M, Yoshihara L, Wall WA (2014) Biotech Bioeng 111:1385
Costerton JW (2007) The Biofilm Primer. Springer, Berlin
Davies DG (2011) Biofilm Highlights In: Flemming H-C, Wingender J, Szewzyk U (eds). Springer, Berlin
Dexter SC (1979) J Coll Int Sci 70:346–353
Doi M, Edwards S (1986) The theory of polymer dynamics. Oxford
Duddu R, Chopp DL, Moran B (2009) Biotech Bioeng 103:92
Dupin HJ, Kitanidis PK, McCarty PL (2001) Wat Res Res 37:2965
Eberl HJ, Sudarsan R (2008) J Theor Biol 253:788
Ehret AE, Böl M (2012) J Roy Soc Int 10:20120676
Espeso DR, Carpio A, Einarsson B (2015) Phys Rev E 91:022710
Evans RML, Tassieri M, Auhl D, Waigh TA (2009) Phys Rev E 80:012501
Fielding SM, Sollich P, Cates ME (2000) J Rheol 44:323
Flemming H-C, Wingender J (2010) Nat Rev Microbiol 8:623
Flemming H-C, Sriyutha Murthy P, Venkatesan R, Cooksey KE (eds) (2009) Marine and industrial biofouling. Springer, Berlin
Fu J, Wang Y-K, Yang MT, Desai RA, Yu X, Liu Z, Chen CC (2011) Nat Meth 7:733–736
Fux CA, Wilson S, Stoodley P (2004) J Bacteriol 186:4486
Galy O, Latour-Lambert P, Zrelli K, Ghigo J-M, Beloin C, Henry N (2012) Biophys J 103:1400
Gilkeson CA, Camargo-Valero MA, Pickin LE, Noakes CJ (2013) Build Env 65:35–48
Guélon T, Mathias J-D, Stoodley P (2011) Biofilm highlights In: Flemming H-C, Wingender J, Szewzyk U (eds). Springer, Berlin
Hall-Stoodley L, Costerton JW, Stoodley P (2004) Nat Rev Microbiol 2:95
He Y, Peterson BW, Jongsma MA, Ren Y, Sharma PK, Busscher HJ, can der Mei HC (2013) PLoS ONE 8:e63750
Hödl I, Mari L, Bertuzzo E, Suweis S, Besemer K, Rinaldo A, Battin TJ (2014) Env Microbiol 16:802
Hohne DN, Younger JG, Solomon MJ (2009) Langmuir 25:7743
Hogg S (2005) Essential microbiology. Wiley, Chichester
Houry A, Gohar M, Deschamps J, Tischenko E, Aymerich S, Gruss A, Briandet R (2012) Proc Nat Acad Sci 109:13088
Howlin RP, Fabbri S, Offin DG, Symonds N, Kiang KS, Knee RJ, Yoganantham DC, Webb JS, Birkin PR, Leighton TG, Stoodley P (2015) J Dent Res 94:1303
Huang Z, McLamore ES, Chuang HS, Zhang W, Werely S, Leon JLC, Banks MK (2013) Biotech Bioeng 110:525
Hwang G, Marsh G, Gao L, Waugh R, Koo H (2015) J Dent Res 94:1310
Katsikogianni MG, Missirlis YF (2010) Acta Biomater 6:1107–1118
Kim MK, Drescher K, Pak OS, Bassler BL, Stone HA (2014) New J Phys 16:065024
Klapper I, Dockery J (2006) Phys Rev E 74:031902
Knutson CE, Werth CJ, Valocchi AJ (2005) Wat Res Res 41:W07007
Landau LD, Lifshitz EM (1986) Theory of elasticity, 3rd edn. Butterworth-Heinemann
Lau PCY, Dutcher JR, Beveridge TJ, Lam JS (2009) Biophys J 96:2935
Lieleg O, Caldara M, Baumgärtel R, Ribbeck K (2011) Soft Matter 7:3307
Lin Y-C, Yao NY, Broedersz CP, Herrmann H, MacKintosh FC, Weitz DA (2010) Phys Rev Lett 104:058101
Marchetti MC, Joanny JF, Ramaswamy S, Liverpool TB, Prost J, Rao M, Simha RA (2013) Rev Mod Phys 85:1143
Marsh PD (1994) Adv Dent Res 8:263
Marsh PD, Martin MV (2009) Oral microbiology, 5th edn. Churchill-Livingstone

Marsh PD, Moter A, Devine D (2011) Perio 2000 55:16
Mason TG (2000) Rheol Acta 39:371
Mason TG, Weitz DA (1995) Phys Rev Lett 74:1250
McDougald D, Rice SA, Barraud N, Steinberg PD, Kjelleberg S (2012) Nat Rev Microbiol 10:39
Mizuno D, Head DA, MacKintosh FC, Schmidt CF (2008) Macromolecules 41:7194
Morse DC (1998) Macromolecules 31:7044
Ochoa J-O, Coufort C, Escudié R, Liné A, Paul E (2007) Chem Eng Sci 62:3672
Ogden RW (1984) Non-linear elastic deformations. Dover, New York
Paramonova E, de Jong ED, Krom BP, van der Mei HC, Busscher HJ, Sharma PK (2007) App Env Microbiol 73:7023
Pavlovsky L, Sturtevant RA, Younger JG, Solomon MJ (2015) Langmuir 31:2036
Pavlovsky L, Younger JG, Solomon MJ (2012) Soft Matter 9:122
Peterson BW, van der Mei HC, Sjollema J, Busscher HJ, Sharma PK (2013) mBio 4:e00497–13
Picioreanu C, van Loosdrecht MCM, Heijnen JJ (2001) Biotech. Bioeng. 72:205
Pritchard RH, Huang YYS, Terentjev EM (2014) Soft Matter 10:1864
Roberts D, Rochas C, Saiani A, Miller AF (2012) Langmuir 28:16196
Rodriguez D, Einarsson B, Carpio A (2012) Phys Rev E 86:061914
Rogers SS, van der Walle C, Waigh TA (2008) Langmuir 24:13549
Rombouts WH, Colomb-Delsuc M, Werten MWT, Otto S, de Wolf FA, van der Gucht J (2013) Soft Matter 9:6936
Rubinstein M, Colby RH (2003) Polymer physics. Oxford
Rupp CJ, Fux CA, Stoodley P (2005) App Env Microbiol 71:2175
Rusconi R, Lecuyer S, Autrusson N, Guglielmini L, Stone HA (2011) Biophys J 100:1392
Seminara A, Angelini TE, Wilking JN, Vlamikas H, Ebrahim S, Kolter R, Weitz DA, Brenner MP (2012) Proc Nat Acad Sci 109:1116
Shaw T, Winston M, Rupp CJ, Klapper I, Stoodley P (2004) Phys Rev Lett 93:098102
Shirtliff M, Leid J (2009) The role of biofilms in device-related infections. Springer, Berlin
Smith SB, Cui Y, Bustamante C (1996) Science 271:795
Song JL, Au KH, Huynh KT, Packman AI (2014) Biotech Bioeng 111:597
Song F, Koo H, Ren D (2015) J Dent Res 94:1027
Stewart PS (2012) Biofouling 28:187
Storm C, Pastore JL, MacKintosh FC, Lubensky TC, Janmey PA (2005) Nature 435:191
Stoodley P, Lewandowski Z, Boyle JD, Lappin-Scott HM (1999a) Env Microbiol 1:447
Stoodley P, Lewandowski Z, Boyle JD, Lappin-Scott HM (1999b) Biotech Bioeng 65:84
Szomolay B, Cogan NG (2015) Env Microbiol 17:1870–1883
Taherzadeh D, Picioreanu C, Horn H (2012) Biophys J 102:1483
Thullner M, Baveye P (2008) Biotech Bioeng 99:1337–1351
Tierra G, Pavissich JP, Nerenberg R, Xu Z, Alber MS (2015) J Roy Soc Interface 12:20150045
Towler BW, Cunningham A, Stoodley P, McKittrick L (2007) Biotech Bioeng 96:259
Tuson HH, Weibel DB (2013) Soft Matter 9:4368
Verkaik MJ, Busscher HJ, Rustema-Abbing M, Slomp AM, Abbas F, van der Mei HC (2010) Clin Oral Invest 14:403–409
Vo GD, Heys J (2011) Biotech Bioeng 108:1893
Vogt SJ, Sanderlin AB, Seymour JD, Codd SL (2013) Biotech Bioeng 110:1366
Walter JT et al (2004) Eur J Oral Sci 112:412–418
Weaver WM, Milisavljevic V, Miller JF, Di Carlo D (2012) App Env Microbiol 78:5890
Wilking JN, Angelini TE, Seminara A, Brenner MP, Weitz DA (2011) MRS Bull 36:385–391
Wilson M, Devine D (2003) Medical implications of biofilms. CUP, Cambridge
Wloka M, Rehage H, Flemming H-C, Wingender J (2004) Coll Poly Sci 282:1067
Xu Z, Meakin P, Tartakovsky A, Scheibe TD (2011) Phys Rev E 83:066702
Zrelli K, Galy O, Latour-Lambert P, Kirwan L, Ghigo JM, Beloin C, Henry N (2013) New J Phys 15:125026

Chapter 9
Designing a Single-Molecule Biophysics Tool for Characterising DNA Damage for Techniques that Kill Infectious Pathogens Through DNA Damage Effects

Helen Miller, Adam J.M. Wollman and Mark C. Leake

Abstract Antibiotics such as the quinolones and fluoroquinolones kill bacterial pathogens ultimately through DNA damage. They target the essential type IIA topoisomerases in bacteria by stabilising the normally transient double-strand break state which is created to modify the supercoiling state of the DNA. Here we discuss the development of these antibiotics and their method of action. Existing methods for DNA damage visualisation, such as the comet assay and immunofluorescence imaging can often only be analysed qualitatively and this analysis is subjective. We describe a putative single-molecule fluorescence technique for quantifying DNA damage via the total fluorescence intensity of a DNA origami tile fully saturated with an intercalating dye, along with the optical requirements for how to implement these into a light microscopy imaging system capable of single-molecule millisecond timescale imaging. This system promises significant improvements in reproducibility of the quantification of DNA damage over traditional techniques.

Keywords Single molecule · YOYO · Quinolone · DNA origami · DNA damage · DNA gyrase

H. Miller (✉) · A.J.M. Wollman · M.C. Leake
Departments of Physics and Biology, Biological Physical Sciences Institute,
University of York, York YO10 5DD, UK
e-mail: hlm531@york.ac.uk

M.C. Leake
e-mail: mark.leake@york.ac.uk

© Springer International Publishing Switzerland 2016

M.C. Leake (ed.), *Biophysics of Infection*, Advances in Experimental Medicine and Biology 915, DOI 10.1007/978-3-319-32189-9_9

Definitions

Intercalating Binding between base pairs.

PCR Polymerase chain reaction; a method to produce multiple copies of a DNA sequence. The DNA to be copied is heated to separate the strands, and then cooled so that short 'primer' sequences bind complementarily at both 5′ ends of the DNA sequence to be copied. A DNA polymerase then extends from the primers to give two copies of the sequence of interest. Multiple repeats of this process produce a high yield of the target sequence.

9.1 Introduction

There exist classes of antibiotics that act to kill bacterial pathogens via DNA damage. These antibiotics and existing methods to measure DNA damage will be discussed, along with considerations for the design of a novel single-molecule fluorescence method of measuring DNA damage.

9.2 Quinolones: Antibiotics that Kill Pathogens via DNA Damage

The quinolones, and more recently fluoroquinolones, are synthetic antibiotics which kill infectious pathogens through DNA damage by stabilising double-strand breaks (DSB) caused by the type IIA topoisomerases. The first quinolone discovered was Nalidixic Acid (Lesher et al. 1962). The quinolone antibiotics are effective against gram-negative bacteria, being bacteriostatic (stopping growth) at low concentrations, and bactericidal at higher concentrations when hydroxyl radicals are formed (Kohanski et al. 2007) (a detailed discussion of the activities of the different drugs is given in Drlica and Zhao 1997). Fluoroquinolones, such as the first one discovered, Norfloxacin, have fluorine attached to the ring system (see Fig. 9.1; for details on the chemical structure of various quinolone and fluoroquinolone compounds see Gootz and Brighty 1996; Appelbaum and Hunter 2000), and have increased activity in gram-positive bacteria compared to the quinolones (Chu and Fernandes 1991).

Fluoroquinolones have several major advantages over quinolones: they generally are more effective against gram-positive bacteria with the same activity against gram-negative bacteria, and they are absorbed well enough and the newer ones have a sufficiently long half-life that a daily dose is possible which increases patient compliability (Appelbaum and Hunter 2000). However, fluoroquinolones can have toxic side effects (for a review see for example Owens and Ambrose 2005) leading to some being abandoned in the late stages of clinical trials or even after marketing (e.g. temafloxacin (Hardy et al. 1987)).

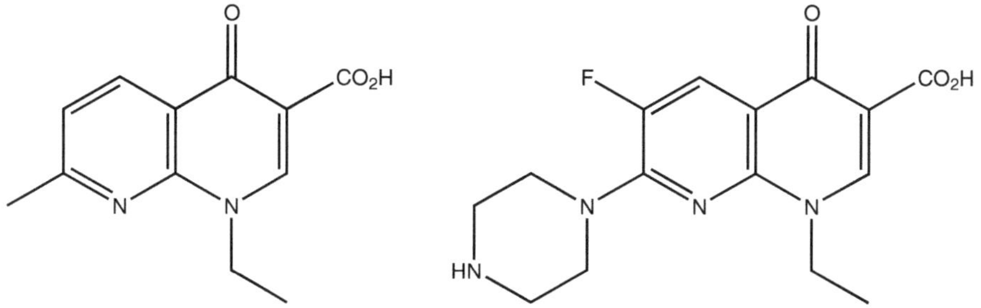

Nalidixic Acid Norfloxacin

Fig. 9.1 The chemical structures of Nalidixic acid and Norfloxacin. Norfloxacin is an example of a fluoroquinolone

There are two main targets of quinolone antibiotics, both of which are type IIA topoisomerases: DNA gyrase (the target in gram-negative bacteria) and topoisomerase IV (the target in gram-positive bacteria). DNA gyrase is a tetrameric protein of the form A_2B_2 (Sugino et al. 1977) which acts to produce negative supercoils in DNA (Gellert et al. 1976). The A subunits bind to DNA whilst the B subunits carry out the ATP hydrolysis (Reece and Maxwell 1991). Topoisomerase IV is a tetrameric protein of the form C_2E_2, which is essential for separating the linked DNA strands produced during chromosome replication (Kato et al. 1990). The C unit binds DNA and the E subunit carries out ATP hydrolysis (Peng and Marians 1993). The C subunit of topoisomerase IV can be thought as analogous to the A subunit of DNA gyrase, E and B likewise. The main difference in the way these two enzymes act is that DNA gyrase wraps the DNA around itself whilst topoisomerase IV does not (Drlica and Zhao 1997).

Type IIA topoisomerases create transient four base pair staggered DSB through which duplex DNA can be passed to modify supercoiling. A covalent bond is formed between the 5′ phosphate of the new DNA terminus and a tyrosine residue in the functional subunit of the type IIA topoisomerase (Sutcliffe et al. 1989; Roca 1995)—to be explicit, in DNA gyrase, the hydroxyl group of Tyrosine-122 and the 5′ DNA terminus form a phosphate ester (Horowitz and Wang 1987; Maxwell 1992).

The bacteriostatic method of action of the quinolones is to stabilise the double-strand break, putatively by causing a conformational change in the enzyme—DNA complex (Orphanides and Maxwell 1994). Studies with Moxifloxacin, an example of a fluoroquinolone, have shown that it is wedge-shaped molecule that intercalates at the points of DNA breakage; the interaction between the antibiotic and the topoisomerase is mediated by a magnesium ion (Wohlkonig et al. 2010). DNA sedimentation experiments on quinolone treated DNA (Chen et al. 1996) indicate that the ends of the DNA are released, possibly by being freed from the quinolone-topoisomerase complex, or by separation of the subunits of DNA gyrase (Drlica and Zhao 1997), this leaves the bacterial chromosome unwound (at the clinical dose levels used mammalian cells are not affected), leaving it too large to fit in the daughter cells, stopping growth (Emmerson and Jones 2003). The bacterial cell starts the SOS response; an error-prone DNA damage repair system which leads to the fast development of antibiotic resistance. The bactericidal drugs also produce hydroxyl radicals which kill the cell (Kohanski et al. 2007). For more detail on the SOS response, see for example Drlica and Zhao (1997), Žgur-Bertok (2013).

9.3 Quantifying DNA Damage

Quantifying the DNA damage required to kill a bacterial pathogen could lead to lower doses of antibiotics being required, which might help with the increasing problem of antibiotic resistance to the quinolones.

The quinolones act by DNA damage, so a method to quantify this potentially gives a way to measure the effectiveness of this class of antibiotics, or to determine the dose required. There are other applications where quantitative measures of DNA damage are required, for example, in studying programmed cell death, or for evaluating the effects of novel cancer treatments. For many medical applications, it is important to know how different cells respond to a new therapy, by cell type or by location relative to the point of application. This is imperative in developing in vivo treatments for cancer, where you only wish to damage the tumour cells and not the healthy tissue cells; you must measure how far from the point of application a cell-killing effect is received—you do not wish to damage healthy tissue. This is important in the application of, for example, medical plasma treatments (Hirst et al. 2015).

9.3.1 Gel Electrophoresis and the Comet Assay

Two common assays for in vivo and in vitro quantification of DNA damage work via electrophoresis. DNA is negatively charged and so will migrate through a porous medium when a potential difference is applied. In electrophoresis, the distance travelled by a particular fragment depends on both its molecular weight and its conformation. For DNA of the same conformation, a strand with lower molecular weight runs further from its starting position than a strand of high molecular weight. DNA of the same molecular weight in supercoiled, circular or linear forms runs to different positions that can be used to identify its structure, and therefore if it is undamaged, has a single-strand break, or a double-strand break (Blazek et al. 1989; O'Connell et al. 2011). The DNA in the gels can be visualised either by adding a DNA intercalating dye to the gel before running electrophoresis or by post staining, and imaging with ultraviolet light. As an example of the information that can be gained from electrophoresis Fig. 9.2 shows a 1 % agarose gel with damage to λ DNA. In Lane 1, the λ DNA is undamaged; its long length (48,502 bp) means the band is smeared from the top, but the clear lower edge shows all the DNA in the band to be the same molecular weight. Lanes 2 and 3

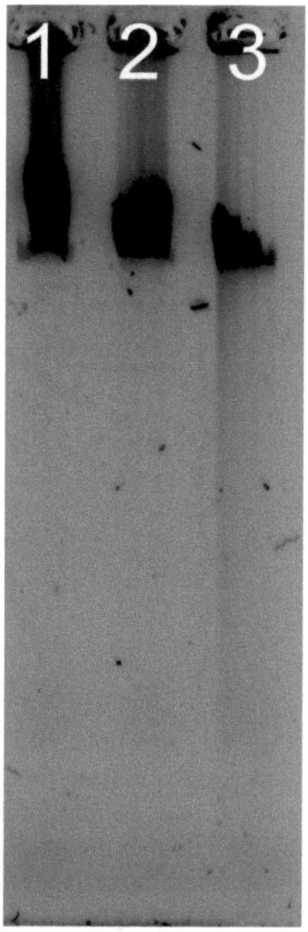

Fig. 9.2 1 % agarose gel showing damage to λ DNA. *Lane 1* Undamaged λ DNA, *Lane 2* λ DNA treated with plasma for 5 s. *Lane 3* λ DNA treated with plasma for 30 s

show λ DNA treated with low-temperature plasma for 5 s and 30 s respectively, this means the DNA is subjected to reactive oxygen and nitrogen species; the shorter damaged λ DNA ran further from the well and shows smearing on the lower edge, indicating DNA of multiple lengths is present (Fig. 9.3).

Agarose gel electrophoresis allows the visualisation of DNA damage from in vitro experiments, or from a cell colony from which the DNA has been collected. This has the advantage that all the DNA starts in the same place in a small well before electrophoresis, so theoretically the sizes of the different fragments can be found based on how far they have run, but it cannot yield information about where within the nucleus of a single cell the damage occurred. For information on the DNA damage within a single cell, the comet assay is used.

The comet assay (first developed by Östling and Johansson (1984) in 1984 and later modified by Singh et al. (1988)) is often used for quantifying DNA damage in single cells. Cells are embedded in low-melting point agarose, cast onto microscope slides, lysed, electrophoresed and stained. As with gel electrophoresis, the DNA then runs from its starting position a distance that depends on its weight and structure, and the resulting shape of DNA gives the technique its name; "comet". The difficulties with this method come in evaluation; the DNA's initial position in the nucleus prior to electrophoresis must be determined, and in samples that have high levels of DNA damage this can be non-trivial and makes analysis subjective.

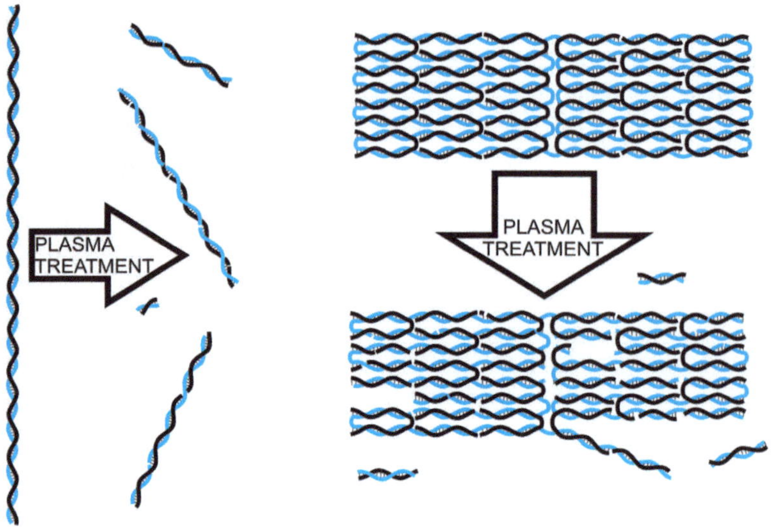

Fig. 9.3 Examples of linear DNA and DNA origami substrates for a fluorescence damage assay, and a possible interpretation of the effects of low-temperature atmospheric plasma on each one (single and double strand breaks). The broken nature of the linear DNA lends itself to gel electrophoresis quantification as the different length pieces of DNA run to different positions. Damage to the origami tile could be quantified by total fluorescence intensity when imaged with an intercalating dye

Additionally, cells and their resulting comets can overlap making analysis difficult or impossible. The major advantage of this assay at the time of development was that compared to other methods it did not require radioactive labelling.

Software such as OpenComet (Gyori et al. 2014), a plug-in for ImageJ (http://imagej.nih.gov/ij/) can reduce the subjectiveness of manual analysis of the comet assay, are fast and in this instance, free and open source (which allows bespoke modification).

DNA damage in vitro obviously has differences to DNA damage in vivo; for one, buffer or free-radical scavenging properties are changed. This, combined with slight differences in the method used to inflict damage make comparisons of the two difficult. For example, with low-temperature plasma treatments the plasma source itself, distance from the plasma source, and detection method used can vary (For example, contrast Hirst et al. 2015; O'Connell et al. 2011). For low-temperature plasma treatments at least, dose times to produce DNA damage generally fall in the range of seconds for low damage levels up to around 10 min for high damage levels, both in vitro and in vivo, indicating that DNA damage occurs on a similar timescale in both methods.

9.3.2 *Single Molecule qPCR*

A technique to quantify RNA transcription is single molecule qPCR. This uses PCR in a single cell to quantitatively analyse transcription products: choosing a cell, and quantitative PCR can be achieved with reasonably high throughput in a microfluidic device (White et al. 2011). This technique is used to analyse the most abundant mRNA in a cell, but with further sensitivity and careful primer choice could plausibly be extended to examine damage to the DNA in the cell, but this is not a reality yet.

9.3.3 *Immunofluorescence*

Immunofluorescence labelling (Coons et al. 1942; Coons and Kaplan 1950) uses fluorescently labelled antibodies to optically detect antigens in tissue. This is a fluorescence technique giving data about single cells, but not quantitatively. Immunofluorescence labelling for visualising DNA damage from a plasma source (Hirst et al. 2014) gives you more spatial information than the comet assay, since the DNA is not moved from its original position. It does, however produce difficulties due to the three-dimensional nature of the cell, causing loss of information out of the focal plane and the potential for vertically overlapping signals being mistaken for single molecules.

9.3.4 *Tunel Staining*

Tunel staining (Gavrieli et al. 1992) is a single cell fluorescence technique that fluorescently labels the 3′ OH of damaged DNA produced by programmed cell death. The method requires staining followed by imaging, and is not without problems (Kraupp et al. 1995). However, the 3′ OH group is not present in drug-induced DNA damage and therefore, although this is a fluorescent method, it will not be further discussed here.

9.4 Putative Fluorescence Single-Molecule DNA Damage Assays

The assays described above based on gel electrophoresis produce a measure of the size of the DNA fragments, but no information on where along the strand or in the chromosome the DNA damage occurred, or whether one region was a particular target. The method of immunofluorescence suffers from uncertainty due to the three-dimensional nature of the cell, and single molecule qPCR for DNA damage has not yet been achieved. Single-molecule assays are preferable for applications where the damage caused is expected to be strongly dependent on distance from an application of a DNA damage source, for example, cells at different distances from the point of application of a cancer treatment. In this case, ensemble measurements are insufficient as they do not give the required level of detail (Wollman et al. 2015a, b). In the following a putative fluorescence, single-molecule DNA damage quantification experiment will be described. Such an experiment requires a suitable substrate and a technique.

9.4.1 Substrate: DNA Origami

Watson-Crick base pairing of DNA can be used to self-assemble a long, single-stranded DNA strand ('scaffold') with complementary short strands ('staples') (Rothemund 2006): For example, a staple which binds to, say, 16 bases in one location on the scaffold, and 16 in another will act to bring those two regions together. By complementing the whole scaffold sequence with staples in this way, any shape can be designed.

Originally the correctly folded yield of DNA origami was very variable (in the Rothemund paper (Rothemund 2006) the 'rectangle' yield was 90 % but the 'square' yield was 13 %). Steps have been taken to increase this; some of these are empirical (see for example, Martin and Dietz 2012), but optimisation of the folding pathway (Dunn et al. 2015) gives the potential for more systematic improvement of yield, potentially making commercial production viable—this would be required for an object to be used as a standard DNA damage probe.

DNA origami might prove to be a good test object for DNA damage due to its two-dimensional nature and well defined shape, thus circumventing the problems of immunofluorescence. Gel electrophoresis has been used as a method of analysing the folding quality of origami in empirical folding optimisation experiments (see for example Schmied et al. 2014). Double-strand breaks may manifest in similar ways to incompletely folded origami (Fig. 9.3). If so, this would mean DNA origami could be used to test a fluorescence single-molecule DNA damage assay, as results could be compared to the results of the established method of gel electrophoresis.

9.4.2 Technique: Single-Molecule Fluorescence Microscopy

Fluorescence microscopy is a relatively non-invasive single-molecule method; small fluorescent markers can be attached to molecules of interest in vitro or in vivo. What can be seen in conventional fluorescence microscopy is restricted: The Abbe diffraction limit gives a theoretical limit on the resolution of a light microscope of around two hundred nanometres, caused by diffraction through the rear aperture of the objective (for a discussion of this see Wollman et al. 2015a, b). This is a problem, since many biological structures of interest are smaller than this, being just a few nanometres in size, so simply cannot be seen by looking down a conventional light microscope. Super-resolution fluorescence microscopy can be achieved by many methods (e.g. STED (Hell and Wichmann 1994), STORM (Rust et al. 2006), PALM (Betzig et al. 2006; Hess et al. 2006)). What all of these methods have in common is they produce images of the point spread function of a single fluorescent molecule, (for example, by spatially separating fluorescent molecules using a low density, or by temporally separating the emission of densely packed molecules). The shape of the emission profile of a single molecule is known to be an Airy function, and by numerically fitting the centre of the distribution the location of the molecule can be found to a precision of nanometres, allowing the observation of interactions or the quantification of the number of molecules present.

A fluorescence microscopy, single-molecule tool for detecting DNA damage that uses a two-dimensional DNA test object would be attractive for quantifying in vitro DNA damage as it removes the need to measure migration of DNA as in gel electrophoresis, and removes the issues inherent in the three-dimensional nature of immunofluorescence. Intercalating dyes such as YOYO-1 increase in brightness intensity on binding to double-stranded DNA by 100–1000 times their typical intensity in solution (Flors 2013). Using an intercalating dye at a concentration to fully saturate the origami tile, the fluorescence intensity of an undamaged tile could be measured. Single-molecule fluorescence microscopy is an increasingly quantitative tool enabling information to be extracted concerning molecular stoichiometry, mobility and co-localization (Leake 2013; Robson et al. 2013; Llorente-Garcia et al. 2014; Leake 2014). By finding the characteristic intensity of a single dye and combining this with measurements of the intensity of dye saturated damaged DNA origami, a measure of the number of double-strand breaks could be made.

A microscope for such work is currently under development, incorporating a total internal reflection fluorescence (TIRF) capability to further reduce background from out of focus dye molecules. Figure 9.4 shows the setup. The excitation light can pass through either of two paths which contain lenses on flip mounts for different excitation modes. One path can be used as a FRAP path or an epifluorescence path, whilst the other can be operated in widefield or Slimfield/narrowfield (Plank et al. 2009; Wollman and Leake 2015), in TIRF or epifluorescence modes. For imaging of DNA origami, illumination is via the narrowfield TIRF mode. Lens L6 is mounted on a translation mount. By positioning L6, the focal length of L7 from L7, L6 lies in a conjugate plane to the focal plane. This means that angle of the beam for TIRF can be adjusted with no lateral deflection of the laser beam in the sample plane. This is beneficial since the physical size of the EMCCD (electron multiplying charge-coupled device) sensor on an Andor Ixon Ultra 897 is $\sim$8.2 mm × 8.2 mm and large shifts would require realignment, making switching between illumination modes non-trivial.

This microscope combines fast high magnification (200X) imaging with bespoke TIRF illumination in a colour channel of choice. This should aid the acquisition of high signal-to-noise ratio images for precise quantification of the number of single

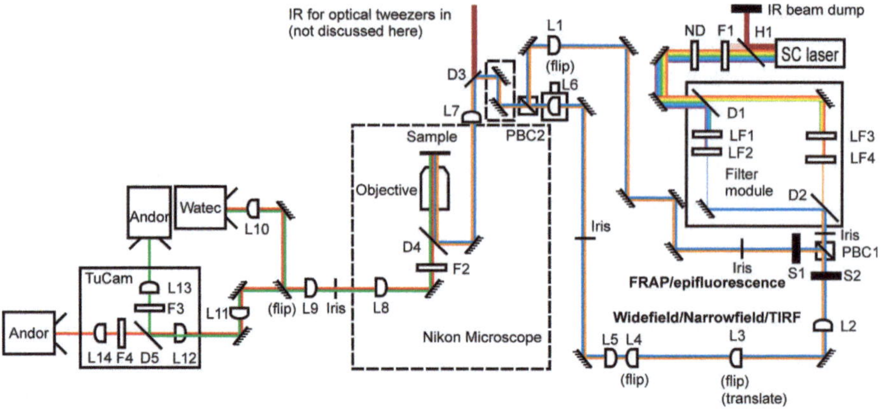

Fig. 9.4 The bespoke fluorescence microscope designed for DNA damage assays. Lens numbers as described in the text. The infrared (IR) beam is part of ongoing development to add optical tweezers to the microscope (Zhou et al. 2015)

dye molecules present on a DNA origami substrate to enable quantitative measures of DNA damage.

A fluorescence microscopy tool still has limitations; the production of free radicals such as molecular oxygen during laser illumination is a well-known factor (Dave et al. 2009). These free radicals cause damage to DNA, for example, in an assay to look at stretched lambda DNA without oxygen scavenging (Miller et al. 2015) DNA strands spontaneously undergo snapping behaviour presumably due to radiative damage. Clearly, a tool to image DNA damage should not itself inflict DNA damage by the same method. Luckily, the effects of illumination can be mitigated somewhat with oxygen scavenging systems; detailed studies have been conducted into how effective these are (see for example Dave et al. 2009).

9.5 Conclusions

There exist bacterial pathogens that act through DNA damage. Whilst several methods exist to quantify damage, none use the exact same target for each treatment. A DNA origami fluorescence microscopy assay would aid reproducibility and has advantages over techniques such as immunofluorescence since the entire DNA tile can be in focus in one image. A fluorescence microscopy assay to image DNA damage on single origami tiles is under development.

Acknowledgements This work was funded by with the assistance of the Biological Physical Sciences Institute (BPSI) and the Royal Society (MCL). Thanks to Katherine Dunn, Adam Hirst and Deborah O'Connell for discussions and technical assistance on DNA origami and plasma treatments respectively.

References

Appelbaum PC, Hunter PA (2000) The fluoroquinolone antibacterials: past, present and future perspectives. Int J Antimicrob Agents 16:5–15. doi:10.1016/S0924-8579(00)00192-8

Betzig E, Patterson GH, Sougrat R et al (2006) Imaging intracellular fluorescent proteins at nanometer resolution. Science 313:1642–1645. doi:10.1126/science.1127344

Blazek ER, Peak JG, Peak MJ (1989) Singlet oxygen induces frank strand breaks as well as alkali—and piperidine-labile sites in supercoiled plasmid DNA. Photochem Photobiol 49:607–613. doi:10.1111/j.1751-1097.1989.tb08431.x

Chen CR, Malik M, Snyder M, Drlica K (1996) DNA gyrase and topoisomerase IV on the bacterial chromosome: quinolone-induced DNA cleavage. J Mol Biol 258:627–637. doi:10.1006/jmbi.1996.0274

Chu DTW, Fernandes PB (1991) Recent developments in the field of quinolone antibacterial agents. Adv Drug Res 21:39–144

Coons AH, Creech HJ, Jones RN, Berliner E (1942) The demonstration of pneumococcal antigen in tissues by the use of fluorescent antibody. J Immunol 45:159–170

Coons AH, Kaplan MH (1950) Localization of antigen in tissue cells: II. Improvements in a method for the detection of antigen by means of fluorescent antibody. J Exp Med 91:1–13. doi:10.1084/jem.91.1.1

Dave R, Terry DS, Munro JB, Blanchard SC (2009) Mitigating unwanted photophysical processes for improved single-molecule fluorescence imaging. Biophys J 96:2371–2381. doi:10.1016/j.bpj.2008.11.061

Drlica K, Zhao X (1997) DNA gyrase, topoisomerase IV, and the 4-quinolones. Microbiol Mol Biol Rev 61:377–392

Dunn KE, Dannenberg F, Ouldridge TE et al (2015) Guiding the folding pathway of DNA origami. Nature. doi:10.1038/nature14860

Emmerson AM, Jones A (2003) The quinolones: decades of development and use. J Antimicrob Chemother 51:13–20. doi:10.1093/jac/dkg208

Flors C (2013) Super-resolution fluorescence imaging of directly labelled DNA: from microscopy standards to living cells. J Microsc 251:1–4. doi:10.1111/jmi.12054

Gavrieli Y, Sherman Y, Ben-Sasson SA (1992) Identification of programmed cell death in situ via specific labeling of nuclear DNA fragmentation. J Cell Biol 119:493–501

Gellert M, Mizuuchi K, O'Dea MH, Nash HA (1976) DNA gyrase: an enzyme that introduces superhelical turns into DNA. Proc Natl Acad Sci U S A 73:3872–3876

Gootz TD, Brighty KE (1996) Fluoroquinolone antibacterials: SAR mechanism of action, resistance, and clinical aspects. Med Res Rev 16:433–486. doi:10.1002/(SICI)1098-1128(199609)16:5<433:AID-MED3>3.0.CO;2-W

Gyori BM, Venkatachalam G, Thiagarajan PS et al (2014) OpenComet: an automated tool for comet assay image analysis. Redox Biol 2:457–465. doi:10.1016/j.redox.2013.12.020

Hardy DJ, Swanson RN, Hensey DM et al (1987) Comparative antibacterial activities of temafloxacin hydrochloride (A-62254) and two reference fluoroquinolones. Antimicrob Agents Chemother 31:1768–1774. doi:10.1128/AAC.31.11.1768

Hell SW, Wichmann J (1994) Breaking the diffraction resolution limit by stimulated emission: stimulated-emission-depletion fluorescence microscopy. Opt Lett 19:780–782

Hess ST, Girirajan TPK, Mason MD (2006) Ultra-high resolution imaging by fluorescence photoactivation localization microscopy. Biophys J 91:4258–4272. doi:10.1529/biophysj.106.091116

Hirst AM, Frame FM, Maitland NJ, O'Connell D (2014) Low temperature plasma causes double-strand break DNA damage in primary epithelial cells cultured from a human prostate tumor. IEEE Trans Plasma Sci 42:2740–2741. doi:10.1109/TPS.2014.2351453

Hirst AM, Simms MS, Mann VM et al (2015) Low-temperature plasma treatment induces DNA damage leading to necrotic cell death in primary prostate epithelial cells. Br J Cancer 112:1536–1545. doi:10.1038/bjc.2015.113

Horowitz DS, Wang JC (1987) Mapping the active site tyrosine of Escherichia coli DNA gyrase. J Biol Chem 262:5339–5344
Kato J, Nishimura Y, Imamura R et al (1990) New topoisomerase essential for chromosome segregation in E. coli. Cell 63:393–404. doi:10.1016/0092-8674(90)90172-B
Kohanski MA, Dwyer DJ, Hayete B et al (2007) A common mechanism of cellular death induced by bactericidal antibiotics. Cell 130:797–810. doi:10.1016/j.cell.2007.06.049
Kraupp BG, Ruttkay-Nedecky B, Koudelka H et al (1995) In situ detection of fragmented dna (tunel assay) fails to discriminate among apoptosis, necrosis, and autolytic cell death: a cautionary note. Hepatology 21:1465–1468. doi:10.1002/hep.1840210534
Leake MC (2013) The physics of life: one molecule at a time. Philos Trans R Soc Lond B Biol Sci 368:20120248. doi:10.1098/rstb.2012.0248
Leake MC (2014) Analytical tools for single-molecule fluorescence imaging in cellulo. Phys Chem Chem Phys 16:12635–12647. doi:10.1039/c4cp00219a
Lesher GY, Froelich EJ, Gruett MD et al (1962) 1,8-Naphthyridine derivatives. A new class of chemotherapeutic agents. J Med Pharm Chem 5:1063–1065. doi:10.1021/jm01240a021
Llorente-Garcia I, Lenn T, Erhardt H et al (2014) Single-molecule in vivo imaging of bacterial respiratory complexes indicates delocalized oxidative phosphorylation. Biochim Biophys Acta 1837:811–824. doi:10.1016/j.bbabio.2014.01.020
Martin TG, Dietz H (2012) Magnesium-free self-assembly of multi-layer DNA objects. Nat Commun 3:1103. doi:10.1038/ncomms2095
Maxwell A (1992) The molecular basis of quinolone action. J Antimicrob Chemother 30:409–414. doi:10.1177/146642407309300601
Miller H, Zhou Z, Wollman AJM, Leake MC (2015) Superresolution imaging of single DNA molecules using stochastic photoblinking of minor groove and intercalating dyes. Methods. doi:10.1016/j.ymeth.2015.01.010
O'Connell D, Cox LJ, Hyland WB et al (2011) Cold atmospheric pressure plasma jet interactions with plasmid DNA. Appl Phys Lett 98:043701. doi:10.1063/1.3521502
Orphanides G, Maxwell A (1994) Evidence for a conformational change in the DNA gyrase–DNA complex from hydroxyl radical footprinting. Nucleic Acids Res 22:1567–1575. doi:10.1093/nar/22.9.1567
Ostling O, Johanson KJ (1984) Microelectrophoretic study of radiation-induced DNA damages in individual mammalian cells. Biochem Biophys Res Commun 123:291–298. doi:10.1016/0006-291X(84)90411-X
Owens RC, Ambrose PG (2005) Antimicrobial safety: focus on fluoroquinolones. Clin Infect Dis 41(Suppl 2):S144–S157. doi:10.1086/428055
Peng H, Marians K (1993) Escherichia coli topoisomerase IV. Purification, characterization, subunit structure, and subunit interactions. J Biol Chem 268:24481–24490
Plank M, Wadhams GH, Leake MC (2009) Millisecond timescale slimfield imaging and automated quantification of single fluorescent protein molecules for use in probing complex biological processes. Integr Biol (Camb) 1:602–612. doi:10.1039/b907837a
Reece RJ, Maxwell A (1991) DNA gyrase: structure and function. Crit Rev Biochem Mol Biol 26:335–375. doi: 10.3109/10409239109114072
Robson A, Burrage K, Leake MC (2013) Inferring diffusion in single live cells at the single-molecule level. Philos Trans R Soc Lond B Biol Sci 368:20120029. doi:10.1098/rstb.2012.0029
Roca J (1995) The mechanisms of DNA topoisomerases. Trends Biochem Sci 20:156–160. doi:10.1016/S0968-0004(00)88993-8
Rothemund PWK (2006) Folding DNA to create nanoscale shapes and patterns. Nature 440:297–302. doi:10.1038/nature04586
Rust MJ, Bates M, Zhuang X (2006) Sub-diffraction-limit imaging by stochastic optical reconstruction microscopy (STORM). Nature 3:793–795. doi:10.1038/NMETH929
Schmied JJ, Raab M, Forthmann C et al (2014) DNA origami-based standards for quantitative fluorescence microscopy. Nat Protoc 9:1367–1391. doi:10.1038/nprot.2014.079

Singh NP, McCoy MT, Tice RR, Schneider EL (1988) A simple technique for quantitation of low levels of DNA damage in individual cells. Exp Cell Res 175:184–191. doi:10.1016/0014-4827(88)90265-0

Sugino A, Peebles CL, Kreuzer KN, Cozzarelli NR (1977) Mechanism of action of nalidixic acid: purification of Escherichia coli nalA gene product and its relationship to DNA gyrase and a novel nicking-closing enzyme. Proc Natl Acad Sci 74:4767–4771. doi:10.1073/pnas.74.11.4767

Sutcliffe JA, Gootz TD, Barrett JF (1989) Biochemical characteristics and physiological significance of major DNA topoisomerases. Antimicrob Agents Chemother 33:2027–2033. doi:10.1128/AAC.33.12.2027

White AK, VanInsberghe M, Petriv OI et al (2011) High-throughput microfluidic single-cell RT-qPCR. Proc Natl Acad Sci U S A 108:13999–14004. doi:10.1073/pnas.1019446108

Wohlkonig A, Chan PF, Fosberry AP et al (2010) Structural basis of quinolone inhibition of type IIA topoisomerases and target-mediated resistance. Nat Struct Mol Biol 17:1152–1153. doi:10.1038/nsmb.1892

Wollman AJM, Leake MC (2015) Millisecond single-molecule localization microscopy combined with convolution analysis and automated image segmentation to determine protein concentrations in complexly structured, functional cells, one cell at a time. Faraday Discuss 184:401–424. doi:10.1039/c5fd00077g

Wollman AJM, Miller H, Zhou Z, Leake MC (2015a) Probing DNA interactions with proteins using a single-molecule toolbox: inside the cell, in a test tube and in a computer. Biochem Soc Trans 43:139–145. doi:10.1042/BST20140253

Wollman AJM, Nudd R, Hedlund EG, Leake MC (2015b) From Animaculum to single molecules: 300 years of the light microscope. Open Biol 5:150019–150019. doi:10.1098/rsob.150019

Žgur-Bertok D (2013) DNA damage repair and bacterial pathogens. PLoS Pathog 9:e1003711. doi:10.1371/journal.ppat.1003711

Zhou Z, Miller H, Wollman A, Leake M (2015) Developing a new biophysical tool to combine magneto-optical tweezers with super-resolution fluorescence microscopy. Photonics 2:758–772. doi:10.3390/photonics2030758

Chapter 10
Bacterial Surfaces: Front Lines in Host–Pathogen Interaction

Jane E. King and Ian S. Roberts

Abstract All bacteria are bound by at least one membrane that acts as a barrier between the cell's interior and the outside environment. Surface components within and attached to the cell membrane are essential for ensuring that the overall homeostasis of the cell is maintained. However, many surface components of the bacterial cell also have an indispensable role mediating interactions of the bacteria with their immediate environment and as such are essential to the pathogenesis of infectious disease. During the course of an infection, bacterial pathogens will encounter many different ecological niches where environmental conditions such as salinity, temperature, pH, and the availability of nutrients fluctuate. It is the bacterial cell surface that is at the front-line of these host–pathogen interactions often protecting the bacterium from hostile surroundings but at the same time playing a critical role in the adherence to host tissues promoting colonization and subsequent infection. To deal effectively with the changing environments that pathogens may encounter in different ecological niches within the host many of the surface components of the bacterial cell are subject to phenotypic variation resulting in heterogeneous subpopulations of bacteria within the clonal population. This dynamic phenotypic heterogeneity ensures that at least a small fraction of the population will be adapted for a particular circumstance should it arise. Diversity within the clonal population has often been masked by studies on entire bacterial populations where it was often assumed genes were expressed in a uniform manner. This chapter, therefore, aims to highlight the non-uniformity in certain cell surface structures and will discuss the implication of this heterogeneity in bacterial–host interaction. Some of the recent advances in studying bacterial surface structures at the single cell level will also be reviewed.

Keywords Bacteria · Cell surface · Heterogeneity · Pathogenicity

---

J.E. King · I.S. Roberts (✉)
Faculty of Life Sciences, University of Manchester, Manchester M13 9PT, UK
e-mail: i.s.roberts@manchester.ac.uk

© Springer International Publishing Switzerland 2016
M.C. Leake (ed.), *Biophysics of Infection*, Advances in Experimental Medicine and Biology 915, DOI 10.1007/978-3-319-32189-9_10

10.1 The Bacterial Surface

Nearly, all bacteria can be classified into two groups based on whether they are able to retain the Gram stain or not (named after Christian Gram who developed the method in 1884). Gram-positive bacteria can retain the stain while Gram-negative bacteria do not. This difference is due to the physical and chemical structure of the cell envelope of these two groups of bacteria (Fig. 10.1) (reviewed by Silhavy et al. 2010). The main difference between the two groups is that Gram-negative bacteria possess two membranes, an outer cell membrane composed mainly of phospholipids and lipopolysaccharide (LPS) and an inner or cell membrane between which is sandwiched a thin peptidoglycan (a polymer of amino acids and sugars) layer. Gram-positive bacteria, on the other hand, have only one cell membrane, which is surrounded by multiple layers of peptidoglycan containing long threads of anionic glycopolymers known as teichoic acid. While these complex multi-layered structures that make up the cell envelope have themselves been shown to have roles in interactions with the host (Davis and Weiser 2011; Rosenfeld and Shai 2006; Raetz and Whitfield 2002; Weidenmaier and Peschel 2008; Brown et al. 2013) other surface component (such as flagella, fimbriae, S-layer and capsule) extend out beyond, or form the outermost layers, of the cell surface in many bacteria and as such will generally be the first points of contact with the external environment and

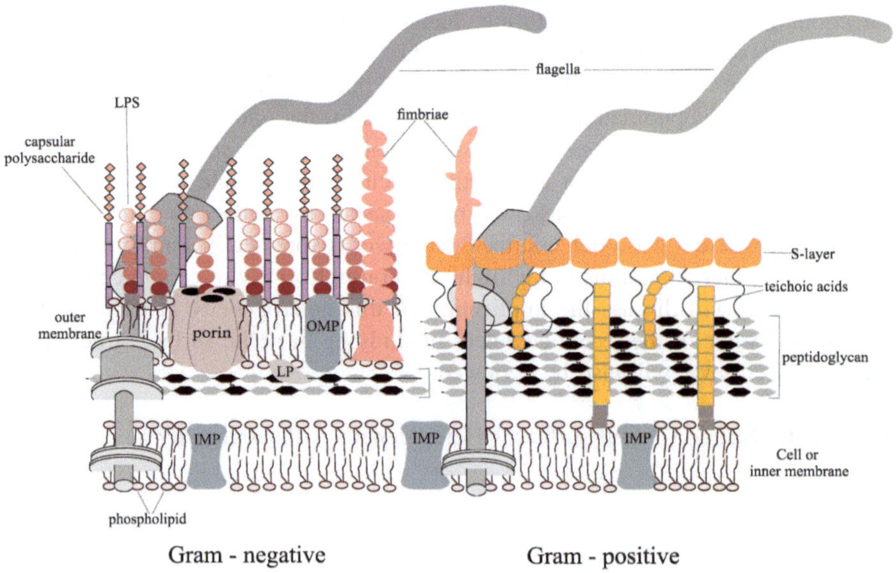

Fig. 10.1 Depiction of the Gram-negative and Gram-positive cell surface. For clarity, the capsule and S-layer are shown only on the Gram-negative and Gram-positive surfaces respectively. However, both structures can be present in both groups of bacteria. *IMP* integral membrane protein; *OMP* outer membrane protein; *LPS* lipopolysaccharide; *LP* lipoprotein

host. These outermost surface components have all been shown to be subject to phenotypic heterogeneity in many bacterial strains and their role(s) in interactions with the host will be discussed in more detail after a brief description of how phenotypic heterogeneity can arise.

10.2 Phenotypic Heterogeneity

Phenotypic variation within the microbial population arises from either a heritable but reversible alteration to the genome or by non-genetic change. The classic example of genotypic heterogeneity is phase or antigenic variation where genetic alterations such as mutation, inversion, rearrangement and methylation (reviewed by van der Woude 2011) results in spontaneous changes in the biosynthesis of many bacterial surface exposed structures. Phase variation generally refers to an on/off mechanism resulting in variation in the levels of a protein (or proteins) between single bacteria in a clonal population. Whereas antigenic variation results in clonal bacteria expressing antigenically distinct moieties on their cell surface. Other types of genetic change such as adaptive mutagenesis (Finkel and Kolter 1999) and the appearance of mutator cells (Travis and Travis 2002) can also lead to the dynamic coexistence of multiple phenotypes. Phenotypic variation can also occur without a change or modification to the DNA. Most gene expression levels in a clonal population display a normal (or unimodal) distribution around the mean because of stochastic fluctuations in transcription and translation. These fluctuations are known as “noise”. The consequences of molecular level noise on gene expression have been well documented (Kaern et al. 2005; Veening et al. 2008). When noise is amplified (for example by a positive feedback loop) it can give rise to another type of phenotypic variation whereby the unimodal population bifurcates into two subsets. This type of gene expression pattern is known as bistability (Dubnau and Losick 2006; Smits et al. 2006). Other nongenetic molecular mechanisms responsible for phenotypic heterogeneity include cellular age, periodic oscillations and cell-to-cell interactions and these mechanisms have been discussed recently elsewhere (Ackermann 2015) Bacteria exploit both genetic and nongenetic mechanisms to generate phenotypic heterogeneity of their surface structures.

10.3 Flagella

The flagellum is primarily the motility organelle of the bacterium that allows movement and chemotaxis—the biophysical properties of this organelle have been reviewed recently (Rossez et al. 2015). Flagella have generally been regarded as virulence factors mainly because of their ability to propel a number of microorganisms through often relatively viscous surroundings (Persat et al. 2015) such as the intestinal mucosa (Eaton et al. 1996; Ormonde et al. 2000; O’Neil and Marquis

2006; Baban et al. 2013). However, scenarios can also be imagined whereby motility of the bacteria is no longer required, for example, for tight adherence to host cells, continuous slow growth and when energy needs to be conserved (flagellar biosynthesis and function requires 2.1 % of total cellular energy Macnab 2003). Indeed in some microorganisms there appears to be a selective pressure against flagellar expression under host conditions in order to prevent unnecessary expenditure of energy. For example, *Yersinia enterocolitica* (Kapatral et al. 1996) uses body temperature (37 °C) to repress flagellum biosynthesis in the host and there is a subset of human pathogens including *Yersinia pestis*, *Shigella* spp., *Bordetella pertussis* and certain *Escherichia coli* strains (Deng et al. 2002; Al Mamun et al. 1996; Leigh et al. 1993; Monday et al. 2004) that have lost motility by mutation in an operon involved in flagellum biosynthesis. In addition, in a number of Gram-positive and Gram-negative bacteria the flagellum has been shown to be involved in host adherence, colonization and invasion (Haiko and Westerlund-Wikstrom 2013; Duan et al. 2013) and recently the bacterial flagellum has also been shown to have a role as a mechanosensor responding to surface contact (Belas 2014).

Antigenic variation of the flagella of *Salmonella enterica* serovar Typhimurium was one of the first examples of bacterial switching of a cell surface structure (van der Woude and Baumler 2004). The microorganism can alternately express two antigenic variants of the flagellin subunit (the principal substituent of the flagella). Mutants unable to switch between the two flagella forms were altered in virulence compared to wild-type cells (Ikeda et al. 2001). Furthermore, it has been shown that one of the flagella forms is more able to induce tumor necrosis factor alpha in a human cell line (Ciacci-Woolwine et al. 1998) indicating flagella variation may facilitate evasion of the host immune response. Other strains of bacteria that show antigenic variation in their flagella production include *E. coli* (Liu et al. 2012) *Helicobacter pylori* (Josenhans et al. 2000) and Campylobacter species (Diker et al. 1992; Hendrixson 2008; Wilson et al. 2010). Phenotypic heterogeneity of Vibrio flagella has also been documented (Chang et al. 2009).

10.4 S-Layers

Bacterial surface (S) layers are found in many Gram-positive and Gram-negative bacteria. When present, they are the outermost proteinaceous layer of the cell envelope and are covered only sometimes by capsule (see later). S-layers are composed of numerous identical (glyco) protein subunits, which form a two-dimensional crystalline array that covers the entire cell surface. S-layers are structurally diverse with oblique, square and hexagonal lattice symmetries and although many functions have been demonstrated for S-layers there is no common function between strains (reviewed by Fagan and Fairweather 2014). A number of possible roles that relate to pathogenicity and host interaction include adhesion (Buck et al. 2005; Sakakibara et al. 2007; Poppinga et al. 2012); resistance to serum killing by impairing complement binding and phagocytosis (Blaser and Pei 1993; Sakakibara et al. 2007; Shimotahira et al. 2013) and the ability to modulate host immunity (Ausiello et al. 2006; Settem et al. 2013). Several bacterial species show antigenic variation in their S-layers (Boot and Pouwels 1996; Scholz et al. 2001; Spigaglia et al. 2011) but possibly the best characterized is that of *Campylobacter fetus* (Tu et al. 2003). *C. fetus* is a small gram-negative bacterium which primarily causes abortion in cattle and sheep. The bacterium can cause bacteraemia in immunocompromised humans leading to meningitis, pericarditis, cellulitis and abortion (Cypierre et al. 2014). The ability of this microorganism to disseminate through the bloodstream without being killed by serum is attributed to its surface layer proteins (SLPs) that prevent components of the complement system such as C3b from binding to the cell surface and causing cell lysis. This lack of deposition of C3b will reduce opsonisation and thereby lower the uptake of these bacteria by phagocytic cells mediated via the C3b receptor (Blaser et al. 1988). The SLPs of *C. fetus* have been shown to be essential for host colonization (Grogono-Thomas et al. 2000) and during an infection *C. fetus* has evolved a mechanism which allows high-frequency variation of its SLPs which helps the microorganism avoid antibody mediated killing. This mechanism is accomplished by DNA inversion and rearrangement events involving eight or nine copies of SLP—encoding genes being shuttled behind one unique promoter on the chromosome. This genomic plasticity results in substantial antigenic variation (with SLPs ranging in mass from 97–149 kDa), which in turn helps the microorganism evade the host immune response (Tu et al. 2003).

10.5 Capsule

The bacterial capsule (when present) forms the outermost layer of the cell envelope (Fig. 10.1). It is made up of long polysaccharide chains (sometimes more than 200 sugars in length), which can extend 100–400 nm from the cell surface (Pelkonen et al. 1988; Kroncke et al. 1990). As such, the capsule is at the very front-line of host pathogen interaction for many bacterial species. For many years, the capsule has been recognized as a major virulence factor—absolutely necessary for systemic infection by a number of bacterial pathogens (Roberts 1995). Indeed, the most common etiological agents of bacterial meningitis, including *Streptococcus pneumoniae*, *Haemophilus influenzae*, *Neisseria meningitidis*, and in neonates, *E. coli K1* are all encapsulated and this encapsulation plays a key role in allowing these microorganisms to breech the blood-brain barrier (Kim 2008). The capsule (like S-layers) has been shown to act in a shielding capacity protecting the bacterium against antimicrobial agents and the immune system during an infection (Burns and Hull 1998; Cunnion et al. 2001; Uria et al. 2008; Campos et al. 2004). However, it is now clear that the capsule plays additional roles in mediating interaction with its host (Corbett and Roberts 2009). Many of these roles are outlined in Table 10.1 and include moderating induction of chemokines and cytokines, perturbation of mucus

Table 10.1 Interactions of the polysaccharide capsule with the host

Organism	Capsule function	References
Streptococcus agalactiae (Group B Strep.)	Interaction with sialic acid-binding immunoglobulin–like lectins (siglecs) on neutrophils and monocytes.	Carlin et al. (2007)
Group A Streptococci	Interaction between hyaluronic acid capsule and CD44 inducing a signaling pathway	Cywes and Wessels (2001)
E. coli K1	Modulates trafficking of vacuoles, enhancing intracellular survival in brain endothelial cells	Kim et al. (2003)
UPEC	Intracellular bacterial community (IBC) formation in bladder cells	Anderson et al. (2010)
N. meningitidis	Intracellular survival	Spinosa et al. (2007)
Staphylococcus aureus	Induce release of Interleukin and TNF α	Soell et al. (1995)
Porphyromonas gingivalis and *Streptococcus suis*	Induces monocyte chemoattractant protein-1 production from monocytes	Graveline et al. (2007), d'Empaire et al. (2006)
Bacteriodes fragilis	Anti-inflammatory properties	Mazmanian et al. (2008)

clearance, interacting with Toll Like receptors and intracellular survival. Some capsular polysaccharides are structurally similar or identical to polysaccharides found on mammalian cells (such as sialic acid and hyaluronan) and this molecular mimicry limits immune recognition allowing the pathogen to multiply in its host "unnoticed" (reviewed by Cress et al. 2014).

For bacterial pathogens the initial stages of pathogenesis involve intimate contact, or adherence, to host cells, As well as allowing bacteria to "stick" to host cells this attachment is also required for the release of toxins and virulence factors necessary for infection and can also lead directly to changes in host cell signaling enabling evasion of host immune responses and hence spread. For intracellular pathogens attachment initiates a series of signaling events that allow entry of the bacterium into the cell. A plethora of specific adhesins are expressed on the bacterial cell surface (see section of fimbriae below) (Kline et al. 2009) which have been shown to be indispensible for this attachment process. Although the *Streptococcus pyogenes* (group A Streptococci) hyaluronic acid capsule has been shown to mediate attachment of this bacterium to pharyngeal cells (Cywes and Wessels 2001) in most cases it appears that the capsule inhibits adhesion and/or uptake into non-mucus producing cells (Runnels and Moon 1984; Schembri et al. 2004; St Geme and Falkow 1991; Virji et al. 1993; Hammerschmidt et al. 2005) presumably by masking or obstructing specific bacterial adhesins. Conversely, there is some evidence that the polysaccharide capsule of certain microorganisms is involved in adherence to, in particular, mucus expressing cells. For example, in *Klebsiella pneumonia* the loss of capsule reduced adhesion to the mucosal surface of the mouse colon (Favre-Bonte et al. 1999b) (but increased adhesion to non-mucous secreting epithelial cells Favre-Bonte et al. 1999a) and in *E. coli* K5 capsule expression has been shown to enhance colonization of this microorganism to the large intestine of a gnotobiotic rat (Herias et al. 1997). It would appear, therefore, that many bacteria need to spatially and temporally regulate their capsule production during infection to be effective pathogens, that is, down regulation or loss of capsule expression to enhance non-mucous producing host cell attachment and entry, followed by an up regulation of capsule expression to shield against the immune system. One way to address this is for the microorganism to phase vary their capsule expression so that at least some of the bacterial population will have the appropriate phenotype for the constantly changing environment and in fact, capsule expression has reported to be phase variable in many bacterial pathogens including *N. meningitidis*, *Campylobacter jejuni*, *H. influenzae* and *Bacteriodes fragilis* (de Vries et al. 1996; Bacon et al. 2001; Roche and Moxon 1995; Krinos et al. 2001).

One bacterium that does not appear to phase vary its capsule production is *E. coli*. The polysaccharide capsule of uropathogenic *E. coli* (UPEC) strain UTI89 has shown to be an important virulence factor during urinary tract infections (UTI) (Goller and Seed 2010). The pathogenesis of UTI has been studied extensively in mice where UPEC initially adhere to the superficial (umbrella) epithelial cells that line the luminal surface of the bladder (Fig. 10.2) (Hung et al. 2009; Eto et al. 2007). Upon infection of these uroepithelial cells, UPEC can either be translocated back out of the host cell, trafficked into late endosomal compartments where they can persist for the long term (quiescent intracellular reservoirs QIRs) or enter a complex developmental pathway involving the formation of intracellular bacterial communities (IBCs), which have biofilm like properties (Anderson et al. 2003). Around 16 h post infection bacteria on the surface of IBCs become motile, detach and escape from the umbrella cell entering into the lumen of the bladder where they can attach naïve epithelial cells and start the cycle again. Exfoliation of infected umbrella cells (presumably in order to eliminate the infecting bacteria) is also observed at this stage. The polysaccharide capsule of UPEC has been shown to be involved in IBC development (Anderson et al. 2010), however, during the initial stages of infection, during growth in urine it is envisaged that the capsule may inhibit attachment of UPEC to uroepithelial cells. Recently we have shown that there is a down regulation of capsule expression during growth in urine of UPEC strain UTI89. We have also shown that the capsule is subject to phenotypic heterogeneity (King et al. 2015) with the observation of two distinct subpopulations (capsulated and un-encapsulated) in the clonal population (Fig. 10.3a). Further studies have revealed that the un-encapsulated bacteria appear to be the initial colonizers to bladder epithelial cells and once internalized the bacteria appear to increase their capsule production presumably to aid in intracellular survival (Fig. 10.3b). It is unclear if the bacteria that efflux from the infected bladder cell during an infection are all encapsulated or not, however we speculate that there still would be a small percentage of un-encapsulated bacteria at this stage ready to go on and restart the infection cycle. This phenotypic heterogeneity, therefore, appears to be a type of the so-called "bet-hedging" mechanism ensuring that at least some of the population has the appropriate phenotype for the environmental situation. The mechanism underpinning this heterogeneity in K1 capsule expression in UPEC remains to be determined.

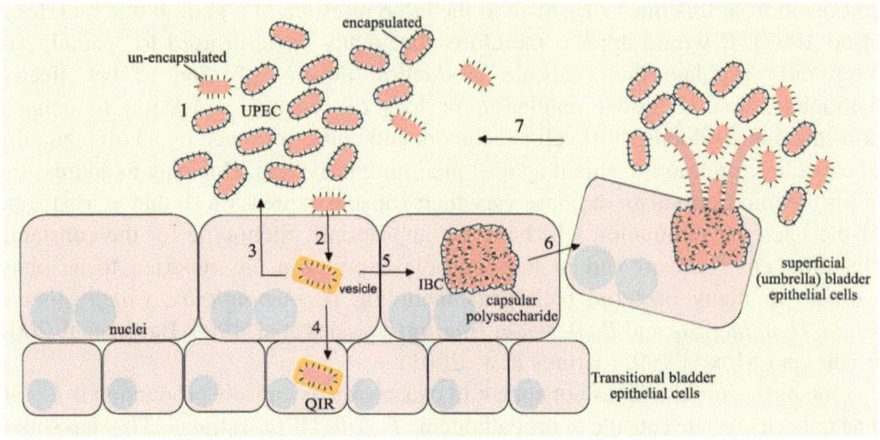

Fig. 10.2 UPEC pathogenic cycle in bladder epithelial cells. *1* During a UTI UPEC replicate in urine in the lumen of the bladder. *2* Un-encapsulated UPEC attach to and invade bladder umbrella cells. *3* The majority of the internalized bacteria are exocytosed. *4* UPEC can persist within vesicles forming quiescent intracellular reservoirs (QIRs) that have been found in the underlying transitional and basal cells. These QIRs are resistant to antibiotics and are thought to be the source of recurrent infections. *5* UPEC can break out from the vesicle and replicate in the host cell cytosol forming large intracellular bacterial communities (IBC) that contain up to 10^5 cells. Polysaccharide capsule has been shown to be important in IBC formation. *6* Bacteria are released from the IBC. Long filamentous bacteria have been observed at this stage, which have been shown to be resistant to neutrophil attack. Infection also triggers the exfoliation of bladder cells presumably in an attempt to eliminate the internalized bacteria. *7* Dispersal of the bacteria reinitiates the infection cycle

10.6 Pili and Fimbriae

As mentioned above, bacteria produce a large number of diverse adhesins on their cell surface that allow specific attachment or colonization to host cells and many of these surface bound proteins show phenotypic heterogeneity. The polysaccharide capsule (see above) that surrounds the cell surface will often mask shorter adhesins (such as the cell aggregation protein Antigen 43) (Schembri et al. 2004). We will focus here, therefore, on the extended polymeric bacterial adhesins that are thought to reach beyond the capsule layer allowing for initial host interactions (Kline et al. 2009). Phenotypic variation of these surface appendages has been well documented and since they have been shown to be excellent immunogens (Giron et al. 1991; Rene et al. 1982) the ability of bacteria to vary production of these appendages will be beneficial—allowing the bacteria to adhere to and invade their host but then turning off the expression to avoid elimination by the hosts innate defenses.

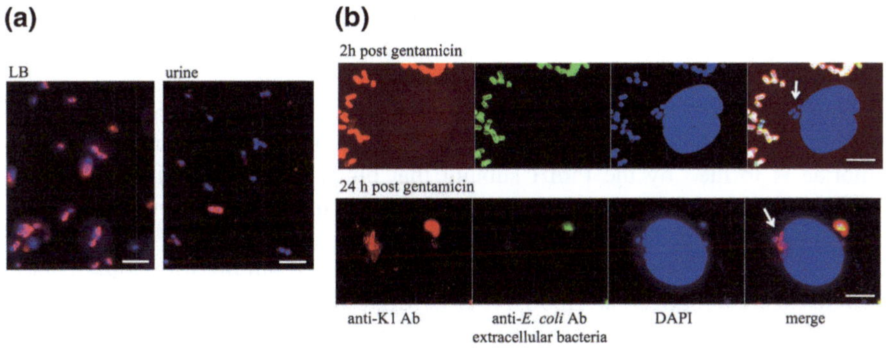

Fig. 10.3 **a** Decrease in the amount of capsule on the bacterial cell surface during growth in urine compared to Luria Bertani (LB) media and evidence of an un-encapsulated subpopulation. UTI89 was grown to mid-exponential phase in LB or urine, fixed on a microscope slide and stained with DAPI, anti-K1 primary antibody (Ab) and Alexa-fluor 594 conjugated secondary Ab. Bacteria expressing capsule appear *pink* while un-encapsulated bacteria appear *blue*. Heterogeneity in capsule expression in both media and a decrease in capsule on the cell surface in urine can clearly be seen. **b** Un-encapsulated UTI89 preferentially adhere/invade bladder epithelial cells before up regulating capsule expression. Bladder cells were infected with UTI89 and at the time points indicated following gentamicin treatment (to kill extracellular bacteria) cells were incubated with anti—*E. coli* Ab before fixation to stain extracellular bacteria. Epithelial cells were then permeabilised and stained with anti-K1 Ab, DAPI and secondary Abs to the *E. coli* and K1 primary Abs. A montage of the individual fluorescent channels is shown alongside the composite image. *Arrows* highlight intracellular bacteria (Scale bars = 10 μm.) (Adapted from King et al. 2015)

10.6.1 Chaperone-Usher Pili/Fimbriae

Pili structures exported by the chaperone-usher pathway include some of the best characterized Gram-negative bacterial adhesins. In this group pili subunits are processed and assembled by two proteins: a periplasmic chaperone and an outer membrane usher protein. Pili/fimbriae are filamentous hair-like organelles of 1–2 μm in length and can be divided into a “tip” and helically wound rod like domain (Wurpel et al. 2013; Hahn et al. 2002). Since certain pili can also be used for the transfer of genetic material during conjugation, the term “fimbriae” is often used to describe pili whose function is devoted to adhesion of bacteria to surfaces but generally the names are interchangeable.

One of the first characterized fimbriae is the P (or PAP—Pylonephritis Associated Pili) fimbriae expressed by UPEC strains and is involved in colonization of the kidney during UTI via the tip adhesion subunit PapG. PapG targets a receptor found primarily on the kidney epithelium (glycosphingolipids containing galabiose Roberts et al. 1994) but variations in PapG can recognize different receptors differentially distributed in the host and this is thought to drive tissue and host specificity (Hultgren et al. 1991). Pap phase variation is one of the most studied phase variation systems (Hernday et al. 2004) and occurs due to methylation-controlled modifications within the *pap* gene operon (this mechanism is described in detail elsewhere Hernday et al. 2004). Another microorganism capable of colonizing the urinary tract is *Proteus mirabilis* and its MR/P fimbriae (similar to PAP) are also subject to phase variation (Li et al. 2002).

Another class of well-characterized fimbriae found on most clinical isolates of *E. coli* is that of type-1 fimbriae (Iida et al. 2001). The adhesion tip of type-1 fimbriae is formed by the FimH subunit that binds mono- and tri-mannose containing glycoproteins on the host cell surface. About 80 % of commensal *E. coli* fecal isolates encode FimH that binds only tri-mannose receptors while about 70 % of UPEC isolates express FimH variants with enhanced affinity for mono-mannose residues (Sokurenko et al. 1995, 1998). Mono-mannose residues are found in abundance on the glycoprotein receptor uroplankin expressed by uroepithelial cells. Hence the mono-mannose binding phenotype enhances UPEC colonization of the urinary tract (Zhou et al. 2001) and indeed it has been shown that Type 1 fimbriae are absolutely essential for UPEC colonization of the bladder (Martinez et al. 2000). As well as adhesion FimH also appears to play important roles in host cell signaling and persistence of UPEC during UTI (Mulvey et al. 1998; Klumpp et al. 2001; Chen et al. 2009). Interestingly, biophysical analysis of type 1 fimbriae has shown that binding of FimH to mannose is via a catch bond mechanism, that is, the strength of the bond is increased by force. So the shear stress induced by urine flow, which may be envisaged as a defense mechanism, actually strengthens the interaction between the bacteria and the host (Zakrisson et al. 2013). Among other interactions, FimH has also been shown to be important in the adherence of associated adherent-invasive *E. coli* to the inflamed epithelium of patients with Crohn's disease (Tawfik et al. 2014) and for the invasion of *E. coli* K1 to brain microvascular endothelial cells (Khan et al. 2007).

The expression of type 1 fimbriae in *E. coli* is phase variable and is transcriptionally controlled by a promoter situated on an invertible element that is able to switch between a "phase–on" (resulting in fimbrial expression) orientation and a "phase-off" (no fimbrial expression) orientation (Eisenstein 1981). The invertible element is controlled by two recombinases FimB and FimE which are able to mediate the switching (Klemm 1986). The FimB recombinase catalyzes switching at low frequency in both directions whereas FimE predominantly causes switching at high frequency from phase-on to phase-off. Several regulators are also involved in the fine-tuning of Type1 fimbriae expression including integration host factor, leucine responsive protein and the histone-like protein H-NS. Several environmental factors are known to affect phase switching and these include temperature, pH, osmolarity and oxygen concentration. Studies using "locked-on" and "locked-off" mutants have shown the importance of phase variability of type -1 fimbriae in colonization of the urinary tract by UPEC (Snyder et al. 2006; Chen et al. 2009). Other microorganisms that contain phase variable fimbrial structures similar to Type-1 in *E. coli* include *S. enterica* serotype Typhimurium (Clegg et al. 1996); *B. pertussis* (Willems et al. 1990) and *H. influenza* (van Ham et al. 1993).

The Afa/Dr adhesins are another class of adhesive structures exported by the chaperone usher pathway and are associated with bacteria causing both diarrheal and UTI (Servin 2005). They can be fimbrial or afimbrial in structure and bind to epithelial cells via the decay accelerating factor (DAF or CD55)—a glycoprotein that regulates complement activation (Nowicki et al. 1993). Some Afa/Dr adhesins can also recognize type IV collagen and carcinoembryonic antigen related cell adhesion molecules (CEACAMs) (Westerlund et al. 1989; Berger et al. 2004). The gene encoding the major adhesive subunit of these adhesins is highly variable and it has been shown that naturally occurring point mutations in these adhesins can affect their adhesive properties (Korotkova et al. 2006, 2007).

10.6.2 Retractile Type IV Pili

Unlike the pili assembled by the chaperone-usher pathway above, type IV pili are polymeric assemblies of the protein pilin which are formed at the cytoplasmic membrane of gram-negative bacteria and the entire organelle is projected across the outer membrane by a type II secretion system (Wolfgang et al. 2000). Type IV pili are a similar length to the type1 pili but appear more flexible and can form bundles at the poles of the bacterial cell (Craig et al. 2004). A major feature of these pili is their ability to retract through the bacterial cell wall while the pilus remains bound to the target surface. This mechanism is required for a type of bacterial movement across semi-solid surfaces (such as the mucosal epithelia) called twitching motility (Wolfgang et al. 1998). Type IV pili are expressed in many Gram-negative pathogens and their structure and role in pathogenicity has been reviewed in depth elsewhere (Craig et al. 2004; Giltner et al. 2012). In the bloodstream, it has been shown that the type IV pili of encapsulated *N. meningitidis* are the only factors that allow initial colonization by direct interaction with the endothelial cells of the central nervous system (Carbonnelle et al. 2009). *N. meningitidis* expresses a wide diversity in the amino acid sequence of its major pilin subunit PilE. This antigenic variation involves recombination events that occur at a high rate in vivo and result in the microorganism being able to express many different pilin variants during colonization of the same host (Gibbs et al. 1989; Omer et al. 2011). This antigenic variation has recently been shown to control host cell responses (Miller et al. 2014).

10.6.3 Gram-Positive Pili

For many years pilus research was dominated by studies on Gram-negative bacteria but more recently pilus structures have been observed in pathogenic strains of Gram-positive bacteria including *Corynebacterium diphtheriae* (Ton-That and Schneewind 2003), group A Streptococci. (Mora et al. 2005), group B Streptococci. (Lauer et al. 2005), *S. pneumonia* (Barocchi et al. 2006) *Enterococcus faecalis* (Nallapareddy et al. 2006), *Clostridium perfringens* and other clostridia (Varga et al. 2006). Two types of pili have been observed in Gram-positive bacteria. The first are composed of covalently linked major pilin subunits (creating a shaft) into which minor pilins are inserted. These are linked together and transported by a sortase—dependent mechanism (Ton-That and Schneewind 2004). In group B Streptococci the major pilin subunit (required for translocation of the pathogen through human epithelial cells) is dispensable for adhesion, with the minor subunits being responsible for adhesion to cervical and lung epithelial cells (Pezzicoli et al. 2008). These minor pili structures have also been shown to be involved in adhesion to and invasion of brain microvascular endothelial cells by group B Streptococci. In *C. diphtheriae*, the minor pilin subunits also appear to play a major role in adhesion and colonization of this bacterium to pharyngeal cells (Rogers et al. 2011). The second type of pili consists of "type IV-like pili", which are similar to the type IV pili of Gram-negative bacteria and are found in a select number of Gram-positive bacteria in particular clostridia (Melville and Craig 2013).

Heterogeneity in expression of gram-positive pili has been reported for a number of strains including, *S. pneumonia* (De Angelis et al. 2011; Basset et al. 2011), *Corynebacterium pilosum* (Hiramune et al. 1991) and *E. faecalis* (Nallapareddy et al. 2006). In *S. pneumonia* expression of its type 1 pilus is known to be bistable and regulated by a transcription factor RlrA. RlrA positively controls its own expression and this positive feedback loop mediates the bistable expression of the pneumococci pilus genes (Basset et al. 2012). The basis for culture heterogeneity of the endocarditis and biofilm associated pilus (Ebp) of *E. faecalis* remains uncertain but also appears to be due to a nongenetic change such as bistability in expression of the pilus gene operon (Tariq et al. 2012). Heterogeneity in the *C. pilosum* pili appears to be by a phase variable mechanism (Hiramune et al. 1991).

Many other types of fimbrial and afimbrial adhesins are involved in the interaction of bacteria with their host which are not addressed here but are discussed in the following reviews (Kline et al. 2009; McWilliams and Torres 2014; Pizarro-Cerda and Cossart 2006; Vengadesan and Narayana 2011).

It is now clear that both the capsule and fimbriae are prominent surface components of many bacterial cells and both play roles in survival and pathogenicity of the microorganism. There have been some conflicting investigations into whether the expression of these surface structures is coordinated, particularly in the opportunistic pathogen *K. pneumonia*. Two studies have suggested an inverse relationship between capsule and type-I fimbriae expression in this microorganism. Encapsulated sepsis isolates appeared to have no type-1 fimbriae whereas the majority of UTI isolates expressed type-1 fimbriae but were un-encapsulated (Matatov et al. 1999; Sahly et al. 2000). However, an earlier report showed that most *K. pneumonia* UTI strains were able to express both capsule and type-1 fimbriae (Tarkkanen et al. 1992). Shembri et al. (2004) also showed fimbrial expression in both encapsulated and un-encapsulated strain backgrounds. However, they also showed that there was a capsule-mediated effect on fimbrial function, which was probably due to physical interference (Schembri et al. 2004). It still remains unclear therefore, whether there is some sort of coordination (or cross-talk) in fimbriae and capsule production or whether the heterogeneity of these cell surface structures is sufficient to ensure that a percentage of the population will always have the appropriate phenotype for the immediate environment.

Although most of these interactions described above are between a specific receptor and ligand, microbial cell adhesion to its host can also be nonspecific, that is, can occur due to hydrogen bonding, hydrophobic, van der Waals, electrostatic and macromolecular forces (Busscher et al. 2008). Recent studies have shown that a variety of mechanical responses are involved in adhesion of a microorganism to a surface (reviewed by Dufrene 2015).

10.7 Surface Structure Glycosylation

Another (often phase variable) mechanism bacteria have evolved which appear to have important roles in pathogenicity is glycosylation of cell surface structures. All of the cell surface structures described in the above sections (flagella, S-layers, capsule and fimbriae) can be glycosylated and this adds yet another level of complexity and heterogeneity to the bacterial cell surface. Bacteria can express a diverse array of glycans on their cell surface, which have been classified on the nature of their glycosidic linkage: N-glycosylation (attachment of the glycan to the amide nitrogen of asparagine), O-glycosylation (attachment to the hydroxyl oxygen of serine or threonine) and S-glycosylation (attachment to the sulfur of cysteine). Bacterial protein glycosyation has been reviewed recently (Tan et al. 2015). As well as having roles in protein stability (Cote et al. 2013) and motility (Asakura et al. 2010) glycosylation has shown to enhance adhesion. For example, glycosylation of pili of *Pseudomonas aeruginosa* was shown to enhance colonization of this microorganism to the mouse lung (Smedley et al. 2005) and the loss of O-liked glycosylation of Type IV pili was shown to reduce internalization of *N. meningitides* into brain endothelial cells (Takahashi et al. 2012). Glycosylation has also been shown to have roles in immune modulation and evasion. For example, purified glycosylated flagellin from the Gram-negative opportunistic pathogen *Burkholderia cenocepacia* was shown to significantly impair the immune response in an in vitro cell reporter system suggesting glycosylation of flagella may provide an evasive strategy for infecting bacteria (Hanuszkiewicz et al. 2014). Some proteins are highly decorated with glycans (e.g., flagellin of certain Campylobacter species McNally et al. 2007) and this abundance of sugars may mask protein epitopes from the immune system. The fact that the glycans of *Campylobacter* and *Neisseria* have themselves been shown to be immunogenic (Logan et al. 2002; Borud et al. 2010) may explain why glycosylation is often phase variable in these species.

The above sections have described certain bacterial surface structures that are involved in the front-line in host–pathogen interaction and have emphasized the heterogeneity and diversity of these cell surface structures. Within a clonal bacterial population therefore there may be a heterogeneous mix of bacteria displaying different cell surface components that can change during the course of an infection. This appears to increase the fitness of the microorganism by allowing increased possibilities to adhere and survive antimicrobials or the immune response. Antibiotics or antimicrobials have been designed to be active against an entire homogeneous population as classical microbiological assays provided information on entire populations. However, with the increasing evidence of cell-to-cell physiological heterogeneity new methods in the study of single cell microbiology are emerging. This in turn should aid in the development of new antimicrobials that acknowledge the existence of previously "disguised" subpopulations (which may form the basis of resistant strains). Some of the recent advances in studying the cell surface of the bacterium at the single cell level will now be reviewed.

10.8 Atomic Force Microscopy

During the past years atomic force microscopy (AFM)—based techniques have been used increasingly for the analysis of structure-function relationships of microbial cell surfaces (reviewed by Dufrene 2014). AFM allows cellular structures to be viewed under near-physiological conditions and is able to measure the adhesion or nanomechanics of a single component of the bacterial cell surface. AFM senses and measures the small forces acting between a sharp probe or tip (the tip is attached to a soft cantilever that deflects and quantifies the force) and the sample surface. The tip can follow the contours of the cell to generate a 3D image of the cell surface (Fig. 10.4). Using AFM imaging in this way has allowed many cell surface components from various species to be observed including the flagella (Chang et al. 2013; Gillis et al. 2012), S-layers (Dupres et al. 2009; Oh et al. 2013), capsule (Beaussart et al. 2014b; Stukalov et al. 2008) and fimbriae/pili (Merz et al. 2000; Miller et al. 2006). In fact, recently AFM has been used to explore the interactions between Type 3 fimbriae and the polysaccharide capsule of *K. pneumonia* during the formation of a biofilm (Wang et al. 2015). This study showed that the fluidity of the capsular polysaccharide was involved in biofilm formation and that the type 3 fimbriae were involved in maintaining this fluidity by minimizing hydrogen bond connections between the capsular polysaccharide molecules (Wang et al. 2015).

Other AFM techniques include single molecule force spectroscopy (SMFS)—this is when the force between the tip and the sample is quantified. With this technique the tip is functionalized with ligands to measure specific receptor-ligand forces and can provide data on the localization, adhesive strength and mechanics of surface molecules (Hinterdorfer and Dufrene 2006; Muller et al. 2009). In this way, SMFS has been used to explore the mechanisms involved in adhesion of bacteria to their hosts. For example, the *B. cenocepacia* trimeric autotransporter adhesion, BCAM0224, was found to act like a spring, capable of withstanding high forces and was also found to form both homomeric and dimeric interactions that enhanced adhesion to immune cells (El-Kirat-Chatel et al. 2013). Similarly the nanomechanical properties of: the *Pseudomonas fluorescens* adhesion LapA (Ivanov et al. 2012); the *E. coli* Ag43 adhesion (Jacquot et al. 2014b) and Type 1 fimbriae attachment to mannose (Jacquot et al. 2014a) have been investigated under different physiological conditions using this system.

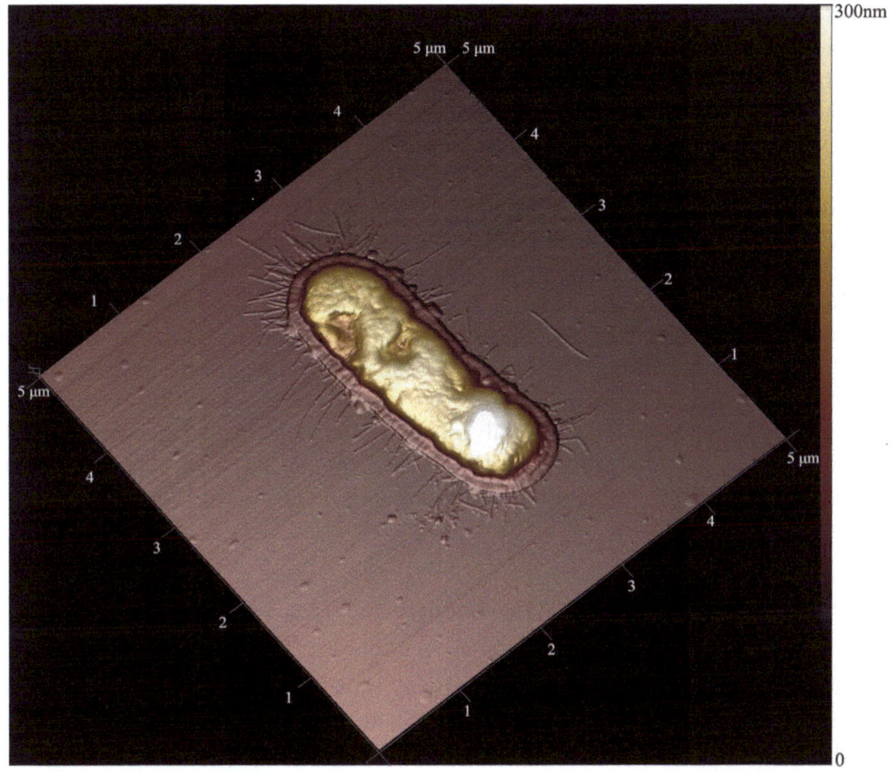

Fig. 10.4 3-D AFM image of a single un-encapsulated *E. coli* bacterium. Fimbriae can clearly be seen around the surface of the cell

A variation of SMFS is single cell force spectroscopy (SCFS) where the tip is replaced by a living cell. This allows not only single cell adhesion forces but also single-adhesion-receptor-ligand interactions to be studied (Helenius et al. 2008). Different techniques have been developed to attach microbial cells to AFM cantilevers and these are reviewed by Dufrene (2015). SCFS techniques have been used to investigate the binding strength and cell surface localization of the *Staphylococcus epidermidis* adhesion SdrG, which binds to the blood plasma protein fibrinogen (Herman et al. 2014). The experiments showed that SdrG mediated very strong (the binding strength of a covalent bond), stable and specific binding to fibrinogen and that during the interaction SdrG proteins accumulated on the bacterial cell surface to form nanoscale domains which were thought to stabilize bacterial adhesion. *S. epidermidis* infections from colonization of medical indwelling devices (such as catheters) are very common (Otto 2009) and the strong binding force exhibited by SdrG presumably allows the microorganism to withstand shear forces and persist on these devices (Herman et al. 2014). SCFS has also been used to study the nanoscale forces guiding the adhesion of Gram-negative *P. aeruginosa* type IV pili to surfaces

(Beaussart et al. 2014a); Gram-positive *Streptococcus mutans* P1 adhesion to target receptors (Sullan et al. 2015) and probiotic *Lactobacillus rhamnosuss* GG to human intestinal cells (Sullan et al. 2014). These single cell analyses have provided novel insights into the mechanisms by which microorganisms colonize surfaces and opens up new avenues in nanomedicine for controlling the adhesion of microbial pathogens. Currently SCFS assays are limited by their low throughput, i.e., only one cell per AFM cantilever, however a recent extension of AFM called fluidicforce microscopy (FluidFM), which uses micro-channeled cantilevers with nano-sized apertures to act as force controlled nanopipettes, should overcome this problem and will allow the isolation of single cells which is important when dealing with the cell-to-cell heterogeneity outlined above (Guillaume-Gentil et al. 2014). Another extension of AFM, suggested by Dufrene (2014, 2015) would be to couple it with super-resolution microscopy (Gitai 2009) which would provide a more powerful structural and functional imaging system for the analyses of specific molecules of the microbial cell surface as they interact with their targets. Recent advances in super-resolution microscopy to study the microbial cell surface will be discussed below.

10.9 Super-Resolution Microscopy

Fluorescent labeling of proteins is proving invaluable for studying real-time dynamics of biological systems (Fernandez-Suarez and Ting 2008; Chudakov et al. 2010). A number of fluorescent super-resolution imaging techniques, capable of breaking the diffraction limit of light and achieving resolution of tens of nanometers, have been developed for imaging live cells (reviewed by Tuson and Biteen 2015). These techniques include: stimulated emission depletion (STED); photo-activated localization microscopy (PALM); stochastic optical reconstruction microscopy (STORM) and points accumulation for imaging in nanoscale topography (PAINT). These techniques have provided insights into basic bacterial cell biology such as the mechanisms of cell division (Buss et al. 2013) and are also allowing the study of bacterial cell surfaces with their extracellular environment (Gunsolus et al. 2014). Although much of the research to date has focused on molecules inside cells, for example, nucleoid association proteins (Wang et al. 2011; Lee et al. 2011) and transcription factors (Elf et al. 2007; Hensel and Xiao 2013) there are examples where these systems have been applied to study bacterial cell membrane proteins. For example, in the human gut symbiont *Bacteroides thetaiotaomicron* the assembly and real-time dynamics of the outer membrane proteins involved in starch utilization were characterized (Karunatilaka et al. 2014). Also, in the human pathogen *Vibrio cholera* molecular scale interactions and dynamics of the membrane localized TcpP protein (a transcriptional activator central to the disease process) have been uncovered (Haas et al. 2015). With the ongoing improvement and development of new fluorophores, specifically for bacterial applications (Wang et al. 2014), we should see advances in this area with more focus on cell surface structures of pathogenic bacteria and their interaction with the host.

10.10 Surface Plasmon Resonance

Surface Plasmon Resonance (SPR) is another technique that is allowing for sensitive, specific, and label free analyses of real-time molecular interactions (Zeidan et al. 2015). This optical technique that probes for refractive index changes within the region of a sensor surface has been used recently alongside nuclear magnetic resonance spectroscopy to gain structural insights into host recognition by aggregative adherence fimbriae of enteroaggregative *E. coli* (Berry et al. 2014). SPR experiments have also shown direct binding between a type II transmembrane glycoprotein (CD98) of intestinal epithelial cells and both enteropathogenic *E. coli* and its murine equivalent *Citrobacter rodentium* (Berry et al. 2014; Charania et al. 2013). Furthermore, Syal et al. recently used a plasmonic imaging technique to study and quantify the interaction of a single cell of the highly virulent food-borne pathogen *E. coli O157:H7* with an antibody (Syal et al. 2015). This technique, as well as providing important information about the interaction kinetics between antigen and antibody, highlighted the physiological heterogeneity in the binding capacity between bacteria that exist in the same microbial population.

As with AFM, it has been suggested that combining SPR with other existing technologies (such as fluorescence and Raman) will provide new and improved platforms from which to study biomolecular interactions (Zeidan et al. 2015).

10.11 Conclusions

Bacteria produce a wide variety of surface structures that help them interact with their hosts. These structures include flagella, S-layers, capsule, and fimbriae among many others. Phenotypic heterogeneity of these structures mediated by genetic or nongenetic mechanisms increases the fitness of the microorganisms allowing them to be ready for any situation that may arise. The cell surface of a single bacterium is therefore a complex dynamic structure that can constantly change over time and space. Emerging and improving technologies for live single—cell analyses is allowing better understanding of how these surface structures change and interact with their host which in turn should lead to the development of better antimicrobials.

Acknowledgments Thanks to Dr. Nigel Hodson, BioAFM Facility, The University of Manchester, for generating the AFM image.

References

Ackermann M (2015) A functional perspective on phenotypic heterogeneity in microorganisms. Nat Rev Microbiol 13(8):497–508. doi:10.1038/nrmicro3491

Al Mamun AA, Tominaga A, Enomoto M (1996) Detection and characterization of the flagellar master operon in the four Shigella subgroups. J Bacteriol 178(13):3722–3726

Anderson GG, Palermo JJ, Schilling JD, Roth R, Heuser J, Hultgren SJ (2003) Intracellular bacterial biofilm-like pods in urinary tract infections. Science 301(5629):105–107. doi:10.1126/science.1084550

Anderson GG, Goller CC, Justice S, Hultgren SJ, Seed PC (2010) Polysaccharide capsule and sialic acid-mediated regulation promote biofilm-like intracellular bacterial communities during cystitis. Infect Immun 78(3):963–975. doi:10.1128/IAI.00925-09

Asakura H, Churin Y, Bauer B, Boettcher JP, Bartfeld S, Hashii N, Kawasaki N, Mollenkopf HJ, Jungblut PR, Brinkmann V, Meyer TF (2010) Helicobacter pylori HP0518 affects flagellin glycosylation to alter bacterial motility. Mol Microbiol 78(5):1130–1144. doi:10.1111/j.1365-2958.2010.07393.x

Ausiello CM, Cerquetti M, Fedele G, Spensieri F, Palazzo R, Nasso M, Frezza S, Mastrantonio P (2006) Surface layer proteins from Clostridium difficile induce inflammatory and regulatory cytokines in human monocytes and dendritic cells. Microbes Infect/Institut Pasteur 8 (11):2640–2646. doi:10.1016/j.micinf.2006.07.009

Baban ST, Kuehne SA, Barketi-Klai A, Cartman ST, Kelly ML, Hardie KR, Kansau I, Collignon A, Minton NP (2013) The role of flagella in Clostridium difficile pathogenesis: comparison between a non-epidemic and an epidemic strain. PLoS ONE 8(9):e73026. doi:10.1371/journal.pone.0073026

Bacon DJ, Szymanski CM, Burr DH, Silver RP, Alm RA, Guerry P (2001) A phase-variable capsule is involved in virulence of Campylobacter jejuni 81–176. Mol Microbiol 40(3):769–777

Barocchi MA, Ries J, Zogaj X, Hemsley C, Albiger B, Kanth A, Dahlberg S, Fernebro J, Moschioni M, Masignani V, Hultenby K, Taddei AR, Beiter K, Wartha F, von Euler A, Covacci A, Holden DW, Normark S, Rappuoli R, Henriques-Normark B (2006) A pneumococcal pilus influences virulence and host inflammatory responses. Proc Natl Acad Sci USA 103(8):2857–2862. doi:10.1073/pnas.0511017103

Basset A, Turner KH, Boush E, Sayeed S, Dove SL, Malley R (2011) Expression of the type 1 pneumococcal pilus is bistable and negatively regulated by the structural component RrgA. Infect Immun 79(8):2974–2983. doi:10.1128/IAI.05117-11

Basset A, Turner KH, Boush E, Sayeed S, Dove SL, Malley R (2012) An epigenetic switch mediates bistable expression of the type 1 pilus genes in Streptococcus pneumoniae. J Bacteriol 194(5):1088–1091. doi:10.1128/JB.06078-11

Beaussart A, El-Kirat-Chatel S, Sullan RM, Alsteens D, Herman P, Derclaye S, Dufrene YF (2014a) Quantifying the forces guiding microbial cell adhesion using single-cell force spectroscopy. Nat Protoc 9(5):1049–1055. doi:10.1038/nprot.2014.066

Beaussart A, Pechoux C, Trieu-Cuot P, Hols P, Mistou MY, Dufrene YF (2014b) Molecular mapping of the cell wall polysaccharides of the human pathogen Streptococcus agalactiae. Nanoscale 6(24):14820–14827. doi:10.1039/c4nr05280c

Belas R (2014) Biofilms, flagella, and mechanosensing of surfaces by bacteria. Trends Microbiol 22(9):517–527. doi:10.1016/j.tim.2014.05.002

Berger CN, Billker O, Meyer TF, Servin AL, Kansau I (2004) Differential recognition of members of the carcinoembryonic antigen family by Afa/Dr adhesins of diffusely adhering Escherichia coli (Afa/Dr DAEC). Mol Microbiol 52(4):963–983. doi:10.1111/j.1365-2958.2004.04033.x

Berry AA, Yang Y, Pakharukova N, Garnett JA, Lee WC, Cota E, Marchant J, Roy S, Tuittila M, Liu B, Inman KG, Ruiz-Perez F, Mandomando I, Nataro JP, Zavialov AV, Matthews S (2014) Structural insight into host recognition by aggregative adherence fimbriae of enteroaggregative Escherichia coli. PLoS Pathog 10(9):e1004404. doi:10.1371/journal.ppat.1004404

Blaser MJ, Pei Z (1993) Pathogenesis of Campylobacter fetus infections: critical role of high-molecular-weight S-layer proteins in virulence. J Infect Dis 167(2):372–377
Blaser MJ, Smith PF, Repine JE, Joiner KA (1988) Pathogenesis of Campylobacter fetus infections. Failure of encapsulated Campylobacter fetus to bind C3b explains serum and phagocytosis resistance. J Clin Investig 81(5):1434–1444. doi:10.1172/JCI113474
Boot HJ, Pouwels PH (1996) Expression, secretion and antigenic variation of bacterial S-layer proteins. Mol Microbiol 21(6):1117–1123
Borud B, Aas FE, Vik A, Winther-Larsen HC, Egge-Jacobsen W, Koomey M (2010) Genetic, structural, and antigenic analyses of glycan diversity in the O-linked protein glycosylation systems of human Neisseria species. J Bacteriol 192(11):2816–2829. doi:10.1128/JB.00101-10
Brown S, Santa Maria JP Jr, Walker S (2013) Wall teichoic acids of gram-positive bacteria. Annu Rev Microbiol 67:313–336. doi:10.1146/annurev-micro-092412-155620
Buck BL, Altermann E, Svingerud T, Klaenhammer TR (2005) Functional analysis of putative adhesion factors in Lactobacillus acidophilus NCFM. Appl Environ Microbiol 71(12):8344–8351. doi:10.1128/AEM.71.12.8344-8351.2005
Burns SM, Hull SI (1998) Comparison of loss of serum resistance by defined lipopolysaccharide mutants and an acapsular mutant of uropathogenic Escherichia coli O75:K5. Infect Immun 66 (9):4244–4253
Buss J, Coltharp C, Xiao J (2013) Super-resolution imaging of the bacterial division machinery. J Visualized Exp (71). doi:10.3791/50048
Busscher HJ, Norde W, van der Mei HC (2008) Specific molecular recognition and nonspecific contributions to bacterial interaction forces. Appl Environ Microbiol 74(9):2559–2564. doi:10.1128/AEM.02839-07
Campos MA, Vargas MA, Regueiro V, Llompart CM, Alberti S, Bengoechea JA (2004) Capsule polysaccharide mediates bacterial resistance to antimicrobial peptides. Infect Immun 72 (12):7107–7114. doi:10.1128/IAI.72.12.7107-7114.2004
Carbonnelle E, Hill DJ, Morand P, Griffiths NJ, Bourdoulous S, Murillo I, Nassif X, Virji M (2009) Meningococcal interactions with the host. Vaccine 27(Suppl 2):B78–B89. doi:10.1016/j.vaccine.2009.04.069
Carlin AF, Lewis AL, Varki A, Nizet V (2007) Group B streptococcal capsular sialic acids interact with siglecs (immunoglobulin-like lectins) on human leukocytes. J Bacteriol 189(4):1231–1237. doi:10.1128/JB.01155-06
Chang C, Jin X, Chaoqun H (2009) Phenotypic and genetic differences between opaque and translucent colonies of Vibrio alginolyticus. Biofouling 25(6):525–531. doi:10.1080/08927010902964578
Chang KC, Cheng SJ, Chen YC, Huang HR, Liou JW (2013) Nanoscopic analysis on pH induced morphological changes of flagella in Escherichia coli. J Microbiol Immunol Infect = Wei mian yu gan ran za zhi 46(6):405–412. doi:10.1016/j.jmii.2012.08.004
Charania MA, Laroui H, Liu H, Viennois E, Ayyadurai S, Xiao B, Ingersoll SA, Kalman D, Merlin D (2013) Intestinal epithelial CD98 directly modulates the innate host response to enteric bacterial pathogens. Infect Immun 81(3):923–934. doi:10.1128/IAI.01388-12
Chen SL, Hung CS, Pinkner JS, Walker JN, Cusumano CK, Li Z, Bouckaert J, Gordon JI, Hultgren SJ (2009) Positive selection identifies an in vivo role for FimH during urinary tract infection in addition to mannose binding. Proc Natl Acad Sci USA 106(52):22439–22444. doi:10.1073/pnas.0902179106
Chudakov DM, Matz MV, Lukyanov S, Lukyanov KA (2010) Fluorescent proteins and their applications in imaging living cells and tissues. Physiol Rev 90(3):1103–1163. doi:10.1152/physrev.00038.2009
Ciacci-Woolwine F, Blomfield IC, Richardson SH, Mizel SB (1998) Salmonella flagellin induces tumor necrosis factor alpha in a human promonocytic cell line. Infect Immun 66(3):1127–1134
Clegg S, Hancox LS, Yeh KS (1996) Salmonella typhimurium fimbrial phase variation and FimA expression. J Bacteriol 178(2):542–545
Corbett D, Roberts IS (2009) The role of microbial polysaccharides in host–pathogen interaction. F1000 biology reports 1:30. doi:10.3410/B1-30

Cote JP, Charbonneau ME, Mourez M (2013) Glycosylation of the Escherichia coli TibA self-associating autotransporter influences the conformation and the functionality of the protein. PLoS ONE 8(11):e80739. doi:10.1371/journal.pone.0080739
Craig L, Pique ME, Tainer JA (2004) Type IV pilus structure and bacterial pathogenicity. Nat Rev Microbiol 2(5):363–378. doi:10.1038/nrmicro885
Cress BF, Englaender JA, He W, Kasper D, Linhardt RJ, Koffas MA (2014) Masquerading microbial pathogens: capsular polysaccharides mimic host-tissue molecules. FEMS Microbiol Rev 38(4):660–697. doi:10.1111/1574-6976.12056
Cunnion KM, Lee JC, Frank MM (2001) Capsule production and growth phase influence binding of complement to Staphylococcus aureus. Infect Immun 69(11):6796–6803. doi:10.1128/IAI.69.11.6796-6803.2001
Cypierre A, Denes E, Barraud O, Jamilloux Y, Jacques J, Durox H, Pinet P, Weinbreck P (2014) Campylobacter fetus infections. Medecine et maladies infectieuses 44(4):167–173. doi:10.1016/j.medmal.2014.02.001
Cywes C, Wessels MR (2001) Group A Streptococcus tissue invasion by CD44-mediated cell signalling. Nature 414(6864):648–652. doi:10.1038/414648a
d'Empaire G, Baer MT, Gibson FC 3rd (2006) The K1 serotype capsular polysaccharide of Porphyromonas gingivalis elicits chemokine production from murine macrophages that facilitates cell migration. Infect Immun 74(11):6236–6243. doi:10.1128/IAI.00519-06
Davis KM, Weiser JN (2011) Modifications to the peptidoglycan backbone help bacteria to establish infection. Infect Immun 79(2):562–570. doi:10.1128/IAI.00651-10
De Angelis G, Moschioni M, Muzzi A, Pezzicoli A, Censini S, Delany I, Lo Sapio M, Sinisi A, Donati C, Masignani V, Barocchi MA (2011) The Streptococcus pneumoniae pilus-1 displays a biphasic expression pattern. PLoS ONE 6(6):e21269. doi:10.1371/journal.pone.0021269
de Vries FP, van Der Ende A, van Putten JP, Dankert J (1996) Invasion of primary nasopharyngeal epithelial cells by Neisseria meningitidis is controlled by phase variation of multiple surface antigens. Infect Immun 64(8):2998–3006
Deng W, Burland V, Plunkett G 3rd, Boutin A, Mayhew GF, Liss P, Perna NT, Rose DJ, Mau B, Zhou S, Schwartz DC, Fetherston JD, Lindler LE, Brubaker RR, Plano GV, Straley SC, McDonough KA, Nilles ML, Matson JS, Blattner FR, Perry RD (2002) Genome sequence of Yersinia pestis KIM. J Bacteriol 184(16):4601–4611
Diker KS, Hascelik G, Akan M (1992) Reversible expression of flagella in Campylobacter spp. FEMS Microbiol Lett 78(2–3):261–264
Duan Q, Zhou M, Zhu L, Zhu G (2013) Flagella and bacterial pathogenicity. J Basic Microbiol 53 (1):1–8. doi:10.1002/jobm.201100335
Dubnau D, Losick R (2006) Bistability in bacteria. Mol Microbiol 61(3):564–572. doi:10.1111/j.1365-2958.2006.05249.x
Dufrene YF (2014) Atomic force microscopy in microbiology: new structural and functional insights into the microbial cell surface. mBio 5(4):e01363–01314. doi:10.1128/mBio.01363-14
Dufrene YF (2015) Sticky microbes: forces in microbial cell adhesion. Trends Microbiol 23 (6):376–382. doi:10.1016/j.tim.2015.01.011
Dupres V, Alsteens D, Pauwels K, Dufrene YF (2009) In vivo imaging of S-layer nanoarrays on Corynebacterium glutamicum. Langmuir: ACS J Surf Colloids 25(17):9653–9655. doi:10.1021/la902238q
Eaton KA, Suerbaum S, Josenhans C, Krakowka S (1996) Colonization of gnotobiotic piglets by Helicobacter pylori deficient in two flagellin genes. Infect Immun 64(7):2445–2448
Eisenstein BI (1981) Phase variation of type 1 fimbriae in Escherichia coli is under transcriptional control. Science 214(4518):337–339
El-Kirat-Chatel S, Mil-Homens D, Beaussart A, Fialho AM, Dufrene YF (2013) Single-molecule atomic force microscopy unravels the binding mechanism of a Burkholderia cenocepacia trimeric autotransporter adhesin. Mol Microbiol 89(4):649–659. doi:10.1111/mmi.12301
Elf J, Li GW, Xie XS (2007) Probing transcription factor dynamics at the single-molecule level in a living cell. Science 316(5828):1191–1194. doi:10.1126/science.1141967

Eto DS, Jones TA, Sundsbak JL, Mulvey MA (2007) Integrin-mediated host cell invasion by type 1-piliated uropathogenic Escherichia coli. PLoS Pathog 3(7):e100. doi:10.1371/journal.ppat.0030100
Fagan RP, Fairweather NF (2014) Biogenesis and functions of bacterial S-layers. Nat Rev Microbiol 12(3):211–222. doi:10.1038/nrmicro3213
Favre-Bonte S, Joly B, Forestier C (1999a) Consequences of reduction of Klebsiella pneumoniae capsule expression on interactions of this bacterium with epithelial cells. Infect Immun 67 (2):554–561
Favre-Bonte S, Licht TR, Forestier C, Krogfelt KA (1999b) Klebsiella pneumoniae capsule expression is necessary for colonization of large intestines of streptomycin-treated mice. Infect Immun 67(11):6152–6156
Fernandez-Suarez M, Ting AY (2008) Fluorescent probes for super-resolution imaging in living cells. Nat Rev Mol Cell Biol 9(12):929–943. doi:10.1038/nrm2531
Finkel SE, Kolter R (1999) Evolution of microbial diversity during prolonged starvation. Proc Natl Acad Sci USA 96(7):4023–4027
Gibbs CP, Reimann BY, Schultz E, Kaufmann A, Haas R, Meyer TF (1989) Reassortment of pilin genes in Neisseria gonorrhoeae occurs by two distinct mechanisms. Nature 338(6217):651–652. doi:10.1038/338651a0
Gillis A, Dupres V, Delestrait G, Mahillon J, Dufrene YF (2012) Nanoscale imaging of Bacillus thuringiensis flagella using atomic force microscopy. Nanoscale 4(5):1585–1591. doi:10.1039/c1nr11161b
Giltner CL, Nguyen Y, Burrows LL (2012) Type IV pilin proteins: versatile molecular modules. Microbiol Mol Biol Rev: MMBR 76(4):740–772. doi:10.1128/MMBR.00035-12
Giron JA, Ho AS, Schoolnik GK (1991) An inducible bundle-forming pilus of enteropathogenic Escherichia coli. Science 254(5032):710–713
Gitai Z (2009) New fluorescence microscopy methods for microbiology: sharper, faster, and quantitative. Curr Opin Microbiol 12(3):341–346. doi:10.1016/j.mib.2009.03.001
Goller CC, Seed PC (2010) Revisiting the Escherichia coli polysaccharide capsule as a virulence factor during urinary tract infection: contribution to intracellular biofilm development. Virulence 1(4):333–337. doi:10.4161/viru.1.4.12388
Graveline R, Segura M, Radzioch D, Gottschalk M (2007) TLR2-dependent recognition of Streptococcus suis is modulated by the presence of capsular polysaccharide which modifies macrophage responsiveness. Int Immunol 19(4):375–389. doi:10.1093/intimm/dxm003
Grogono-Thomas R, Dworkin J, Blaser MJ, Newell DG (2000) Roles of the surface layer proteins of Campylobacter fetus subsp. fetus in ovine abortion. Infect Immun 68(3):1687–1691
Guillaume-Gentil O, Potthoff E, Ossola D, Franz CM, Zambelli T, Vorholt JA (2014) Force-controlled manipulation of single cells: from AFM to FluidFM. Trends Biotechnol 32 (7):381–388. doi:10.1016/j.tibtech.2014.04.008
Gunsolus IL, Hu D, Mihai C, Lohse SE, Lee CS, Torelli MD, Hamers RJ, Murhpy CJ, Orr G, Haynes CL (2014) Facile method to stain the bacterial cell surface for super-resolution fluorescence microscopy. The Analyst 139(12):3174–3178. doi:10.1039/c4an00574k
Haas BL, Matson JS, DiRita VJ, Biteen JS (2015) Single-molecule tracking in live Vibrio cholerae reveals that ToxR recruits the membrane-bound virulence regulator TcpP to the toxT promoter. Mol Microbiol 96(1):4–13. doi:10.1111/mmi.12834
Hahn E, Wild P, Hermanns U, Sebbel P, Glockshuber R, Haner M, Taschner N, Burkhard P, Aebi U, Muller SA (2002) Exploring the 3D molecular architecture of Escherichia coli type 1 pili. J Mol Biol 323(5):845–857
Haiko J, Westerlund-Wikstrom B (2013) The role of the bacterial flagellum in adhesion and virulence. Biology 2(4):1242–1267. doi:10.3390/biology2041242
Hammerschmidt S, Wolff S, Hocke A, Rosseau S, Muller E, Rohde M (2005) Illustration of pneumococcal polysaccharide capsule during adherence and invasion of epithelial cells. Infect Immun 73(8):4653–4667. doi:10.1128/IAI.73.8.4653-4667.2005
Hanuszkiewicz A, Pittock P, Humphries F, Moll H, Rosales AR, Molinaro A, Moynagh PN, Lajoie GA, Valvano MA (2014) Identification of the flagellin glycosylation system in

Burkholderia cenocepacia and the contribution of glycosylated flagellin to evasion of human innate immune responses. J Biol Chem 289(27):19231–19244. doi:10.1074/jbc.M114.562603
Helenius J, Heisenberg CP, Gaub HE, Muller DJ (2008) Single-cell force spectroscopy. J Cell Sci 121(Pt 11):1785–1791. doi:10.1242/jcs.030999
Hendrixson DR (2008) Restoration of flagellar biosynthesis by varied mutational events in Campylobacter jejuni. Mol Microbiol 70(2):519–536. doi:10.1111/j.1365-2958.2008.06428.x
Hensel Z, Xiao J (2013) Single-molecule methods for studying gene regulation in vivo. Pflugers Arch 465(3):383–395. doi:10.1007/s00424-013-1243-y
Herias MV, Midtvedt T, Hanson LA, Wold AE (1997) Escherichia coli K5 capsule expression enhances colonization of the large intestine in the gnotobiotic rat. Infect Immun 65(2):531–536
Herman P, El-Kirat-Chatel S, Beaussart A, Geoghegan JA, Foster TJ, Dufrene YF (2014) The binding force of the staphylococcal adhesin SdrG is remarkably strong. Mol Microbiol 93 (2):356–368. doi:10.1111/mmi.12663
Hernday A, Braaten B, Low D (2004) The intricate workings of a bacterial epigenetic switch. Adv Exp Med Biol 547:83–89
Hinterdorfer P, Dufrene YF (2006) Detection and localization of single molecular recognition events using atomic force microscopy. Nat Methods 3(5):347–355. doi:10.1038/nmeth871
Hiramune T, Onishi K, Kikuchi N, Yanagawa R (1991) Phase variation of pili of Corynebacterium pilosum. Zentralbl Veterinarmed [A] J Vet Med Ser B 38(4):303–305
Hultgren SJ, Normark S, Abraham SN (1991) Chaperone-assisted assembly and molecular architecture of adhesive pili. Annu Rev Microbiol 45:383–415. doi:10.1146/annurev.mi.45.100191.002123
Hung CS, Dodson KW, Hultgren SJ (2009) A murine model of urinary tract infection. Nat Protoc 4(8):1230–1243. doi:10.1038/nprot.2009.116
Iida K, Mizunoe Y, Wai SN, Yoshida S (2001) Type 1 fimbriation and its phase switching in diarrheagenic Escherichia coli strains. Clin Diagn Lab Immunol 8(3):489–495. doi:10.1128/CDLI.8.3.489-495.2001
Ikeda JS, Schmitt CK, Darnell SC, Watson PR, Bispham J, Wallis TS, Weinstein DL, Metcalf ES, Adams P, O'Connor CD, O'Brien AD (2001) Flagellar phase variation of Salmonella enterica serovar Typhimurium contributes to virulence in the murine typhoid infection model but does not influence Salmonella-induced enteropathogenesis. Infect Immun 69(5):3021–3030. doi:10.1128/IAI.69.5.3021-3030.2001
Ivanov IE, Boyd CD, Newell PD, Schwartz ME, Turnbull L, Johnson MS, Whitchurch CB, O'Toole GA, Camesano TA (2012) Atomic force and super-resolution microscopy support a role for LapA as a cell-surface biofilm adhesin of Pseudomonas fluorescens. Res Microbiol 163 (9–10):685–691. doi:10.1016/j.resmic.2012.10.001
Jacquot A, Sakamoto C, Razafitianamaharavo A, Caillet C, Merlin J, Fahs A, Ghigo JM, Beloin C, Duval JF, Francius G (2014a) Dynamic modulation of fimbrial extension and FimH-mannose binding force on live bacteria under pH changes: a molecular atomic force microscopy analysis. J Biomed Nanotechnol 10(11):3361–3372
Jacquot A, Sakamoto C, Razafitianamarahavo A, Caillet C, Merlin J, Fahs A, Ghigo JM, Duval JF, Beloin C, Francius G (2014b) The dynamics and pH-dependence of Ag43 adhesins' self-association probed by atomic force spectroscopy. Nanoscale 6(21):12665–12681. doi:10.1039/c4nr03312d
Josenhans C, Eaton KA, Thevenot T, Suerbaum S (2000) Switching of flagellar motility in Helicobacter pylori by reversible length variation of a short homopolymeric sequence repeat in fliP, a gene encoding a basal body protein. Infect Immun 68(8):4598–4603
Kaern M, Elston TC, Blake WJ, Collins JJ (2005) Stochasticity in gene expression: from theories to phenotypes. Nat Rev Genet 6(6):451–464. doi:10.1038/nrg1615
Kapatral V, Olson JW, Pepe JC, Miller VL, Minnich SA (1996) Temperature-dependent regulation of Yersinia enterocolitica Class III flagellar genes. Mol Microbiol 19(5):1061–1071
Karunatilaka KS, Cameron EA, Martens EC, Koropatkin NM, Biteen JS (2014) Superresolution imaging captures carbohydrate utilization dynamics in human gut symbionts. mBio 5(6): e02172. doi:10.1128/mBio.02172-14

Khan NA, Kim Y, Shin S, Kim KS (2007) FimH-mediated Escherichia coli K1 invasion of human brain microvascular endothelial cells. Cell Microbiol 9(1):169–178. doi:10.1111/j.1462-5822.2006.00779.x

Kim KJ, Elliott SJ, Di Cello F, Stins MF, Kim KS (2003) The K1 capsule modulates trafficking of E. coli-containing vacuoles and enhances intracellular bacterial survival in human brain microvascular endothelial cells. Cell Microbiol 5(4):245–252

Kim KS (2008) Mechanisms of microbial traversal of the blood-brain barrier. Nat Rev Microbiol 6 (8):625–634. doi:10.1038/nrmicro1952

King JE, Aal Owaif HA, Jia J, Roberts IS (2015) Phenotypic Heterogeneity in Expression of the K1 Polysaccharide Capsule of Uropathogenic Escherichia coli and Downregulation of the Capsule Genes during Growth in Urine. Infect Immun 83(7):2605–2613. doi:10.1128/IAI.00188-15

Klemm P (1986) Two regulatory fim genes, fimB and fimE, control the phase variation of type 1 fimbriae in Escherichia coli. EMBO J 5(6):1389–1393

Kline KA, Falker S, Dahlberg S, Normark S, Henriques-Normark B (2009) Bacterial adhesins in host-microbe interactions. Cell Host Microbe 5(6):580–592. doi:10.1016/j.chom.2009.05.011

Klumpp DJ, Weiser AC, Sengupta S, Forrestal SG, Batler RA, Schaeffer AJ (2001) Uropathogenic Escherichia coli potentiates type 1 pilus-induced apoptosis by suppressing NF-kappaB. Infect Immun 69(11):6689–6695. doi:10.1128/IAI.69.11.6689-6695.2001

Korotkova N, Cota E, Lebedin Y, Monpouet S, Guignot J, Servin AL, Matthews S, Moseley SL (2006) A subfamily of Dr adhesins of Escherichia coli bind independently to decay-accelerating factor and the N-domain of carcinoembryonic antigen. J Biol Chem 281 (39):29120–29130. doi:10.1074/jbc.M605681200

Korotkova N, Chattopadhyay S, Tabata TA, Beskhlebnaya V, Vigdorovich V, Kaiser BK, Strong RK, Dykhuizen DE, Sokurenko EV, Moseley SL (2007) Selection for functional diversity drives accumulation of point mutations in Dr adhesins of Escherichia coli. Mol Microbiol 64(1):180–194. doi:10.1111/j.1365-2958.2007.05648.x

Krinos CM, Coyne MJ, Weinacht KG, Tzianabos AO, Kasper DL, Comstock LE (2001) Extensive surface diversity of a commensal microorganism by multiple DNA inversions. Nature 414 (6863):555–558. doi:10.1038/35107092

Kroncke KD, Golecki JR, Jann K (1990) Further electron microscopic studies on the expression of Escherichia coli group II capsules. J Bacteriol 172(6):3469–3472

Lauer P, Rinaudo CD, Soriani M, Margarit I, Maione D, Rosini R, Taddei AR, Mora M, Rappuoli R, Grandi G, Telford JL (2005) Genome analysis reveals pili in Group B Streptococcus. Science 309(5731):105. doi:10.1126/science.1111563

Lee SF, Thompson MA, Schwartz MA, Shapiro L, Moerner WE (2011) Super-resolution imaging of the nucleoid-associated protein HU in Caulobacter crescentus. Biophys J 100(7):L31–L33. doi:10.1016/j.bpj.2011.02.022

Leigh AF, Coote JG, Parton R, Duggleby CJ (1993) Chromosomal DNA from both flagellate and non-flagellate Bordetella species contains sequences homologous to the Salmonella H1 flagellin gene. FEMS Microbiol Lett 111(2–3):225–231

Li X, Lockatell CV, Johnson DE, Mobley HL (2002) Identification of MrpI as the sole recombinase that regulates the phase variation of MR/P fimbria, a bladder colonization factor of uropathogenic Proteus mirabilis. Mol Microbiol 45(3):865–874

Liu B, Hu B, Zhou Z, Guo D, Guo X, Ding P, Feng L, Wang L (2012) A novel non-homologous recombination-mediated mechanism for Escherichia coli unilateral flagellar phase variation. Nucleic Acids Res 40(10):4530–4538. doi:10.1093/nar/gks040

Logan SM, Kelly JF, Thibault P, Ewing CP, Guerry P (2002) Structural heterogeneity of carbohydrate modifications affects serospecificity of Campylobacter flagellins. Mol Microbiol 46(2):587–597

Macnab RM (2003) How bacteria assemble flagella. Annu Rev Microbiol 57:77–100. doi:10.1146/annurev.micro.57.030502.090832

Martinez JJ, Mulvey MA, Schilling JD, Pinkner JS, Hultgren SJ (2000) Type 1 pilus-mediated bacterial invasion of bladder epithelial cells. EMBO J 19(12):2803–2812. doi:10.1093/emboj/19.12.2803

Matatov R, Goldhar J, Skutelsky E, Sechter I, Perry R, Podschun R, Sahly H, Thankavel K, Abraham SN, Ofek I (1999) Inability of encapsulated Klebsiella pneumoniae to assemble functional type 1 fimbriae on their surface. FEMS Microbiol Lett 179(1):123–130

Mazmanian SK, Round JL, Kasper DL (2008) A microbial symbiosis factor prevents intestinal inflammatory disease. Nature 453(7195):620–625. doi:10.1038/nature07008

McNally DJ, Aubry AJ, Hui JP, Khieu NH, Whitfield D, Ewing CP, Guerry P, Brisson JR, Logan SM, Soo EC (2007) Targeted metabolomics analysis of Campylobacter coli VC167 reveals legionaminic acid derivatives as novel flagellar glycans. J Biol Chem 282(19):14463–14475. doi:10.1074/jbc.M611027200

McWilliams BD, Torres AG (2014) EHEC Adhesins. Microbiol Spectr 2(2):EHEC00032013. doi:10.1128/microbiolspec.EHEC-0003-2013

Melville S, Craig L (2013) Type IV pili in Gram-positive bacteria. Microbiol Mol Biol Rev: MMBR 77(3):323–341. doi:10.1128/MMBR.00063-12

Merz AJ, So M, Sheetz MP (2000) Pilus retraction powers bacterial twitching motility. Nature 407 (6800):98–102. doi:10.1038/35024105

Miller E, Garcia T, Hultgren S, Oberhauser AF (2006) The mechanical properties of E. coli type 1 pili measured by atomic force microscopy techniques. Biophys J 91(10):3848–3856. doi:10.1529/biophysj.106.088989

Miller F, Phan G, Brissac T, Bouchiat C, Lioux G, Nassif X, Coureuil M (2014) The hypervariable region of meningococcal major pilin PilE controls the host cell response via antigenic variation. mBio 5(1):e01024-01013. doi:10.1128/mBio.01024-13

Monday SR, Minnich SA, Feng PC (2004) A 12-base-pair deletion in the flagellar master control gene flhC causes nonmotility of the pathogenic German sorbitol-fermenting Escherichia coli O157:H- strains. J Bacteriol 186(8):2319–2327

Mora M, Bensi G, Capo S, Falugi F, Zingaretti C, Manetti AG, Maggi T, Taddei AR, Grandi G, Telford JL (2005) Group A Streptococcus produce pilus-like structures containing protective antigens and Lancefield T antigens. Proc Natl Acad Sci USA 102(43):15641–15646. doi:10.1073/pnas.0507808102

Muller DJ, Krieg M, Alsteens D, Dufrene YF (2009) New frontiers in atomic force microscopy: analyzing interactions from single-molecules to cells. Curr Opin Biotechnol 20(1):4–13. doi:10.1016/j.copbio.2009.02.005

Mulvey MA, Lopez-Boado YS, Wilson CL, Roth R, Parks WC, Heuser J, Hultgren SJ (1998) Induction and evasion of host defenses by type 1-piliated uropathogenic Escherichia coli. Science 282(5393):1494–1497

Nallapareddy SR, Singh KV, Sillanpaa J, Garsin DA, Hook M, Erlandsen SL, Murray BE (2006) Endocarditis and biofilm-associated pili of Enterococcus faecalis. J Clin Investig 116 (10):2799–2807. doi:10.1172/JCI29021

Nowicki B, Hart A, Coyne KE, Lublin DM, Nowicki S (1993) Short consensus repeat-3 domain of recombinant decay-accelerating factor is recognized by Escherichia coli recombinant Dr adhesin in a model of a cell-cell interaction. J Exp Med 178(6):2115–2121

O'Neil HS, Marquis H (2006) Listeria monocytogenes flagella are used for motility, not as adhesins, to increase host cell invasion. Infect Immun 74(12):6675–6681. doi:10.1128/IAI.00886-06

Oh YJ, Sekot G, Duman M, Chtcheglova L, Messner P, Peterlik H, Schaffer C, Hinterdorfer P (2013) Characterizing the S-layer structure and anti-S-layer antibody recognition on intact Tannerella forsythia cells by scanning probe microscopy and small angle X-ray scattering. J Mol Recogn: JMR 26(11):542–549. doi:10.1002/jmr.2298

Omer H, Rose G, Jolley KA, Frapy E, Zahar JR, Maiden MC, Bentley SD, Tinsley CR, Nassif X, Bille E (2011) Genotypic and phenotypic modifications of Neisseria meningitidis after an accidental human passage. PLoS ONE 6(2):e17145. doi:10.1371/journal.pone.0017145

Ormonde P, Horstedt P, O'Toole R, Milton DL (2000) Role of motility in adherence to and invasion of a fish cell line by Vibrio anguillarum. J Bacteriol 182(8):2326–2328
Otto M (2009) Staphylococcus epidermidis–the 'accidental' pathogen. Nat Rev Microbiol 7 (8):555–567. doi:10.1038/nrmicro2182
Pelkonen S, Hayrinen J, Finne J (1988) Polyacrylamide gel electrophoresis of the capsular polysaccharides of Escherichia coli K1 and other bacteria. J Bacteriol 170(6):2646–2653
Persat A, Nadell CD, Kim MK, Ingremeau F, Siryaporn A, Drescher K, Wingreen NS, Bassler BL, Gitai Z, Stone HA (2015) The mechanical world of bacteria. Cell 161(5):988–997. doi:10.1016/j.cell.2015.05.005
Pezzicoli A, Santi I, Lauer P, Rosini R, Rinaudo D, Grandi G, Telford JL, Soriani M (2008) Pilus backbone contributes to group B Streptococcus paracellular translocation through epithelial cells. J Infect Dis 198(6):890–898. doi:10.1086/591182
Pizarro-Cerda J, Cossart P (2006) Bacterial adhesion and entry into host cells. Cell 124(4):715–727. doi:10.1016/j.cell.2006.02.012
Poppinga L, Janesch B, Funfhaus A, Sekot G, Garcia-Gonzalez E, Hertlein G, Hedtke K, Schaffer C, Genersch E (2012) Identification and functional analysis of the S-layer protein SplA of Paenibacillus larvae, the causative agent of American Foulbrood of honey bees. PLoS Pathog 8(5):e1002716. doi:10.1371/journal.ppat.1002716
Raetz CR, Whitfield C (2002) Lipopolysaccharide endotoxins. Annu Rev Biochem 71:635–700. doi:10.1146/annurev.biochem.71.110601.135414
Rene P, Dinolfo M, Silverblatt FJ (1982) Serum and urogenital antibody responses to Escherichia coli pili in cystitis. Infect Immun 38(2):542–547
Roberts IS (1995) Bacterial polysaccharides in sickness and in health. The Fleming Lecture. Microbiology 141(Pt 9):2023–2031
Roberts JA, Marklund BI, Ilver D, Haslam D, Kaack MB, Baskin G, Louis M, Mollby R, Winberg J, Normark S (1994) The Gal(alpha 1-4)Gal-specific tip adhesin of Escherichia coli P-fimbriae is needed for pyelonephritis to occur in the normal urinary tract. Proc Natl Acad Sci USA 91(25):11889–11893
Roche RJ, Moxon ER (1995) Phenotypic variation of carbohydrate surface antigens and the pathogenesis of Haemophilus influenzae infections. Trends Microbiol 3(8):304–309
Rogers EA, Das A, Ton-That H (2011) Adhesion by pathogenic corynebacteria. Adv Exp Med Biol 715:91–103. doi:10.1007/978-94-007-0940-9_6
Rosenfeld Y, Shai Y (2006) Lipopolysaccharide (Endotoxin)-host defense antibacterial peptides interactions: role in bacterial resistance and prevention of sepsis. Biochim Biophys Acta 1758 (9):1513–1522. doi:10.1016/j.bbamem.2006.05.017
Rossez Y, Wolfson EB, Holmes A, Gally DL, Holden NJ (2015) Bacterial flagella: twist and stick, or dodge across the kingdoms. PLoS Pathog 11(1):e1004483. doi:10.1371/journal.ppat.1004483
Runnels PL, Moon HW (1984) Capsule reduces adherence of enterotoxigenic Escherichia coli to isolated intestinal epithelial cells of pigs. Infect Immun 45(3):737–740
Sahly H, Podschun R, Oelschlaeger TA, Greiwe M, Parolis H, Hasty D, Kekow J, Ullmann U, Ofek I, Sela S (2000) Capsule impedes adhesion to and invasion of epithelial cells by Klebsiella pneumoniae. Infect Immun 68(12):6744–6749
Sakakibara J, Nagano K, Murakami Y, Higuchi N, Nakamura H, Shimozato K, Yoshimura F (2007) Loss of adherence ability to human gingival epithelial cells in S-layer protein-deficient mutants of Tannerella forsythensis. Microbiology 153(Pt 3):866–876. doi:10.1099/mic.0.29275-0
Schembri MA, Dalsgaard D, Klemm P (2004) Capsule shields the function of short bacterial adhesins. J Bacteriol 186(5):1249–1257
Scholz HC, Riedmann E, Witte A, Lubitz W, Kuen B (2001) S-layer variation in Bacillus stearothermophilus PV72 is based on DNA rearrangements between the chromosome and the naturally occurring megaplasmids. J Bacteriol 183(5):1672–1679. doi:10.1128/JB.183.5.1672-1679.2001

Servin AL (2005) Pathogenesis of Afa/Dr diffusely adhering Escherichia coli. Clin Microbiol Rev 18(2):264–292. doi:10.1128/CMR.18.2.264-292.2005

Settem RP, Honma K, Nakajima T, Phansopa C, Roy S, Stafford GP, Sharma A (2013) A bacterial glycan core linked to surface (S)-layer proteins modulates host immunity through Th17 suppression. Mucosal Immunol 6(2):415–426. doi:10.1038/mi.2012.85

Shimotahira N, Oogai Y, Kawada-Matsuo M, Yamada S, Fukutsuji K, Nagano K, Yoshimura F, Noguchi K, Komatsuzawa H (2013) The surface layer of Tannerella forsythia contributes to serum resistance and oral bacterial coaggregation. Infect Immun 81(4):1198–1206. doi:10.1128/IAI.00983-12

Silhavy TJ, Kahne D, Walker S (2010) The bacterial cell envelope. Cold Spring Harb Perspect Biol 2(5):a000414. doi:10.1101/cshperspect.a000414

Smedley JG 3rd, Jewell E, Roguskie J, Horzempa J, Syboldt A, Stolz DB, Castric P (2005) Influence of pilin glycosylation on Pseudomonas aeruginosa 1244 pilus function. Infect Immun 73(12):7922–7931. doi:10.1128/IAI.73.12.7922-7931.2005

Smits WK, Kuipers OP, Veening JW (2006) Phenotypic variation in bacteria: the role of feedback regulation. Nat Rev Microbiol 4(4):259–271. doi:10.1038/nrmicro1381

Snyder JA, Lloyd AL, Lockatell CV, Johnson DE, Mobley HL (2006) Role of phase variation of type 1 fimbriae in a uropathogenic Escherichia coli cystitis isolate during urinary tract infection. Infect Immun 74(2):1387–1393. doi:10.1128/IAI.74.2.1387-1393.2006

Soell M, Diab M, Haan-Archipoff G, Beretz A, Herbelin C, Poutrel B, Klein JP (1995) Capsular polysaccharide types 5 and 8 of Staphylococcus aureus bind specifically to human epithelial (KB) cells, endothelial cells, and monocytes and induce release of cytokines. Infect Immun 63 (4):1380–1386

Sokurenko EV, Courtney HS, Maslow J, Siitonen A, Hasty DL (1995) Quantitative differences in adhesiveness of type 1 fimbriated Escherichia coli due to structural differences in fimH genes. J Bacteriol 177(13):3680–3686

Sokurenko EV, Chesnokova V, Dykhuizen DE, Ofek I, Wu XR, Krogfelt KA, Struve C, Schembri MA, Hasty DL (1998) Pathogenic adaptation of Escherichia coli by natural variation of the FimH adhesin. Proc Natl Acad Sci USA 95(15):8922–8926

Spigaglia P, Galeotti CL, Barbanti F, Scarselli M, Van Broeck J, Mastrantonio P (2011) The LMW surface-layer proteins of Clostridium difficile PCR ribotypes 027 and 001 share common immunogenic properties. J Med Microbiol 60(Pt 8):1168–1173. doi:10.1099/jmm.0.029710-0

Spinosa MR, Progida C, Tala A, Cogli L, Alifano P, Bucci C (2007) The Neisseria meningitidis capsule is important for intracellular survival in human cells. Infect Immun 75(7):3594–3603. doi:10.1128/IAI.01945-06

St Geme JW, 3rd, Falkow S (1991) Loss of capsule expression by Haemophilus influenzae type b results in enhanced adherence to and invasion of human cells. Infect Immun 59(4):1325–1333

Stukalov O, Korenevsky A, Beveridge TJ, Dutcher JR (2008) Use of atomic force microscopy and transmission electron microscopy for correlative studies of bacterial capsules. Appl Environ Microbiol 74(17):5457–5465. doi:10.1128/AEM.02075-07

Sullan RM, Beaussart A, Tripathi P, Derclaye S, El-Kirat-Chatel S, Li JK, Schneider YJ, Vanderleyden J, Lebeer S, Dufrene YF (2014) Single-cell force spectroscopy of pili-mediated adhesion. Nanoscale 6(2):1134–1143. doi:10.1039/c3nr05462d

Sullan RM, Li JK, Crowley PJ, Brady LJ, Dufrene YF (2015) Binding forces of Streptococcus mutans P1 adhesin. ACS Nano 9(2):1448–1460. doi:10.1021/nn5058886

Syal K, Wang W, Shan X, Wang S, Chen HY, Tao N (2015) Plasmonic imaging of protein interactions with single bacterial cells. Biosens Bioelectron 63:131–137. doi:10.1016/j.bios.2014.06.069

Takahashi H, Yanagisawa T, Kim KS, Yokoyama S, Ohnishi M (2012) Meningococcal PilV potentiates Neisseria meningitidis type IV pilus-mediated internalization into human endothelial and epithelial cells. Infect Immun 80(12):4154–4166. doi:10.1128/IAI.00423-12

Tan FY, Tang CM, Exley RM (2015) Sugar coating: bacterial protein glycosylation and host-microbe interactions. Trends Biochem Sci 40(7):342–350. doi:10.1016/j.tibs.2015.03.016

Tariq M, Bruijs C, Kok J, Krom BP (2012) Link between culture zeta potential homogeneity and Ebp in Enterococcus faecalis. Appl Environ Microbiol 78(7):2282–2288. doi:10.1128/AEM.07618-11

Tarkkanen AM, Allen BL, Williams PH, Kauppi M, Haahtela K, Siitonen A, Orskov I, Orskov F, Clegg S, Korhonen TK (1992) Fimbriation, capsulation, and iron-scavenging systems of Klebsiella strains associated with human urinary tract infection. Infect Immun 60(3):1187–1192

Tawfik A, Flanagan PK, Campbell BJ (2014) Escherichia coli-host macrophage interactions in the pathogenesis of inflammatory bowel disease. World J Gastroenterol: WJG 20(27):8751–8763. doi:10.3748/wjg.v20.i27.8751

Ton-That H, Schneewind O (2003) Assembly of pili on the surface of Corynebacterium diphtheriae. Mol Microbiol 50(4):1429–1438

Ton-That H, Schneewind O (2004) Assembly of pili in Gram-positive bacteria. Trends Microbiol 12(5):228–234. doi:10.1016/j.tim.2004.03.004

Travis JM, Travis ER (2002) Mutator dynamics in fluctuating environments. Proc Biol Sci/Royal Soc 269(1491):591–597. doi:10.1098/rspb.2001.1902

Tu ZC, Wassenaar TM, Thompson SA, Blaser MJ (2003) Structure and genotypic plasticity of the Campylobacter fetus sap locus. Mol Microbiol 48(3):685–698

Tuson HH, Biteen JS (2015) Unveiling the inner workings of live bacteria using super-resolution microscopy. Anal Chem 87(1):42–63. doi:10.1021/ac5041346

Uria MJ, Zhang Q, Li Y, Chan A, Exley RM, Gollan B, Chan H, Feavers I, Yarwood A, Abad R, Borrow R, Fleck RA, Mulloy B, Vazquez JA, Tang CM (2008) A generic mechanism in Neisseria meningitidis for enhanced resistance against bactericidal antibodies. J Exp Med 205 (6):1423–1434. doi:10.1084/jem.20072577

van der Woude MW (2011) Phase variation: how to create and coordinate population diversity. Curr Opin Microbiol 14(2):205–211. doi:10.1016/j.mib.2011.01.002

van der Woude MW, Baumler AJ (2004) Phase and antigenic variation in bacteria. Clin Microbiol Rev 17(3):581–611, table of contents. doi:10.1128/CMR.17.3.581-611.2004

van Ham SM, van Alphen L, Mooi FR, van Putten JP (1993) Phase variation of H. influenzae fimbriae: transcriptional control of two divergent genes through a variable combined promoter region. Cell 73(6):1187–1196

Varga JJ, Nguyen V, O'Brien DK, Rodgers K, Walker RA, Melville SB (2006) Type IV pili-dependent gliding motility in the Gram-positive pathogen Clostridium perfringens and other Clostridia. Mol Microbiol 62(3):680–694. doi:10.1111/j.1365-2958.2006.05414.x

Veening JW, Smits WK, Kuipers OP (2008) Bistability, epigenetics, and bet-hedging in bacteria. Annu Rev Microbiol 62:193–210. doi:10.1146/annurev.micro.62.081307.163002

Vengadesan K, Narayana SV (2011) Structural biology of Gram-positive bacterial adhesins. Protein Sci: Publ Protein Soc 20(5):759–772. doi:10.1002/pro.613

Virji M, Makepeace K, Ferguson DJ, Achtman M, Moxon ER (1993) Meningococcal Opa and Opc proteins: their role in colonization and invasion of human epithelial and endothelial cells. Mol Microbiol 10(3):499–510

Wang H, Wilksch JJ, Strugnell RA, Gee ML (2015) Role of Capsular Polysaccharides in Biofilm Formation: an AFM nanomechanics Study. ACS Appl Mater Interfaces 7(23):13007–13013. doi:10.1021/acsami.5b03041

Wang S, Moffitt JR, Dempsey GT, Xie XS, Zhuang X (2014) Characterization and development of photoactivatable fluorescent proteins for single-molecule-based superresolution imaging. Proc Natl Acad Sci USA 111(23):8452–8457. doi:10.1073/pnas.1406593111

Wang W, Li GW, Chen C, Xie XS, Zhuang X (2011) Chromosome organization by a nucleoid-associated protein in live bacteria. Science 333(6048):1445–1449. doi:10.1126/science.1204697

Weidenmaier C, Peschel A (2008) Teichoic acids and related cell-wall glycopolymers in Gram-positive physiology and host interactions. Nat Rev Microbiol 6(4):276–287. doi:10.1038/nrmicro1861

Westerlund B, Kuusela P, Risteli J, Risteli L, Vartio T, Rauvala H, Virkola R, Korhonen TK (1989) The O75X adhesin of uropathogenic Escherichia coli is a type IV collagen-binding protein. Mol Microbiol 3(3):329–337

Willems R, Paul A, van der Heide HG, ter Avest AR, Mooi FR (1990) Fimbrial phase variation in Bordetella pertussis: a novel mechanism for transcriptional regulation. EMBO J 9(9):2803–2809

Wilson DL, Rathinam VA, Qi W, Wick LM, Landgraf J, Bell JA, Plovanich-Jones A, Parrish J, Finley RL, Mansfield LS, Linz JE (2010) Genetic diversity in Campylobacter jejuni is associated with differential colonization of broiler chickens and C57BL/6 J IL10-deficient mice. Microbiology 156(Pt 7):2046–2057. doi:10.1099/mic.0.035717-0

Wolfgang M, Park HS, Hayes SF, van Putten JP, Koomey M (1998) Suppression of an absolute defect in type IV pilus biogenesis by loss-of-function mutations in pilT, a twitching motility gene in Neisseria gonorrhoeae. Proc Natl Acad Sci USA 95(25):14973–14978

Wolfgang M, van Putten JP, Hayes SF, Dorward D, Koomey M (2000) Components and dynamics of fiber formation define a ubiquitous biogenesis pathway for bacterial pili. EMBO J 19 (23):6408–6418. doi:10.1093/emboj/19.23.6408

Wurpel DJ, Beatson SA, Totsika M, Petty NK, Schembri MA (2013) Chaperone-usher fimbriae of Escherichia coli. PLoS ONE 8(1):e52835. doi:10.1371/journal.pone.0052835

Zakrisson J, Wiklund K, Axner O, Andersson M (2013) The shaft of the type 1 fimbriae regulates an external force to match the FimH catch bond. Biophys J 104(10):2137–2148. doi:10.1016/j.bpj.2013.03.059

Zeidan E, Kepley CL, Sayes C, Sandros MG (2015) Surface plasmon resonance: a label-free tool for cellular analysis. Nanomedicine 10(11):1833–1846. doi:10.2217/nnm.15.31

Zhou G, Mo WJ, Sebbel P, Min G, Neubert TA, Glockshuber R, Wu XR, Sun TT, Kong XP (2001) Uroplakin Ia is the urothelial receptor for uropathogenic Escherichia coli: evidence from in vitro FimH binding. J Cell Sci 114(Pt 22):4095–4103

Chapter 11
Biophysical Approaches to Bacterial Gene Regulation by Riboswitches

Cibran Perez-Gonzalez, Jonathan P. Grondin, Daniel A. Lafontaine and J. Carlos Penedo

Abstract The last decade has witnessed the discovery of a variety of non-coding RNA sequences that perform a broad range of crucial biological functions. Among these, the ability of certain RNA sequences, so-called riboswitches, has attracted considerable interest. Riboswitches control gene expression in response to the concentration of particular metabolites to which they bind without the need for any protein. These RNA switches not only need to adopt a very specific tridimensional structure to perform their function, but also their sequence has been evolutionary optimized to recognize a particular metabolite with high affinity and selectivity. Thus, riboswitches offer a unique opportunity to get fundamental insights into RNA plasticity and how folding dynamics and ligand recognition mechanisms have been efficiently merged to control gene regulation. Because riboswitch sequences have been mostly found in bacterial organisms controlling the expression of genes associated to the synthesis, degradation or transport of crucial metabolites for bacterial survival, they offer exciting new routes for antibiotic development in an era where bacterial resistance is more than ever challenging conventional drug discovery strategies. Here, we give an overview of the architecture, diversity and regulatory mechanisms employed by riboswitches with particular emphasis on the biophysical methods currently available to characterise their structure and functional dynamics.

C. Perez-Gonzalez · J. Carlos Penedo
SUPA School of Physics and Astronomy, University of St Andrews,
St Andrews, Fife KY16 9SS, UK

J.P. Grondin · D.A. Lafontaine (✉)
Department of Biology, Faculty of Science, RNA Group,
Université de Sherbrooke, Sherbrooke, QC J1K 2R1, Canada
e-mail: Daniel.Lafontaine@USherbrooke.ca

J. Carlos Penedo (✉)
Biomedical Sciences Research Complex, University of St Andrews,
St Andrews, Fife KY16 9ST, UK
e-mail: jcp10@st-andrews.ac.uk

© Springer International Publishing Switzerland 2016
M.C. Leake (ed.), *Biophysics of Infection*, Advances in Experimental Medicine and Biology 915, DOI 10.1007/978-3-319-32189-9_11

Keywords Non-coding RNA · Gene regulation · RNA aptamers · Metabolite sensing · Antibiotic resistance · Bacterial pathogens

11.1 Introduction

The discovery that RNA can catalyze chemical reactions changed the perception of RNA from being a rather passive and disposable carrier of the genetic information to become a true player enabling biological functions previously thought to be exclusive to proteins. The discovery of these RNA enzymes, or ribozymes, not only challenged the supremacy of proteins as the only molecular effectors within the cell but also inspired the RNA World hypothesis which emerged as a potential solution to the 'chicken and egg' question of how a translation system could have evolved without proteins initially in place (Lilley and Sutherland 2011). Since the discovery of ribozymes, a continuously growing number of biological roles have been found to be controlled through RNA-based mechanisms. Its widespread involvement in many fundamental cellular processes such as protein synthesis, self-editing and gene regulation not only seems to reinforce RNA as a candidate for the 'Original pre-Life Molecule', but also has triggered an interest in the exploration of RNA structures as targets for drug development (Breaker 2012). In this context metabolite-sensing mRNAs, so-called riboswitches, that are able to respond to the cellular concentration of a particular metabolite to control gene expression have received considerable attention in recent years as potential targets for antibiotic development (Serganov and Nudler 2013). Riboswitches are widespread in bacterial organisms, with only one example binding thiamine pyrophosphate (TPP) found in plants, and they control the expression of genes responsible for synthesizing or transporting essential metabolites for cellular viability following a feedback mechanism (Serganov and Patel 2012). The fact that the ligand recognition element of riboswitches, the aptamer domain, has been evolutionary optimized to respond to a very specific small ligand suggests that small-molecule ligand analogues could be designed through medicinal chemistry approaches to efficiently bind the riboswitch sequence (Blouin et al. 2009a). In this chapter, we summarize the current knowledge on the antimicrobial action of riboswitches with specific emphasis on the biophysical strategies available to characterise the structure, dynamics and function of the RNA–ligand complex and how this knowledge might empower a more efficient use of these RNA elements as antibiotic targets.

11.2 Riboswitches as Antibacterial Drug Targets

Some riboswitches classes such as those recognizing flavin mononucleotide (FMN) and thiamine pyrophosphate (TPP) are widely spread between bacterial species that present great clinical threats such as *Staphylococcus aureus, Pseudomonas aeruginosa* and *Klebsiella pneumoniae* (Blount and Breaker 2006) (Table 11.1) and in

Table 11.1 RNA riboswitches from 17 different classes present in 19 pathogens

		Acinetobacter baumannii	*Baciuls anthracis*	*Brucella melitensis*	*Clostridium difficile*	*Enterococcus faecalis*	*Escherichia coli*	*Francisella tularensis*	*Hoemophilus influenzoe*	*Helicobacter pylori*	*Klebsiella pneumoniae*	*Listeria monocytogenes*	*Mycobacterium tuberculosis*	*Pseudomonos aeruginosa*	*Salmonella enterica*	*Shigella flexneri*	*Stophylococcus aureus*	*Streptococcus pneumoniae*	*Vibrio cholerae*	*Yersinia pestis*
Amino acids	Gly	1	1	1	2				1			1	2				1	1	1	
	Lys		4		1	1	1		1		1	1				1	2		3	
Coenzymes	AdoCbl	1	1	4	3	2	1				2	2	2	5	2	1			1	1
	FMN	1	2	1	2	1	1	1	1		1	1		1	1	1	2	2	1	1
	Moco						1		1		1				1	1			1	1
	SAM-I		17		5							7					4			
	SAM-II			1				1												
	SAM-III					1														
	SAM-IV												1							
	SAH													1						
	THF					1														
	TPP	1	7	2	2	2	3	1	3	1	4	2	2	1	3	3	2	4	2	3
Nucleotides and nucleobases	c-di-AMP		4										1							
	c-di-GMP		1		12															
	PreQ1-I		2			2		1				1					1			
	Purine		6		3	1						2					1	1	2	
Sugars	GlcN6P		1		1	1						1					1			

The table shows, for each class, the number of riboswitches present in each species. Updated from Blount and Breaker (2006), Lee et al. (2011)

some bacterial species they control over 2 % of the genome (Winkler and Breaker 2005). Because riboswitch representative have not been found in humans, riboswitches hold the potential to be pathogen-selective therapeutic targets (Lünse et al. 2014).

To date, the potential of several riboswitch classes has been investigated including the TPP (Thore et al. 2006), lysine (Blount et al. 2007), FMN (Grill et al. 2008; Lee et al. 2009; Kim et al. 2009) and guanine families (Batey et al. 2004; Mulhbacher et al. 2010). Interestingly, these reports have demonstrated previously identified antibacterial compounds for which the mechanism of action was unknown indeed target riboswitch structures (Ott et al. 2009). For instance, pyrithiamine, an analogue of thiamine, was determined to be toxic to bacteria and fungi (Woolley and White 1943) but the exact mechanism of pyrithiamine toxicity remained a mystery for years. The discovery of the TPP riboswitch led to the hypothesis that pyrithiamine inhibition of bacterial growth could be due to pyrithiamine-targeting TPP riboswitches and thus inhibiting thiamine biosynthesis (Sudarsan et al. 2005). Using pyrithiamine-resistant strains of *E. coli* and ligand binding assays, it was confirmed that pyrithiamine inhibition of bacterial growth was, at least partially, due to its interaction with one or more TPP riboswitches. The reason why pyrithiamine competes so well with the natural ligand was clarified by comparing the X-ray structures of wild-type ligand-bound complexes (Edwards and Ferré-D'Amaré 2006; Serganov et al. 2006) and the structure of the riboswitch from *Arabidopsis thaliana* bound to pyrithiamine pyrophosphate (Thore et al. 2008). It was found that the thiazole group of the TPP ligand is only weakly recognized by the RNA, thus explaining why the pyrithiamine analogue can be easily accommodated with minimal disruption of the binding pocket.

The presence of guanine riboswitches in some important pathogenic bacteria including *S. aureus* and *B. anthracis* combined with a wealth of structural and functional data makes them attractive drug targets. The identity of a single nucleotide within the ligand-recognition element determines whether the riboswitch responds to guanine or adenine (Mandal and Breaker 2004a, b; Tremblay et al. 2011). X-ray analysis of guanine and adenine-responsive riboswitches (Mandal and Breaker 2004a, b) have shown the ligand to be completely engaged in interactions with the RNA binding pocket. Despite of this tight RNA–ligand interaction ($K_D \sim 5$ nM for guanine), several studies demonstrated that modifications of the natural ligand are well accommodated within the RNA pocket, particularly modifications at positions C2 and C6 of the purine ring led to analogues with K_D values ranging from 8 to 500 nM (Kim et al. 2009). In another study, a pyrimidine compound, 2,5,6-triaminopyrimidin-4-one (PC1), lacking one aromatic ring exhibited a binding affinity ($K_D \sim 100$ nM) similar to that of hypoxanthine, which is a naturally occurring guanine analogue (Mulhbacher et al. 2010). PC1 showed a marked antibacterial action in 9 g-positive bacteria including multiple drug resistance (MDR) strains and the *Clostridium difficile* CD6 isolate (Mulhbacher et al. 2010). Importantly, it was found that all PC1-responsive strains had the *guaA* gene under riboswitch control, whereas the unresponsive ones did not use a riboswitch element to regulate *guaA*. The dose-response inhibitory action of PC1 on *S. aureus* showed a minimum inhibitory concentration (MIC) of 0.625 mg/mL and when compared to known antibiotics such as ciprofloxacin, PC1 showed similar antibactericidal action over a 4-h period. When the same experiment was repeated by adding GMP or AMP, bacterial growth was only restored in the presence of GMP, confirming that PC1 inhibition of bacterial growth was specific to guanine metabolism. The proof-of-concept for the antibacterial efficiency of PC1 was further confirmed using a murine model of *S. aureus*-induced mastitis providing the first evidence that antibiotics targeting riboswitches constitute a viable route to kill bacterial pathogens in mammalian infection models. Crucially, the study suggested that the development of resistance toward PC1 was unlikely (Mulhbacher et al. 2010).

FMN riboswitches are not as common as members of the TPP family but they still appear frequently in numerous bacterial pathogens (Barrick and Breaker 2007). For instance, *E. coli* has a single FMN representative located in the 5′ untranslated region (UTR) of the *ribB* gene (Winkler et al. 2002). Gram-negative bacteria such as *B. subtilis* have two FMN riboswitches, one located in the 5′ UTR of an operon that encodes proteins involved in riboflavin synthesis and the other located in the 5′ UTR of a riboflavin transporter protein (*ypaA*) (Vogl et al. 2007). Interestingly, it has been found that nature has already optimized a FMN analogue, roseoflavin. Roseoflavin (RoF) is a pigment synthesized by *Streptomyces davawensis* that was known for long time to exhibit antimicrobial activity (Otani et al. 1974). The chemical structure of roseoflavin only differs from FMN by the lack of a phosphate group and the presence of a single dimethylamino group. As expected, in-line probing demonstrated that FMN riboswitches bind RoF less tightly than FMN, with K_D values of 100 nM and 5 nM, respectively; but more tightly than riboflavin ($K_D \sim 3$ μM), which only lacks the phosphate group (Lee et al. 2009; Winkler et al. 2002). It has been found that bacteria resistant to RoF overproduce riboflavin and have mutations in the 5′ UTR region corresponding to FMN riboswitch sequences (Ott et al. 2009). This increase in riboflavin production was demonstrated to be caused by the inability of the mutant species to bind FMN and thus to turn on the transcription of riboflavin synthesis genes (Lee et al. 2009).

The potential of the FMN riboswitch as antibiotic target has been recently demonstrated with the discovery and characterization of ribocil, a synthetic chemical modulator of FMN riboswitch function that is not structurally related to the natural ligand (Howe et al. 2015). Ribocil was discovered from a library of 57,000 antibacterial compounds using a phenotypic assay that selected those molecules whose growth-inhibitory effects could be reversed by supplementing the *E. coli* cultures with riboflavin, thus ensuring the selected compounds were target-specific inhibitors of the riboflavin biosynthetic pathway. The X-ray structure of ribocil bound to the *Fusobacterium nucleatum impX* FMN aptamer showed that ribocil adopts a U-shaped conformation positioned inside the RNA junction (Howe et al. 2015) as observed for the natural FMN ligand (Serganov et al. 2009). Importantly, it was found that the biologically active form of ribocil corresponds to the (S)-isomer with K_D and IC_{50} values of 6.6 nM and 0.13 μM, respectively, whereas the (R)-isomer showed K_D and IC_{50} values of 10^4 nM and 26 μM. The antimicrobial activity of the (S)-isomer was subsequently tested on a murine *E. coli septicaemia* model of infection. It was found that ribocil treatment of infected mice reduced the concentration of pathogenic bacteria by more than 1000-fold (Hermann 2015; Howe et al. 2015).

The lysine riboswitch recognizes lysine and represses the expression of the *lysC* gene in response to high lysine concentrations (Grundy et al. 2003; Sudarsan et al. 2003). Repression of *lysC,* that encodes for aspartokinase II that catalyzes the first step in lysine biosynthesis, was known to be lysine dependent but the underlying regulation mechanism was unknown until the discovery of the lysine riboswitch (Patte 1996). The widespread distribution of lysine riboswitches and their role in lysine biosynthesis has triggered strong interest in their use as potential antibiotic targets and a number of lysine analogues have been found in vitro with K_D values ranging between 1 and 13 μM (Blount et al. 2007; Sudarsan et al. 2003). Interestingly, analysis of riboswitch variant sequences from *B. subtilis* that induced resistance to some of the ligand analogues investigated only affected the ligand binding affinity in vitro very modestly (Blount et al. 2007), suggesting that these mutations confer resistance by altering other aspects of the RNA–ligand interaction than ligand affinity. Moreover, the concentration of ligand required to induce half-maximal transcription termination was higher for the variants than for the wild type. This was interpreted as evidence that for some riboswitches the binding thermodynamics in vivo may not be as important as the kinetic aspects of ligand binding and RNA folding, and highlights the importance of understanding these features to develop successful antibiotic approaches (Wickiser et al. 2005).

11.3 Riboswitch Anatomy

Most biological functions performed by proteins and nucleic acids rely on acquiring a very specific tridimensional structure following a process known as folding. This relationship between structure and function constitutes one of the fundamental principles of molecular biology and considerable efforts have directed over the years to generate sufficient knowledge about the structure–function relationship to generate folding models. However, in addition to structural information, accurate models need to take into account the dynamics of the folding process and the time window in which this process takes place. Riboswitches constitute excellent platforms to investigate the structure–dynamics–function relationship because not only the structure but also the time scale of the folding and ligand recognition processes determines the regulatory fate of the gene or genes under control.

As previously mentioned, riboswitches are *cis* acting regulatory elements normally found in the 5′ UTR region of mRNA and are typically composed of two domains. As the nascent RNA emerges from the RNA polymerase complex, the first structure to be formed is the aptamer domain (Mandal and Breaker 2004b) (Fig. 11.1). The aptamer domain is the sensory element of the riboswitch that has been evolutionarily optimized to recognize its corresponding target metabolite with high specificity. It is the evolutionary pressure to optimize the aptamer domain for a

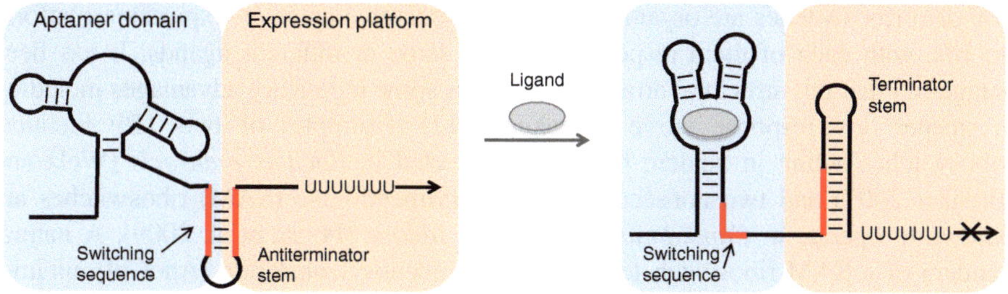

Fig. 11.1 Ligand-free and ligand-bound conformations of a negatively controlled riboswitch acting at the transcriptional level. When the ligand is not bound, sequences from the aptamer (switching sequence) and the expression platform form a secondary structure, called antiterminator stem, that allows the transcription to continue. The binding of the ligand promotes the folding of the aptamer domain and the formation of a repressor terminator stem that causes the premature end of the transcription

specific metabolite that makes the sequence of the aptamer structure to be usually highly conserved and it is the sequence conservation that is commonly used to identify specific riboswitch classes (Blouin et al. 2009a).

After the aptamer domain has been transcribed, the second domain to be formed is the expression platform (Fig. 11.1). The expression platform is located downstream of the aptamer domain, but in some cases the two domains shared a common sequence and they overlap to some extent. The role of the expression platform is to transduce the ligand binding event into a gene regulation event. This signalling process involves a conformational change in the expression platform between two mutually exclusive structures. In contrast to the high conservation observed for the aptamer domain, the expression platform varies substantially between even closely related organisms. Importantly, the physical separation between the ligand-recognition element (aptamer domain) and the signalling module (expression platform) is a beneficial approach as it allows the modular combination of similar ligand-binding domains with different expression platforms to regulate gene expression at different levels (Henkin 2008).

Although the combination of a single aptamer domain with a downstream positioned expression platform is the most typical organization of riboswitches, there are some exceptions to this pattern. The first is exemplified by RNA thermosensors, which constitute the simplest class of RNA switches (Narberhaus et al. 2006). RNA thermosensors regulate gene expression in response to a change in temperature usually by sequestering the Shine-Dalgarno (SD) sequence. These RNA structures do not require a sensing domain and the regulatory response depends on the stability of a hairpin structure. Usually, at low temperature the helical structure is stable and the SD sequence is sequestered within the helical stem resulting in translation inhibition. At temperatures above the melting temperature of the helix, the SD sequence becomes exposed and available to initiate translation.

The second exception to the single aptamer–expression platform rule includes riboswitch structures organized in a tandem configuration (Sudarsan et al. 2006).

Tandem riboswitches are organized as two consecutive aptamer–expression platform motifs, with each of them responding to the same or different ligands. It has been suggested that this structural arrangement offers some regulatory advantages including a steeper dose-response curve (Breaker 2012). Examples of two TPP identical riboswitches acting in tandem have been detected in *Bacillus anthracis* (Welz and Breaker 2007) and two consecutive S-adenosylmethionine (SAM) riboswitches are known to operate in *Candidatus pelagibacter ubique* (Poiata et al. 2009). A natural tandem of a SAM riboswitch followed by an adenosylcobalamin (AdoCbl) aptamer has been observed in *metE* mRNA from Bacillus clausii (Sudarsan et al. 2006). The tandem structure acts as a two-input NOR Boolean logic gate, where the binding of SAM or AdoCbl causes transcription termination before the coding region of the mRNA is formed. A different tandem organization with two consecutives aptamers controlling a single downstream expression platform has been observed in the glycine riboswitch (Huang et al. 2010; Mandal et al. 2004a). In this case, each aptamer recognizes the same ligand, glycine, but instead of operating independently, the two aptamer domains interact with each other to adopt the native state. Early studies on the *Vibrio cholera* (VC) glycine riboswitch reported that the ligand binding to the downstream aptamer allows tertiary interactions between both aptamer domains that facilitate binding of a separate glycine molecule to the first aptamer with Hill coefficients of 1.5–1.6 (Mandal et al. 2004; Trausch et al. 2011). This cooperative model for ligand binding was further confirmed using biochemical (Kwon and Strobel 2008), small angle X-ray scattering (Lipfert et al. 2007, 2010) as well as crystallographic data (Butler et al. 2011; Huang et al. 2010). However, more recently, separate studies using RNA-structure modelling tools (Kladwang et al. 2012), nucleotide analogue interference mapping (NAIM) and in-line probing (Sherman et al. 2012) have demonstrated that the linker between both aptamers instead of being single-stranded RNA adopts a highly conserved k-turn structure (Schroeder et al. 2011). The presence of this k-turn motif enhanced the inter-aptamer interaction, increased the binding affinity by 4.5–86-fold depending on the specific construct and abolished cooperativity, suggesting that previously observed cooperativity was most likely an artefact resulting from using a truncated construct.

11.4 Structural Diversity and Ligand Recognition Mechanisms

Currently, there are 20 classes of riboswitches. In addition to the already mentioned mRNAs sensing SAM, TPP, glycine and AdoCbl (Vitamin B12), riboswitch structures responsive to nucleotides such as adenine and guanine (Fig. 11.2); second messenger molecules such as c-di-GMP (Sudarsan et al. 2008) and c-di-AMP (Nelson et al. 2013); amino acids such as lysine (Fig. 11.2) (Grundy et al. 2003), glycine (Mandal et al. 2004) and glutamine (Ames and Breaker 2011) have already been identified. Additional small molecules including preQ1 (Roth et al. 2007),

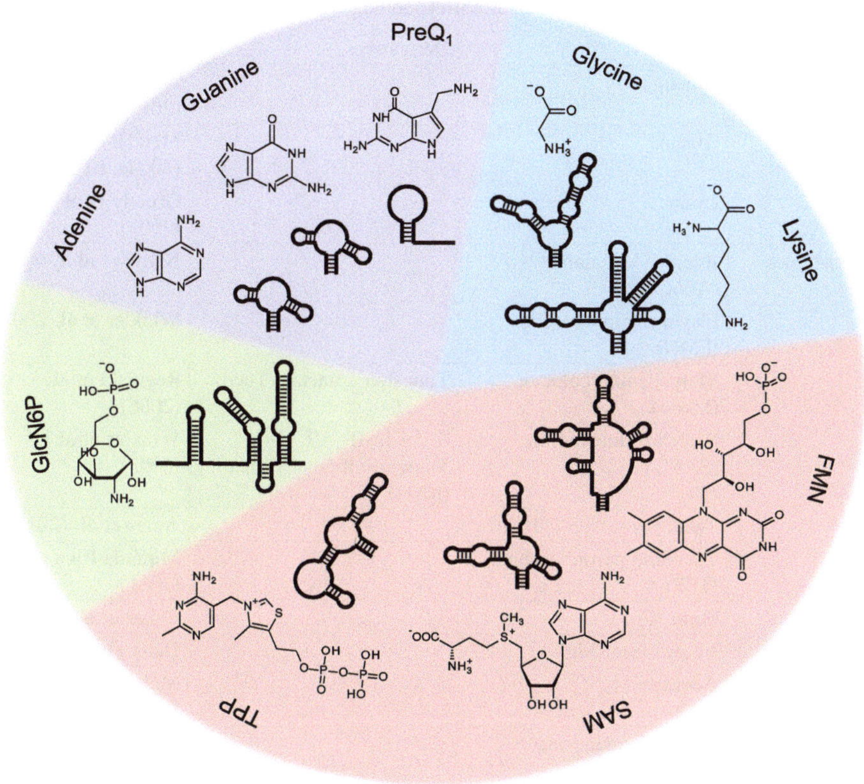

Fig. 11.2 Aptamer organization according to the nature of its associated ligand. *Purple* Nucleotides and nucleobases: adenine, guanine and preQ $_1$ riboswitches. *Blue* Amino acid ligands: glycine and lysine riboswitches. *Red* Coenzymes: flavin mononucleotide (FMN), S-adenosylmethionine (SAM-I) and thiamine pyrophosphate (TPP) riboswitches. *Green* Sugar: glucosamine-6-phosphate (glmS) riboswitch

tetrahydrofolate (THF) (Ames et al. 2010), glucosamine-6-phosphate (GlcN6P) (Jansen et al. 2006), metal ions (Dann et al. 2007, Furukawa et al. 2015), fluoride (Baker et al. 2011) and putative S-adenosylhomocysteine (SAH) (Wang et al. 2008), and molybdenum cofactor (Regulski et al. 2008) riboswitches have also been observed along with some orphan riboswitches whose cognate metabolite has not yet been identified. The *glms* riboswitch sensing GlcN6P is an atypical riboswitch in that it is also a ribozyme and ligand binding to the aptamer domain induces a 1000-fold increase in self-cleaving efficiency (Winkler et al. 2004) (Fig. 11.2). A list of riboswitches recognizing small molecules is given in Table 11.2.

High-resolution crystal structures of the aptamer–ligand complex have been determined for most riboswitch classes (Serganov and Patel 2012). With the exception of the purine class of riboswitches that can switch its specificity from guanine to adenine by a single point mutation (C-to-U change) (Mandal and

Table 11.2 List of riboswitches and their corresponding ligands

	Ligand	Derivatives	References
Amino acids	Glutamine		Ames and Breaker (2011)
	Glycine		Mandal et al. (2004a, b)
	Lysine		Grundy et al. (2003)
Coenzymes	Adenosylcobalamin (AdoCbl)		Nahvi et al. (2002)
	Flavin mononucleotide (FMN)		Winkler et al. 2002
	Molybdenum cofactor (Moco)	Tungsten cofactor (Tuco)	Regulski et al. (2008)
	S-adenosylmethionine (SAM)	SAM-I, -II, -III, -IV. S-adenosylhomocysteine (SAH)	Winkler et al. (2003)
	Tetrahydrofolate (THF)		Ames et al. (2010)
	Thiamine pyrophosphate (TPP)		Miranda-Rios et al. (2001)
Ions	Fluor (F^-)		Baker et al. (2011)
	Magnesium (Mg^{2+})		Dann et al. (2007)
Nucleotides and nucleobases	Adenine		Mandal and Breaker (2004a, b)
	Cyclic di-adenosine monophosphate (c-di-AMP)		Nelson et al. (2013)
	Cyclic di-guanosine monophosphate (c-di-GMP)	c-di-GMP-I, -II	Sudarsan et al. (2008)
	Guanine	2′-deoxyguanosine	Mandal et al. (2003)
	Prequeuosine-1 (preQ1)	PrEquation $_1$-I, -II	Roth et al. (2007)
Sugars	Glucosamine-6-phosphate (GlcN6P)		Jansen et al. (2006)

Breaker 2004a, b), each type of ligand is recognized by a different RNA sequence and folding structure. Crystallographic and biochemical data have demonstrated that riboswitch aptamers achieve a high level of specificity towards their cognate ligand by interacting with nearly all available functional groups (Roth and Breaker 2009). For instance, adenine and lysine-sensing riboswitches almost completely encapsulate their ligands leaving only <15 % of the ligand surface accessible to the solvent (Serganov and Patel 2012). To achieve this high degree of interaction between the RNA and the ligand, the aptamer needs to be sufficiently flexible to allow access of the ligand to the binding pocket.

On the basis of available crystal structures for aptamer–ligand complexes, it has been proposed that riboswitches can be classified into two distinct types according to their binding pocket architecture and the conformational changes taking place in the aptamer domain induced by ligand binding (Serganov and Nudler 2013; Montange and Batey 2008). The first type (Type I) is exemplified by the purine class and includes riboswitch aptamers folded into multi-helical junctions. The binding pocket and the tertiary structure in these structures is largely pre-organized and conformational changes induced by ligand binding are limited to the aptamer core from where the helical stems irradiate and regions adjacent to the ligand binding pocket. Recently, the Type I group has been divided in two subgroups (Ia and Ib) (Serganov and Nudler 2013) using as classification criteria the relative positioning of the multi-helical junction with respect to the P1 stem. The P1 stem is a common element found in most riboswitches that closes the metabolite sensing domain and shares a common sequence, the switching sequence, with the expression platform (Fig. 11.1) (Serganov and Patel 2012). As exemplified by the purine riboswitch class, in the Type Ia, the helical junction is centrally positioned, adjacent to the P1 stem, and acts as a sequence joining the regulatory P1 stem to the rest of the helical stems. These peripheral stems interact with each other through long-range interactions that stabilize the global structure. In the Type Ib, a key junction is located far from the P1 stem and ligand binding at the junction influences the formation of the P1 stem through tertiary interactions. Examples of Type Ib are the THF-sensing riboswitch (Trausch et al. 2011) and the Mg^{2+} class (Dann et al. 2007). Lastly, the Type II includes those riboswitch classes whose global architecture is based on the formation of pseudoknot structures (base-pairing structures formed between a loop of an RNA stem-loop and another region resulting in a knot-shaped conformation (Lilley and Eckstein 2008). Examples of Type II riboswitches include the SAM-II (Gilbert et al. 2008) and the fluoride-sensing aptamers (Ren et al. 2012).

The mechanisms of ligand recognition have been extensively studied by both X-ray crystallography and RNA biochemical probing assays (Garst et al. 2011). Although the emerging picture arising from these studies is that there is no a uniform recognition mechanism, several common and general features have been observed. For instance, ligand binding pockets are usually formed by conserved nucleotides and the recognition of most metabolites involves the formation of specific hydrogen bonds and electrostatic interactions with the RNA. Those ligands with planar groups are additionally stabilized by stacking interactions with purine bases (G and A) forming the ligand-binding pocket and brought into close proximity to the ligand by local or long-range tertiary interactions. As expected for RNA in general, mono- and divalent metal ions such as Na^+, K^+ and Mg^{2+} play a fundamental role in riboswitch folding and ligand recognition (Draper et al. 2005). In addition to diffusive metal ions while screening the negative charge of the RNA phosphate groups and the negative charge of some functional groups within the ligand (carboxylate, phosphate), specifically trapped Mg^{2+} ions are commonly observed in the crystal structures of many riboswitches (Draper et al. 2005). These locally bound metal ions mediate interactions between different nucleobases or between the RNA backbone and the ligand either directly or coordinated with water molecules (Garst et al. 2011).

Depending on the role played by the ligand in the folding and formation of the binding pocket, two mechanisms have been proposed for ligand recognition: induced-fit and conformational selection. Ligand-binding following an induced-fit mechanism is exemplified by the adenine- and guanine-sensing riboswitches. The intrinsic flexibility of the ligand-free state makes difficult to obtain crystal structures and normally they are studied using NMR and in-line probing techniques. NMR studies of the *pbuE* and *xpt-pbuX* riboswitch aptamer suggested that at physiological concentrations of Mg^{2+}, the core of the three-way junction is disordered in the absence of ligand but becomes ordered upon ligand binding (St-Pierre et al. 2014). The local reorganization of the aptamer core induced by the ligand has been interpreted as evidence for an induced-fit mechanism (Gilbert et al. 2006; Noeske et al. 2005, 2007; Sharma et al. 2009). In contrast, SAM-I (Heppell et al. 2011) and SAM-II (Haller et al. 2011a) have been shown to recognize the ligand following a conformational capture mechanism. In the absence of ligand, these aptamers are sampling a range of equilibrium conformations and the ligand selectively binds one of these conformations, thus shifting the population equilibrium towards the ligand-bound state. A detailed description of both models is beyond the scope of this paper but for further information, we refer the reader to an excellent review by Haller et al. that had focus on this particular aspect of the ligand–aptamer interaction (Haller et al. 2011a).

11.5 Riboswitch Gene Regulation Mechanisms

The vast majority of riboswitches have been found to control either mRNA levels or translation initiation. For most riboswitches, ligand binding results in the repression of gene expression. However, for both glycine and adenine-sensing riboswitches, it has been shown that ligand binding promotes the positive expression at the transcription and translation levels, respectively (Mandal and Breaker 2004a, b; Mandal et al. 2004). Furthermore, additional mechanisms have been documented implicating riboswitches in modulating ribozyme activity, mRNA splicing, RNase activity, antisense production (*cis* and *trans* locations relative to the RNA target) and protein sequestration (Debroy et al. 2014; Mellin et al. 2014; Serganov and Nudler 2013).

Riboswitches have been shown to control transcription termination either using an intrinsic stem loop terminator or by employing the transcriptional Rho factor (Serganov and Nudler 2013). For both mechanisms, ligand binding to the aptamer induces a conformational change in the riboswitch resulting in the reorganization of the expression platform. In the case of Rho-independent terminators, the presence of a uridine stretch stalls the polymerase, eventually causing the release of the prematurely terminated nascent RNA. However, for Rho-dependent terminators, ligand binding has been shown to terminate transcription using stem loop structure without the presence of the U-rich region (Fig. 11.3a) (Hollands et al. 2012, 2014). Riboswitch expression platforms controlling translation initiation modulate the

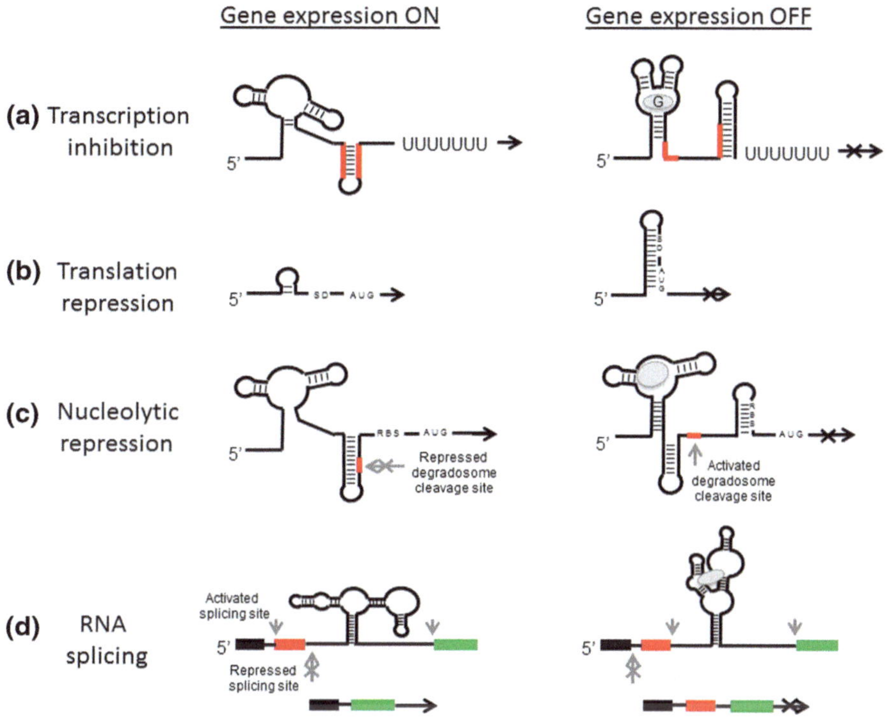

Fig. 11.3 Riboswitch-mediated gene regulation mechanisms. **a** Guanine riboswitch is shown as an example of transcription inhibition, in which a high concentration of guanine ligand favours the formation of the terminator stem, thus causing the premature end of the transcription. **b** The translation repression is exemplified with the miniROSE RNA, a small RNA thermosensor operating at the translational level. The miniROSE element aborts the translation initiation with the sequestration of the Shine-Dalgarno sequence (SD) and the starting AUG codon. **c** The nucleolytic repression mechanism is found on the *lysC* riboswitch, which operates at the translational level. In the absence of ligand, the cleavage site (*red box*) is sequestered in a stable secondary structure with a region of the aptamer, thus inhibiting its recognition by the cleavage machinery. High concentrations of lysine ligand induce the folding of the riboswitch in a structure that liberates the cleavage site and sequesters the ribosomal binding site (RBS) sequence, inhibiting the translation initiation. **d** RNA splicing is mechanism followed by the thiamine pyrophosphate (TPP) riboswitch in the *NMT1* gene regulation. A sequence of the TPP aptamer interacts with a non-coding region in the proximity of a splicing site, making it inaccessible to the splicing machinery and leading to cleavage at a distal splice site. The conformational change induced upon ligand binding breaks the previous interaction since the rearrangement of the stem involved is required for the correct folding. The proximal splice site becomes then active and produces a longer mRNA, incorporating an extra sequence (in *red*), that inhibits the gene expression

accessibility of Shine-Dalgarno site (SD) and AUG start codon (Fig. 11.3b). In absence of ligand, the SD is accessible and this allows the ribosome to bind to the SD to initiate translation. However, upon ligand binding, the SD is sequestrated in a stem loop and the ribosome cannot bind the SD thus repressing translation initiation. Due to the presence of cellular ribonucleases and the transcription factor Rho, it is difficult to envisage that such riboswitches may strictly control translation initiation without significantly involving regulatory mechanisms altering mRNA levels upon translation inhibition. For example, it was recently reported that the *E. coli lysC* lysine riboswitches controls both mRNA levels and translation initiation (Caron et al. 2012). In the absence of ligand, the riboswitch adopts the ON state and sequesters an RNase E cleavage site located in the expression platform, thus allowing mRNA accumulation and translation initiation (Fig. 11.3c). However, upon ligand binding, the riboswitch folds into conformation which frees the RNase E cleavage site and sequesters the SD, resulting in both mRNA degradation and the repression of translation initiation (Caron et al. 2012). Whether this particular regulation mechanism is shared by other riboswitches will require further investigation.

The TPP riboswitch is particular as it is not only found in bacteria but also in archaea and eukaryotes. In fungi, this riboswitch has been found to control mRNA splicing by modulating the access of splice sites (Serganov and Nudler 2013). The aptamer contains two 5′ splice sites and a single 3′ splice site. In the absence of TPP, the proximal 5′ splice site is sequestered allowing the activation of the distal 5′ splice site. However, upon ligand binding, the structural conformation adopted by the riboswitch promotes the use of the 5′ proximal site in the splicing regulation, thereby repressing the 5′ distal splice (Fig. 11.3d).

Clearly, it is very likely that riboswitches employ additional mechanisms since riboswitch expression platforms exhibit low sequence conservation, and thus potentially a large structural complexity. While most obvious regulatory mechanisms have been uncovered, more subtle or complex mechanisms will be deciphered in the future due to the ever expanding power of biocomputational approaches and the development of more sophisticated biophysical techniques able to monitor RNA conformational changes in real time.

11.6 Biophysical Characterization of Riboswitch Structure, Dynamics and Function

Crystallographic studies of riboswitch aptamers constitute the main source of information for the structure of the ligand-bound complex. However, these studies only provide a snapshot of the most stable ligand-bound conformation, and thus information about the dynamics of aptamer folding and ligand binding needs to be extracted from additional solution-based methods. Moreover, due to the intrinsic flexibility of the ligand-free state, it is very difficult to obtain crystal structures of unbound aptamers. Biophysical methods such as NMR spectroscopy (Noeske et al. 2007), small angle scattering (SAXS) (Lipfert and Doniach 2007) and chemical probing methods such as nucleotide analogue interference mapping (NIAM) (Suydam and Strobel 2009) and selective 2′-hydroxyl acylation analyzed by primer extension (SHAPE) have been widely used to investigate the structure of the both ligand-free and ligand-bound states. Fluorescence studies using 2-aminopurine (2AP), which is an intrinsically fluorescent analogue of adenine, have been extensively employed to monitor local conformational changes within the aptamer sequence due to Mg^{2+}-induced folding and ligand binding (St-Pierre et al. 2014).

Usually, the folding and ligand-recognition mechanisms in riboswitch structures composed of multi-helical junctions require the formation of long-range tertiary interactions between peripheral RNA motifs. These long-range interactions are commonly studied using Förster resonance energy transfer (FRET) (Selvin 2000). FRET allows to measure the distance between two points within the riboswitch structure and how this distance vector changes as the structure progresses from the ligand-free to the ligand-bound state or as a consequence of the folding process. FRET requires the use of RNA constructs labelled with a donor and an acceptor fluorescent dye, positioned at those locations for which the distance needs to be extracted. The transfer of energy from the directly excited donor to the ground state acceptor is highly dependent on the distance between both dyes and results in an excited acceptor that emits at longer wavelength than the donor. By monitoring the ratio between the donor and acceptor emission, it is possible to extract FRET efficiency values, which can then be transformed in distance values (Blouin et al. 2009b).

Importantly, in recent years the development of methods that can monitor conformational changes in single RNA structures using FRET has revolutionized our understanding of bacterial gene regulation and many other biological processes. The main advantage of single-molecule FRET (smFRET) compared to traditional 'ensemble-averaging' methods, which monitor simultaneously many RNA molecules, lies on the ability to determine the number of structural states coexisting in a mixture and their relative populations (Ha 2001). Moreover, by targeting one RNA molecule at a time, these methods provide information about low-populated and transient states that will remain hidden in ensemble methods. Importantly, smFRET offers the possibility of monitoring these conformational changes in real time, and because the time-dependent trajectory of each RNA molecule is recorded individually, there is no need for synchronization (Ha 2001; Blouin et al. 2009b). In parallel to the development and application of smFRET, a range of techniques using optical force (smForce) to directly stretch the RNA structure, or indirectly sense conformational changes via variations in RNA length, have emerged as complementary techniques to smFRET that can extract information about variations in the RNA conformational landscape even at co-transcriptional level (Frieda and Block 2012). In the following sections, we focus on the application of these fluorescence-based methods as indicators of RNA structural changes at local level using intrinsically fluorescent nucleotides and to monitor the formation of long-range interactions using smFRET and smForce.

11.6.1 Understanding Ligand-Recognition Mechanisms Using 2-Aminopurine Fluorescence

The need to develop new fluorescence methods to investigate local conformational changes in nucleic acids led to the introduction of the 2-aminopurine (2AP), an analogue of adenine that also forms hydrogen bonds with uracil and thiamine. It is the presence of a band with maximum at 305 nm, outside the absorption region of the DNA and RNA bases, what makes possible to selectively excite the 2AP without influencing the nucleic acids. Despite the low extinction coefficient of the 305 nm band ($\sim$6000 M^{-1} cm^{-1}), the strong dependence of the 2AP quantum yield, lifetime and anisotropy with the surrounding medium make 2AP a valuable probe for the detection of local changes in DNA and RNA structures (Holmen et al. 1997; Albert and Taguchi 1973; Wilhelmsson 2010). 2AP-based assays exploit the fact that the high fluorescence quantum yield of 2AP free in solution (0.68) is strongly quenched upon stacking between the base pairs of a nucleic acid sequence (Wilhelmsson 2010).

The sensitivity and selectivity of 2AP emission to variations in the local environment and the structure of the RNA molecule makes 2AP substitutions an ideal probe to study folding and ligand-binding mechanisms in riboswitches. As an example, 2AP labelling of purine riboswitches has been used to study the folding pathways of the aptamer domain of the adenosine deaminase *add* riboswitch from *Vibrio vulnificus* (Rieder et al. 2007). The response of the aptamer domain from the *add* riboswitch to the addition of Mg^{2+} and adenine ligand was studied for two different 2AP substitutions on the aptamer core (U48AP, U24AP) and two substitutions on the P2 loop (A35AP and G38AP/C60U) (Fig. 11.4a). The U48AP variant showed little response to Mg^{2+} and only the addition of adenine induced a significant increase of the 2AP fluorescence. The emission of the A24AP variant, on the contrary, was reduced to one-third of the initial signal upon addition of Mg^{2+}. The presence of a Mg^{2+}-binding site next to the A24 may stabilize this region of the aptamer, where the 2AP is mainly stacked between G72 and A73 (Serganov et al. 2004). The A35AP and G38AP/C60U substitutions at the P2 loop were introduced to sense the P2–P3 loop–loop interaction in the ligand-free and ligand-bound states. One-third of the initial emission of the A35AP variant was quenched upon addition of Mg^{2+}, which may confirm a preorganization of stems P2 and P3 in the absence of ligand.

The incorporation of 2AP to specific positions of the aptamer was extended to monitor the folding of the full-length *add* riboswitch in its U48AP variant, a mRNA acting at the translational level, and the G48AP variant of the *pbuE* riboswitch from *Bacillus subtilis* acting at the transcriptional level (Rieder et al. 2007). The *add* riboswitch experienced a significant increase of the 2AP fluorescence in the presence of adenine in a similar way to the aptamer sequence (Fig. 11.4b), whilst the full-length *pbuE* riboswitch remained almost invariant to the addition of Mg^{2+} and

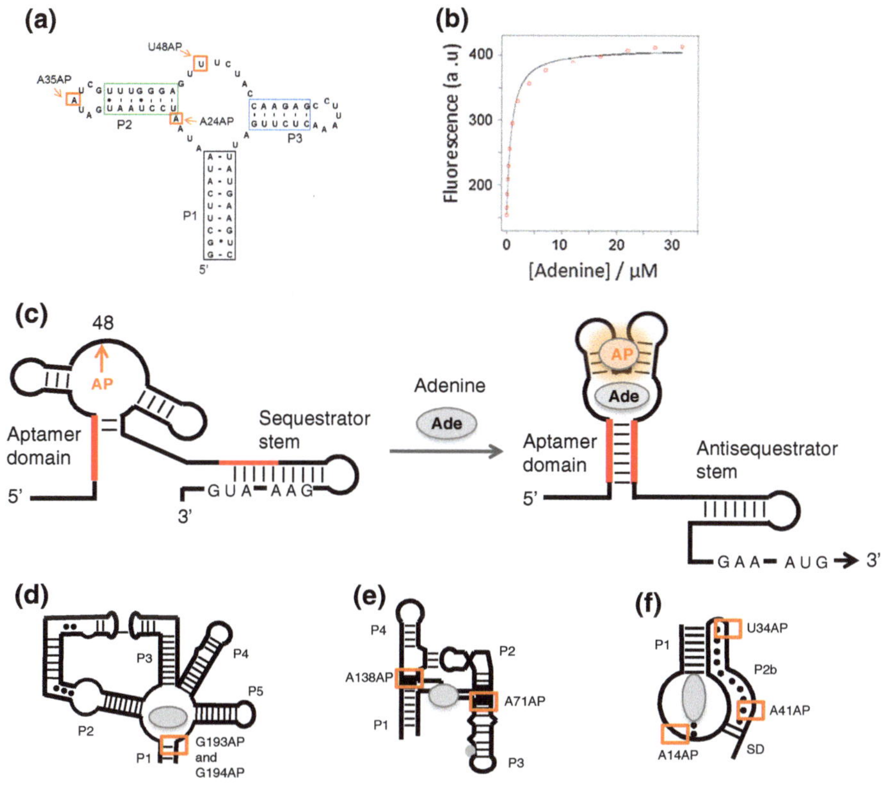

Fig. 11.4 **a** Secondary structure of the adenine riboswitch aptamer showing the A24, A35 and U48 positions of the 2-aminopurine substitutions. **b** Variations in fluorescence intensity of the *add* U48AP variant of as a function of adenine ligand added and determination of the dissociation constant by fitting to a single-binding site model, obtaining a K_D of 680 nM (Rieder et al. 2007). **c** Schematic representation of the full-length *add* riboswitch on the U48AP variant. Ligand binding promotes a solvent exposed conformation of the 2AP, which can be observed as an increase of its fluorescence. **d** Ligand-bound state of the lysine riboswitch showing the G193AP and G194AP substitutions. **e** Ligand-bound state of the SAM-I riboswitch highlighting the A71AP (P2-P3 stack) and the A138AP (P1–P4 stack) substitutions. **f** Ligand-bound state of the SAM-II riboswitch and positions for the A14AP, U34AP and A41AP substitutions

ligand. This was explained by the presence of the shorter terminator stem on the *add* riboswitch, which allows a certain organization of the aptamer even when the sequestrator is formed (Fig. 11.4c). The *pbuE* riboswitch, with a longer terminator stem formed by complementary sequences from the expression platform and from the P1 and P3 stems of the aptamer, remained unchanged in the presence of adenine, most likely due to the high stability of the repressor sequence (Rieder et al. 2007; Lemay et al. 2006).

2AP labelling was also used to study the Mg^{2+}-induced folding and ligand binding of purine (Tremblay et al. 2011), lysine (Blouin et al. 2010), preQ1 (Rieder et al. 2010; Souliere et al. 2011), SAM-I (Heppell et al. 2011; Heppell and Lafontaine 2008; Eschbach et al. 2012), SAM-II (Haller et al. 2011; Souliere et al. 2011) and TPP (Lang et al. 2007) riboswitches. The lysine *lysC* riboswitch from *Bacillus subtilis* was studied by incorporating 2AP at the non-conserved G194 position (Fig. 11.4d). This conformer exhibited a notable increase in the fluorescence after the addition of ligand due to the transition of 2AP from a stacked configuration in the unfolded state to a more exposed conformation in the ligand-bound complex (Blouin et al. 2010). Further studies on the G193AP variant of the *lysC* riboswitch focused their attention on the binding of other amino acids and lysine analogues (Garst et al. 2012). The *lysC* riboswitch exhibits a >1000-fold specificity for lysine over any other amino acid investigated. Most of them, such as alanine, asparagine and threonine showed no increase in the fluorescence of the 2AP. Only serine showed some affinity ($K_D \sim 20$ mM), although this value is three orders of magnitude greater than its intracellular concentration ~70 μM (Bennett et al. 2009). The position of the charged ε-amino group on the side chain was thought to play a key role in the high specificity for lysine over other metabolites. The analysis of the G193AP variant showed that an analogue with a positive charge on the side chain only present a four- to fivefold higher affinity than two similar compounds lacking this charge. The small differences in the affinity for the ligand, and the fact that the neutral analogues should also be affected by a greater steric bulk, suggest that the charged ε-amino group is important for the ligand binding (Garst et al. 2012).

The SAM-I and SAM-II riboswitches were also investigated by intercalating a 2AP on key positions within the aptamer domain. SAM-I riboswitch is a four-way junction with two pairs of stacked stems, one between helices P1 and P4, and the second involving P2 and P3. The role of each of the stacks was investigated with two 2AP variants. A 2AP substitution was located at position A71 between the P2 and P3 stem, and a second substitution was located at position A138, which lies at the P1–P4 interface (Fig. 11.4e) (Heppell et al. 2011). The initial emission of the A71AP variant in the presence of Na^+ was quenched to a half of its initial signal upon addition of Mg^{2+} and no change was observed after the addition of SAM ligand. In contrast, the A138AP variant was unresponsive to the addition of Mg^{2+} and the addition of ligand induced a strong quenching of the 2AP fluorescence. This significant response was removed by incorporating the same substitution in the U107C mutant of the SAM-I riboswitch, which is known to be unable to bind the ligand (Heppell et al. 2011). Thus, the data from the 2AP experiments for the SAM-I riboswitch suggested the formation of a P2–P3 stack induced by Mg^{2+} ions and that the formation of the P1–P4 helical stacking was exclusively formed upon ligand binding (Heppell et al. 2011; St-Pierre et al. 2014).

The structure of the SAM-II riboswitch differs from the four-way junction of the SAM-I. The SAM-II riboswitch is composed of a H-type pseudoknot with two paired helices (P1 and P2a) and two loops (L1 and L3) (Gilbert et al. 2008). Three different variants were investigated using the potential of the 2AP labelling:

A14AP, to monitor the folding of the P2a stem and the sequestration of the Shine-Dalgarno (SD) sequence; U34AP, to monitor the P1/L3 folding; and A41, in the proximity of the ligand-binding site (Fig. 11.4f) (Haller et al. 2011b). The three variants showed similar results with an increase in 2AP fluorescence after addition of Mg^{2+} and a further increase upon with the addition of SAM ligand. However, kinetic assays showed that the rate constant for ligand binding was 1.8 times faster for A14AP than for the U34AP and A41AP variants, indicating that local changes in these regions are rate-limiting. Thus, the data from the thermodynamics and the kinetics of the SAM-II riboswitch suggest that the folding of the L1 loop starts the formation of the pseudoknot and allows ligand binding, with the subsequent formation of the L3 loop and the P2a helix (Haller et al. 2011b).

The use of 2AP for the study of riboswitches is not only restricted to substitutions within the mRNA sequence. Since the 2AP is a fluorescent analogue of adenine, it can be used as an intrinsically fluorescent ligand to characterise the RNA folding and ligand binding of adenine-sensing riboswitches in a similar way to the intrinsically fluorescent FMN ligand. The potential application of 2AP as a fluorescent sensor was effectively demonstrated for different wild-type adenine aptamers. The total emission of the 2AP was progressively quenched in the bound complex and reached a plateau at saturating aptamer concentrations (Lemay et al. 2006, 2009; Lemay and Lafontaine 2007; Wickiser et al. 2005; Gilbert et al. 2006). The data obtained for the *add* riboswitch from *Vibrio vulnificus*, and the *add* riboswitch for the *Clostridium perfringens* corroborate the ability of 2AP to act as an effective sensor for ligand binding. Interestingly, the aptamer from *V. vulnificus* with only two base pairs on its P1 stem showed a 54-fold less affinity for the ligand than the full length. This was taken as evidence that the length of the P1 stem is key to stabilize the ligand-bound complex (Lemay and Lafontaine 2007).

11.6.2 Single-Molecule FRET Studies of Riboswitch Function

11.6.2.1 Principle of Single-Molecule FRET

FRET is based on the non-radiative transfer of energy between a donor and an acceptor that are very close to one another (Stryer 1978). The transfer of resonant energy involves a weak interaction between the oscillating dipole moments of the donor and acceptor molecules and it requires a significant overlap between the absorption spectrum of the donor and the emission spectrum of the acceptor. The transfer of energy from the donor to the acceptor constitutes an additional deactivation pathway of the donor from the excited state that competes with the deactivation via radiative (fluorescence emission) and non-radiative pathways

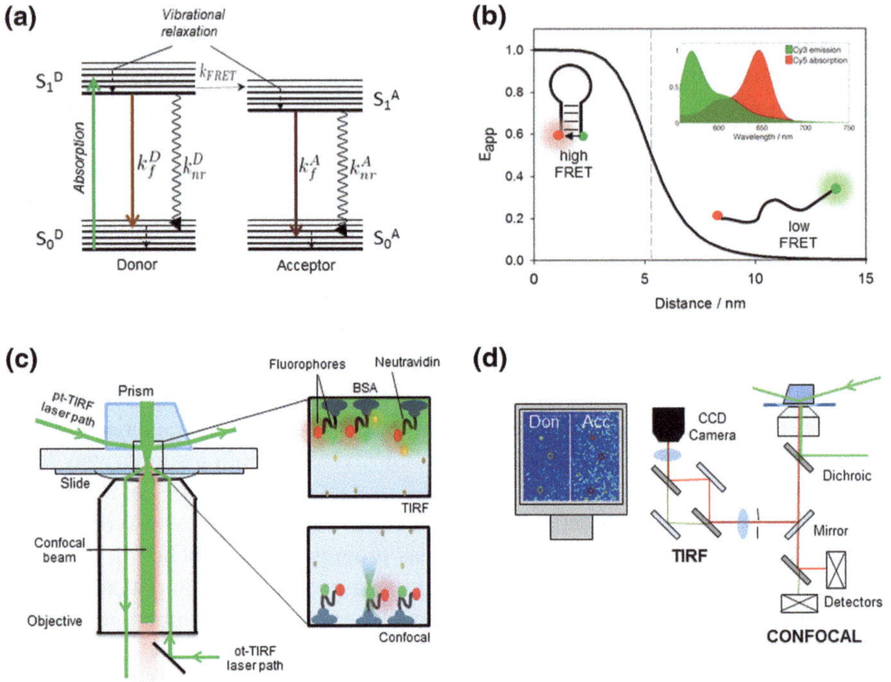

Fig. 11.5 **a** Jablonski diagram illustrating the FRET process. Absorption of light excites the donor molecule, which reaches its first excited electronic state, S_1^D. The molecule can then relax through fluorescence radiative emission (k_f^D), via non-radiative processes or undergo resonance energy transfer (k_{FRET}) to an acceptor in close proximity. The acceptor, on its excited state S_1^A, can also reach the ground state, S_0^A, by radiative (k_f^A) and non-radiative processes (k_{nr}^A). **b** FRET efficiency as a function of the separation between the popular Cy3 (donor)/Cy5 (acceptor) FRET pair. The FRET efficiency depends inversely on the sixth power of the distance (r^6) between the two dyes. *Right inset* Cy3 emission (*green*) and Cy5 absorption (*red*) spectra. The FRET efficiency directly depends on the spectral overlap between the selected dyes. **c** Diagram for single-molecule confocal and TIR microscopy on its objective- (ot-TIRF) and prism-type (pt-TIRF) variants. TIR is a wide-field technique in which a evanescent wave is created on the of the sample channel with a penetration depth of ~150 nm, whilst confocal microscopy excites one molecule at a time at the focal spot of the laser. **d** Excitation and emission pathways for TIR and confocal FRET assays. The emission signal in a TIR setup is collected with an electron-multiplying CCD camera and for confocal techniques avalanche photodiodes are employed providing a faster integration time

(Fig. 11.5a). The rate of energy transfer depends inversely on the sixth power of the distance between the donor and the acceptor (R) (Fig. 11.5b), making the technique very sensitive to small changes in inter-dye distance (Blouin et al. 2009b) (Eq. 11.1).

$$E = \frac{1}{1 + \left(\frac{R}{R_o}\right)^6} \tag{11.1}$$

R_0 is the Förster radius and represents the distance between the donor and the acceptor dyes resulting in 50 % of energy transfer efficiency. The R_0 value is characteristic for a given FRET pair and it can be calculated from the expression (Eq. 11.2):

$$R_0^6 = \frac{9000 ln 10\, \kappa^2 Q_D}{128\pi^5 N n^4} \int_0^\infty F_F(\lambda)\varepsilon_A(\lambda)\lambda^4 d\lambda \tag{11.2}$$

where Q_D represents the quantum yield of the donor, k is a dipole orientation factor which is normally assumed to have a value of two-third for isotropic orientation of the dyes, n is the refractive index of the medium and N is the Avogadro's number. The integral term in this expression represents the overlap between the emission and the acceptor and includes the normalized emission spectrum of the donor (F_D) and the absorption spectrum of the acceptor expressed in units of extinction coefficient (ε_A). Overall, the efficiency of the FRET process depends on the distance between the donor and the acceptor, their spectral overlap, the relative orientation of their dipole moments and the quantum yield of the donor (Heppell et al. 2009; Blouin et al. 2009b; Penedo et al. 2004).

The basic idea underlying any FRET experiment designed to study the tertiary folding pathway of a riboswitch aptamer relies on the site-specific labelling of the RNA structure with a donor and an acceptor fluorophore that can have efficient energy transfer between them. The most commonly used FRET pair for ensemble studies involve fluorescein acting as a donor and Cy3 acting as acceptor ($R_0 \sim 56$ Å). For single-molecule studies and due to the low photostability of fluorescein, the fluorescein/Cy3 FRET pair is replaced by Cy3 acting as donor and Cy5 acting as acceptor ($R_0 \sim 53$ Å). These fluorescent reporters can be incorporated at specific locations of the RNA sequence either during the solid-phase synthesis of the RNA sequence of post-synthetically using the coupling reaction between an N-hydroxysuccinimide ester (NHS) group of the dye or an amino linker group placed at the desired position on the RNA (Blouin et al. 2009b; McCluskey et al. 2014). The labelling step is followed by a purification step using high performance liquid chromatography (HPLC) or polyacrylamide gel electrophoresis (PAGE) to separate labelled from unlabelled RNA and also the unreacted dye (Blouin et al. 2009b). It is important to make sure that the fluorescent labelling of the riboswitch aptamer does not affect its function and structure. If available, the crystal structures of the RNA–ligand complex are particularly useful as they allow to determine those nucleotides that are fully exposed to the solvent and therefore potential good candidates for labelling.

The majority of single-molecule FRET studies involving riboswitch aptamers have been performed using total-internal reflection (smTIR) illumination methods and surface-immobilization techniques to allow the observation of folding trajectories and binding events for long periods of time (Fig. 11.5c, d). The most common method to immobilize the RNA aptamer on the surface of a quartz slide exploits the use of biotin–streptavidin interactions (McCluskey et al. 2014; Shaw et al. 2014). For this, an additional functionalization of the RNA sequence on the 5′ or 3′ end terminal of the RNA with a biotin groups needs to be carried out and this is usually incorporated during the solid-phase synthesis of the RNA. A detailed description of the immobilization methods and the wide-field single-molecule setup required to measure FRET using a single-molecule TIR approach has been extensively covered in many reviews (McCluskey et al. 2014; Perez-Gonzalez and Penedo 2015; Roy et al. 2008). A combination of ensemble-FRET and single-molecule FRET was used to study the folding pathway of some bacterial riboswitches including purine (Brenner et al. 2010; Dalgarno et al. 2013; Lemay et al. 2006), SAM-I (Heppell et al. 2011; Eschbach et al. 2012), SAM-II (Haller et al. 2011b), lysine (Fiegland et al. 2012), TPP (Haller et al. 2013), c-di-GMP (Wood et al. 2012) and preQ1 (Suddala et al. 2013) riboswitches. In the next section, we will focus on how the use of smFRET has contributed to unravel the folding pathways and ligand binding mechanisms using the adenine aptamer as a model example, but the reader is referred to recent reviews on the field for more detailed information regarding single-molecule studies on the remaining RNA aptamers (St-Pierre et al. 2014; Boudreault et al. 2015; Haller et al. 2011a; Savinov et al. 2014).

11.6.2.2 SmFRET Studies of Adenine-Sensing Riboswitch Aptamers

Adenine aptamers are organized as a three-way junction connecting three helical domains (P1, P2 and P3), as shown in Figs. 11.4a and 11.6a. Crystallographic (Montange and Batey 2006), biochemical and NMR data on the *xpt* adenine aptamer indicated the formation of a long-range tertiary interaction between the loops located in the P2 and P3 stems. To investigate this loop–loop interaction and the factors modulating it, a Cy3/Cy5 pair was located at positions U27 of the P2 loop and U53 of the P3 loop of the *Vibrio vulnificus add* aptamer and the FRET efficiency was measured in single immobilized aptamers as a function of Mg^{2+} ions and ligand (Fig. 11.6a) (Lemay et al. 2006, 2009; Tremblay et al. 2011).

The single-molecule histograms obtained as a function of increasing concentrations of Mg^{2+} showed a shift from a low-FRET state centred at $E_{app} \sim 0.3$ (unfolded state, U) in the absence of Mg^{2+} to a value of $\sim$0.85 at 2 mM concentration of Mg^{2+} ions (ligand-free folded state, F_{LF}) (Lemay et al. 2006). This shortening of the distance between the dyes provided strong evidence for the formation of the loop–loop interaction in the absence of ligand and confirmed that divalent metal ions are sufficient to promote this long-range tertiary interaction. Interestingly, in addition to the U and F_{LF} states, it was found that the folding pathway between these states takes place via an intermediate transient state (I) with

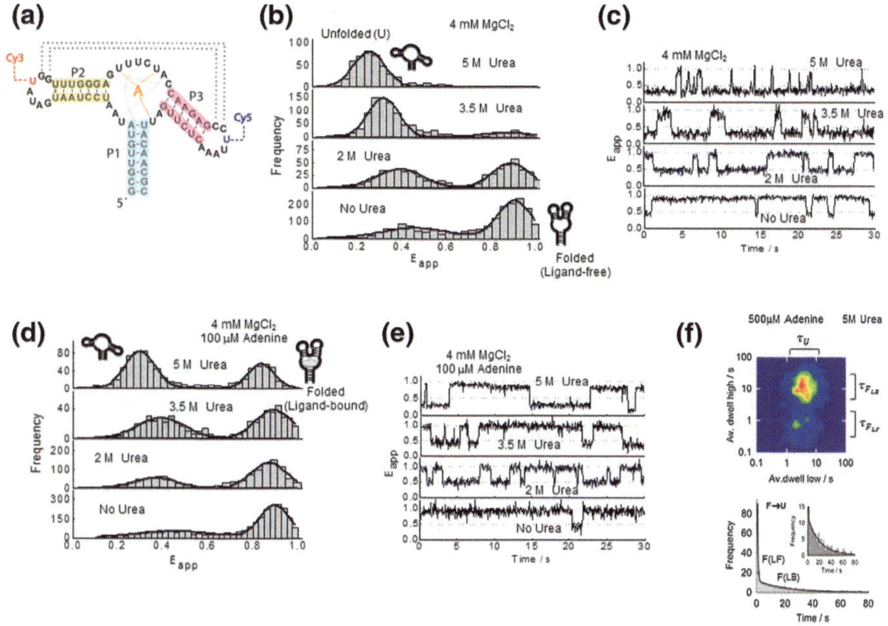

Fig. 11.6 **a** Schematic of the adenine aptamer showing the aptamer nucleotides involved in ligand recognition (*orange*) and stabilization of the ligand-bound conformation (*grey*). The dye-labelling positions and base-pair interactions involved in the formation of the loop–loop interaction are also shown (*dotted line*). **b** Single-molecule FRET histograms for the adenine riboswitch as a function of concentration of urea in a background of 4 mM $MgCl_2$. **c**.Representative FRET trajectories of the adenine aptamer in the presence of 4 mM $MgCl_2$ and increasing concentrations of urea illustrating the transition from the high-FRET to the low-FRET upon addition of the denaturant. **d** Single-molecule FRET histograms for the adenine aptamer in a background of 4 mM $MgCl_2$ and 0.1 mM adenine ligand. **e** Representative FRET trajectories of the adenine aptamer in the presence of 4 mM $MgCl_2$ and 0.1 mM adenine ligand for increasing concentrations of urea. **f** (*Top*) Two-dimensional contour plot for the adenine riboswitch representing the averaged dwell times of the folded (in both ligand-free and ligand-bound states) and unfolded conformation at 4 mM $MgCl_2$, 0.5 mM adenine and 5 M urea. *Bottom* Dwell-time histograms of the unfolding events fitted to a bi-exponential function (*solid line*), reporting a $2.1 \pm 0.1\ s^{-1}$ and $0.045 \pm 0.003\ s^{-1}$ rate constants for the ligand-free and ligand-bound states, respectively.

an $E_{app} \sim 0.5$ (Lemay et al. 2006). Transitions between the three states (U, I and F_{LF}) were observed within single aptamers, most easily at relatively low concentrations of Mg^{2+} ions (~20–100 μM). An analysis of 846 single-molecule FRET trajectories demonstrated that only a 6 % of them showed direct transitions from the unfolded to the folded conformation, suggesting that the intermediate structure constitutes and obligatory step in the folding pathway. By comparing the dwell times in the folded and unfolded states (U and I) with and without ligand added, which correspond to the inverse of the unfolding and folding rates, respectively, the authors uncovered a significant decrease in the dynamic heterogeneity of the riboswitch aptamer induced by ligand binding. In the absence of ligand, the average folding and unfolding rates varied from RNA aptamer to RNA aptamer by more than two orders of magnitude. In contrast, the addition of 50 μM concentration of adenine ligand reduced this heterogeneity in both rates by tenfold. Both findings, the discovery of an intermediate folding state as an obligatory step and the observed heterogeneity, were only possible using a single-molecule approach and highlight the ability of these techniques to get insights that otherwise will remain hidden in conventional ensemble measurements.

In this study, the fact that the ligand-free and the ligand-bound states exhibited identical values of FRET efficiency precluded a further investigation into the ligand-binding dynamics and the influence of the ligand in the folding mechanism. This was later addressed in a study reporting, for the first time, the combination of smFRET with chemical denaturation (Dalgarno et al. 2013). It is important to emphasize that although the use of chemical denaturants such as urea and guanidinium chloride constitutes a common procedure to investigate protein unfolding (Nienhaus 2009; Schuler and Eaton 2008), its application in the RNA folding field was mostly unexploited. smFRET applied to investigate RNA conformational changes has focused entirely on understanding the influence on mono- and divalent metal ions as folding enhancers and stabilizers of the tertiary structure (Bokinsky and Zhuang 2005). Using a similar labelling strategy of the *add* adenine aptamer to monitor the formation of the loop–loop interaction, it was shown that the use of concentrations as high as 5 M urea does not disrupt the biotin–streptavidin interaction used for immobilization strategies and the small change in refraction index compared to aqueous solution is still compatible with TIR illumination. By placing the immobilized aptamers in the folded conformation using a 4 mM concentration of Mg^{2+} ions, the addition of urea induced the transition to an unfolded state with an identical FRET value ($E_{app} \sim 0.3$) to that observed in the absence of Mg^{2+} ions (Fig. 11.6b, c). Importantly, chemical denaturation of the tertiary structure was a reversible process and immobilized aptamers could be repetitively folded by addition of metal ions and unfolded by injection of urea in a cyclic manner. Moreover, the presence of an intermediate state during urea-induced unfolding, with a FRET efficiency also similar to that observed for the folding transition, was also observed. These findings confirmed that, at least for the adenine aptamer, urea-stimulated unfolding was a mirror image of the Mg^{2+}-induced folding process, with the same number and structural intermediates being involved in both processes.

Intuitively, the formation of the RNA–ligand complex should stabilize the tertiary folded structure, and in fact this constitutes the basis of the regulation mechanism (Fig. 11.1). However, due to the high stability of the ligand-bound complex in most riboswitch aptamers, recording a significant number of transitions from this state to the ligand-free unfolded state before the donor or the acceptor photobleach has been a difficult task. As a result, quantifying the degree of stabilization due to ligand binding by comparing the unfolding rates, with and without ligand, has not been easily accessible using conventional smFRET. Dalgarno and co-workers demonstrated that the different response of the ligand-free and the ligand-bound states towards urea-induced denaturation can be used to determine the degree of stabilization induced by ligand-binding and to differentiate both states even when they have identical FRET values (Fig. 11.6d, e) (Dalgarno et al. 2013; Shaw et al. 2014).

In a mixture of 4 mM Mg^{2+}, 5 M urea and 0.5 mM adenine ligand, a biexponential decay was observed with two unfolding rates (Fig. 11.6f). The fast component ($\sim 2\ s^{-1}$) was assigned to urea-induced unfolding events taking place form the ligand-free state, whereas the slow component ($\sim 0.045\ s^{-1}$), whose contribution increased as the ligand concentration was increased, was assigned to urea-induced unfolding events taking place from the ligand-bound state (Fig. 11.6f). The ratio between both rates provided a 50-fold value for the degree of stabilization of the adenine aptamer due to ligand binding, which is very similar to the value obtained by single-molecule force methods (Neupane et al. 2011; Greenleaf et al. 2008).

When the urea-induced unfolding rates allowed to determine the ligand-induced stabilization of the aptamer tertiary structure, a similar investigation on the influence of urea on the folding rates provided insights into the nature of the rate-limiting folding step, with and without ligand, and the influence of ligand binding at early stages of the folding process. For the ligand-free state, the rate for the formation of the loop–loop interaction decreased moderately ($\sim$2–3 fold) with the addition of urea and the relative change in the presence and absence of urea was independent on the concentration of Mg^{2+} ions. The influence of urea on the loop–loop interaction was similar to that found for the hairpin ribozyme that also relies on the docking of two peripheral loops for efficient folding. It was hypothesized that in both RNA elements, productive encounters between both loops leading to the formation of key tertiary contacts may constitute the rate-limiting step for ligand-free folding. In the presence of ligand, urea also decreased the rate for the loop–loop docking, but in contrast to the observations in the absence of ligand, the influence of urea on the docking rate was strongly dependent on the concentration of Mg^{2+} ions and more pronounced ($\sim$eightfold) at 0.1 mM than at 4 mM ($\sim$twofold) concentration of Mg^{2+} ions. Because Mg^{2+} ions trapped at specific positions have been observed in the crystal structure of the RNA–ligand complex, two of them located near the ligand-binding pocket (Serganov et al. 2004), it was suggested that the folding pathway in the presence of ligand involves a rate-limiting step that requires the assistance of specifically positioned Mg^{2+} ions to stabilize the native state.

The adenine riboswitch constitutes a good model example to describe how the application of smFRET and its combination with chemical denaturation can probe the details of ligand binding on the folding dynamics. However, because smFRET is limited to monitoring conformational changes within a relatively small distance range (2–8 nm), it is not possible to quantify the changes in the entire RNA structure as it folds. Here, single-molecule techniques capable of monitoring variations in total length of the RNA structure as it progressively folds offer a complementary approach. In the next section, we describe the application of single-molecule force (SMF) microscopy to extract information about the complete RNA folding landscape, the influence of ligand binding and how this may influence the position and high of the energy barriers involved in the transition from the unstructured, unliganded aptamer to the native state.

11.6.3 Single-Molecule Force Studies of Riboswitch Folding

11.6.3.1 Principle of Single-Molecule Force-Based Techniques

The principle of single-molecule force (SMF) techniques is based on the concept that when a small transparent dielectric particle of glass or polystyrene interacts with an intense diffraction-limited focused laser beam, usually on the near infrared region, significant forces can be generated to trap the object. The tightly focused laser beam creates a steep intensity gradient and a particle in the immediate vicinity of this trap experiences a force pulling the particle towards the brighter region of the intensity gradient. The theory, analysis and technical aspects of optical trapping and its application to single molecules has been reviewed detailed in several excellent papers (Dudko et al. 2008; Neuman and Block 2004). For a particle trapped in the highest intensity region of the beam, the optical trap behaves like a spring and thus, after careful calibration of the spring constant, it is possible to apply forces and induced particle displacements in a known and controlled manner. The two most common SMF setups are single-beam and dual-beam trap. In a single-beam setup, the micrometre sized trapped particle is functionalized with a single-stranded nucleic acid handle can hybridize with a specific terminal region of the RNA or DNA under study. The other end of the RNA or DNA is attached to the quartz slide using biotin–streptavidin interactions or alternative methods. In the dual-beam setup, the RNA sample is placed between two optically trapped beads using handle with different sequences at each particle. In an SMF experiment, variations in the end-to-end distance of the trapped RNA as a function of applied force can be used to track the folding pathway and reveal intermediate states and their lifetime. The assignation of the observed states to specific RNA tertiary and secondary structures can be done based on their different extension. Species with identical extension such as the ligand-bound and ligand-free forms of the adenine aptamer can be differentiated using a similar concept to the chemical denaturation methods, but in this case replacing the different stability against urea by the different resistance to unfolding by an applied force.

Two different kinds of experiments can be performed using single-molecule force: near equilibrium measurements of extension under constant-force and non-equilibrium experiments using force ramps (Savinov et al. 2014). The first involves applying a constant force and measuring the fluctuations in extension between different conformational states of the RNA and their respective lifetimes. By varying the applied force, in known steps, all the intermediate structures present along the unfolding landscape is explored. In the force-ramp experiment, a monotonically increasing force is applied to the sample over time (Woodside and Block 2014). The result is a force-extension (FEC) curve that shows abrupt discontinuities corresponding to sudden unfolding events (i.e. the disruption of a helical stem). The smooth portions of the FEC curve can be fitted to a polymer elasticity model such as the worm-like chain (WLC) to extract the change in extension and the number of base pairs disrupted associated within that specific unfolding event. To date, single-molecule force analysis of riboswitch structure has been performed on the adenine- (Frieda and Block 2012; Greenleaf et al. 2008; Neupane et al. 2011) and TPP-sensing riboswitches (Anthony et al. 2012).

11.6.3.2 Single-Molecule Force Studies on the Adenine Aptamer

The application of single-molecule optical force to investigate the folding landscape of a riboswitch was pioneered by S. Block and co-workers using the *pbuE* aptamer from *B. subtillis* (Greenleaf et al. 2008) under constant-force and force-ramp regimes. In their initial configuration, they used a dual-beam optical trap to monitor the conformational landscape of an add riboswitch being transcribed in situ. The RNA polymerase from *E. coli* was attached to one of the beads and transcriptionally stalled downstream of promoter site. Whilst stalled, the initial 29 nt transcript emerging from the RNAP was hybridized to a DNA handle attached to the other bead, thus providing a dual-beam 'dumbbell' geometry to monitor FEC between the RNAP and the transcribed RNA once the transcription was restarted by introducing nucleoside triphosphates. To avoid disassembly of the complex once the aptamer sequence was transcribed, a road block consisting of a streptavidin molecule bound to a biotin group placed at the 5′ termini of the DNA template.

Using a combination of constant-force and force-ramp methods, a transcriptional landscape consisting of five distinct states was observed. In addition to demonstrating that the formation of the P2 and P3 loops takes place before the assembly of the ligand-binding pocket, they uncovered a ligand-binding intermediate state, termed A-comp, formed prior to the native fully folded state. A later study of the *add* aptamer from *V. vulnificus* (Neupane et al. 2011) also revealed a similar five-state free energy folding landscape and an intermediate conformation, termed P1_U, located before the fully folded state and after the formation of the P2 and P3 stems. It was suggested that this P1_U state may arise from the docking of the P2 and P3 stem loops and the RNA junction partially organized but with the P1 stem still unfolded. To date, it remains to be determined whether the intermediate state (I) observed by smFRET (Lemay et al. 2006) and this P1_U state observed by single-molecule force are somehow structurally related. The single-molecule free energy folding trajectory obtained for both aptamers suggests that weak tertiary interactions are a key to promote the long-range conformational changes needed to close the P1 switching stem. Thus, both aptamers can be added to the growing number of examples challenging the generality of a strictly hierarchical RNA folding model, where secondary structures are typically formed first and followed by the formation of tertiary contacts (Cruz and Westhof 2009).

It is interesting to note that these single-molecule force experiments detected significant differences for the transition from a pre-organized P1_U (*add* aptamer) or A-comp (*pbuE* aptamer) to the fully folded native state. These differences were not only on the degree of stabilization of the fully folded state provided by ligand-binding, which was 4 kcal mol^{-1} for the *pbuE* aptamer compared to

8 kcal mol^{-1} for the *add*, but also on the nature of key interactions involved in transition from P1_U or A-comp to the ligand-bound state, whereas for the *pbuE* aptamer, a GC base pair at the base of the P1 stem is critical for junction stabilization and once disrupted the entire aptamer falls apart, for the *add* aptamer the position of the transition state suggested that the structural keystone may involve a UA-A triplex. These results suggest that aptamer sequences may have evolutionary evolved not only to discriminate their cognate ligands with high specificity, but also carefully optimized to modulate the stability of the P1 switching sequence and fine-tune the regulatory response.

A latter study using single-molecule force has analyzed, for the first time, the co-transcriptional folding landscape of a riboswitch element (Frieda and Block 2012). Using a similar dumbbell geometry as previously described, with the RNAP attached to one of the beads and the nascent RNA hybridized to a handle that was attached to the other bead, the authors recorded the variations in the distance between the two beads as a function of time as a way to investigate the co-transcriptional folding landscape of the pbuE riboswitch from *B. subtillis*. In the context of co-transcriptional folding, if the *pbuE* aptamer binds adenine, an antiterminator stem is formed and a run-through product is formed. Alternatively, if the aptamer–ligand complex is not formed, transcription of the expression platform will result in the formation of a terminator hairpin structure that will stop transcription. The authors observed at sufficiently high forces (> 8 pN) that preclude the formation of the ligand binding pocket, the time-extension curves (TECs) were the same in the presence or absence of adenine. However, at lower loads (~5.8 pN) and in the presence of adenine, most TECs exhibited a characteristic aptamer folding event that appeared as a large decrease in extension and 95 % of these TECs displayed run-through events. In the absence of adenine, the probability of transcriptional termination when the aptamer-fold signature absent was ~85 % and run-through events in these conditions exhibited only 15 % probability. Importantly, this study confirmed that the regulatory function of the *pbuE* riboswitch is kinetically controlled because aptamers were observed to fold and bind the adenine ligand only once, or not at all. If the ligand-bound aptamer did not form by the time the RNAP reached the decision point, the terminator hairpin was formed and transcription was terminated. Although this work constitutes, to date, the only co-transcriptional study of riboswitch function, the methodology and procedures used in this work could be adapted to investigate a variety of other co-transcriptional biological processes including RNA-protein interactions and chaperone-assisted RNA folding.

11.7 Future Perspectives

The discovery of riboswitches has shown, once again, that RNA can perform biological functions that previously were thought to be exclusive to proteins. Without doubt, this has changed our perception of non-coding RNA and has triggered a drastic increase in our efforts to decipher the complexity and plasticity of functional RNA sequences. More recently, the discovery of riboswitches working in tandem as logic gates to control gene expression has expanded even further our interest on using these structures for biotechnological applications. However, for this field to crystallise, our understanding of riboswitch function still faces many challenges that need to be addressed. Crucially, many putative riboswitch sequences have been identified for which no metabolite has been found. Here, significant efforts need to be directed towards novel biochemical strategies for metabolite identification.

At the fundamental level, the fact that riboswitch function relies on the balance between multiple factors including transcription rate, ligand binding kinetics and folding pathway requires biophysical and biochemical techniques that can interrogate the full-length riboswitch in a co-transcriptional context. As we have shown, single-molecule techniques are well suited for this task but so far the two most commonly used techniques, smFRET and smForce, have been applied separately. New hybrid methodologies combining both are needed to monitor simultaneously global changes in end-to-end distance (smForce) and assign these changes to specific conformations (smFRET). Another important aspect is the development of fluorescently labelled ligands to allow the direct monitoring of ligand binding dynamics so that essential knowledge during the sensing phase of riboswitch regulation can be gathered. Future advances in these areas should enable the development of more detailed models of riboswitch function and structure; a knowledge that will help to explore and deliver a new range of antibiotics and biotechnological applications.

Acknowledgments The authors thank all members of Penedo's and Lafontaine's groups for critical reading of the manuscript and for helpful discussions.

References

Albert A, Taguchi H (1973) Ultraviolet spectral correlation between the cation of an amino-N-heteroaromatic compound and the neutral species of the corresponding oxo-analogue. J Chem Soc Perkin Trans 1101–1103

Ames TD, Breaker RR (2011) Bacterial aptamers that selectively bind glutamine. RNA Biol 8:82–89

Ames TD, Rodionov DA, Weinberg Z, Breaker RR (2010) A eubacterial riboswitch class that senses the coenzyme tetrahydrofolate. Chem Biol 17:681–685

Anthony PC, Perez CF, Garcia-Garcia C, Block SM (2012) Folding energy landscape of the thiamine pyrophosphate riboswitch aptamer. Proc Natl Acad Sci USA 109:1485–1489

Baker JL, Sudarsan N, Weinberg Z, Roth A, Stockbridge RB, Breaker RR (2011) Widespread genetic switches and toxicity resistance proteins for fluoride. Science 335:233–235

Barrick JE, Breaker RR (2007) The distributions, mechanisms, and structures of metabolite-binding riboswitches. Genome Biol 8:R239

Batey RT, Gilbert SD, Montange RK (2004) Structure of a natural guanine-responsive riboswitch complexed with the metabolite hypoxanthine. Nature 432:411–415

Bennett BO, Kimball EH, Gao M, Osterhout R, Van Dien SJ, Rabinowitz JD (2009) Absolute metabolite concentrations and implied enzyme active site occupancy in *Escherichia coli*. Nat Chem Biol 5:593–599

Blouin S, Mulhbacher J, Penedo JC, Lafontaine DA (2009a) Riboswitches: ancient and promising genetic regulators. ChemBioChem 10:400–416
Blouin S, Craggs TD, Lafontaine DA, Penedo JC (2009b) Functional studies of DNA-protein interactions using FRET techniques. Methods Mol Biol 543:475–502
Blouin S, Chinnappan R, Lafontaine DA (2010) Folding of the lysine riboswitch: importance of peripheral elements for transcriptional regulation. Nucleic Acids Res 39:3373–3387
Blount KF, Breaker RR (2006) Riboswitches as antibacterial drug targets. Nat Biotechnol 24:1558–1564
Blount KF, Wang JX, Lim J, Sudarsan N, Breaker RR (2007) Antibacterial lysine analogues that target lysine riboswitches. Nat Chem Biol 3:44–49
Bokinsky G, Zhuang X (2005) Single-molecule RNA folding. Acc Chem Res 38:566–573
Boudreault J, Perez-Gonzalez DA, Penedo JC, Lafontaine DA (2015) Single-molecule approaches for the characterization of riboswitch folding mechanisms. Methods Mol Biol 1334:101–107
Breaker RR (2012) Riboswitches and the RNA World. Cold Spring Harb Perspect Biol 4:003566
Brenner MD, Scanlan MS, Nahas MK, Ha T, Silverman SK (2010) Multivector fluorescence analysis of the xpt guanine riboswitch aptamer domain and the conformational role of guanine. Biochemistry 49:1596–1605
Butler EB, Xiong Y, Wang J, Strobel SA (2011) Structural basis of cooperative ligand binding by the glycine riboswitch. Chem Biol 18:293–298
Caron MP, Bastet L, Lussier A, Simoneau-Roy M, Massé E, Lafontaine DA (2012) Dual-acting riboswitch control of translation initiation and mRNA decay. Proc Natl Acad Sci USA 109: E3444–E3453
Cruz JA, Westhof E (2009) The dynamic landscapes of RNA architecture. Cell 136:604–609
Dalgarno PA, Bordello J, Morris R, St-Pierre P, Dubé A, Samuel IDW, Lafontaine DA, Penedo JC (2013) Single-molecule chemical denaturation of riboswitches. Nucleic Acids Res 41:4253–4265
Dann CE 3rd, Wakemann CA, Sieling CL, Baker SC, Irnov I, Winkler WC (2007) Structure and mechanism of a metal-sensing regulatory RNA. Cell 130:878–892
DebRoy S, Gebbie M, Ramesh A, Goodson JR, Cruz MR, van Hoof A, Winkler WC, Garsin DA (2014) A riboswitch-containing sRNA controls gene expression by sequestration of a response regulator. Science 345:937–940
Draper DE, Grilley D, Soto AM (2005) Ions and RNA folding. Annu Rev Biophys Biomol Struct 34:221–243
Dudko OK, Hummer G, Szabo A (2008) Theory, analysis, and interpretation of single-molecule force spectroscopy experiments. Proc Natl Acad Sci USA 105:15755–15760
Edwards TE, Ferré d'Amaré AR (2006) Crystal structures of the thi-box riboswitch bound to thiamine pyrophosphate analogues reveal adaptive RNA-small molecule recognition. Structure 14:1459–1468
Eschbach SH, St-Pierre P, Penedo JC, Lafontaine DA (2012) Folding of the SAM-I riboswitch: a tale with a twist. RNA Biol 9:535–541
Fiegland LR, Garst AD, Batey RT, Nesbitt DJ (2012) Single-molecule studies of the lysine riboswitch reveal effector-dependent conformational dynamics of the aptamer domain. Biochemistry 51:9223–9233
Frieda KL, Block SM (2012) Direct observation of cotranscriptional folding in an adenine riboswitch. Science 338:397–400
Furukawa K, Ramesh A, Zhou Z, Weinberg Z, Vallery T, Winkler WC, Breaker RR (2015) Baterial riboswitches bind Ni^{2+} or Co^{2+} ions and control expression of heavy metal transporters. Mol Cell 57:1088–1098
Garst AD, Edwards AL, Batey RT (2011) Riboswitches: structures and mechanisms. Cold Spring Harb Perspect Biol 4:a003533
Garst AD, Porter EB, Batey RT (2012) Insights into the regulatory landscape of the lysine riboswitch. J Mol Biol 423:17–33
Gilbert SD, Stoddard CD, Wise SJ, Batey RT (2006) Thermodynamic and kinetic characterization of ligand binding to the purine riboswitch aptamer domain. J Mol Biol 359:754–768

Gilbert SD, Rambo RP, Van Tyne D, Batey RT (2008) Structure of the SAM-II riboswitch bound to S-adenosylmethionine. Nat Struct Mol Biol 15:177–182

Greenleaf WJ, Frieda KL, Foster DAN, Woodside MT, Block SM (2008) Direct observation of hierarchical folding in single riboswitch aptamers. Science 319:630–633

Grill S, Busenbender S, Pfeiffer M, Köhler U, Mack M (2008) The bifunctional flavokinase/flavin adenine dinucleotide synthetase from Streptomyces davawensis produces inactive flavin cofactors and is not involved in resistance to the antibiotic roseoflavin. J Bacteriol 190:1546–1553

Grundy FJ, Lehman SC, Henkin TM (2003) The L box regulon: lysine sensing by leader RNAs of bacterial lysine biosynthesis genes. Proc Natl Acad Sci USA 100:12057–12062

Ha T (2001) Single-molecule fluorescence energy transfer. Methods 25:78–86

Haller A, Rieder U, Aigner M, Blanchard SC, Micura R (2011a) Conformational capture of the SAM-II riboswitch. Nat Chem Biol 7:393–400

Haller A, Soulière MF, Micura R (2011b) The dynamic nature of RNA as key to understanding riboswitch mechanisms. Acc Chem Res 44:1339–1348

Haller A, Altman RB, Souliere MF, Blanchard SC, Micura R (2013) Folding and ligand recognition of the TPP riboswitch aptamer at single-molecule resolution. Proc Natl Acad USA 110:4188–4193

Henkin TM (2008) Riboswitch RNAs: using RNA to sense cellular metabolism. Genes Dev 22:3383

Heppell B, Lafontaine DA (2008) Folding of the SAM aptamer is determined by the formation of a K-turn dependent pseudoknot. Biochemistry 47:1490–1499

Heppell B, Mulhbacher J, Penedo JC, Lafontaine DA (2009) Application of fluorescence measurements for the characterization of riboswitch-ligand interactions. Methods Mol Biol 540:25–37

Heppell B, Blouin S, Dussault AM, Mulhbacher J, Ennifar E, Penedo JC, Lafontaine DA (2011) Molecular insights into the ligand-controlled organization of the SAM-I riboswitch. Nat Chem Biol 7:384–392

Hermann T (2015) Non-coding RNA: antibiotic tricks a switch. Nature 526:650–651

Hollands K, Proshkin S, Sklyarova S, Epshtein V, Mironov A, Nudler E, Groisman EA (2012) Riboswitch control of Rho-dependent transcription termination. Proc Natl Acad Sci USA 109:5376–5381

Hollands K, Sevostiyanova A, Groisman EA (2014) Unusually long-lived pause required for regulation of a Rho-dependent transcription terminator. Proc Natl Acad Sci USA 111:E1999–E2007

Holmén A, Nordén B, Albinsson B (1997) Electronic transition moments of 2-aminopurine. J Am Chem Soc 119:3114–3121

Howe JA, Wang H, Fischmann TO et al (2015) Selective small-molecule inhibition of an RNA structural element. Nature 526:672–677

Huang L, Serganov A, Patel DJ (2010) Structural insights into ligand recognition by a sensing domain of the cooperative glycine riboswitch. Mol Cell 40:774–786

Jansen JA, McCarthy TJ, Soukup GA, Soukup JK (2006) Backbone and nucleobase contacts to glusocamine-6-phosphate in the glmS ribozyme. Nat Struct Mol Biol 13:517–523

Kim JN, Blount KF, Puskarz I, Lim J, Link KH, Breaker RR (2009) Design and antimicrobial action of purine analogues that bind Guanine riboswitches. ACS Chem Biol 4:915–927

Kladwang W, Chou FC, Das R (2012) Automated RNA structure prediction uncovers a kink-turn linker in double glycine riboswitches. J Am Chem Soc 134:1404–1407

Kwon M, Strobel SA (2008) Chemical basis of glycine riboswitch cooperativity. RNA 14:25–34

Lang K, Rieder R, Micura R (2007) Ligand-induced folding of the thiM TPP riboswitch investigated by a structure-based fluorescence spectroscopic approach. Nucleic Acids Res 35:5370–5378

Lee ER, Blount KF, Breaker RR (2009) Roseoflavin is a natural antibacterial compound that binds to FMN riboswitches and regulates gene expression. RNA Biol 6:187–194

Lee ER, Blount KF, Breaker RR (2011) Metabolite-sensing riboswitches as antibacterial drug targets. In: Miller AA, Miller PF (eds) Emerging trends in antibacterial discovery: answering the call to arms. Caister Academic Press, Norfolk, pp 107–130

Lemay JF, Lafontaine DA (2007) Core requirements of the adenine riboswitch aptamer for ligand binding. RNA 13:339–350

Lemay JF, Penedo JC, Tremblay R, Lilley DM, Lafontaine DA (2006) Folding of the adenine riboswitch. Chem Biol 13:857–868

Lemay JF, Penedo JC, Mulhbacher J, Lafontaine DA (2009) Molecular basis of RNA-mediated gene regulation on the adenine riboswitch by single-molecule approaches. Methods Mol Biol 540:65–76

Lilley DMJ, Eckstein F (eds) (2008) Ribozymes and RNA catalysis. Royal Society of Chemistry

Lilley DMJ, Sutherland J (2011) The chemical origins of life and its early evolution: an introduction. Philos Trans R Soc Lond B Biol Sci 366:2853–2856

Lipfert J, Doniach S (2007) Small-angle X-ray scattering from RNA, proteins, and protein complexes. Annu Rev Biophys Biomol Struct 36:307–327

Lipfert J, Das R, Chu VB, Kudaravalli M, Boyd N, Herschlag D, Doniach S (2007) Structural transitions and thermodynamics of a glycine-dependent riboswitch from *Vibrio cholerae*. J Mol Biol 365:1393–1406

Lipfert J, Sim AY, Herschlag D, Doniach S (2010) Dissecting electrostatic screening, specific ion binding, and ligand binding in an energetic model for glycine riboswitch folding. RNA 16:708–719

Lünse CE, Schüller A, Mayer G (2014) The promise of riboswitches as potential antibacterial drug targets. Int J Med Microbiol 304:79–92

Mandal M, Breaker RR (2004a) Adenine riboswitches and gene activation by disruption of a transcription terminator. Nat Struct Mol Biol 11:29–35

Mandal M, Breaker RR (2004b) Gene regulation by riboswitches. Nat Rev Mol Cell Biol 5:451–463

Mandal M, Boese B, Barrick JE, Winkler WC, Breaker RR (2003) Riboswitches control fundamental biochemical pathways in *Bacillus subtilis* and other bacteria. Cell 113:577–586

Mandal M, Lee M, Barrick JE, Weinberg Z, Emilsson GM, Ruzzo WL, Breaker RR (2004) A glycine-dependent riboswitch that uses cooperative binding to control gene expression. Science 306:275–279

McCluskey K, Shaw E, Lafontaine DA, Penedo JC (2014) Single-molecule fluorescence of nucleic acids. Methods Mol Biol 1076:759–791

Mellin JR, Koutero M, Dar D, Nahori MA, Sorek R, Cossart P (2014) Sequestration of a two-component response regulator by a riboswitch-regulated noncoding RNA. Science 345:940–943

Miranda-Rios J, Navarro M, Soberón (2001) A conserved RNA structure (thi box) is involved in regulation of thiamine biosynthetic gene expression in bacteria. Proc Natl Acad Sci 98:9736–9741

Montange RK, Batey RT (2006) Structure of the S-adenosylmethionine riboswitch regulatory mRNA element. Nature 441:1172–1175

Montange RK, Batey RT (2008) Riboswitches: emerging themes in RNA structure and function. Ann Rev Biophys 37:117–133

Mulhbacher J, Brouillete E, Allard M, Fortier LC, Malouin F, Lafontaine DA (2010) Novel riboswitch ligand analogues as selective inhibitors of guanine-related metabolic pathways. PLoS Pathog 6:e1000865

Nahvi A, Sudarsan N, Ebert MS, Zou X, Brown KL, Breaker RR (2002) Genetic control by a metabolite binding mRNA. Chem Biol 9:1043–1049

Narberhaus F, Waldminghaus T, Chowdhury S (2006) RNA thermometers. FEMS Microbiol Rev 30:3–16

Nelson JW, Sudarsan N, Furukawa K, Weinberg Z, Wang JX, Breaker RR (2013) Riboswitches in eubacteria sense the second messenger c-di-AMP. Nat Chem Biol 9:834–839

Neuman KC, Block SM (2004) Optical trapping. Rev Sci Instrum 75:2787–2809

Neupane K, Yu H, Foster DAN, Wang F, Woodside MT (2011) Single-molecule force spectroscopy of the add adenine riboswitch relates folding to regulatory mechanism. Nucleic Acids Res 39:7677–7687
Nienhaus GU (2009) Single-molecule fluorescence studies of protein folding. Methods Mol Biol 490:311–337
Noeske J, Richter C, Grundi MA, Nasiri HR, Schwalbe H, Wohnert J (2005) An intermolecular base triple as the basis of ligand specificity and affinity in the guanine- and adenine-sensing riboswitch RNAs. Proc Natl Acad Sci USA 102:1372–1377
Noeske J, Schwalbe H, Wohnert J (2007) Metal-ion binding and metal-ion induced folding of the adenine-sensing riboswitch aptamer domain. Nucleic Acids Res 35:5262–5273
Otani S, Takatsu M, Nakano M, Kasai S, Miura R (1974) Letter: roseoflavin, a new antimicrobial pigment from *Streptomyces*. J Antibiot (Tokyo) 27:88–89
Ott E, Stolz J, Lehmann M, Mack M (2009) The RFN riboswitch of *Bacillus subtilis* is a target for the antibiotic roseoflavin produced by *Streptomyces davawensis*. RNA Biol 6:276–280
Patte JC (1996) Biosynthesis of threonine and lysine. In: *Escherichia coli* and *Salmonella typhimurium* (Neidhardt, FC, ed), vol 2. American Society for Microbiology, Washington, pp 528–541
Penedo JC, Wilson TJ, Jayasena S, Khvorova A, Lilley DMJ (2004) Folding of the natural hammerhead is enhanced by interaction of auxiliary elements. RNA 10:880–888
Perez-Gonzalez DC, Penedo JC (2015) Single-molecule strategies for DNA and RNA diagnostics. In: Erdmann DA, Jurga S, Barciszweski J (eds) RNA and DNA Diagnostics. Springer, pp 297–332
Poiata E, Meyer MM, Ames TD, Breaker RR (2009) A variant riboswitch apatamer class for S-adenosylmethionine common in marine bacteria. RNA 15:2046–2056
Regulski EE, Moy RH, Weinberg Z, Barrick JE, Yao Z, Ruzzo WL, Breaker RR (2008) A widespread riboswitch candidate that controls bacterial genes involved in molybdenum cofactor and tungsten cofactor metabolism. Mol Microbiol 68:918–932
Ren A, Rajashankar KR, Patel DJ (2012) Fluoride ion encapsulation by Mg2+ ions and phosphates in a fluoride riboswitch. Nature 486:85–89
Rieder R, Lang K, Graber D, Micura R (2007) Ligand-induced folding of the adenosine deaminase A-riboswitch and implications on riboswitch translational control. ChemBioChem 8:896–902
Rieder U, Kreutz C, Micura R (2010) Folding of a transcriptionally acting preQ1 riboswitch. Proc Natl Acad Sci USA 107:10804–10809
Roth A, Breaker RR (2009) The structural and functional diversity of metabolite-binding riboswitches. Annu Rev Biochem 78:305–334
Roth A, Winkler WC, Regulski EE, Lee BW, Lim J, Jona I, Barrick JE, Ritwik A, Kim JN, Welz R, Iwata-Reuyl D, Breaker RR (2007) A riboswitch selective for the queuosine precursor preQ1 contains an unusually small aptamer domain. Nat Struct Mol Biol 14:308–317
Roy R, Hohng S, Ha T (2008) A practical guide to single-molecule FRET. Nat Methods 5:507–516
Savinov A, Perez CF, Block SM (2014) Single-molecule studies of riboswitch folding. Biochim Biophys Acta 1839:1030–1045
Schroeder KT, Daldrop P, Lilley DMJ (2011) RNA tertiary interactions in a riboswitch stabilize the structure of a kink turn. Structure 19:1233–1240
Schuler B, Eaton WA (2008) Protein folding studied by single-molecule FRET. Curr Opin Struct Biol 18:16–26
Selvin PR (2000) The renaissance of fluorescence resonance energy transfer. Nat Struct Biol 7:730–734
Serganov A, Nudler E (2013) A decade of riboswitches. Cell 152:17–24
Serganov A, Patel DJ (2012) Metabolite recognition principles and molecular mechanisms underlying riboswitch function. Annu Rev Biophys 41:343–370
Serganov A, Yuan YR, Pikovskaya O, Polonskaia A, Malinina L, Phan AT, Hobartner C, Micura R, Breaker RR, Patel DJ (2004) Structural basis for discriminative regulation of gene expression by adenine- and guanine-sensing mRNAs. Chem Biol 11:1729–1741

Serganov A, Polonskaia A, Phan AT, Breaker RR, Patel DJ (2006) Structural basis for gene regulation by a thiamine pyrophosphate-sensing riboswitch. Nature 441:1167–1171

Serganov A, Huang L, Patel DJ (2009) Coenzyme recognition and gene regulation by a flavin mononucleotide riboswitch. Nature 458:233–237

Sharma M, Bulusu G, Mitra A (2009) MD simulations of ligand-bound and ligand-free aptamer: molecular level insights into the binding and switching mechanism of the add A-riboswitch. RNA 15:1673–1692

Shaw E, St-Pierre P, McCluskey K, Lafontaine DA, Penedo JC (2014) Using smFRET and denaturants to reveal folding landscapes. Methods Enzymol 549:313–341

Sherman EM, Esquiaqui J, Elsayed G, Ye JD (2012) An energetically beneficial leader-linker interaction abolishes ligand-binding cooperativity in glycine riboswitches. RNA 18:496–507

Soulière MF, Haller A, RIeder R, Micura R (2011) A powerful approach for the selection of 2-aminopurine substitution sites to investigate RNA folding. J Am Chem Soc 133:16161–16167

St-Pierre P, McCLuskey K, Shaw E, Penedo JC, Lafontaine DA (2014) Fluorescence tools to investigate riboswitch structural dynamics. Biochim Biophys Acta 1839:1005–1019

Stryer L (1978) Fluorescence energy transfer as a spectroscopic ruler. Annu Rev Biochem 47:819–846

Sudarsan N, Wickiser JK, Nakamura S, Ebert MS, Breaker RR (2003) An mRNA structure in bacteria that controls gene expression by binding lysine. Genes Dev 17:2688–2697

Sudarsan N, Cohen-Chalamish S, Nakamura S, Emilsson GM, Breaker RR (2005) Thiamine pyrophosphate riboswitches are targets for the antimicrobial compound pyrithiamine. Chem Biol 12:1325–1335

Sudarsan N, Hammond MC, Block KF, Welz R, Barrick JE, Roth A, Breaker RR (2006) Tandem riboswitch architectures exhibit complex gene control functions. Science 314:300–304

Sudarsan N, Lee ER, Weinberg Z, Moy RH, Kim JN, Link KH, Breaker RR (2008) Riboswitches in eubacteria sense the second messenger cyclic di-GMP. Science 321:411–413

Suddala KC, Rinaldi AJ, Feng J, Mustoe AM, Eichhorn CD, Liberman JA, Wedekind JE, Al-Hashimi HM, Brooks CL 3rd, Walter NG (2013) Single transcriptional and translational preQ1 riboswitches adopt similar pre-folded ensembles that follow distinct folding pathways into the same ligand-bound structure. Nucleic Acids Res 41:10462–10475

Suydam IT, Strobel SA (2009) Nucleotide analogueinterference mapping. Methods Enzymol 468:3–30

Thore S, Leibundgut M, Ban N (2006) Structure of the eukaryotic thiamine pyrophosphate riboswitch with its regulatory ligand. Science 312:1208–1211

Thore S, Frick C, Ban N (2008) Structural basis of thiamine pyrophosphate analogues binding to the eukaryotic riboswitch. J Am Chem Soc 130:8116–8117

Trausch JJ, Ceres P, Reyes FE, Batey RT (2011) The structure of a tetrahydrofolate-sensing riboswitch reveals two ligand binding sites in a single aptamer. Structure 19:1413–1423

Tremblay R, Lemay JF, Blouin S, Mulhbacher J, Bonneau É, Legault P, Dupont P, Penedo JC, Lafontaine DA (2011) Constitutive regulatory activity of an evolutionarily excluded riboswitch variant. J Biol Chem 286:27406–27415

Vogl C, Grill S, Schilling O, Stülke J, Mack M, Stolz J (2007) Characterization of riboflavin (Vitamin B2) transport proteins from *Bacillus subtilis* and *Corynebacterium glutamicum.* J Bacteriol 189:7367–7375

Wang JX, Lee ER, Morales DR, Lim J, Breaker RR (2008) Riboswitches that sense S-adenosylhomocysteine and activate genes involved in coenzyme recycling. Mol Cell 29:691–702

Welz R, Breaker RR (2007) Ligand binding and gene control characteristics of tandem riboswitches in *Bacillus anthracis*. RNA 13:573–583

Wickiser JK, Cheah MT, Breaker RR, Crothers DM (2005) The kinetics of ligand binding by an adenine-sensing riboswitch. Biochemistry 44:13404–13414

Wilhelmsson LM (2010) Fluorescent nucleic acid base analogues. Q Rev Biophys 43:159–183

Winkler WC, Breaker RR (2005) Regulation of bacterial gene expression by riboswitches. Annu Rev Microbiol 59:487–517
Winkler WC, Cohen-Chalamish S, Breaker RR (2002) An mRNA structure that controls gene expression by binding FMN. Proc Natl Acad Sci 99:15908–15913
Winkler WC, Nahvi A, Sudarsan N, Barrick JE, Breaker RR (2003) An mRNA structure that controls gene expression by binding S-adenosylmethionine. Nat Struct Biol 10:701–707
Winkler WC, Nahvi A, Roth A, Collins JA, Breaker RR (2004) Control of gene expression by a natural metabolite-responsive ribozyme. Nature 428:281–286
Wood S, Ferré-D`Amaré AR, Rueda D (2012) Allosteric tertiary interactions preorganize the c-di-GMP riboswitch and accelerate ligand binding. ACS Chem Biol 7:920–927
Woodside MT, Block SM (2014) Reconstructing folding energy landscapes by single-molecule force spectroscopy. Annu Rev Biophys 43:19–39
Woolley DW, White AGC (1943) Selective reversible inhibition of microbial growth with pyrithiamine. J Exp Med 78:489–497

Chapter 12
Bugs on a Slippery Plane

Understanding the Motility of Microbial Pathogens with Mathematical Modelling

Dmitri O. Pushkin and Martin A. Bees

Abstract Many pathogenic microorganisms live in close association with surfaces, typically in thin films that either arise naturally or that they themselves create. In response to this constrained environment, the cells adjust their behaviour and morphology, invoking communication channels and inducing physical phenomena that allow for rapid colonization of biomedically relevant surfaces or the promotion of virulence factors. Thus, it is very important to measure and theoretically understand the key mechanisms for the apparent advantage obtained from swimming in thin films. We discuss experimental measurements of flows around a peritrichously flagellated bacterium constrained in a thin film, derive a simplified mathematical theory and Green's functions for flows in a thin film with general slip boundary conditions, and establish connections between theoretical and experimental results. This article aims to highlight the importance of mathematics as a tool to unlock qualitative mechanisms associated with experimental observations in the medical and biological sciences.

Keywords Flow fields · Thin-film flow · Microorganisms · Biofilms · Mathematical modelling

12.1 Introduction

Single-celled microorganisms such as bacteria and algae live in microscopic fluid environments, where frictional forces dominate and inertia is negligible. As individuals, they employ a range of techniques to move planktonically in fluid or in close association with a surface (Harshey 2003) ranging from swimming with flagella or body deformations (Turner et al. 2000), to gliding by extending and

D.O. Pushkin · M.A. Bees (✉)
Department of Mathematics, University of York, York YO10 5DD, UK
e-mail: martin.bees@york.ac.uk

D.O. Pushkin
e-mail: mitya.pushkin@york.ac.uk

© Springer International Publishing Switzerland 2016

M.C. Leake (ed.), *Biophysics of Infection*, Advances in Experimental Medicine and Biology 915, DOI 10.1007/978-3-319-32189-9_12

contracting pili (Skerker and Berg 2001), or otherwise, such as propulsion using directed exudate. By swimming or attempting to swim, they generate intricate fluid flow structures that are affected by and influence the orientation and motion of other swimming cells.

Frequently and fruitfully microorganisms are found to colonise surfaces, where they are able to exert a measure of control over their local environment with a subtle balance of growth, organisation and expansion. These dynamics are coupled to nutrient availability, signal-molecule-mediated intercellular communication, extracellular secretions and the biophysics of interfaces and contact lines (Bees et al. 2000). As a colony, they may swim in an existing surface film, forming diffusive rings, or make use of signal molecules, surfactants and wetting agents to expand the perimeter of a thin fluid drop on a surface. Many of these regulatory pathways give rise to or are implicated in biofilm formation, when cells (mono- or multi-species) adhere to one another and to a surface (Verstraeten et al. 2008). Generally, cells in a biofilm are embedded within a structure of extracellular polymeric substance (EPS; mostly proteins, polysaccharides and DNA) that they produce. Biofilms are as important as they are ubiquitous, having the potential to resist antimicrobial treatment whilst expressing virulence factors.

Therefore, it is of fundamental importance to study the physical mechanisms for surface colonisation, swimming in a constrained environment and the earliest stages of biofilm formation. To this end, we introduce a simple mathematical model of flows due to swimming in a thin film. While similar modelling is standard in the fluid dynamics literature, we do not know of its application to the motility of microorganisms.

12.2 The Need for Mechanistic Mathematical Modelling

Modelling can provide a means to test mechanistic understanding of a biological or physical system not easily accessible with experiments. It is the most advantageous when there is a close marriage of experiment and theory such that quantitative or qualitative predictions can be made based on experimental data and tested. In most cases, there are no valid models, only models that are good enough to explain the phenomena at hand without having so many unmeasured or unmeasurable free parameters that any behaviour can be fitted, a concept closely related to that of Occam's razor. Thus, simplification is prerequisite to any successful theory, especially in the biological sciences where the complexity can quickly become excessive, obscuring the wood for the trees.

The equations of fluid dynamics are immensely successful, describing flows in fluids (including gases) across many spatial and temporal scales with no free parameters. Their application to modelling swimming microorganisms is of much recent interest (Lauga and Powers 2009) as is the measurement of flows around cells, planktonic (Guasto et al. 2011) and sessile (Cisneros et al. 2008). However, their swimming apparatus can be complex, consisting, for example, of multiple interacting helical or plane-wave flagella. Hence, simplification is not only necessary to obtain analytical or computational solutions but desirable for clarity of the mechanisms.

The aim of this article is to provide a minimal model of swimming in a constrained environment that can be employed analytically, rather than a highly complicated description incorporating many features of the morphology of the microorganism that would necessitate complex numerical investigation. We will show that due to its simplicity, it provides an insight into the highly nontrivial structure of flow fields around microorganisms observed in experiments (Cisneros et al. 2008; Guasto et al. 2011).

12.3 Morphology

Of significance to this article is the swimming of microorganism in preexisting thin films. The responses of microorganisms to their environment, such as due to the availability of nutrients, via quorum sensing, cell specialisation and spatial self-organisation, are fundamentally different for planktonic cells suspended in a liquid medium than for cells inhabiting wet surfaces and liquid films (Donlan 2002). This is particularly true for motile cells that swim using flagella (Harshey 2003).

A striking example is the change in cell morphology when bacteria grow within a wet surface colony. For example, the bacterium *Vibrio parahaemolyticus* (found in saltwater and capable of causing gastroenteritis in humans) grows longer and sprouts many long lateral flagella in addition to the polar flagellum used for swimming in liquid medium (McCarter 1999). This transformation is orchestrated with the production of extracellular surfactants and wetting agents that allow it to develop a new type of collective motility, swarming, and to form differentiated multicellular communities. Remarkably, most bacterial cells capable of swarming, such as *Escherichia*, *Salmonella*, *Proteus* and *Serratia spp.*, undergo similar (reversible) morphological changes to form elongated, aseptate, multinucleated, hyperflagellated cells. This remarkable fact seems to suggest that a distinct evolutionary advantage must be conferred by such morphology on cells inhabiting wet surfaces and films. There is no doubt that longer and more numerous flagella are capable of creating stronger flows in films. However, swarming cells cannot exert sufficient force to counteract surface tension or contact line dynamics, instead they change the physics with wetting agents and surfactants that can even allow for rapid (even exponential) expansion of the colony radius (Bees et al. 2002). But the swimming characteristics of the individuals do have an effect: the morphology of a cell can induce complex flow fields, especially in constrained environments, and collective motion can arise (e.g. Sokolov et al. 2007) potentially adjusting effective mechanical properties of the suspension and influencing fluid–substrate interactions and dispersion of signal molecules.

The enormous medical impact of thin-film infections of pathogenic bacteria, such as *Pseudomonas aeruginosa* and *Staphylococcus aureus*, is long-established and well-documented (Costerton et al. 1999). These and many other bacteria are responsible for lung and airway infections, and nosocomial and catheter-associated bloodstream infections. Typically, the pathogens initially are associated with surfaces, living within a thin film of fluid upon a solid substrate. The upper boundary may be either a liquid–air or liquid–liquid interface. In such a system, it is important to disentangle coupled biological and physical mechanisms for the establishment of biofilms; fluid flow around one or more stuck cells may entrain other cells in the vicinity and allow for sufficiently high local concentrations to upregulate quorum sensing activity.

The investigation of swimming within thin films may also have application in other areas, such as biofuels, where an efficient exchange of gases between the suspension and the air is desirable (Bees and Croze 2014).

12.4 Measuring the Flow field

Whilst there have been many numerical studies of the interaction between swimming microorganisms, there are only a few measurements of the flow around individuals. Drescher et al. (2010b) measured the three-dimensional mean flow around freely swimming cells of *Chlamydomonas reinhardtii* and *Volvox carterii*, averaged over many flagella beats and cells. They found the expected far-field stresslet (force-dipole) behaviour in three dimensions, with the magnitude of the velocity decaying approximately as the reciprocal of the radius squared for a buoyant free-swimmer, and Stokeslet behaviour (a point force), reciprocal of the radius, for a negatively buoyant swimmer. Guasto et al. (2011) extended these results by measuring the flow around freely swimming cells of *Chlamydomonas reinhardtii*, temporally resolving the flow as it conducts its 50 Hz flagellar beat. They performed these experiments in a pseudo-two-dimensional flow in a film with two stress-free boundaries (a bubble), revealing the complexities of the near-field flows. These flows can be compared with flow fields from numerical computations (O'Malley and Bees 2012). Drescher et al. (2010a) measured the flow around freely swimming bacteria. These studies again revealed stresslet behaviour in the far field (decaying approximately as the reciprocal of the radius squared) although they reported the overwhelming importance of stochastic effects in cell–cell and cell–boundary interactions over larger distances.

However, prior to these articles Cisneros et al. (2008) measured an altogether different flow for bacteria constrained in thin films between a no-slip and stress-free boundary, as depicted in Fig. 12.1d, in contrast to what they observed for freely swimming cells. One might expect that cells would generate a flow reminiscent of a point force singularity due to the bundled flagella pushing against frictional forces due to the no-slip boundary, especially in the far field. Yet, this was not what was observed. Instead, the dominant flow over the observable range was one that rotated about the cell in the plane of the film.

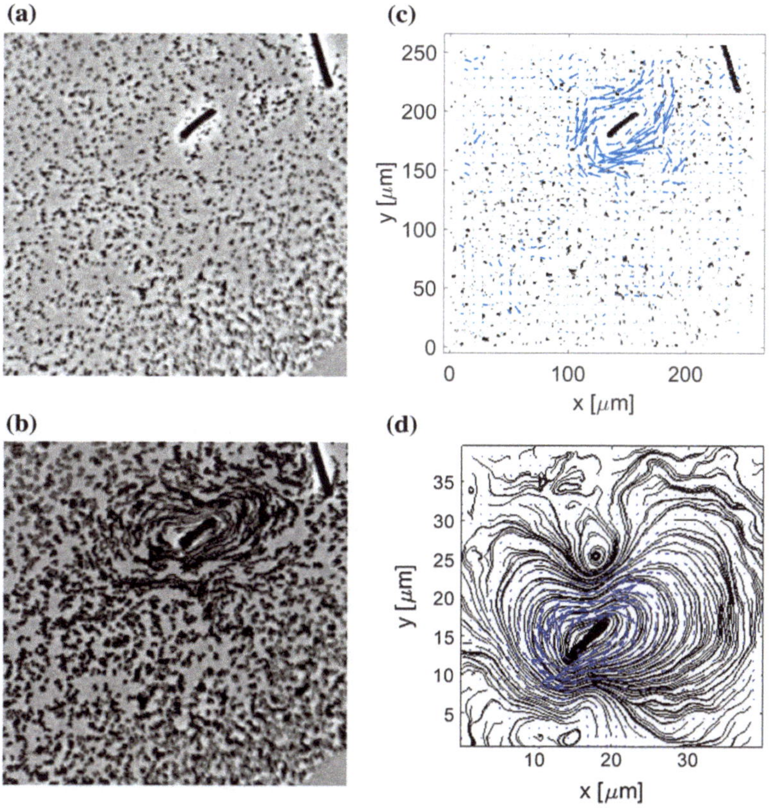

Fig. 12.1 **a** One frame of a series of 1500: the large particle is the body (4 μm × 0.8 μm) of a peritrichously flagellated (6–15 flagella; 10 μm long and 20 nm diameter; 3 μm pitch; rotating at 100 Hz) cell of *Bacillus subtilis*. The *small dots* are tracer particles of diameter 210 nm, and the resolution is 256 × 256 pixels. **b** A processed image of the minimum value at each pixel over 30 frames (at 60 fps) around a stuck cell of *Bacillus subtilis*: the motion of the tracer particles illustrates the flow around the cell body. **c** The *arrows* indicate the vector field around the stuck cell obtained with PIV (ImageJ plugin, Q. Tseng). The small dots are the tracer particles. **d** Streamlines and flow field vectors about a second cell. (Subfigure **d** is replotted, with permission, from data used in Cisneros et al. 2008 using similar methods.)

Experimental methods: Liquid cultures of *Bacillus subtilis* 1085B were grown in Ezmix Terrific Broth (TB) for 18 h in a shaker bath at 37 °C and then a 1:50 dilution was prepared in TB and left for an additional 5 h (see Cisneros et al. 2008 for full culture details). This process induced the formation of long, healthy motile cells (typically 4 μm × 0.8 μm). After a further 1:100 dilution with TB, carboxylate-modified 0.21 μm microspheres (Molec. Probes: F8809—FluoSpheres) were added to act as tracer particles so that the fluid velocity could be observed. A drop of the suspension was placed upon a cover slip that was glued to the outside of a square aperture cut into the base of a Petri dish. A water reservoir and lid eliminated potential for evaporative flows. The drop wetted the surface and fluid was removed by pipette to leave a thin layer of cells. (Not all cover slips allowed thin films to persist; dewetting occurs for some glassware.) Some motile cells swam into shallow regions near the contact line and became wedged between the upper and lower boundaries (several body lengths from the contact line) in a film depth of approximately 1 μm. See Cisneros et al. (2008) for more experimental details. Figure 12.1 illustrates the method to extract the flow field using PIV: Fig. 12.1a is one frame out of 1500 (at 60 fps) with a (large) cell body and (small) tracer particles; Fig. 12.1b demonstrates that the flow is mostly close to the cell body by plotting simply the minimum value at each pixel over 30 frames; Fig. 12.1c uses PIV (ImageJ plugin, Q. Tseng) to generate the flow field; Fig. 12.1d presents reprocessed data used in Cisneros et al. (2008) with streamlines. These authors report 25 observations of rotating flows from 27 isolated and wedged cells. Cisneros et al. (2008) also present data for the much smaller non-rotating flow around freely swimming cells.

Discussion of experimental results: Clearly, the results indicate that the flow rotates around the mostly stationary cells (the cells rotate stochastically by a couple of degrees about their mean orientations over the course of each experiment). Moreover, it is apparent that the typically 10 μm long flagella are not bundled with any particular orientation. Cisneros et al. (2008) argue that the rotating flagella interact more strongly with the lower no-slip boundary than the upper stress-free boundary leading to flagella wrapping around the cell, forming a continuously evolving envelope and driving rotating flow in the plane of the film. A conclusion supported by the absence of rotation when the boundaries are symmetric. Furthermore, one can measure the magnitude, U, of the flow over several directions as a function of distance, r, from the cell (in μm, units of the cell diameter or units of film thickness). Surprisingly, Cisneros et al. (2008) find that the flow decays over a couple of body lengths (10 body widths) as $U \propto e^{-kr}$, where $k \approx \frac{1}{4}$ μm^{-1}, in stark contrast to the expected logarithmic or reciprocal decay expected of a point force in two or three-dimensional non-constrained flow, respectively. This observation should be explained.

12.5 Mathematics of Modelling Swimming in a Thin Film

In this section, we shall describe some approximations and solutions for swimming cells in thin films. Mathematically inclined readers will discover in this section that swimming in a constrained environment in a thin film that allows for asymmetry in the boundary conditions leads to near-field behaviour that is distinct from swimming in simple two-dimensional or three-dimensional fluid. The transition between near- and far-field effects will be explored. Mathematically disinclined readers can skip this section in favour of the discussion of results and comparison with experiments in Sect. 12.6.

For simple fluids at low Reynolds number, the fluid motion is governed by the Stokes equations, a set of linear equations that possess a time-reversal symmetry. The implication in this microscopic environment is that to move, cells need to adopt a swimming stroke that looks different when played in reverse (allowing for an arbitrary strictly increasing nonlinear scaling of time).

Many biological fluids have complex viscoelastic and rheological properties (mucus, blood, etc.). Here, however, we consider the simplest case of a Newtonian fluid with viscosity μ (i.e. we assume that the fluid possesses negligible viscoelasticity). We assume the film to be infinite and flat and to have a constant thickness h. Furthermore, and significantly, we assume the film to be thin. This assumption means that flows generated by swimmers quickly become (quasi) two-dimensional at distances r much larger than h, such that $r \gg h$.

Boundary conditions at small spatial scales can be complex. For microscopic scales, a no-slip condition is commonly used. However, biological and complex fluids often show significant apparent slip in a variety of situations, e.g. when polymers dissolved in the fluid are poorly adsorbed by the wall, or the wall-adsorbed polymer chains undergo a stretch–coil transition (Lauga et al. 2007). Other factors leading to a significant slip include surface roughness, hydrophobicity, electric charge and dissolved gases (Tretheway and Meinhart 2002; Zhu and Granick 2002; Lauga et al. 2007). Therefore, we employ a general *slip* boundary condition at the bottom and the top of the film, such that

$$\mathbf{v} - l\frac{\partial \mathbf{v}}{\partial n} = 0, \qquad (12.1)$$

where $\mathbf{v}$ is the (three-dimensional) fluid velocity, the derivative is taken along the normal to the film boundary with coordinate n and the slip length l may be different for the top and the bottom film boundaries. For $l = 0$, this boundary condition corresponds to the *no-slip* condition requiring that the fluid velocity vanishes on a fluid–solid interface; for $l = \infty$, the condition describes a stress-free interface. Intermediate values of l are typical for fluid–fluid interfaces. For example, if another fluid layer is sandwiched between the film and the solid substrate $l \sim h_1\mu/\mu_1$, where μ_1 and h_1 are the viscosity and the thickness of the sandwiched fluid layer, respectively.

Within the thin-film approximation the velocity component normal to the film is negligible, and it can be shown that the two-dimensional fluid velocity $\mathbf{u}(x, y)$, defined as the local velocity $\mathbf{v}$ averaged over the film thickness, is governed by

$$-\nabla p + \mu\nabla^2\mathbf{u} - \kappa\mathbf{u} = 0, \qquad (12.2)$$

where $p(x, y)$ is the fluid pressure and κ is the local friction coefficient. While the first two terms describing the balance of the pressure and viscous stresses between neighbouring layers of fluid are identical to the terms of the Stokes equation, the last term describes the damping effect of the film boundaries. The local friction coefficient can be related to the film thickness and the slip lengths l_1 and l_2 are for the bottom and the top of the film, respectively:

$$\kappa = \mu/\lambda^2, \quad \lambda^2 = \frac{h(h^2 + 4h(l_1 + l_2) + 12 l_1 l_2)}{h + l_1 + l_2}. \tag{12.3}$$

In particular, for a channel that is formed by two parallel solid planes with no-slip boundary conditions (a Hele-Shaw cell) $\lambda = h$. For a stress-free top boundary $\lambda = 2\sqrt{h(h + 3l_1)}$. If, in addition, the bottom boundary is no-slip then $\lambda = 2h$; alternatively, if $l_1 \gg h$, then $\lambda = 2\sqrt{3hl_1}$. Equation (12.2) should be supplemented by the fluid incompressibility condition

$$\nabla \cdot \mathbf{u} = 0. \tag{12.4}$$

The thin-film flow generated by a point force $\mathbf{F}$ applied at the origin must satisfy

$$-\nabla p + \mu \nabla^2 \mathbf{u} - \kappa \mathbf{u} = -\rho \mathbf{F} \delta(\mathbf{r}), \tag{12.5}$$

where $\mathbf{r}$ is the position vector and ρ is the fluid density. It turns out that the exact solution of (12.4) and (12.5) can be found analytically, such that

$$\mathbf{u}_\mathbf{f}(\mathbf{r}, \mathbf{F}) = \mathbf{G}_\mathbf{f} \cdot \frac{\mathbf{F}}{2\pi\nu}, \quad \mathbf{G}_\mathbf{f} = f_1(r/\lambda)\mathbf{I} + f_2(r/\lambda)\hat{\mathbf{r}}\,\hat{\mathbf{r}}, \tag{12.6}$$

$$f_1(x) = K_0(x) - \left(x^{-2} + x^{-1} K_0'(x)\right), \tag{12.7}$$

$$f_2(x) = 2x^{-2} - x\left(x^{-1} K_0'(x)\right)', \tag{12.8}$$

where $\nu = \mu/\rho$ is the kinematic viscosity of the fluid, $\mathbf{I}$ is the identity tensor, $r = \|r\|r$, $\hat{\mathbf{r}} = \mathbf{r}/r$ and $K_0(x)$ is the modified Bessel function of the second kind. As the governing equations are linear, this solution may be used to construct arbitrary (quasi-two-dimensional) flows in thin films using the Green's functions formalism (Pozrikidis 1992). Therefore, we will refer to $\mathbf{u}_\mathbf{f}$ as the Green's function for a thin-film slip flow.

In order to understand this solution, let us first consider its asymptotes:

$$x \to 0: \quad f_1 \sim \frac{1}{2}\ln(2x^{-1}) - \gamma,\ f_2 \sim \frac{1}{2}; \tag{12.9}$$

$$x \to \infty: \quad f_1 \sim -x^{-2},\ f_2 \sim 2x^{-2}. \tag{12.10}$$

Here, $\gamma \approx 0.577$ is the Euler constant. The first asymptote describes the flow generated by a concentrated force at length scales much less than the slip length λ, $r = \lambda$. Hence,

$$\mathbf{u}_\mathbf{f} \approx \mathbf{u}_\text{free} = \left[\left(\ln\frac{2\lambda}{r} - \gamma\right)\mathbf{I} + \hat{\mathbf{r}}\,\hat{\mathbf{r}}\right] \cdot \frac{\mathbf{F}}{4\pi\nu}, \quad r \ll \lambda. \tag{12.11}$$

The Green's function $\mathbf{u}_{free}$ describes hydrodynamic interactions in a film with no friction at the film boundaries. The structure of this flow is presented in Fig. 12.2a.

The expression $\mathbf{u}_{free}$ can be obtained directly from the Stokes equations (i.e. Eq. (12.2) with $\kappa = 0$; Pozrikidis 1992) in two dimensions. Note that the first term in the brackets contains a logarithm, which diverges for $r \to 0$ and $r \to \infty$. The short-distance divergence is natural as we assume a point force; in real flows it will be cured by the finite size of the microorganism. The large-distance divergence is a signature of the *Stokes paradox*, which is explicitly resolved here by the requirement $r \ll \lambda$; the (weak) dependence of the argument of the logarithm on λ remains its only vestige. The Green's function analogous to $\mathbf{u}_{\text{free}}$ has been used recently to quantify hydrodynamic interactions in free fluid films (Di Leonardo et al. 2008); note, however, that we obtain somewhat different numerical values for the constants in the prefactor of $\mathbf{I}$ in (12.11).

The asymptote (12.10) describes the flow produced by a concentrated force far beyond the slip length λ, yielding

$$\mathbf{u} \approx \mathbf{u}_{SD} = -\frac{1}{r^2}\left[\mathbf{I} - 2\hat{\mathbf{r}}\,\hat{\mathbf{r}}\right] \cdot \frac{\rho}{\kappa}\frac{\mathbf{F}}{2\pi}, \quad r \gg \lambda. \tag{12.12}$$

This flow can be obtained as the balance of the friction force at the boundaries and the pressure, which enforces incompressibility of the fluid. Importantly, this flow can be rewritten as

$$\mathbf{u}_{SD} = -2\nabla\nabla \ln r \cdot \frac{\lambda^2 \mathbf{F}}{4\pi\nu}, \tag{12.13}$$

which means that this singular flow can be considered as a two-dimensional source doublet flow (Pozrikidis 1992). Figure 12.2 shows the streamlines of this flow. Notably, the streamlines form closed loops. For a film with no-slip boundaries $\lambda = h$; this particular source doublet flow has been used recently to describe flows created in Hele-Shaw cells by microorganisms and by moving droplets (e.g. see Pepper et al. 2009; Brotto et al. 2013). We predict that the effect of fluid slip at large distances is to enhance the source dipole strength by a factor of $(\lambda/h)^2$.

Thus, the flow $\mathbf{u_f}$ has a very different structure for $r \ll \lambda$ and $r \gg \lambda$. The structure of the full flow is shown in Fig. 12.2c, d using the system of units with $\lambda = 1$. The region dominated by the logarithmic singularity continuously merges with the recirculatory flow region. This picture illustrates the utility of the Green's function $\mathbf{u_f}$, which may be used to describe thin-film flows in the whole domain.

Finally, we would like to comment on the behaviour of the fluid vorticity in slip flows. By contrast with the Stokes equations, for which the vorticity diffuses infinitely fast throughout the fluid (Acheson 2002), the vorticity in thin-film flows with slip boundary conditions can diffuse only a distance of order λ from the vorticity

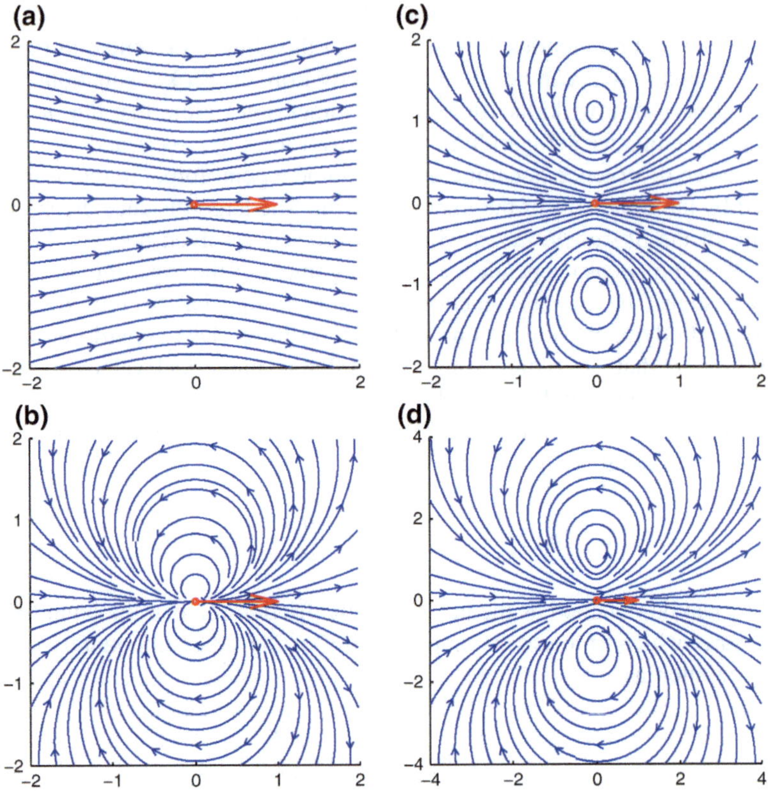

Fig. 12.2 Fluid flows due to a unit point force in thin films with different boundary conditions. The direction of the force is indicated by the unit *red arrow*. **a** The flow in a free-standing film is unidirectional, dominated by a (weak) logarithmic singularity at $r = 0$. **b** The flow in a film with no-slip boundaries (Hele-Shaw cell) shows characteristic closed-loop streamlines. The structure of the slip flow $\mathbf{u_f}$ (6) with $\lambda = 1$ is given in (**c**) and (**d**) for the box sizes $L_{box} = 2$ and $L_{box} = 4$, respectively. It is clear that the region dominated by the logarithmic singularity continuously merges with the recirculatory flow region. These figures illustrate the utility of the Green's function $\mathbf{u_f}$, which may be used to describe thin-film flows in the whole domain

source. Indeed, taking the curl of (12.2) results in an equation for the (only non-zero) vorticity component normal to the film ω, satisfying

$$\nabla^2\omega - \frac{1}{\lambda^2}\omega = 0. \tag{12.14}$$

Its bounded axisymmetric solution is given by

$$\omega = \text{const} \times K_0(r/\lambda) \sim e^{-r/\lambda}\sqrt{\frac{\lambda}{r}}, \quad r \gg \lambda, \tag{12.15}$$

and the flow circulation around the contour $|\mathbf{r}| = r$ varies with distance as

$$\Gamma(r) \sim e^{-r/\lambda}\sqrt{\frac{r}{\lambda}}, \quad r \gg \lambda. \tag{12.16}$$

Hence, the decay of the circulation is approximately exponential.

12.6 Discussion and Concluding Remarks

Thin-film flows are very typical of environments where pathogens are seen to thrive, with rapid colonisation of the film and other tissues, going on to form troublesome antimicrobial resistant biofilms with enhanced virulence. We have described how mathematical modelling can be used as a tool to explore mechanisms behind experimental observations. A proper understanding of the fluid dynamics and coupling with biological and physical processes will reveal why it is that microorganisms gain such an advantage when colonies are associated with surfaces. It is likely that there is not one main reason, but a combination of factors.

We have shown from experimental measurements and mathematical analysis of a simplified fluid system that the flow due to a swimming microorganism in a confined environment, such as within a thin film of fluid can be very different to the flow due to an unconstrained swimmer, particularly when there are asymmetric boundary conditions (e.g. a film on a substrate with a liquid–liquid or liquid–gas upper interface).

We have found a Green's function for a thin film in the presence of two general boundary conditions for the film that is valid over the whole domain. In particular, the results make clear how near-field logarithmic dynamics merge into a recirculatory flow regime. We can tune our slip parameters for symmetric and asymmetric cases (see (12.2)), and have observed that the circulation decays exponentially with distance, with a parameter that can relate the slip length to experimental measurements of flow fields.

Finally, it is tempting to consider how the local flows observed in experiments (see Fig. 12.1d) could be constructed from the fundamental Green's solutions. In fact, a similar flow arises very naturally. In Fig. 12.3 we plot the flow field due to a force couple, or, in other words, to two forces associated with the principal flagella bundles located either side of a rod-shaped bacterium pointing in opposing directions. The resemblance of Figs. 12.1d and 12.3 is striking; each has a region of pinched closed orbits about the body or origin that are clasped between two eddies (at the top and bottom of each picture). Therefore, it would appear that the complex system of a cell with several interacting helical flagella that are attached to the body via a flexible joint in a confined three-dimensional fluid between two general boundaries may be reduced to a system of two point force solutions for an asymptotically equivalent two-dimensional system.

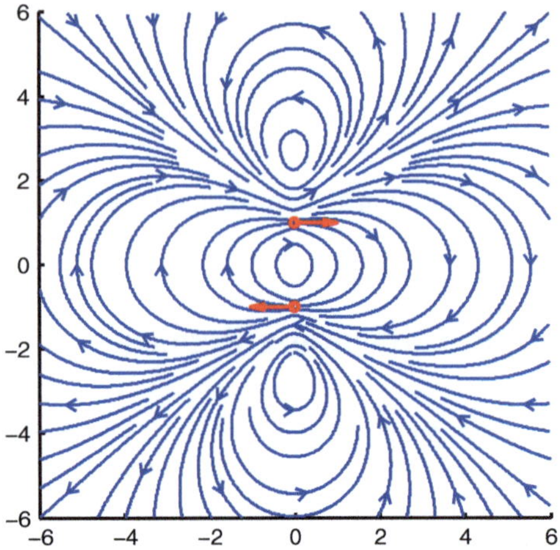

Fig. 12.3 Slip flow in a thin film due to a force couple; two point forces (*red arrows*) pointing in positive and negative x-directions separated by a distance of 2 μm in the y-direction

We plan to explore this system further and investigate interactions between swimmers, but for this we need first to consolidate and challenge the mathematical simplification and results with new experiments.

References

Acheson DJ (2002) Elementary fluid dynamics. Oxford University Press

Bees MA, Andresén P, Mosekilde E, Giskov M (2000) The interaction of thin-film flow, bacterial swarming and cell differentiation in colonies of *Serratia liquefaciens*. J Math Biol 40:27–63

Bees MA, Andresén P, Mosekilde E, Giskov M (2002) Quantitative effects of medium hardness and nutrient availability on the swarming motility of *Serratia liquefaciens*. Bull Math Biol 64:565–587

Bees MA, Croze OA (2014) Mathematics for streamlined biofuel production from unicellular algae. Biofuels 5(1):53–65

Brotto T, Caussin J-B, Lauga E, Bartolo D (2013) Hydrodynamics of confined active fluids. Phys Rev Lett 110(3):038101

Cisneros LH, Kessler JO, Ortiz R, Cortez R, Bees MA (2008) Unexpected bipolar flagellar arrangements and long-range flows driven by bacteria near solid boundaries. Phys Rev Lett 101:168102

Costerton JW, Stewart PS, Greenberg EP (1999) Bacterial biofilms: a common cause of persistent infections. Science 284(5418):1318–1322

Di Leonardo R, Keen S, Ianni F, Leach J, Padgett MJ, Ruocco G (2008) Hydrodynamic interactions in two dimensions. Phys Rev E 78(3):031406

Donlan RM (2002) Biofilms: microbial life on surfaces. Emerg. Infect. Dis. 8(9)

Drescher K, Dunkel J, Cisneros LH, Gangulya S, Goldstein RE (2010a) Fluid dynamics and noise in bacterial cell–cell and cell–surface scattering. Proc Natl Acad Sci USA **108**(27):10940–10945

Drescher K, Goldstein RE, Michel N, Polin M, Tuval I (2010b) Direct measurement of the flow field around swimming microorganisms. Phys Rev Lett **105:**168101

Guasto JS, Johnson KA, Gollub JP (2011) Measuring oscillatory velocity fields due to swimming algae. Phys Fluids 23:091112
Harshey RM (2003) Bacterial motility on a surface: many ways to a common goal. Annu Rev Microbiol 57:249–273
Lauga E, Brenner M, Stone H (2007) Microfluidics: the no-slip boundary condition. In: Springer handbook of experimental fluid mechanics. Springer, pp 1219–1240
Lauga E, Powers TR (2009) The hydrodynamics of swimming microorganisms. Rep. Prog. Phys. **72**:096601, 36pp
McCarter L (1999) The multiple identities of *Vibrio parahaemolyticus*. J Mol Microbiol Biotechnol 1(1):51–57
O'Malley SO, Bees MA (2012) The orientation of swimming bi-flagellates in shear flows. Bull Math Biol 74:232–255
Pepper RE, Roper M, Ryu S, Matsudaira P, Stone HA (2009) Nearby boundaries create eddies near microscopic filter feeders. J R Soc Interface 12:1–12
Pozrikidis C (1992) Boundary integral and singularity methods for linearized viscous flow. Cambridge University Press
Skerker J, Berg HC (2001) Direct observation of extension and retraction of type iv pili. Proc Natl Acad Sci USA 98:6901–6904
Sokolov A, Aranson IS, Kessler JO, Goldstein RE (2007) Concentration dependence of the collective dynamics of swimming bacteria. Phys Rev Lett 98:158102
Tretheway DC, Meinhart CD (2002) Apparent fluid slip at hydrophobic microchannel walls. Phys Fluids 14:L9–L12
Turner L, Ryu WS, Berg HC (2000) Real-time imaging of fluorescent flagellar filaments. J Bacteriol 182(10):2793–2801
Verstraeten N, Braeken K, Debkumari B, Fauvart M, Fransaer J, Vermant J, Michiels J (2008) Living on a surface: swarming and biofilm formation. Trends Microbiol 16:496–506
Zhu Y, Granick S (2002) Limits of the hydrodynamic no-slip boundary condition. Phys Rev Lett 88(10):106102

Chapter 13
Transcription Regulation and Membrane Stress Management in Enterobacterial Pathogens

Nan Zhang, Goran Jovanovic, Christopher McDonald, Oscar Ces, Xiaodong Zhang and Martin Buck

Abstract Transcription regulation in a temporal and conditional manner underpins the lifecycle of enterobacterial pathogens. Upon exposure to a wide array of environmental cues, these pathogens modulate their gene expression via the RNA polymerase and associated sigma factors. Different sigma factors, either involved in general 'house-keeping' or specific responses, guide the RNA polymerase to their cognate promoter DNAs. The major alternative sigma54 factor when activated helps pathogens manage stresses and proliferate in their ecological niches. In this chapter, we review the function and regulation of the sigma54-dependent Phage shock protein (Psp) system—a major stress response when Gram-negative pathogens encounter damages to their inner membranes. We discuss the recent development on mechanisms of gene regulation, signal transduction and stress mitigation in light of different biophysical and biochemical approaches.

Keywords Transcription · RNA polymerase · Sigma factors · Psp · Membrane stress

---

N. Zhang and G. Jovanovic contributed equally to this work.

N. Zhang · G. Jovanovic · M. Buck (✉)
Division of Cell and Molecular Biology, Imperial College London,
Sir Alexander Fleming Building, Exhibition Road, London SW7 2AZ, UK
e-mail: m.buck@imperial.ac.uk

C. McDonald · O. Ces
Division of Chemistry, Imperial College London,
Exhibition Road, London SW7 2AZ, UK

X. Zhang
Department of Medicine, Imperial College London,
Sir Ernst Chain Building, Exhibition Road, London SW7 2AZ, UK

© Springer International Publishing Switzerland 2016

M.C. Leake (ed.), *Biophysics of Infection*, Advances in Experimental Medicine and Biology 915, DOI 10.1007/978-3-319-32189-9_13

13.1 Introduction

Transcription of the Psp regulon in many important bacterial pathogens depends upon the major variant form of RNA polymerase (RNAP) containing the sigma54 factor as the dissociable promoter specificity factor. Working with *Yersinia enterocolitica*, Miller and colleagues identified *psp* genes as being virulence determinants in the mouse model, through using signature tag mutagenesis (Darwin and Miller 1999). In addition to directing RNAP to promoter sequences characterised by the consensus sequences TGCA around −12 and CTGGCAC around −24, the interactions between sigma54 and the core RNAP set the dependency of gene expression at the level of promoter DNA opening in the transition from a closed promoter complex (RP_C) to an open promoter complex (RP_O). The transition from RP_C to RP_O is catalysed by the ATPase function of transcription activators such as PspF acting directly upon the sigma54 Region I to relieve the inhibition imposed upon the −12 DNA fork junction (Bose et al. 2008; Morris et al. 1994). The rearranged RNAP subsequently accepts the single-stranded template DNA at the active site for RNA synthesis. Thus, the formation of RP_O at, for example the *pspA* and *pspG* promoters leads to expression of sets of *psp* genes used to manage plasma (inner) membrane stress.

13.2 Implications of Sigma54 in Pathogenicity

The major alternative sigma54 factor is present in almost all diderm species with an outer membrane (OM) mainly consisted of lipopolysaccharides (Francke et al. 2011). It has been found to cross-regulate genes with other sigma factors in *Pseudomonas aeruginosa* (Schulz et al. 2015). Under stress conditions, sigma54 may repress sigma70-dependent transcription by blocking either a proximal sigma70 binding site or through interacting at the intragenic regions (Schafer et al. 2015).

Historically, sigma54 has been associated with nitrogen metabolism, alternative carbon source utility and motility (Buck et al. 2000; Studholme and Dixon 2003). Recent studies have broadly expanded its functional repertoire in pathogenicity: biofilm formation and toxin production in *Bacillus cereus* ATCC 14579 (Hayrapetyan et al. 2015); internal colonisation in *Vibrio parahaemolyticus* (Whitaker et al. 2014); antibiotics sensitivity in *Escherichia coli* (Liu et al. 2010); acid resistance in enterohemorrhagic *E. coli* (Mitra et al. 2014); dental pulp infection in *Lactobacillus rhamnosus* (Nadkarni et al. 2014); protein secretion and invasion in *Campylobacter jejuni* (Carrillo et al. 2004; Hendrixson and DiRita 2004); intracellular trafficking within macrophages in *Burholderia cenocepacia* (Saldias et al. 2008); osmotolerance in *Listeria monocytogenes* (Okada et al. 2006); lipoprotein biosynthesis and virulence in *Borrelia burgdorferi* (Fisher et al. 2005);

stress adaptation and virulence in *Edwardsiella tarda* (Liu et al. 2014); autolysis and biofilm formation in *Enterococcus faecalis* (Iyer and Hancock 2012); and type I and type IV pilli biogenesis in *Xylella fastidiosa* (da Silva Neto et al. 2008). Computational synteny analyses suggest that genes under regulation of sigma54 conform to a common theme, i.e. the transport and biosynthesis of molecules that constitute the bacterial exterior (extracellular polysaccharides, flagella, lipopolysaccharides, lipoprotein and the peptidoglycan cell wall, Francke et al. 2011).

13.3 Domain Reorganisation of Sigma54 During Transcription Activation

A combination of single-molecule biophysical methods has been used to probe the events taking the RP_C to RP_O at a few sigma54 test promoters. Kinetic studies on the *glnAP2* promoter using COSMO methodologies indicate that the transition from RP_C to transcript generation takes around 90 s, and activators interact with both RP_C and RP_O (Friedman and Gelles 2012). The Stockley and Tuma labs addressed the domain movement of sigma54 in relation to the promoter DNA and ATP hydrolysis using single-molecule FRET analysis (Sharma et al. 2014). At the point of ATP hydrolysis, sigma54 Region I moves nearly 30 Å towards the leading edge of the transcription bubble (approximately the same distance as from −12 to +1, Fig. 13.1). This downstream movement may have two functional consequences: First, it may facilitate the removal of inhibitory interactions formed around the −12 fork junction DNA by the 'power stroke' action of multiple L1 loops on the PspF activator hexamer (Fig. 13.1). Secondly, the downstream movement brings sigma54 Region I in close proximity to the +1 site where it potentially constitutes an interaction network with the bridge helix and switch regions (Fig. 13.1b, Zhang et al. 2015). Deletion of sigma54 Region I has been shown to rescue activation defects of certain bridge helix variants, reverse the regulatory effects of DksA bound to the secondary channel (Zhang et al. 2015), and bypass the activator requirement on several promoters in vivo (Schafer et al. 2015).

13.4 New Structural Insights of the RNAP-Sigma54 Transcription Complex

The structure of the sigma54-containing RNAP and its co-complexes with promoter DNA and with the ATPase domain of PspF has been determined using combinations of single-particle cryo-electron microscopy and X-ray crystallography. It is clear that the closed or intermediate promoter complexes are maintained at the downstream promoter −12 by contacts made between several L1 loops and sigma54

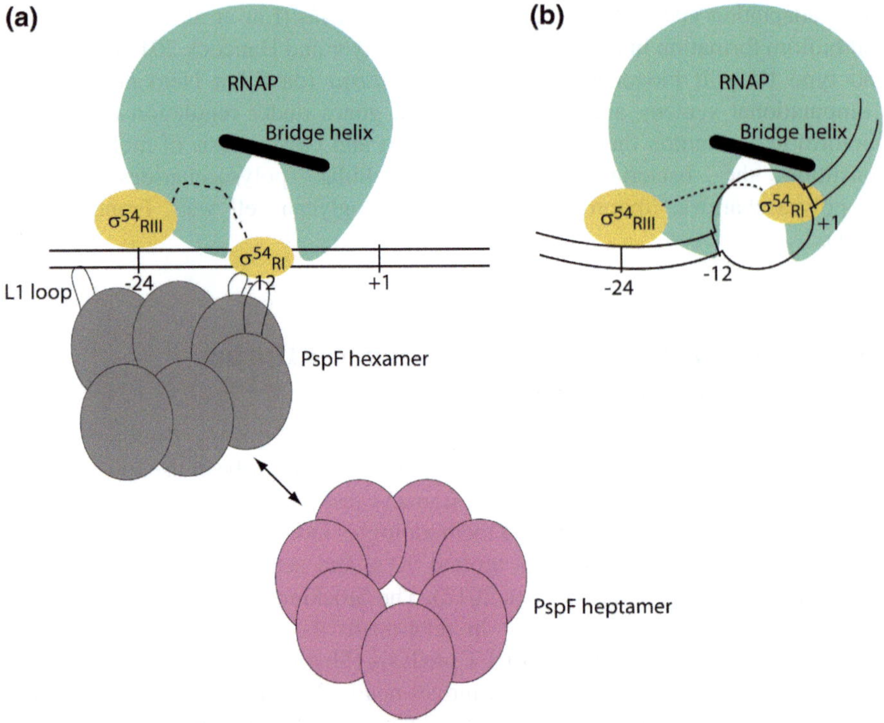

Fig. 13.1 Structural reorganisation of a sigma54-dependent transcription complex. **a** The apo AAA^+ domain of the sigma54 activator (PspF) can form planar heptamers (and some hexamers) with low affinity for nucleotides (*pink*), but assume an open spiral hexameric configuration (*grey*) during ATP hydrolysis and its binding to RP_C. The sigma54 Region I (σ^{54} RI) in the RP_C binds to the −12 inhibitory fork junction and blocks the RNAP DNA-binding channel for accessing the RNAP active site, while the sigma54 Region III (σ^{54} RIII) binds to the −24 element. ATP binding and hydrolysis drives at least two adjacent L1 loops to engage σ^{54} RI at the downstream end and one L1 loop to engage the non-template strand of the promoter DNA at the −30 upstream region. **b** Possibly via a 'power stroke' action, the PspF L1 loop moves σ^{54} RI from the −12 fork junction to the +1 site. This domain reorganisation brings σ^{54} RI in close proximity to the bridge helix. Both sigma54 and PspF L1 loop may contribute to the DNA melting from −10 to −1 (bubble). For simplicity, the components are not drawn to scale

Region I and at the upstream end by an L1 loop and the non-template −30 promoter DNA (Fig. 13.1a, Bose et al. 2008; Rappas et al. 2005; Zhang et al. 2012). Recent structural elucidation of the RNAP-sigma54 holoenzyme by the Zhang lab shed new light on the sigma54 inhibitory mechanism (Yang et al. 2015). In order for the DNA template strand to be loaded into the active channel, the blocking sigma54 Regions I–III 'gate' must be shifted (Fig. 13.2a). In contrast, the template strand can pass through the 'V'-shaped wedge formed between sigma70 regions 2 and 3 (Fig. 13.1c) or between the TFIIB core and linker regions (Yang et al. 2015). In line with the structural data, sigma54 variants that disrupt the Regions I–III 'gate'—such as deletion of Region I or mutations of Region III residue R336—can spontaneously bypass the activation energy barrier (Schafer et al. 2015; Chaney et al. 2001; Wang et al. 1995, 1997). The aromatic residues in sigma54 Regions I and III and the PspF L1 'GAFTGA' motif could potentially facilitate the DNA melting process.

As shown in Fig. 13.2, the downstream DNA channel is blocked by both sigma54 Region II and sigma70 region 1.1 (Fig. 13.2b, d). The core RNAP binding domain (CBD) on sigma54 contacts the β flap feature and blocks the RNA exit

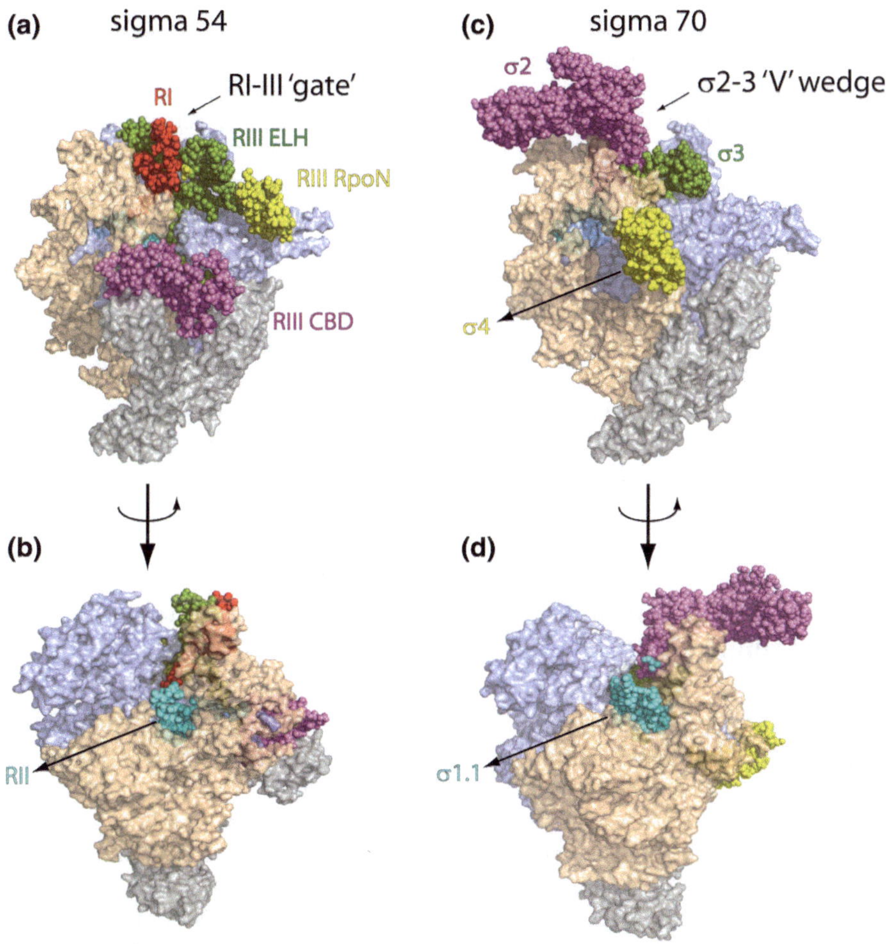

Fig. 13.2 Comparisons of sigma54 and sigma70 holoenzymes. **a**, **b** RNAP-sigma54 with RNAP displayed as surface and sigma54 as spheres. The sigma54 Region I (RI), Region II (RII), Region III Extra Long Helix (RIII ELH), RpoN, Core Binding Domain (CBD) are highlighted in *red*, *cyan*, *green*, *yellow* and *pink* respectively. The RNAP β, β′ and α subunits are highlighted in *blue*, *wheat* and *grey* respectively. **c**, **d** RNAP-sigma70. The sigma70 regions 1.1, 2, 3, and 4 (σ 1.1–4) are highlighted in *cyan*, *pink*, *green* and *yellow* respectively. The same colour code is assigned to domains with similar functions between RNAP-sigma54 and RNAP-sigma70

channel, and suggests that sigma54 must dissociate to allow the nascent RNA to extend beyond 8 nts (Fig. 13.2a). This obligatory dissociation of sigma54 in early transcription is notable, as it is not found with sigma70 where the CBD (σ2 region) binds to the surface of the core RNAP to allow its retention during elongation (Fig. 13.2c). Comparisons between sigma54 and sigma70 holoenzyme structures suggest that the two sigma factors employ different functional domains to contact similar regions on the core RNAP (Fig. 13.2).

A transcription inhibitor—Gp2—encoded by the bacteriophage T7 has been shown to significantly affect RP_O formation in the sigma70-containing but not the sigma54-containing RNAP complexes (Wigneshweraraj et al. 2004). Recent structural elucidation of the Gp2-sigma70 holoenzyme suggests that Gp2 bridges between the β' jaw domain and sigma70 region 1.1 through electrostatic interactions, thereby preventing the egress of region 1.1 from the active channel and blocking the promoter DNA from entering (Bae et al. 2013). In comparison, it is unclear if Gp2 binds to Region II of sigma54, which is located in a similar position to that of region 1.1 of sigma70. Furthermore, the β' jaw domain in the sigma54-containing RNAP changes its conformation upon activator binding and ATP hydrolysis (Wigneshweraraj et al. 2004). Both could contribute to its insensitivity towards any Gp2 inhibition.

Describing the conformational change pathway to RP_O will require further structural determinations, and in particular complexes with fully and partially opened promoter DNA templates, with and without the bound activators and at high resolution. Recent advances in detector technology and image processing algorithms promise a major continuing role for cryo-electron microscopy in such structural studies.

13.5 Introduction of the Psp Regulon and Its Inducing Stimuli

The sigma54-dependent Psp regulon in enterobacterial pathogens is triggered by extracytoplasmic stresses that damage the inner membrane (IM) integrity, and is recognised as of major importance in biofilm formation, virulence, macrophage infection, antibiotics resistance and persistence (Darwin 2013; Joly et al. 2010; Rowley et al. 2006; Wallrodt et al. 2014). A large variety of stimuli can cause IM stresses and lead to dissipation of the proton motive force (pmf) and changes in the redox states (Darwin 2013; Joly et al. 2010; Rowley et al. 2006). These include mislocalisation of OM secretins (filamentous phage protein pIV and its bacterial homologues, such as YscC, PulD, OutD) into the IM, malfunction of the protein translocation systems, blockage of phospholipids and peptidoglycan biogenesis, antimicrobials that target membrane biosynthesis and exposure to bile salts or the intracellular environment of phagocytic cells. The Psp response is also induced upon exposure to organic solvents (e.g., ethanol and methanol), hyperosmotic shock, extreme temperatures as well as incorporations of large protein assemblies, protonophores and fatty acids into the IM. These stimuli may affect the IM potential, the net charge on anionic lipids (redox states), membrane fluidity/rigidity, phospholipids packing and the stored curvature elastic (SCE) stress.

In *E. coli*, there are seven genes present in the Psp regulon (Fig. 13.3a): the *pspABCDE* operon at 29.44 min (under the *pspA* promoter), divergently oriented *pspF* (*pspF* promoter) and *pspG* at 91.84 min (*pspG* promoter). Transcription from the *pspA* and *pspG* promoters is dependent on sigma54 and the PspF transcription activator (Jovanovic et al. 1996; Lloyd et al. 2004; Weiner et al. 1991). Notably, the two promoters possess an altered −12 recognition element (GT or GA *in Y. enterocolitica*) in place of the usual consensus GC element in enterobacteria (Huvet et al. 2011). The PspF activators hexamerise at their cognate upstream activating

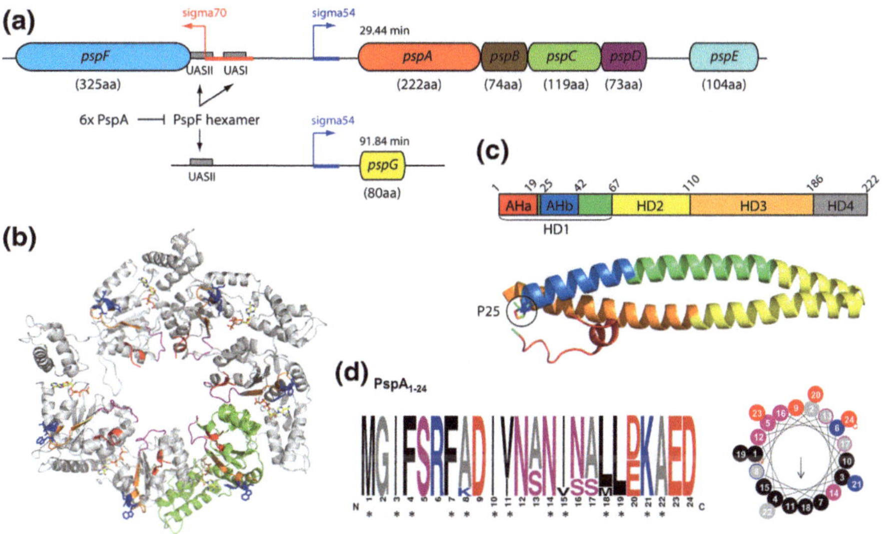

Fig. 13.3 Schematic representation of the Psp regulon. **a** There are seven *psp* genes located on two loci in *E. coli*. The promoter control and the number of amino acid residues (aa) of each protein are indicated. **b** The PspF hexamer model (based on energy minimisation) is depicted in *grey* ribbons, with one subunit highlighted in *green* to delineate the boundaries of the self-association interface. Key residues (the 'YLW' patch, *blue*) in the W56 loop (*orange*), L1 loop (*red*), Linker 1 (*pink*), and ATP (*rainbow*) are highlighted in the hexamer. **c** The predicted helical domain (HD) organisation and an available crystal structure of PspA$_{1\text{-}144}$ (PDB entry 4WHE). The helix break residue P25 (*rainbow*) that connects the two amphipathic helices (AHa, *red* and AHb, *blue*) is highlighted. The AHb has been shown to interact with PspF and the AHa folds from a random structure to an α-helix upon lipid binding and membrane engagement. **d** Sequence alignment and the helical wheel of the PspA AHa domain (residues 1–24) from Gram-negative bacteria, including *Escherichia coli, Shigella flexeri, Salmonella enterica, Yerssinia enterocolitica, Erwinia piriflorinigrans, Vibrio cholera, Shewanella oneidensis, Photobacterium profundum, Idiomarina loihiensis* and *Desulfovibrio fructosivorans*. Residues are colour-coded for their properties: negatively charged—*red*, positively charged—*blue*, hydrophobic—*black*, polar—*purple* and neutral—*grey*. On the helical wheel, the *core colour* corresponds to the primary conserved residue property while the *border colour* to the secondary conserved property

sequence (UAS) and are brought to close proximity to the sigma54-containing RNAP by DNA looping as facilitated by the integration host factor (IHF) (Lloyd et al. 2004; Jovanovic and Model 1997). In contrast to the single UAS site present on the *pspG* promoter (between −78 and −92), the *pspA* promoter contains two UAS sites (between −80 and −126, Fig. 13.3a). The greater transcription strength of the *pspA* promoter may be particularly suited for controlling PspA levels to achieve its negative repressor function (acting on PspF) as well as its effector function (acting on the IM) under stress conditions (Lloyd et al. 2004; Seo et al. 2007). Binding of the PspF hexamers to the UAS sites is a prerequisite for activation of the *pspA* operon. However, it collides with *pspF* gene transcription in the opposite orientation (Fig. 13.3a). So PspF expression is under control of an autonomous negative feedback loop (Jovanovic et al. 1997). Congruent with this observation, microarray analyses in *E. coli*, *Salmonella typhimurium* and *Y. enterocolitica* indicate that under IM stresses (pIV secretin production) the *pspA* and *pspG* promoters are strongly induced while the *pspF* promoter remains largely unaffected (Lloyd et al. 2004; Seo et al. 2007).

13.6 Activation of the Psp Regulon by PspF

The PspF activator unlike many other activators of the sigma54-RNAP lacks an N-terminal regulatory domain and so it is active *per* default (Jovanovic et al. 1996). Rather, the catalytic AAA$^+$ ATPase domain of PspF is negatively regulated *in trans* by PspA to achieve control of transcription outputs (Joly et al. 2009). The concentration-dependent oligomerisation of PspF plays a key role in both ATP hydrolysis and energy coupling to the RP_C remodelling event. The hydrolytic state of ATP bound at the hexameric interface is sensed by the Walker motifs and arginine hand, and relayed to L1 loop movement by the 'Glutamate Switch' pair (E108-N64 in PspF) via the W56 loop and Linker 1 (Fig. 13.3b, Joly et al. 2012; Zhang and Wigley 2008). Disruption of the 'Glutamate Switch' leads to phenotypes that decouple the inhibitory interactions from DNA melting (Darbari et al. 2014). After hydrolysis, the 'ADP-switch' pair (E43-Y126 in PspF) is thought to facilitate the ADP release (Joly et al. 2012).

In the absence of ATP, the AAA$^+$ domain of PspF self-associates into heptamers with low nucleotide-binding affinities (Fig. 13.1a, Zhang et al. 2014). The cooperative binding of nucleotides in at least two adjacent subunits causes the L1 loops from these subunits to engage sigma54 Region I (Fig. 13.1a). This functional asymmetry in PspF subunits may well correlate with a heterogeneity in nucleotide occupancy and be key to achieving an ATPase driven remodelling of RP_C (Joly et al. 2006). The 'GAFTGA' motif in PspF L1 loop, along with aromatic residues in sigma54 Regions I and III, might be presented in a way to facilitate promoter DNA melting. The shedding of the seventh PspF subunit and the opening of the PspF hexameric ring might occur in the transition state (Fig. 13.1a, Zhang et al. 2014). The physiological relevance of any heptameric assembly is a matter of debate not only for the AAA$^+$ activators of sigma54 but for AAA$^+$ proteins in general, and so any roles in vivo are debatable. The observed nucleotide-driven functional asymmetry and partial sequential hydrolysis model may conform to a ubiquitous mechanism as employed by the sigma54 activator family, or at least by PspF and NtrC1 (Zhang et al. 2014; Sysoeva et al. 2013).

13.7 Negative Regulator Function of PspA

PspA is a bifunctional IM binding protein: it negatively regulates the PspF activator under non-stress conditions and switches to an effector function when the IM is stressed (Joly et al. 2010). Like its homologue Vipp1 that is essential in thylakoid biogenesis and photosynthesis in cyanobacteria, green algae and plants (Vothknecht et al. 2012; Westphal et al. 2001; Zhang and Sakamoto 2013), PspA is composed of helical domains (HD 1–4, Fig. 13.3c). The last helical domain (HD4, residues 187–222) has been implicated in higher order oligomerisation associated with PspA's effector function (Joly et al. 2009). Deletion of this domain does not seem to affect PspA's ability to bind and inhibit PspF in vitro (Elderkin et al. 2005). Recent structural elucidation of a PspA fragment (residues 1–144, Fig. 13.3c) revealed a coiled-coil topology similar to the M-domain of the unfoldase ClpB (Osadnik et al. 2015). The HD1 (residues 1–64) of PspA contains two N-terminal amphipathic helices (AHa, residues 2–19; AHb, residues 25–42, Fig. 13.3c) (Jovanovic et al. 2014). It has been proposed that the AHa domain can sense anionic lipids and lipid-packing defects caused by membrane SCE stress or curvature. AHa may then insert its hydrophobic face (Fig. 13.3d) into the lipid bilayer to alleviate the membrane stress via a wedge effect (McDonald et al. 2015). The AHb domain of PspA has been implicated in and binding to the 'YLW56' hydrophobic patch on the surface of PspF, so shutting down the ATPase activity and uncoupling across the hexameric AAA$^+$ ring the hydrolysis-driven L1 loop movement (Jovanovic et al. 2014; Zhang et al. 2013).

The dynamics of PspA and PspF was studied in vitro and in live cells using single-molecule tracking and photobleaching approaches. Under non-stress conditions, up to six PspA monomers are recruited to inhibit one PspF hexamer at the nucleoid (Joly et al. 2009; Osadnik et al. 2015; Jovanovic et al. 2014; Mehta et al. 2013). Occasionally, the PspA-F inhibitory complexes are transiently recruited to the IM cardiolipin-rich polar regions by PspB and PspC for stress sensing—this leads to a partial release of PspF and thus a basal level of *psp* gene expression. Indeed, over-production of PspB and PspC induces the Psp regulon in the absence of stress signals (Joly et al. 2010; Darwin 2005). The PspA-F inhibitory complexes or PspF hexamers rarely occupy both *pspA* and *pspG* promoters simultaneously (Mehta et al. 2013). The heterogeneous promoter occupancy is likely due to the limited amount of intracellular PspF activators. Such limitation could potentially lead to variations in *psp* gene expression—hence differential stress responses—within a population of cells. Joly et al. (2009) showed that the PspA-F inhibitory complex can still engage sigma54. How PspA inhibits the ATPase activity of PspF is not known in detail, but mutational studies suggest the interaction of PspA at the W56 loop causes changes at the ATPase active site (Joly et al. 2009). Inhibition of transcription activation amy additionally then involve a loss of nucleotide-dependent co-ordination between the subunits of PspF (Zhang et al. 2014). It is possible that the elongated PspA folds back on the outer rim of the PspF hexamer so as to vacate for sigma54 interactions. Similar domain folding events have been observed in the N-terminal regulatory domains of other sigma54 activators such as NtrC and NorR (Bush and Dixon 2012; De Carlo et al. 2006).

13.8 Effector Function of PspA

Under membrane stress conditions, PspA dissociates from the inhibitory complexes made with PspF to form higher order oligomers (≥36 mers) with a calculated mass of 1034 kDa, thus releasing PspF for transcription activation (Fig. 13.4, McDonald

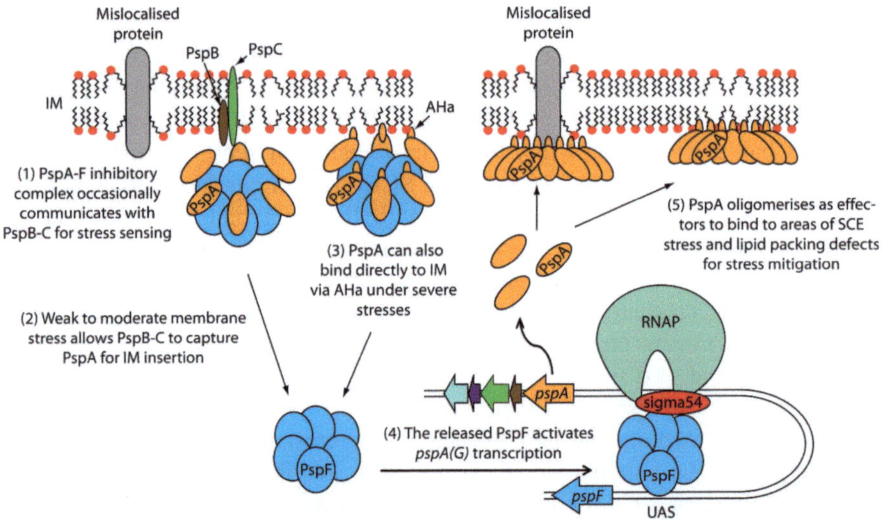

Fig. 13.4 The proposed model of the Psp response in pathogens. Under non-stress conditions, the PspA-F inhibitory complex is transiently recruited to the PspB and PspC sensors located in the IM curved polar regions. Under weak to moderate stress conditions, the PspB-C active complex stably engages PspA for its subsequent membrane binding via the AHa domain (see text for detail). This step releases PspF for activation of the Psp regulon. PspA produced in a large quantity oligomerises (up to 36 mers) to target IM areas of SCE stress and lipid-packing defects. Such defects are proposed to exist around IM mislocalised proteins such as the secretins of secretion systems. The *psp* induction also leads to a high level expression of PspB and PspC which in their active states keeps PspA from switching back to the negative regulator function. Under severe stress conditions, the PspA-F complex or PspA alone can directly bind to the IM, bypassing the PspB-C requirement

et al. 2015; Yamaguchi et al. 2013). Such large PspA assemblies have been studied using cryo-electron microscopy and by single-molecule photobleaching (Jovanovic et al. 2014; Hankamer et al. 2004; Lenn et al. 2011). They are proposed to bind to liposomes, thereby, directly suppressing a proton leakage (Kobayashi et al. 2007). Binding of the PspA higher order oligomers to the IM is facilitated by its AHa domain (Jovanovic et al. 2014) that upon membrane association folds from a random coil into an α-helix (C.M., G.J., O.C., M.B., unpublished data, Figs. 13.3c, d, 13.4). The PspA-IM interaction via AHa and the PspA-F interaction via AHb (i.e., effector function vs. negative regulator function) are mutually exclusive. This functional exclusivity is thought to be maintained by a highly conserved P25 helix-breaking residue that separates the two amphipathic helices in PspA (Fig. 13.3c, Jovanovic et al. 2014). It seems that the higher order oligomeric state is an intrinsic property of PspA when PspF is not present. PspA has been shown to form active oligomeric effectors in heterologous systems (such as bacteria, archaea and plants) that carry only PspA homologues (Otters et al. 2013; Wolf et al. 2010).

Work with purified components has established that SCE stresses and the presence of anionic phospholipids are important for membrane binding and the effector function of PspA (McDonald et al. 2015). Increasing the SCE stress in vesicle membranes by varying their phospholipid composition causes a release of PspA-F inhibitory complexes and elevates PspA-phospholipid interactions via AHa. Depolarisation of the plasma membrane may induce a nano-scale reorganisation of the negatively charged anionic lipids microdomains containing phosphatidyl-glycerol (PG) and phosphatidyl-serine (PS) and trigger a nano-clustering of membrane-targeting PspA proteins via electrostatic interactions. SCE stress and a negatively charged lipid bilayer drive the association of PspA to the IM independently, and they do not seem to reinforce one another. For instance, the negative charge on anionic lipids dampens the SCE stress sensing by PspA (McDonald et al. 2015).

PspA higher order oligomers may also play a key role in the maintenance of torque tension stored in the IM within a critical range thus preventing phase transition into a porous state upon IM stress. The presence of ordered anionic lipid microdomains restricts lateral phospholipid diffusion and imparts high stresses within the IM. It has been shown that insertions of randomly aggregated synthetic PspA AHa peptides could further elevate stress and eventually lead to a membrane phase transition to a porous state (C.M., G.J., O.C., M.B., unpublished data). However, the full-length PspA protein or the high-order PspA scaffold may provide a regular dispersion of the AHa domains and limit IM contacts.

The effects of anionic lipids and SCE stress observed with PspA also apply to Vipp1 (a PspA homologue in cyanobacteria) and its AHa domain. However, differences in effector functions between PspA and Vipp1 have been observed. For instance, Vipp1 can substitute for PspA in *E. coli*, but PspA can only partially replace Vipp1's function (DeLisa et al. 2004). Vipp1 is primarily involved in anionic lipids interactions, which is consistent with its proposed function of thylakoid membrane fusion (McDonald et al. 2015; Hennig et al. 2015). This preference to anionic lipids recognition is likely to be attributed to a higher percentage of cationic residues in Vipp1's AHa domain (McDonald et al. 2015). PspA primarily senses the SCE stress with less of an involvement in anionic lipids interactions (McDonald et al. 2015). Despite these differences, it is clear that a ubiquitous membrane stress response amongst functional homologues is employed across multiple domains of life, from enterobacteria to plants.

In *E. coli*, the PspA higher order oligomers move along a helical path and evidence from interaction studies suggests this may well occur by forming direct complexes with MreB (Engl et al. 2009). The actin-like MreB is essential in maintaining peptidoglycan biosynthesis and cell morphology and is implicated in diffusional restriction along the plasma membrane microdomains (Dominguez-Escobar et al. 2011; Strahl et al. 2014; van den Ent et al. 2010). The MreB protein governs the dynamics of PspA along the IM to deliver it to distinct membrane regions. Many of these regions are marked by flotillin YqiK and RodZ and contain the peptidoglycan biosynthesis machineries (Jovanovic et al. 2014). Apparently, addition of amphipathic lipid II to the bilayer during peptidoglycan biosynthesis may cause SCE stress in membrane microdomains containing the peptidoglycan biosynthesis machinery and so be the target of PspA effectors. The restricted diffusion, along with the transient protein–protein and protein–lipid interactions, within these lipid microdomains may be important for IM engagement by PspA effectors and the SCE stress mitigation process. In *Bacillus subtilis*, the PspA homologue LiaH displays MreB-independent dynamics in the cytoplasm prior to stress and assembles with LiaI to form static complexes along the membrane upon membrane stresses (Dominguez-Escobar et al. 2014).

13.9 Signal Transduction via PspB and PspC

PspB and PspC are IM proteins that positively regulate the Psp regulon and under microaerobic growth link the control of the Psp regulon to respiration and the ArcAB two-component system (Joly et al. 2010). They sense and transduce the IM stress signals to PspA following a direct and stable engagement with the PspA-F inhibitory complex (Fig. 13.4). The co-localisation and function of PspB and PspC in the IM polar regions depend on flotillin YqiK and anionic phospholipid cardiolipin (which is preferentially found in the curved polar IM regions) (Jovanovic et al. 2014; Yamaguchi et al. 2013). Changes in membrane potential alone, following stimulation, dissipation of pmf and metabolic shutdown, is not sufficient to significantly induce the Psp response in vivo (Engl et al. 2011; Wang et al. 2010). However, membrane depolarisation in conjunction with membrane distortion might be the signals required for strong psp induction. The PspB and PspC sensors located in the polar curved IM regions are sensitive to changes in membrane potential and tension, and well placed to recognise IM damages for subsequent signal integration. PspC is a polytopic protein and its C-terminus (Ct) appears to be able to undergo topological changes upon membrane stresses to facilitate its role in stress signaling (Flores-Kim and Darwin 2012, 2015; Jovanovic et al. 2010). One possible scenario is that the PspC Ct domain stays in the periplasm under non-stress conditions or flips from the periplasm to the cytoplasm for PspB interaction upon minor changes in the head-group charges of the anionic lipids located in the IM inner leaflet. This inactive complex of PspB-C only allows transient interactions with the PspA-F inhibitory complex and permits a basal level of *psp* expression. Upon experiencing a second stress signal—related to changes in physico-chemical characteristics of the IM phospholipids—the PspC Ct domain switches to stably interact with PspA, possibly via the AHb domain. The stable binding of PspA to PspC Ct has two functional consequences (Fig. 13.4): (1) The PspA-F inhibitory complex is either dismantled or cannot reform to allow a high level of *psp* gene expression, and (2) PspA is encouraged to bind to the curved polar IM regions and form higher order oligomers (36 mers) as effectors. It is likely that bacteria have developed a 'double checkpoint' mechanism via the PspB and PspC sensors in dealing with low to moderate IM stresses.

The signal transduction pathway leading to increased *psp* promoter activity may occur in a PspB/C-dependent or independent manner that varies with the severity of IM stresses (Joly et al. 2010; Darwin 2005; Model et al. 1997). For instance, if the stress signal is weak to moderate, such as the IM mislocalisation of pIV secretins, the PspB and PspC sensors are targeted for *psp* gene activation. In contrast, severe signals that significantly depolarise the IM or greatly increase the membrane SCE stress (e.g., extreme temperature, hyperosmotic or ethanol shock) can be directly recognised by the PspA-F inhibitory complex without the intermediate step of PspB/C consultation (Fig. 13.4). This PspB/C-bypass response also occurs when the secretin IM mislocalisation is combined with a blockage in peptidoglycan biosynthesis or when the quantity of secretin pIV and fatty acids are elevated above a threshold in anaerobiosis (Jovanovic et al. 2009, 2014). Overall, the PspA-F inhibitory complex may serve as a minimal regulatory unit and an emergency shortcut in dealing with severe IM stresses in enterobacteria.

13.10 Additional Psp Effectors to Mitigate IM Stresses

In *Y. enterocolitia*, the PspB and PspC sensors function as effectors in dealing with secretin-induced membrane stresses (Gueguen et al. 2009; Horstman and Darwin 2012). In *Salmonella enterica*, the loss of PspB function causes a reduction in virulence (Wallrodt et al. 2014). In *E. coli*, members of the Psp regulon also participate in the conservation of energy and pmf. PspD is a peripheral IM protein that can support and to some extents substitute PspA for the effector function (Jovanovic et al. 2006). PspE is a *bona fide* periplasmic rhodanase (Adams et al. 2002; Cheng et al. 2008; Chng et al. 2012). Under IM stress conditions, it may function to repair the damaged Fe–S clusters present in the IM respiratory enzymes and/or support the disulfide bond formation in the increasingly oxidative environment in the periplasm. PspG interacts with PspC in the static polar complexes and migrates to specific membrane regions in an MreB-dependent fashion as seen for PspA (Engl et al. 2009; Jovanovic et al. 2010). The effector function of PspG is needed to fully ameliorate the IM stress in *E. coli* (Lloyd et al. 2004; Engl et al. 2009; Jovanovic et al. 2006, 2010). When over-produced, PspG and PspA jointly down-regulate H^+ mediated transport via IM and cell motility that consumes the pmf (Jovanovic et al. 2006). They also bring down the glycerol shift in order to promote the synthesis of glycerol-3-phosphate which can be used in phospholipids biogenesis and potential replacement of damaged anionic lipids during membrane stresses (Jovanovic et al. 2006; Bury-Mone et al. 2009). In addition, they up-regulate the polyamine spermidine production by nearly 70-fold in order to slow down protein production and elevate the formate-nitrate respiration mode to conserve the pmf.

13.11 Psp Proteins in Pathogenicity and Antimicrobial Responses

The Psp response was initially discovered in enterobacteria following the production of filamenous phage secretin pIV. Since then, it has been implicated in support of cellular growth, adaptation and delivery of toxins and virulence factors in many pathogens (Darwin 2013). Many of the OM associated secretion systems, such as T2/3/4 SS, are affected by the Psp response genes (Joly et al. 2010). For enterobacterial pathogens, surviving the antibacterial action of bile salts is of the uppermost importance. *E. coli* mutants that lack the *psp* genes are not resistant to bile salts and show impaired growth in the stationary phase (Adams et al. 2003). When *Salmonella* consumes phospholipids in bile salts, production of PspA protects its IM and potentially elevates the synthesis of phospholipids (Antunes et al. 2011). The Psp response, including PspA and its homologues, is also required for intracellular adaptation, virulence and survival of proteobacteria in phagocytic cells (Darwin 2013; Mavromatis et al. 2015; Southern et al. 2015), particularly important in mycobacteria infection and survival inside the macrophages (Datta et al. 2015; White et al. 2011).

The Psp response genes are involved in biofilm formation and chronic infections of multi-drug resistant enterobacterial persister cells (Dhamdhere and Zgurskaya 2010; Keren et al. 2004; Ma et al. 2010; Shah et al. 2006). Persistence can be achieved by abolishing the glycerol shift (aerobic respiration) and increasing phospholipids synthesis in glycerol-3-phosphate dehydrogenase (*glpD*) mutants (Girgis et al. 2012). This type of persistence is congruent with the metabolic fine-tuning of Psp effectors (see also above, Sect. 13.10). The indol-induced formation of persister cells depends on the induction of *psp* and *oxyR* genes, when the target protein Flu—a membrane antigen responsible for biofilm formation—is absent (Kint et al. 2012; Vega et al. 2012, 2013). Notably, an increased production of PspA and PspG leads to a fivefold down-regulation of Flu and a 13-fold up-regulation of YodA (another OxyR target, Jovanovic et al. 2006). Psp (in the absence of sigmaE) and YodA are important in cadmium-resistance, divalent cation transport and virulence in *S. typhimurium* (Becker et al. 2005; Wang and Crowley 2005). In addition, synthesis of PspA and PspG up-regulates the transport and biosynthesis of polyamine spermidine in order to arrest cells in a metabolically stagnant state and polyamines are required for virulence of e.g., *S. enterica* (Jelsbak et al. 2012).

The PspA homologue LiaH is present in Gram-positive pathogens and directly protects cells from antibiotics that work on cell walls or peptidoglycan (e.g. bacitracin, nisin, ramoplanin, vancomycin and cationic peptides (Joly et al. 2010; Wolf et al. 2010; Jordan et al. 2008). LiaH confers resistance to daptomycin (a class of antibiotics that binds to phosphatidyl-glycerol and reorganises the membrane architecture Hachmann et al. 2009; Pogliano et al. 2012), vancomycin and nisin (both are lipid II interacting compounds, Breukink and de Kruijff 2006). Given the structural similarity between the AHas of LiaH and PspA, it is possible that LiaH remodels the membrane using this domain to counter the effects of daptomycin. Interestingly, LiaH is strongly induced by biosurfactants (rhamnolipids, Wecke et al. 2011). Biosurfactants are produced by Gram-negative bacteria to assist growth competition, and by inference, the Psp response might be involved in intercellular competition and contact-dependent growth inhibition.

The highly hydrophobic IbsC peptides that are toxic for cells can induce the Psp response in *E. coli* (Fozo et al. 2008). After penetrating the membrane, the cationic antimicrobial peptides sequester anionic phospholipids and cause displacement of membrane proteins required for biogenesis and respiration (Epand et al. 2011) and so the specialised microdomains harbouring PG, PS and CL and Psp proteins might serve as potential targets for treating chronic persistent infections (Allison et al. 2011; Hurdle et al. 2011). In addition, the vesicle destabilising properties of PspA and Vipp1 AHa-derived peptides, in particular the Vipp1 AHa peptide (residues 1–24), could be used specifically for antimicrobial therapies, as the eukaryotic membranes carry a much lower net negative charge (C.M., G.J., O.C., M.B., unpublished data). Finally, the mutants of key Psp response genes/determinants might help in development of attenuated vaccine against bacterial pathogens, including enterobacteria and mycobacteria.

13.12 Potential Cross-Talks with Other Cell Envelope Protection Networks

The integrity of the cell envelope is constantly monitored by at least five multi-component signal transduction systems; alongside the Psp system, there are Cpx, sigmaE, Rcs and Bae systems (Flores-Kim and Darwin 2014). Although these systems respond to stresses acquired in different envelope compartments, it is reasonable to speculate a level of their interplay might well occur when common cellular processes—such as motility, biofilm formation, virulence and antimicrobial resistance—are targeted. The Cpx system responds to accumulation of misfolded cell envelope proteins in the periplasm. Recently, it has been associated with sensing the IM surface charge as influenced by altered lipid composition and perturbations in the IM (Keller et al. 2015). In addition, the Cpx response leads to changes in the peptidoglycan structure and fosfomycin resistance (Bernal-Cabas et al. 2015; Kurabayashi et al. 2014). Similar to the Psp system, it regulates chemotaxis, motility, adhesion, biofilm formation, pathogenicity, resistance to antimicrobials, and T3/4 secretion systems (Raivio 2014). The sigmaE system mainly controls the OM homeostasis. The extracytoplasmic sigmaE regulates the system which senses the accumulation of unassembled OM proteins (e.g., porins) and LPS in the periplasm and regulates cellular responses that ameliorate envelope perturbations (Lima et al. 2013). It has been established that under severe cell envelop damages, all three Psp, Cpx and sigmaE systems are induced and their responses are likely to be interlinked in order to minimise the envelope damage (Joly et al. 2010). The Rcs system responds to damages in the OM and peptidoglycan and its action might depend on the Cpx system (Evans et al. 2013). Activation of the Rcs response down-regulates gene expression in motility and virulence but is required for normal biofilm formation and periplasmic space content quality control. Under osmotic stress conditions, the Rcs system enhances capsule production to protect the OM while the Psp system protects the IM (Laubacher and Ades 2008). The Bae system specifically responds to alterations and damages of the envelope caused by toxic agents and so as to up-regulate efflux pumps to remove the toxins. It also target drug efflux systems and so is implicated in multi-drug transport and resistance (Raivio 2005).

13.13 Conclusions

The major alternative sigma54 factor is a global regulator of genes involved in metabolism and virulence in pathogens. Sigma54-dependent transcription requires activation by ATPase activators that form homo- or hetero-hexamers in respond to their cognate regulators. These regulators are often part of a two-component system that helps cells sense environmental cues. The Psp response we described in this chapter is an example of how pathogens employ the sigma54-dependent transcription system to perceive and ameliorate various membranes stresses. The success of this system lies in a coordinated cascade of protein expressions and a delicate balance between regulator and effector functions. Depending on the severity of membrane stresses, the Psp system can fine tune the signal transduction and response or take an emergency shortcut by bypassing several sensors and effectors. This autonomous pathway selection represents a level of cellular intelligence and is vital for the survival of pathogens in their niches. Understanding the function of the Psp response, and by extension the sigma54-dependent transcription system, is of broad interest not only for detecting the complex cellular regulatory circuitries but also for developing novel antimicrobial targets and therapeutic cationic peptides with bactericidal properties and has been greatly advanced through findings arising from biophysical and structural biology.

Acknowledgements The work conducted in our laboratory was supported by the Wellcome Trust, Biotechnology and Biological Sciences Research Council (BBSRC) research grants (BB/J002828/1, BB/G001278/1 and BB/H012249/1), and the Leverhulme Trust Research project grant (RPG-2012-705).

References

Adams H, Teertstra W, Koster M, Tommassen J (2002) PspE (phage-shock protein E) of Escherichia coli is a rhodanese. FEBS Lett 518(1–3):173–176

Adams H, Teertstra W, Demmers J, Boesten R, Tommassen J (2003) Interactions between phage-shock proteins in Escherichia coli. J Bacteriol 185(4):1174–1180

Allison KR, Brynildsen MP, Collins JJ (2011) Metabolite-enabled eradication of bacterial persisters by aminoglycosides. Nature 473(7346):216–220. doi:10.1038/nature10069

Antunes LC, Andersen SK, Menendez A, Arena ET, Han J, Ferreira RB, Borchers CH, Finlay BB (2011) Metabolomics reveals phospholipids as important nutrient sources during Salmonella growth in bile in vitro and in vivo. J Bacteriol 193(18):4719–4725. doi:10.1128/JB.05132-11

Bae B, Davis E, Brown D, Campbell EA, Wigneshweraraj S, Darst SA (2013) Phage T7 Gp2 inhibition of Escherichia coli RNA polymerase involves misappropriation of sigma70 domain 1.1. Proc Natl Acad Sci USA 110(49):19772–19777. doi:10.1073/pnas.1314576110

Becker LA, Bang IS, Crouch ML, Fang FC (2005) Compensatory role of PspA, a member of the phage shock protein operon, in rpoE mutant Salmonella enterica serovar Typhimurium. Mol Microbiol 56(4):1004–1016. doi:10.1111/j.1365-2958.2005.04604.x

Bernal-Cabas M, Ayala JA, Raivio TL (2015) The Cpx envelope stress response modifies peptidoglycan cross-linking via the L, D-transpeptidase LdtD and the novel protein YgaU. J Bacteriol 197(3):603–614. doi:10.1128/JB.02449-14

Bose D, Pape T, Burrows PC, Rappas M, Wigneshweraraj SR, Buck M, Zhang X (2008) Organization of an activator-bound RNA polymerase holoenzyme. Mol Cell 32(3):337–346

Breukink E, de Kruijff B (2006) Lipid II as a target for antibiotics. Nat Rev Drug Discov 5(4):321–332. doi:10.1038/nrd2004

Buck M, Gallegos MT, Studholme DJ, Guo Y, Gralla JD (2000) The bacterial enhancer-dependent sigma(54) (sigma(N)) transcription factor. J Bacteriol 182(15):4129–4136

Bury-Mone S, Nomane Y, Reymond N, Barbet R, Jacquet E, Imbeaud S, Jacq A, Bouloc P (2009) Global analysis of extracytoplasmic stress signaling in Escherichia coli. PLoS Genet 5(9): e1000651. doi:10.1371/journal.pgen.1000651

Bush M, Dixon R (2012) The role of bacterial enhancer binding proteins as specialized activators of sigma54-dependent transcription. Microbiol Mol Biol Rev: MMBR 76(3):497–529. doi:10.1128/MMBR.00006-12

Carrillo CD, Taboada E, Nash JH, Lanthier P, Kelly J, Lau PC, Verhulp R, Mykytczuk O, Sy J, Findlay WA, Amoako K, Gomis S, Willson P, Austin JW, Potter A, Babiuk L, Allan B, Szymanski CM (2004) Genome-wide expression analyses of Campylobacter jejuni NCTC11168 reveals coordinate regulation of motility and virulence by flhA. J Biol Chem 279(19):20327–20338. doi:10.1074/jbc.M401134200

Chaney M, Grande R, Wigneshweraraj SR, Cannon W, Casaz P, Gallegos MT, Schumacher J, Jones S, Elderkin S, Dago AE, Morett E, Buck M (2001) Binding of transcriptional activators to sigma 54 in the presence of the transition state analog ADP-aluminum fluoride: insights into activator mechanochemical action. Genes Dev 15(17):2282–2294. doi:10.1101/gad.205501

Cheng H, Donahue JL, Battle SE, Ray WK, Larson TJ (2008) Biochemical and genetic characterization of PspE and GlpE, two single-domain sulfur transferases of Escherichia coli. Open Microbiol J 2:18–28. doi:10.2174/1874285800802010018

Chng SS, Dutton RJ, Denoncin K, Vertommen D, Collet JF, Kadokura H, Beckwith J (2012) Overexpression of the rhodanese PspE, a single cysteine-containing protein, restores disulphide bond formation to an Escherichia coli strain lacking DsbA. Mol Microbiol 85(5):996–1006. doi:10.1111/j.1365-2958.2012.08157.x

da Silva Neto JF, Koide T, Abe CM, Gomes SL, Marques MV (2008) Role of sigma54 in the regulation of genes involved in type I and type IV pili biogenesis in Xylella fastidiosa. Arch Microbiol 189(3):249–261. doi:10.1007/s00203-007-0314-x

Darbari VC, Lawton E, Lu D, Burrows PC, Wiesler S, Joly N, Zhang N, Zhang X, Buck M (2014) Molecular basis of nucleotide-dependent substrate engagement and remodeling by an AAA+ activator. Nucleic Acids Res 42(14):9249–9261. doi:10.1093/nar/gku588

Darwin AJ (2005) The phage-shock-protein response. Mol Microbiol 57(3):621–628

Darwin AJ (2013) Stress relief during host infection: the phage shock protein response supports bacterial virulence in various ways. PLoS Pathog 9(7):e1003388. doi:10.1371/journal.ppat.1003388

Darwin AJ, Miller VL (1999) Identification of Yersinia enterocolitica genes affecting survival in an animal host using signature-tagged transposon mutagenesis. Mol Microbiol 32(1):51–62

Datta P, Ravi J, Guerrini V, Chauhan R, Neiditch MB, Shell SS, Fortune SM, Hancioglu B, Igoshin OA, Gennaro ML (2015) The Psp system of Mycobacterium tuberculosis integrates envelope stress-sensing and envelope-preserving functions. Mol Microbiol 97(3):408–422. doi:10.1111/mmi.13037

De Carlo S, Chen B, Hoover TR, Kondrashkina E, Nogales E, Nixon BT (2006) The structural basis for regulated assembly and function of the transcriptional activator NtrC. Genes Dev 20 (11):1485–1495. doi:10.1101/gad.1418306 20/11/1485 [pii]

DeLisa MP, Lee P, Palmer T, Georgiou G (2004) Phage shock protein PspA of Escherichia coli relieves saturation of protein export via the Tat pathway. J Bacteriol 186(2):366–373

Dhamdhere G, Zgurskaya HI (2010) Metabolic shutdown in Escherichia coli cells lacking the outer membrane channel TolC. Mol Microbiol 77(3):743–754. doi:10.1111/j.1365-2958.2010.07245.x

Dominguez-Escobar J, Chastanet A, Crevenna AH, Fromion V, Wedlich-Soldner R, Carballido-Lopez R (2011) Processive movement of MreB-associated cell wall biosynthetic complexes in bacteria. Science 333(6039):225–228. doi:10.1126/science.1203466

Dominguez-Escobar J, Wolf D, Fritz G, Hofler C, Wedlich-Soldner R, Mascher T (2014) Subcellular localization, interactions and dynamics of the phage-shock protein-like Lia response in Bacillus subtilis. Mol Microbiol 92(4):716–732. doi:10.1111/mmi.12586

Elderkin S, Bordes P, Jones S, Rappas M, Buck M (2005) Molecular determinants for PspA-mediated repression of the AAA transcriptional activator PspF. J Bacteriol 187(9):3238–3248. doi:10.1128/JB.187.9.3238-3248.2005 187/9/3238 [pii]

Engl C, Jovanovic G, Lloyd LJ, Murray H, Spitaler M, Ying L, Errington J, Buck M (2009) In vivo localizations of membrane stress controllers PspA and PspG in Escherichia coli. Mol Microbiol 73(3):382–396. doi:10.1111/j.1365-2958.2009.06776.x

Engl C, Beek AT, Bekker M, de Mattos JT, Jovanovic G, Buck M (2011) Dissipation of proton motive force is not sufficient to induce the phage shock protein response in Escherichia coli. Curr Microbiol 62(5):1374–1385. doi:10.1007/s00284-011-9869-5

Epand RF, Mor A, Epand RM (2011) Lipid complexes with cationic peptides and OAKs; their role in antimicrobial action and in the delivery of antimicrobial agents. Cell Mol Life Sci: CMLS 68 (13):2177–2188. doi:10.1007/s00018-011-0711-9

Evans KL, Kannan S, Li G, de Pedro MA, Young KD (2013) Eliminating a set of four penicillin binding proteins triggers the Rcs phosphorelay and Cpx stress responses in Escherichia coli. J Bacteriol 195(19):4415–4424. doi:10.1128/JB.00596-13

Fisher MA, Grimm D, Henion AK, Elias AF, Stewart PE, Rosa PA, Gherardini FC (2005) Borrelia burgdorferi sigma54 is required for mammalian infection and vector transmission but not for tick colonization. Proc Natl Acad Sci USA 102(14):5162–5167. doi:10.1073/pnas.0408536102

Flores-Kim J, Darwin AJ (2012) Phage shock protein C (PspC) of Yersinia enterocolitica is a polytopic membrane protein with implications for regulation of the Psp stress response. J Bacteriol 194(23):6548–6559. doi:10.1128/JB.01250-12

Flores-Kim J, Darwin AJ (2014) Regulation of bacterial virulence gene expression by cell envelope stress responses. Virulence 5(8):835–851. doi:10.4161/21505594.2014.965580

Flores-Kim J, Darwin AJ (2015) Activity of a bacterial cell envelope stress response is controlled by the interaction of a protein binding domain with different partners. J Biol Chem 290 (18):11417–11430. doi:10.1074/jbc.M114.614107

Fozo EM, Kawano M, Fontaine F, Kaya Y, Mendieta KS, Jones KL, Ocampo A, Rudd KE, Storz G (2008) Repression of small toxic protein synthesis by the Sib and OhsC small RNAs. Mol Microbiol 70(5):1076–1093. doi:10.1111/j.1365-2958.2008.06394.x

Francke C, Groot Kormelink T, Hagemeijer Y, Overmars L, Sluijter V, Moezelaar R, Siezen RJ (2011) Comparative analyses imply that the enigmatic Sigma factor 54 is a central controller of the bacterial exterior. BMC Genom 12:385. doi:10.1186/1471-2164-12-385

Friedman LJ, Gelles J (2012) Mechanism of transcription initiation at an activator-dependent promoter defined by single-molecule observation. Cell 148(4):679–689. doi:10.1016/j.cell.2012.01.018 S0092-8674(12)00083-9 [pii]

Girgis HS, Harris K, Tavazoie S (2012) Large mutational target size for rapid emergence of bacterial persistence. Proc Natl Acad Sci USA 109(31):12740–12745. doi:10.1073/pnas.1205124109

Gueguen E, Savitzky DC, Darwin AJ (2009) Analysis of the Yersinia enterocolitica PspBC proteins defines functional domains, essential amino acids and new roles within the phage-shock-protein response. Mol Microbiol 74(3):619–633. doi:10.1111/j.1365-2958.2009.06885.x

Hachmann AB, Angert ER, Helmann JD (2009) Genetic analysis of factors affecting susceptibility of Bacillus subtilis to daptomycin. Antimicrob Agents Chemother 53(4):1598–1609. doi:10.1128/AAC.01329-08

Hankamer BD, Elderkin SL, Buck M, Nield J (2004) Organization of the AAA(+) adaptor protein PspA is an oligomeric ring. J Biol Chem 279(10):8862–8866. doi:10.1074/jbc.M307889200M307889200

Hayrapetyan H, Tempelaars M, Nierop Groot M, Abee T (2015) Bacillus cereus ATCC 14579 RpoN (Sigma 54) is a pleiotropic regulator of growth, carbohydrate metabolism, motility, biofilm formation and toxin production. PLoS ONE 10(8):e0134872. doi:10.1371/journal.pone.0134872

Hendrixson DR, DiRita VJ (2004) Identification of Campylobacter jejuni genes involved in commensal colonization of the chick gastrointestinal tract. Mol Microbiol 52(2):471–484. doi:10.1111/j.1365-2958.2004.03988.x

Hennig R, Heidrich J, Saur M, Schmuser L, Roeters SJ, Hellmann N, Woutersen S, Bonn M, Weidner T, Markl J, Schneider D (2015) IM30 triggers membrane fusion in cyanobacteria and chloroplasts. Nat Commun 6:7018. doi:10.1038/ncomms8018

Horstman NK, Darwin AJ (2012) Phage shock proteins B and C prevent lethal cytoplasmic membrane permeability in Yersinia enterocolitica. Mol Microbiol 85(3):445–460. doi:10.1111/j.1365-2958.2012.08120.x

Hurdle JG, O'Neill AJ, Chopra I, Lee RE (2011) Targeting bacterial membrane function: an underexploited mechanism for treating persistent infections. Nat Rev Microbiol 9(1):62–75. doi:10.1038/nrmicro2474

Huvet M, Toni T, Sheng X, Thorne T, Jovanovic G, Engl C, Buck M, Pinney JW, Stumpf MP (2011) The evolution of the phage shock protein response system: interplay between protein function, genomic organization, and system function. Mol Biol Evol 28(3):1141–1155. doi:10.1093/molbev/msq301

Iyer VS, Hancock LE (2012) Deletion of sigma(54) (rpoN) alters the rate of autolysis and biofilm formation in Enterococcus faecalis. J Bacteriol 194(2):368–375. doi:10.1128/JB.06046-11

Jelsbak L, Thomsen LE, Wallrodt I, Jensen PR, Olsen JE (2012) Polyamines are required for virulence in Salmonella enterica serovar Typhimurium. PLoS ONE 7(4):e36149. doi:10.1371/journal.pone.0036149

Joly N, Schumacher J, Buck M (2006) Heterogeneous nucleotide occupancy stimulates functionality of phage shock protein F, an AAA+ transcriptional activator. J Biol Chem 281(46):34997–35007. doi:10.1074/jbc.M606628200 M606628200 [pii]

Joly N, Burrows PC, Engl C, Jovanovic G, Buck M (2009) A lower-order oligomer form of phage shock protein A (PspA) stably associates with the hexameric AAA(+) transcription activator protein PspF for negative regulation. J Mol Biol 394(4):764–775. doi:10.1016/j.jmb.2009.09.055 S0022-2836(09)01194-2 [pii]

Joly N, Engl C, Jovanovic G, Huvet M, Toni T, Sheng X, Stumpf MP, Buck M (2010) Managing membrane stress: the phage shock protein (Psp) response, from molecular mechanisms to physiology. FEMS Microbiol Rev 34(5):797–827. doi:10.1111/j.1574-6976.2010.00240.x FMR240 [pii]

Joly N, Zhang N, Buck M (2012) ATPase site architecture is required for self-assembly and remodeling activity of a hexameric AAA+ transcriptional activator. Mol Cell 47(3):484–490. doi:10.1016/j.molcel.2012.06.012

Jordan S, Hutchings MI, Mascher T (2008) Cell envelope stress response in Gram-positive bacteria. FEMS Microbiol Rev 32(1):107–146. doi:10.1111/j.1574-6976.2007.00091.x

Jovanovic G, Model P (1997) PspF and IHF bind co-operatively in the psp promoter-regulatory region of Escherichia coli. Mol Microbiol 25(3):473–481

Jovanovic G, Weiner L, Model P (1996) Identification, nucleotide sequence, and characterization of PspF, the transcriptional activator of the Escherichia coli stress-induced psp operon. J Bacteriol 178(7):1936–1945

Jovanovic G, Dworkin J, Model P (1997) Autogenous control of PspF, a constitutively active enhancer-binding protein of Escherichia coli. J Bacteriol 179(16):5232–5237

Jovanovic G, Lloyd LJ, Stumpf MP, Mayhew AJ, Buck M (2006) Induction and function of the phage shock protein extracytoplasmic stress response in Escherichia coli. J Biol Chem 281 (30):21147–21161. doi:10.1074/jbc.M602323200

Jovanovic G, Engl C, Buck M (2009) Physical, functional and conditional interactions between ArcAB and phage shock proteins upon secretin-induced stress in Escherichia coli. Mol Microbiol 74(1):16–28. doi:10.1111/j.1365-2958.2009.06809.x

Jovanovic G, Engl C, Mayhew AJ, Burrows PC, Buck M (2010) Properties of the phage-shock-protein (Psp) regulatory complex that govern signal transduction and induction of the Psp response in Escherichia coli. Microbiology 156(Pt 10):2920–2932. doi:10.1099/mic.0.040055-0

Jovanovic G, Mehta P, McDonald C, Davidson AC, Uzdavinys P, Ying L, Buck M (2014a) The N-terminal amphipathic helices determine regulatory and effector functions of phage shock protein A (PspA) in Escherichia coli. J Mol Biol 426(7):1498–1511. doi:10.1016/j.jmb.2013.12.016

Jovanovic G, Mehta P, Ying L, Buck M (2014b) Anionic lipids and the cytoskeletal proteins MreB and RodZ define the spatio-temporal distribution and function of membrane stress controller PspA in Escherichia coli. Microbiology 160(Pt 11):2374–2386. doi:10.1099/mic.0.078527-0

Keller R, Arioz C, Hansmeier N, Stenberg-Bruzell F, Burstedt M, Vikstrom D, Kelly A, Wieslander A, Daley DO, Hunke S (2015) The Escherichia coli envelope stress sensor CpxA responds to changes in lipid bilayer properties. Biochemistry 54(23):3670–3676. doi:10.1021/acs.biochem.5b00242

Keren I, Shah D, Spoering A, Kaldalu N, Lewis K (2004) Specialized persister cells and the mechanism of multidrug tolerance in Escherichia coli. J Bacteriol 186(24):8172–8180. doi:10.1128/JB.186.24.8172-8180.2004

Kint CI, Verstraeten N, Fauvart M, Michiels J (2012) New-found fundamentals of bacterial persistence. Trends Microbiol 20(12):577–585. doi:10.1016/j.tim.2012.08.009

Kobayashi R, Suzuki T, Yoshida M (2007) Escherichia coli phage-shock protein A (PspA) binds to membrane phospholipids and repairs proton leakage of the damaged membranes. Mol Microbiol 66(1):100–109. doi:10.1111/j.1365-2958.2007.05893.x MMI5893 [pii]

Kurabayashi K, Hirakawa Y, Tanimoto K, Tomita H, Hirakawa H (2014) Role of the CpxAR two-component signal transduction system in control of fosfomycin resistance and carbon substrate uptake. J Bacteriol 196(2):248–256. doi:10.1128/JB.01151-13

Laubacher ME, Ades SE (2008) The Rcs phosphorelay is a cell envelope stress response activated by peptidoglycan stress and contributes to intrinsic antibiotic resistance. J Bacteriol 190 (6):2065–2074. doi:10.1128/JB.01740-07

Lenn T, Gkekas CN, Bernard L, Engl C, Jovanovic G, Buck M, Ying L (2011) Measuring the stoichiometry of functional PspA complexes in living bacterial cells by single molecule photobleaching. Chem Commun 47(1):400–402. doi:10.1039/c0cc01707h

Lima S, Guo MS, Chaba R, Gross CA, Sauer RT (2013) Dual molecular signals mediate the bacterial response to outer-membrane stress. Science 340(6134):837–841. doi:10.1126/science.1235358

Liu A, Tran L, Becket E, Lee K, Chinn L, Park E, Tran K, Miller JH (2010) Antibiotic sensitivity profiles determined with an Escherichia coli gene knockout collection: generating an antibiotic bar code. Antimicrob Agents Chemother 54(4):1393–1403. doi:10.1128/AAC.00906-09

Liu E, Ye J, Song S, Wang K, Zhang Y, Zhang H (2014) Impact of co-deficiency of RpoN and RpoS on stress tolerance, virulence and gene regulation in Edwardsiella tarda. J Basic Microbiol 54(7):678–687. doi:10.1002/jobm.201300622

Lloyd LJ, Jones SE, Jovanovic G, Gyaneshwar P, Rolfe MD, Thompson A, Hinton JC, Buck M (2004) Identification of a new member of the phage shock protein response in Escherichia coli, the phage shock protein G (PspG). J Biol Chem 279(53):55707–55714. doi:10.1074/jbc.M408994200

Ma C, Sim S, Shi W, Du L, Xing D, Zhang Y (2010) Energy production genes sucB and ubiF are involved in persister survival and tolerance to multiple antibiotics and stresses in Escherichia coli. FEMS Microbiol Lett 303(1):33–40. doi:10.1111/j.1574-6968.2009.01857.x

Mavromatis CH, Bokil NJ, Totsika M, Kakkanat A, Schaale K, Cannistraci CV, Ryu T, Beatson SA, Ulett GC, Schembri MA, Sweet MJ, Ravasi T (2015) The co-transcriptome of uropathogenic Escherichia coli-infected mouse macrophages reveals new insights into host-pathogen interactions. Cell Microbiol 17(5):730–746. doi:10.1111/cmi.12397

McDonald C, Jovanovic G, Ces O, Buck M (2015) Membrane stored curvature elastic stress modulates recruitment of maintenance proteins PspA and Vipp1. mBio 6(5): e01188–15. doi:10.1128/mBio.01188-15

Mehta P, Jovanovic G, Lenn T, Bruckbauer A, Engl C, Ying L, Buck M (2013) Dynamics and stoichiometry of a regulated enhancer-binding protein in live Escherichia coli cells. Nat Commun 4:1997. doi:10.1038/ncomms2997

Mitra A, Fay PA, Vendura KW, Alla Z, Carroll RK, Shaw LN, Riordan JT (2014) sigma(N) -dependent control of acid resistance and the locus of enterocyte effacement in enterohemorrhagic Escherichia coli is activated by acetyl phosphate in a manner requiring flagellar regulator FlhDC and the sigma(S) antagonist FliZ. MicrobiologyOpen 3(4):497–512. doi:10.1002/mbo3.183

Model P, Jovanovic G, Dworkin J (1997) The Escherichia coli phage-shock-protein (psp) operon. Mol Microbiol 24(2):255–261

Morris L, Cannon W, Claverie-Martin F, Austin S, Buck M (1994) DNA distortion and nucleation of local DNA unwinding within sigma-54 (sigma N) holoenzyme closed promoter complexes. J Biol Chem 269(15):11563–11571

Nadkarni MA, Chen Z, Wilkins MR, Hunter N (2014) Comparative genome analysis of Lactobacillus rhamnosus clinical isolates from initial stages of dental pulp infection: identification of a new exopolysaccharide cluster. PLoS ONE 9(3):e90643. doi:10.1371/journal.pone.0090643

Okada Y, Okada N, Makino S, Asakura H, Yamamoto S, Igimi S (2006) The sigma factor RpoN (sigma54) is involved in osmotolerance in Listeria monocytogenes. FEMS Microbiol Lett 263(1):54–60. doi:10.1111/j.1574-6968.2006.00405.x

Osadnik H, Schopfel M, Heidrich E, Mehner D, Lilie H, Parthier C, Risselada HJ, Grubmuller H, Stubbs MT, Bruser T (2015) The PspF-binding domain PspA and the PspA.F complex: new insights into the coiled-coil dependent regulation of AAA+ proteins. Mol Microbiol. doi:10.1111/mmi.13154

Otters S, Braun P, Hubner J, Wanner G, Vothknecht UC, Chigri F (2013) The first alpha-helical domain of the vesicle-inducing protein in plastids 1 promotes oligomerization and lipid binding. Planta 237(2):529–540. doi:10.1007/s00425-012-1772-1

Pogliano J, Pogliano N, Silverman JA (2012) Daptomycin-mediated reorganization of membrane architecture causes mislocalization of essential cell division proteins. J Bacteriol 194(17):4494–4504. doi:10.1128/JB.00011-12

Raivio TL (2005) Envelope stress responses and Gram-negative bacterial pathogenesis. Mol Microbiol 56(5):1119–1128. doi:10.1111/j.1365-2958.2005.04625.x

Raivio TL (2014) Everything old is new again: an update on current research on the Cpx envelope stress response. Biochim Biophys Acta 1843(8):1529–1541. doi:10.1016/j.bbamcr.2013.10.018

Rappas M, Schumacher J, Beuron F, Niwa H, Bordes P, Wigneshweraraj S, Keetch CA, Robinson CV, Buck M, Zhang X (2005) Structural insights into the activity of enhancer-binding proteins. Science 307(5717):1972–1975. doi:10.1126/science.1105932 307/5717/1972 [pii]

Rowley G, Spector M, Kormanec J, Roberts M (2006) Pushing the envelope: extracytoplasmic stress responses in bacterial pathogens. Nat Rev Microbiol 4(5):383–394. doi:10.1038/nrmicro1394

Saldias MS, Lamothe J, Wu R, Valvano MA (2008) Burkholderia cenocepacia requires the RpoN sigma factor for biofilm formation and intracellular trafficking within macrophages. Infect Immun 76(3):1059–1067. doi:10.1128/IAI.01167-07

Schafer JE, Engl C, Zhang N, Buck M (2015) Genome wide interactions of wildtype and activator bypass forms of σ54. Unpublished data

Schulz S, Eckweiler D, Bielecka A, Nicolai T, Franke R, Dotsch A, Hornischer K, Bruchmann S, Duvel J, Haussler S (2015) Elucidation of sigma factor-associated networks in Pseudomonas aeruginosa reveals a modular architecture with limited and function-specific crosstalk. PLoS Pathog 11(3):e1004744. doi:10.1371/journal.ppat.1004744

Seo J, Savitzky DC, Ford E, Darwin AJ (2007) Global analysis of tolerance to secretin-induced stress in Yersinia enterocolitica suggests that the phage-shock-protein system may be a remarkably self-contained stress response. Mol Microbiol 65(3):714–727. doi:10.1111/j.1365-2958.2007.05821.x

Shah D, Zhang Z, Khodursky A, Kaldalu N, Kurg K, Lewis K (2006) Persisters: a distinct physiological state of E. coli. BMC Microbiol 6:53. doi:10.1186/1471-2180-6-53

Sharma A, Leach RN, Gell C, Zhang N, Burrows PC, Shepherd DA, Wigneshweraraj S, Smith DA, Zhang X, Buck M, Stockley PG, Tuma R (2014) Domain movements of the enhancer-dependent sigma factor drive DNA delivery into the RNA polymerase active site: insights from single molecule studies. Nucleic Acids Res. doi:10.1093/nar/gku146

Southern SJ, Male A, Milne T, Sarkar-Tyson M, Tavassoli A, Oyston PC (2015) Evaluating the Role of Phage-Shock Protein A in Burkholderia pseudomallei. Microbiology. doi:10.1099/mic.0.000175

Strahl H, Burmann F, Hamoen LW (2014) The actin homologue MreB organizes the bacterial cell membrane. Nat Commun 5:3442. doi:10.1038/ncomms4442

Studholme DJ, Dixon R (2003) Domain architectures of sigma54-dependent transcriptional activators. J Bacteriol 185(6):1757–1767

Sysoeva TA, Chowdhury S, Guo L, Nixon BT (2013) Nucleotide-induced asymmetry within ATPase activator ring drives sigma54-RNAP interaction and ATP hydrolysis. Genes Dev 27 (22):2500–2511. doi:10.1101/gad.229385.113

van den Ent F, Johnson CM, Persons L, de Boer P, Lowe J (2010) Bacterial actin MreB assembles in complex with cell shape protein RodZ. EMBO J 29(6):1081–1090. doi:10.1038/emboj.2010.9

Vega NM, Allison KR, Khalil AS, Collins JJ (2012) Signaling-mediated bacterial persister formation. Nat Chem Biol 8(5):431–433. doi:10.1038/nchembio.915

Vega NM, Allison KR, Samuels AN, Klempner MS, Collins JJ (2013) Salmonella typhimurium intercepts Escherichia coli signaling to enhance antibiotic tolerance. Proc Natl Acad Sci USA 110(35):14420–14425. doi:10.1073/pnas.1308085110

Vothknecht UC, Otters S, Hennig R, Schneider D (2012) Vipp1: a very important protein in plastids?! J Exp Bot 63(4):1699–1712. doi:10.1093/jxb/err357 err357 [pii]

Wallrodt I, Jelsbak L, Thomsen LE, Brix L, Lemire S, Gautier L, Nielsen DS, Jovanovic G, Buck M, Olsen JE (2014) Removal of the phage-shock protein PspB causes reduction of virulence in Salmonella enterica serovar Typhimurium independently of NRAMP1. J Med Microbiol 63(Pt 6):788–795. doi:10.1099/jmm.0.072223-0

Wang A, Crowley DE (2005) Global gene expression responses to cadmium toxicity in Escherichia coli. J Bacteriol 187(9):3259–3266. doi:10.1128/JB.187.9.3259-3266.2005

Wang JT, Syed A, Hsieh M, Gralla JD (1995) Converting Escherichia coli RNA polymerase into an enhancer-responsive enzyme: role of an NH2-terminal leucine patch in sigma 54. Science 270(5238):992–994

Wang JT, Syed A, Gralla JD (1997) Multiple pathways to bypass the enhancer requirement of sigma 54 RNA polymerase: roles for DNA and protein determinants. Proc Natl Acad Sci USA 94(18):9538–9543

Wang P, Kuhn A, Dalbey RE (2010) Global change of gene expression and cell physiology in YidC-depleted Escherichia coli. J Bacteriol 192(8):2193–2209. doi:10.1128/JB.00484-09

Wecke T, Bauer T, Harth H, Mader U, Mascher T (2011) The rhamnolipid stress response of Bacillus subtilis. FEMS Microbiol Lett 323(2):113–123. doi:10.1111/j.1574-6968.2011.02367.x

Weiner L, Brissette JL, Model P (1991) Stress-induced expression of the Escherichia coli phage shock protein operon is dependent on sigma 54 and modulated by positive and negative feedback mechanisms. Genes Dev 5(10):1912–1923

Westphal S, Heins L, Soll J, Vothknecht UC (2001) Vipp1 deletion mutant of Synechocystis: a connection between bacterial phage shock and thylakoid biogenesis? Proc Natl Acad Sci USA 98(7):4243–4248. doi:10.1073/pnas.061501198

Whitaker WB, Richards GP, Boyd EF (2014) Loss of sigma factor RpoN increases intestinal colonization of Vibrio parahaemolyticus in an adult mouse model. Infect Immun 82(2):544–556. doi:10.1128/IAI.01210-13

White MJ, Savaryn JP, Bretl DJ, He H, Penoske RM, Terhune SS, Zahrt TC (2011) The HtrA-like serine protease PepD interacts with and modulates the Mycobacterium tuberculosis 35-kDa antigen outer envelope protein. PLoS ONE 6(3):e18175. doi:10.1371/journal.pone.0018175

Wigneshweraraj SR, Burrows PC, Nechaev S, Zenkin N, Severinov K, Buck M (2004) Regulated communication between the upstream face of RNA polymerase and the beta' subunit jaw domain. EMBO J 23(21):4264–4274. doi:10.1038/sj.emboj.7600407

Wolf D, Kalamorz F, Wecke T, Juszczak A, Mader U, Homuth G, Jordan S, Kirstein J, Hoppert M, Voigt B, Hecker M, Mascher T (2010) In-depth profiling of the LiaR response of Bacillus subtilis. J Bacteriol 192(18):4680–4693. doi:10.1128/JB.00543-10

Yamaguchi S, Reid DA, Rothenberg E, Darwin AJ (2013) Changes in Psp protein binding partners, localization and behaviour upon activation of the Yersinia enterocolitica phage shock protein response. Mol Microbiol 87(3):656–671. doi:10.1111/mmi.12122

Yang Y, Darbari VC, Zhang N, Lu D, Glyde R, Wang YP, Winkelman JT, Gourse RL, Murakami KS, Buck M, Zhang X (2015) TRANSCRIPTION. Structures of the RNA polymerase-sigma54 reveal new and conserved regulatory strategies. Science 349(6250):882–885. doi:10.1126/science.aab1478

Zhang L, Sakamoto W (2013) Possible function of VIPP1 in thylakoids: protection but not formation? Plant Signal Behav 8(2):e22860. doi:10.4161/psb.22860

Zhang X, Wigley DB (2008) The 'glutamate switch' provides a link between ATPase activity and ligand binding in AAA+ proteins. Nat Struct Mol Biol 15(11):1223–1227. doi:10.1038/nsmb.1501 nsmb.1501 [pii]

Zhang N, Joly N, Buck M (2012) A common feature from different subunits of a homomeric AAA + protein contacts three spatially distinct transcription elements. Nucleic Acids Res 40 (18):9139–9152. doi:10.1093/nar/gks661 gks661 [pii]

Zhang N, Simpson T, Lawton E, Uzdavinys P, Joly N, Burrows P, Buck M (2013) A key hydrophobic patch identified in an AAA(+) protein essential for its in trans inhibitory regulation. J Mol Biol 425(15):2656–2669. doi:10.1016/j.jmb.2013.04.024

Zhang N, Gordiyenko Y, Joly N, Lawton E, Robinson CV, Buck M (2014) Subunit dynamics and nucleotide-dependent asymmetry of an AAA(+) transcription complex. J Mol Biol 426(1):71–83. doi:10.1016/j.jmb.2013.08.018

Zhang N, Schafer J, Sharma A, Rayner L, Zhang X, Tuma R, Stockley P, Buck M (2015) Mutations in RNA polymerase bridge helix and switch regions affect active-site networks and transcript-assisted hydrolysis. J Mol Biol. doi:10.1016/j.jmb.2015.09.005

Chapter 14
How Biophysics May Help Us Understand the Flagellar Motor of Bacteria Which Cause Infections

Matthew A.B. Baker

Abstract Motor proteins are molecules which convert chemical energy to mechanical work and are responsible for motility across all levels: for transport within a cell, for the motion of an individual cell in its surroundings, and for movement in multicellular aggregates, such as muscles. The bacterial flagellar motor is one of the canonical examples of a molecular complex made from several motor proteins, which self-assembles on demand and provides the locomotive force for bacteria. This locomotion provides a key aspect of bacteria's prevalence. Here, we outline the biophysics behind the assembly, the energetics, the switching and the rotation of this remarkable nanoscale electric motor that is Nature's first wheel.

Keywords Molecular motors · Bacterial motility · Flagellar motor · Self-assembly · Chemotaxis · Ultrasensitivity

14.1 Introduction

While motility provides a significant evolutionary advantage, it is by no means essential to bacteria, even all pathogenic bacteria. For example, both streptococcus and shigella do not swim, but can reproduce and invade a host causing infections. Likewise, yeast, although a eukaryote, causes infection in humans, and is not motile. However, other highly pathogenic bacteria, such as *E. coli* O157:H7 and *Salmonella enterica*, are some of our world's best swimmers. They can swim up to 40 body-lengths in a second, change direction in only a handful of milliseconds, and swim to where their life may be better. However, not necessarily the life of their host.

The motor that is responsible for this swimming motility is highly conserved across bacterial species. It is a remarkable piece of nanotechnology—at only 40 nm in diameter, it can rotate at 100,000 Hz (Sowa and Berry 2008; Xue et al. 2015),

M.A.B. Baker (✉)
EMBL Australia Node for Single Molecule Science,
The University of New South Wales, NSW, Australia
e-mail: matthew.baker@unsw.edu.au

© Springer International Publishing Switzerland 2016

M.C. Leake (ed.), *Biophysics of Infection*, Advances in Experimental Medicine and Biology 915, DOI 10.1007/978-3-319-32189-9_14

respond to changes in load (Tipping et al. 2013), and dynamically adapt its structure during rotation to retain ultrasensitivity across a large operating range (Lele et al. 2013). This natural technology is the result of evolution, and as such, it is a research target so that we can learn from bacteria how to design bespoke biotechnology that can self-assemble and adapt during function, and rotate at high speed.

14.2 Physics of Bacterial Swimming Motility

Generating locomotion is quite different for a swimming bacteria, at the micron scale, than for a mammal at the meter scale. For a bacterium, the surrounding environment is comparatively very viscous, and it has very little mass. The ratio of inertia to viscous drag is known as the Reynolds Number, *R*, and is typically low for bacterial swimmers ($\sim 10^{-4}$). As a result, inertial considerations are inconsequential, and when a bacterium stops actively propelling itself it stops nearly immediately. Thus, the type of swimming mechanism it uses to propel itself through the viscous environment must be nonreciprocal. That is, propulsion requires a time-irreversible swimming motion, which in a bacterium is typically provided by the helical nature of the filament, acting as a propellor. Second, in order to move, in the big picture, to where there is more nutrient, a bacterium must be able to swim fast enough to outrun diffusion. If it does not, the bacteria does not in fact swim anywhere at all, relative to its nutrient source (Purcell 1977).

To navigate, the motor biases its random walk such that it spends more time swimming toward areas of high nutrients than randomly reorienting its direction. Each individual BFM can rotate in two directions, counterclockwise (CCW—a run) and clockwise (CW—a tumble). When all the motors on an *E. coli* cell spin CCW, its flagellar filaments bundle to form a helical propeller that pushes the cell steadily forward (run state); when one or more of the motors switch to spin CW, the flagellar bundle breaks apart and splays and the cell then tumbles randomly due to diffusion (tumble state). By controlling the proportion of time spent swimming vs randomly tumbling, the bacterium is able to repeat a 'run'-'tumble'-'run' pattern that moves, on average, to where there is a higher concentration of attractant (Berg 2003).

Bacterial response to chemical stimuli is known as chemotaxis, first demonstrated by T.W. Engelmann in 1881 (Adler 1966). The BFM thus not only drives bacterial locomotion, it also plays a crucial role in bacterial chemotaxis by controlling the ratio between CCW and CW rotation, and thus the duration of run and tumble events, and how quickly bacteria are able to move from regions of low nutrients to regions containing attractants.

14.3 Structure of the Flagellar Motor

What is underneath the hood of this molecular machine? Single particle cryoEM has revealed in considerable detail the architecture of an assembled BFM in *Salmonella sp* (DeRosier 2006). The BFM in total consists of ~45 proteins, of which nearly 24 different proteins form the intact complex, and only 6 are directly involved in rotation. These include four proteins in the rotor (Fig. 14.1: green and blue) FliF, FliG, FliM, and FliN and two proteins MotA and MotB forming the MotAB stator complex (Fig. 14.1: pink/orange). Energy for torque generation comes from an electrochemical gradient that drives an influx of cations, which is mediated by the stators (Sowa and Berry 2008). This influx of cations is converted into mechanical rotation by a rotor substructure known as the switch complex, which is responsible for torque generation, motor reversal, and regulating rotation speed (Sowa and Berry 2008; Baker and Berry 2009). The switch complex is comprised of dozens of copies of FliG, FliM, and FliN, which form a large bell-like structure with ~45 nm diameter. This is attached to the base of the MS-ring (Fig. 14.1: green). The MS-ring has a diameter of 30 nm and 26-fold symmetry while the larger C-ring has a diameter of 45 nm and 34-fold symmetry—herein lies the symmetry mismatch in the flagellar motor that is a signature of biological machinery (Sowa and Berry 2008). FliG is the rotor protein most directly involved in torque generation and hence motor switching (Fig. 14.1: light blue). Charged residues on or around a single helix in the C-terminal domain of FliG (the torque

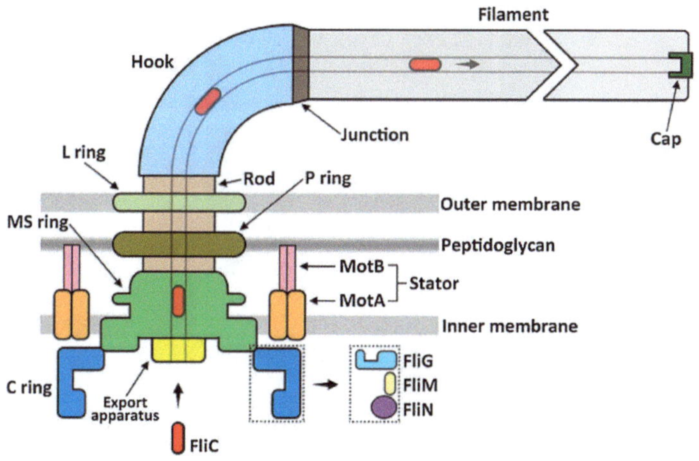

Fig. 14.1 A schematic showing the major structural components of the bacterial flagellum in *E. coli*. The site of torque generation is the interaction between the stator component MotB (*orange*) with the top of the rotor (*blue*), in particular with the protein FliG (*light blue*). Chemotaxis is controlled by the binding of CheY-P (not shown) to the bottom section of the rotor, in particular FliM (*yellow*). Hook and filament are exported by a type-III secretion apparatus which exports hook proteins and filament proteins through the hollow interior of these structures, which then self-assemble at the tip. From (Xue et al. 2015)

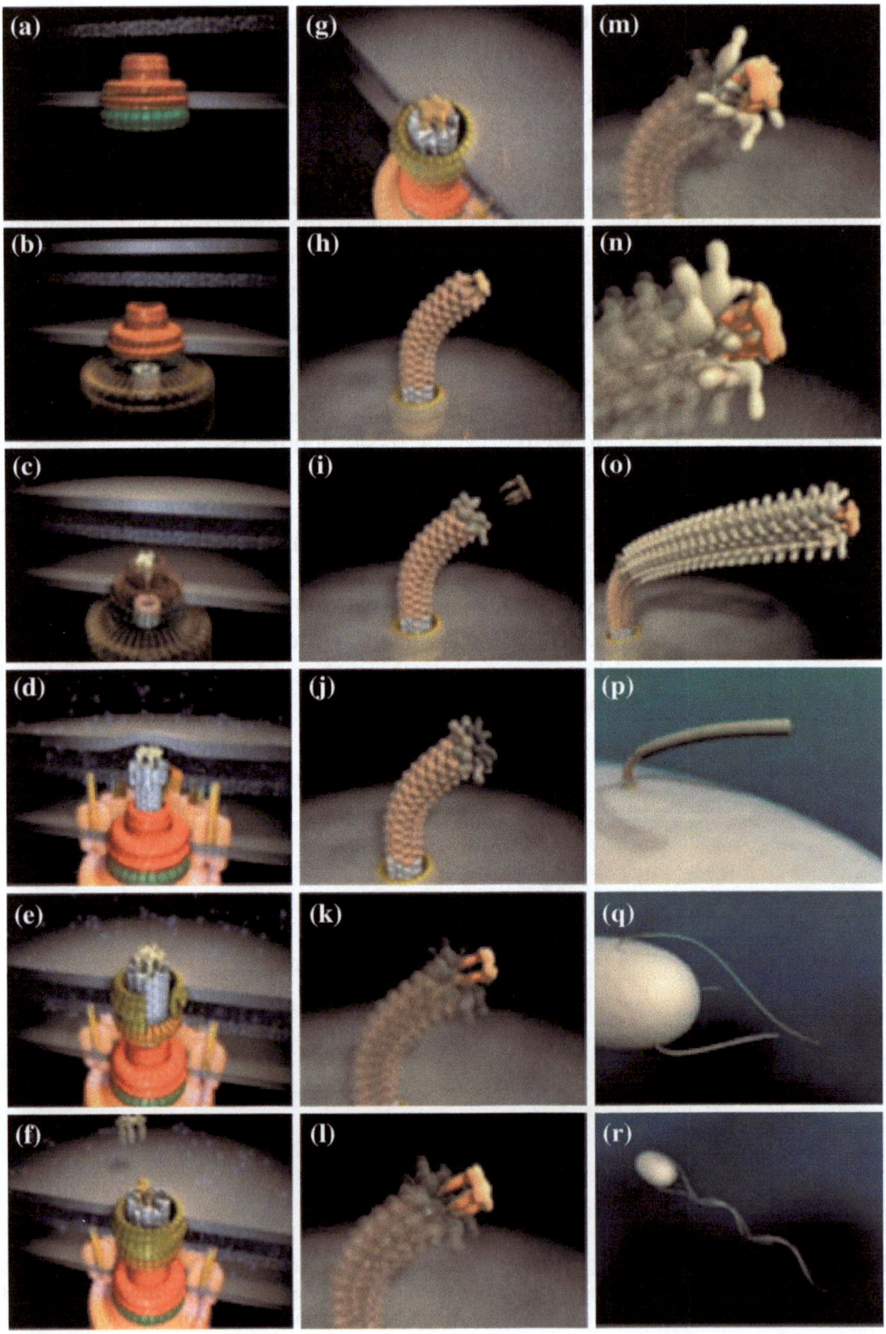
(a)
(b)
(c)
(d)
(e)
(f)
(g)
(h)
(i)
(j)
(k)
(l)
(m)
(n)
(o)
(p)
(q)
(r)

◀ **Fig. 14.2** Steps in the assembly of the bacterial flagellum from *top left* to *bottom right*. Assembly begins with the formation of the MS-ring in the cytoplasmic membrane as the anchor point for the rest of the motor and the scaffold onto which it will assemble (**a**). Afterward, the cytoplasmic C-ring consisting of FliG, FliM, and FliN is attached to the MS-ring and the flagellar-specific type-III secretion apparatus assembles within a central pore of the MS-ring to prepare for export of the hook and filament itself (**b**). Flagellar secretion substrates are now secreted specifically via the type-III secretion apparatus. Flagellar proteins are exported through the hollow interior and when they reach the far end of the growing filament, self-assemble onto the existing structure with the help of cap proteins (**c**–**o**). Taken from (Erhardt et al. 2010)

helix) interact with at least two residues on the stator protein MotA to generate torque (Sowa and Berry 2008).

Assembly starts with the formation of the FliF scaffold in the inner membrane (the MS-ring), followed by the attachment of the rotor (the C-ring) in the cytoplasmic face of the MS-ring. Then a type-III secretion apparatus assembles in a central pore of the MS-ring, which then assembles a driveshaft beneath a rod-cap, which can penetrate the peptidoglycan layer (Erhardt et al. 2010). Once this layer is penetrated, the external structures of the flagellar motor, such as the hook and filament, are exported in an unfolded state through the hollow interior of the driveshaft. When these proteins reach the end of the growing structure, they self-assemble onto the existing structure with the help of a cap protein, which guides filamentous assembly (Fig. 14.2).

14.4 Energy Source

Most linear molecular motors are powered by ATP hydrolysis, such as myosin on actin filaments, or kinesin and dynein on microtubules. In contrast, the BFM is powered by the free energy released from the transit of protons across the electrochemical gradient maintained over the cell membrane. This energy source is termed the proton motive force (PMF), and is maintained by the electron transport chain and the ATPase involved in metabolic processes. Other bacteria, especially those living in marine or high-pH environments are powered by alternate cations, such as Na^+ rather than H^+ (Yorimitsu and Homma 2001). One such marine bacteria, *Vibrio alginolyticus*, is sodium powered, and has been used to genetically engineer chimeric hybrid stators that function in *E. coli* but are powered by sodium motive force (Asai et al. 2003).

The PMF consists of two parts (Eq. 14.1): The first contribution is from the transmembrane electrical potential gradient, or the membrane potential. This arises from the electric field generated by different concentrations of cations and anions across the membrane. The second component consists of the entropic force arising from transmembrane concentration differences:

$$\mathrm{PMF} = \Delta \mathrm{p} = \Delta \psi + 2.3 \frac{k_B T}{e} \Delta pH. \tag{14.1}$$

Here, $\Delta\psi$ represents the electrical potential across cytoplasmic membrane, ΔpH denotes the change in proton concentration across the membrane, k_B is the Boltzmann constant, T the absolute temperature, and e the proton charge. For further detail on how biophysics can enable us to measure the proton motive force, consult Chap. 20 of this volume by C.J. Lo.

14.5 Cooperative Switching of the Bacterial Flagellar Motor

The process of chemotaxis and interaction with the environment begins on the surface of the bacteria where arrays of chemoreceptors are located. Information from environmental changes are translated into phosphorylation levels of a signaling molecule, CheY, through a series of biochemical reactions (Sowa and Berry 2008). This is communicated to the motor directly by the binding of phosphorylated CheY-P to FliM, located at the bottom of the rotor. When the concentration of CheY-P in the cytoplasm is low, the rotor is more likely to rotate counterclockwise (CCW); while the concentration is higher, the reverse is true, and the motors predominantly rotate in the clockwise direction (CW) (Cluzel et al. 2000; Bray 2002; Sourjik 2002). Binding of CheY-P to FliM is believed to trigger conformational changes in FliM, which is coupled to conformational changes in the FliG protein (Toker and Macnab 1997; Lee et al. 2010). Conformational rearrangement of individual FliG proteins generates the two opposite directions of rotation when torque is delivered from the stator units (Lee et al. 2010) (Fig. 14.1). Strategic mutation of FliG can affect switch bias, and deletion of key residues can result in motors that have their rotational direction locked (Togashi et al. 1997).

The rapid switching of the BFM between the "CCW" and "CW" states leads to the transition between "runs" and "tumbles" and forms the basis of bacterial chemotaxis. We can measure the duration of switch events, and the dwell time between switches by measuring the rotation of a spherical bead attached to the stub of a filament, close to the motor (Chen and Berg 2000). The position and rotational direction of this bead can be monitored, and thus we can acquire statistical measurements for the kinetics of a switch event. This kinetic data inform models for how a switch of the rotational direction is driven by the underlying spread over the whole rotor of the conformational change in individual rotor subunits.

Bai et al. measured that intervals between switch events were distributed exponentially (Bai et al. 2010). This demonstrated that switching was well approximated by a two-state Poisson process, despite the underlying multistate nature of the switch. Bai et al. showed that a model of conformational spread well explained these switch distributions and their data showed quantitative agreement

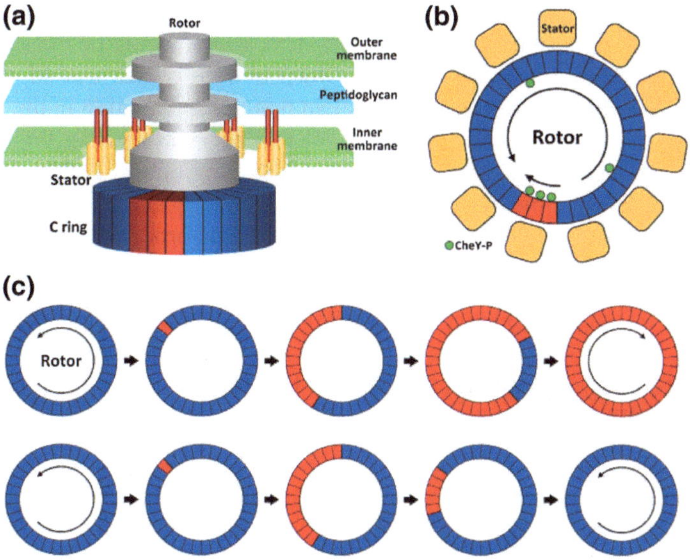

Fig. 14.3 Conformational spread as a mechanism for ultrasensitivity in the flagellar motor switch. **a** In this figure some RSUs (*red*) are in the CW state while the others (*blue*) in the CCW state. **b** Schematic top view of the motor, comprising a ring of 34 protomers and 11 stator units. **c** The same-state domain may grow to encompass the entire ring (*top*), known as conformational spread, or shrink and disappear as the motor restores its previous state (*bottom*). From (Xue et al. 2015)

with a model of allosteric cooperativity. Their model for conformational spread was based on consideration of neighbor-neighbor interaction energies, with stochastic growth and shrinkage of domains in adjacent subunits that share a particular conformational state. Fluctuations in these domains were observed as transient speed fluctuations in otherwise stable rotations. This mechanism of a global conformational spread based upon near–neighbor interactions enabled the amplification of a signal and the cooperative spread of a conformational change across the switch complex (Fig 14.3).

To remove any complications from filament transition during switching, and to examine the kinetics of the switching mechanism directly at the motor, Yuan et al. extended this study to examine the dynamics of switching at low load (Yuan et al. 2009). Using gold nanoparticles as previously, they observed that switching rates increased as viscosity was increased. This indicated that switching rates were correlated with load and that the switch complex itself also sensed stator–rotor interactions (Yuan et al. 2009). Wang et al. saw recently that switch intervals at low load were distributed exponentially, as with high load (Wang et al. 2014). Wang et al. also observed an additional curve in the semilog plots away from exponential behavior. On long recordings of 30 min, greater fluctuation in switching bias was observed than obtained in simulations derived from a two-state Poisson process. This fluctuation was not observed at high loads, where the number of stators was large, but only at low load and low stator number. This confirmed that fluctuation was not a result of variation in the level of CheY, which would be the same at high stator number. Additionally, when stator proteins were overexpressed, this fluctuation was somewhat suppressed. This implied that it was variation in the number of stators, which has greater effect at low stator number, was causing the heightened fluctuation in switching bias over relatively long time scales. Otherwise, high-spatial resolution measurements at low load, directly at the motor's drive shaft confirmed that motor switching was a two-state Poisson process, in accordance with previous measurements.

14.6 Structural Adaptivity of the Bacterial Flagellar Motor

What is remarkable about the sensitivity of the flagellar motor's switch complex is that it maintains a high sensitivity across a very large operating range. This sensitivity arises due to the conformational spread as detailed above, but the operating range can be adjusted by directly adapting the structure of the rotor (Lele et al. 2012; Yuan et al. 2012). Yuan et al. showed that the number of FliM units is able to change in response to levels of CheY-P, increasing the number of binding sites and thus sensitivity when CheY-P concentrations are low. Furthermore, Lele et al. showed that FliM binding directly strengthened in response to the rotational direction of the motor, and that it was the direction of rotation that mattered not only CheY-P binding. Indeed, both the stators and the rotor proteins at the bottom of the rotor, FliM and FliN, have been shown to turnover during the motor operation (Delalez and Armitage 2009; Delalez et al. 2010, 2014). This turnover is a property of bionanotechnology that lies in stark contrast to man-made technology, in that the structure of the motor can adaptively remodel during operation. As remarkable as this is from a macroscopic machinery perspective, this seems to have evolved as a common theme throughout biology (Yuan et al. 2012; Loquet et al. 2012; Olia et al. 2011; Stock et al. 1999). Thus far, turnover in the topmost rotor protein FliG has not been observed (Fukuoka et al. 2014), presumably due to the role in torque generation and in the conformational change that dictates a switch event (Bai et al. 2010). This leads to a further perplexing symmetry mismatch between the top and bottom of the C-ring: FliG should be 34-fold in *Salmonella enterica* (DeRosier 2006), and while FliM is ~34-fold in an exclusively clockwise rotating motor, it can be as many as 44-fold in an exclusively counterclockwise rotating motor (Lele et al. 2012). Additionally, the scaffold to which all the rotor proteins are attached, exists at 26-fold symmetry, so natural machinery seems to tolerate, perhaps even require, symmetry mismatch, in order to function.

It has also been shown recently that the motor adapts in response to increased load (Lele et al. 2013). Lele et. al showed there that the stators themselves dynamically respond to load, and this response was consistently observed even when chemotaxis and rotational switching were removed—it is the stators themselves which engage or disengage with changes in load. This is presumably because at low load only few stators are required to drive rotation and additional stators simply waste ions (Yuan et al. 2012). However, the mechanism for how the stators sense force remains unknown.

14.7 Diversity

Studies of the model *E. coli* motor have yielded great insight into the structure, function, energy source, and switching method of the motor. However, bacteria exhibit large diversity, and the frontier of flagellar research has now moved onto studying how motors differ in different species (Chen et al. 2011; Zhao et al. 2014; Beeby et. al. 2016). Bacteria have evolved to live in diverse environments, and this diversity is reflected phenotypically, and structurally, in the flagellar motor. We will consider two test cases, looking at the flagellar motor in two pathogenic species of bacteria, *Borrelia burgdorferi*, a species of spirochaetes responsible for Lyme disease, and *Salmonella enterica*, analogous to *E. coli*, and responsible for typhoid fever and enterocolitis.

Borrelia burgdorferi is a bacteria in a taxa known as spirochetes. These are the bacteria that are responsible for Lyme disease and syphilis, and thus are of interest as a research target to prevent their pathogenic effects. Lyme disease is a strongly emerging tropical disease accounting for 90 % of all vector-borne disease in the United States, which was named after an epidemic break in the 1970s clustered around the town of Lyme in southeastern Connecticut, USA (Steere et al. 1977). Lyme disease spirochetes have a dual-host relationship with ticks and humans. These spirochetes disseminate in their reservoir hosts, such as a mouse or a human, and then are acquired by larval ticks taking a blood meal from these reservoir hosts. The bacteria are in fact attracted to chemotactic stimuli exported during larval feeding. *Borrelia burgdorferi* then matures in the tick gut, and is transferred to other reservoir hosts, typically deposited into the bite wound from ticks along with tick saliva during tick feeding. Ingredients of tick saliva impede various mammalian responses, such as activation of complement, chemotaxis of neutrophils, and antibody-mediated killing. After a delay of ~2 days, *B. burgdorferi* begins spreading to distant tissues, penetrating the matrix between cells and entering capillary beds. The health complications from borrelia arise due to inflammation and immunological reactions, which can even cause neurological disorders in 10 % of cases (Auwaerter et al. 2004).

Visually, one can see immediately that *B. burgdorferi* is a very different bacterium to *E. coli* (Fig. 14.4). Spirochetes in general are tightly coiled bacteria that are only 5 um wide but up to 250 um long. They are coiled into tight spirals, reminiscent of a helical flagellar bundle. In fact, uniquely, their flagella run inside the periplasm between the inner and outer membrane, wrapping around the inner membrane in a right-handed helix, twisting the shell shape. When the motors at the base of the flagellar filaments rotate, this twists the entire surface of the cell and results in a

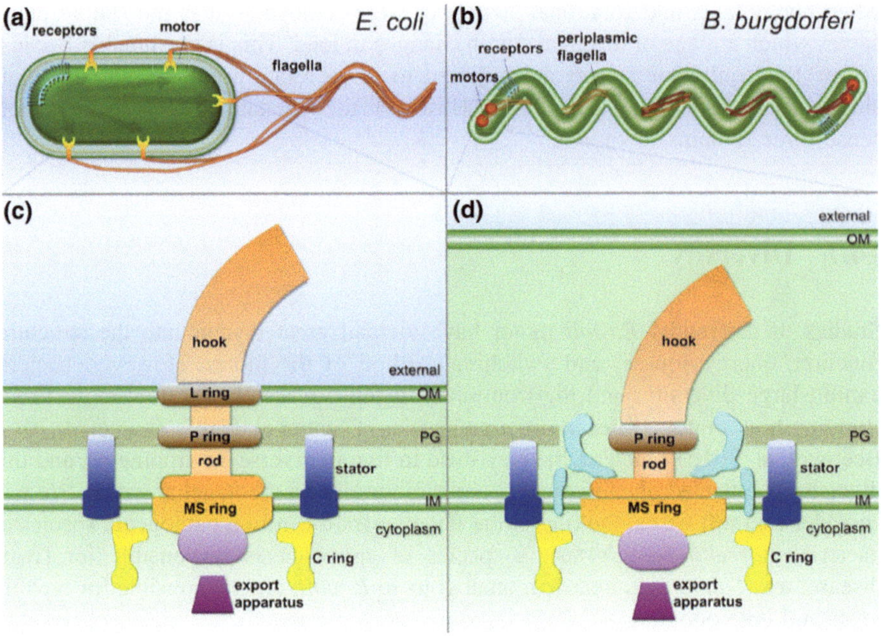

Fig. 14.4 Schematic models of the flagellar motor in *E. coli*. (*left*) and *B. burgdorferi* (*right*). In *E. coli* (**a** and **c**), the flagella are external, and bundle to form a helical filament for propulsion when the motors all spin in the same CCW direction. Salmonella motors are analogous to *E. coli* in their structure. In *B. burgdorferi*, the flagella sit in the periplasm and wind around the cell. *B. burgdorferi* swims by twisting the cell body, and as a result, the flagellar motors are considerably larger. The core structure is comparable, with shared structures including the MS ring, the C ring, the rod and the hook. Notable differences are the lack of an L-ring in the *B. burgdorferi* motor, and the presence of a periplasmic collar around the P-ring of the motor in *B. burgdorferi*. Taken from (Zhao et al. 2014)

writhing motility. Borrelia has 7–11 periplasmic flagella inserted subterminally at both poles. These winds around the cylindrical protoplasmic cylindrical center of the cell, and are coordinated to rotate in opposite directions during translational motion, and in the same direction in the 'tumbling' equivalent of spirochetes. Rotation of the flagellar causes a snake-like movement of the whole cell body, which enables them to very efficiently penetrate viscous media and tissues.

Interestingly, at the site of the rotor, *B. burgdorferi* have a larger rotor, $\sim$60 nm in diameter, with $\sim$44 switching subunits (FliG/M/N). As per the physics of rotating objects, a larger rotor radius results in a larger lever arm and the ability to exert a greater torque. Thus, borrelia's larger rotor facilitates swimming in viscous environments, which is critical to borrelia dissemination.

S. enterica causes as many as 1.3 billion cases of disease annually (Coburn et al. 2006). *S. enterica* typically enters the host orally and cause one of four syndromes: enteric fever (typhoid), diarrhea, bacteremia, and chronic asymptomatic carriage. The manifestation of the disease depends on the host susceptibility and the infectious serovar that is present. The most common forms of the disease are human typhoid fever and intestinal/diarrheal disease, with typhoid the most common and thus the most studied. Following oral inoculation, virulent serovar Typhimurum can survive gastric acidity and colonize the ileum and cecum outcompeting local microflora. Via invasion of the phagocytic epithelial cells, and through uptake by dendritic cells, the bacteria gain access to the host circulation. From there, the bacteria can take up residence in various splenocytes, mostly macrophages, as well as hepatocytes in the liver. *S. enterica* are internalized and take up residence inside a membrane-bound compartment termed the *Salmonella* containing vacuole (SCV). In humans, typhoid disease manifests within 2 weeks of inoculation with general fever and abdominal pain.

Enterocolitis and diarrhea cause 3 million deaths each year worldwide with the incidence being highest in the developing world. Disease typically follows the ingestion of 50,000 bacteria in contaminated food or water, with symptoms occurring between 6 and 72 h after consumption. Symptoms typically resolve in the absence of treatment after 5–7 days, but antimicrobial treatment is required in the presence of positive signs of invasive disease, or in children. For a thorough review of Salmonella's role in disease, see Coburn et al. (2006).

The flagellar motor of *S. enterica,* in contrast to *B. burgdorferi*, is highly structurally analogous to *E. coli*. The rotor has only 34 units, on average, and a rotor diameter of 40 nm. Like *E. coli*, salmonella rotates peritrichous flagellar that bundle to form a helical propellor when driving bacteria forward, or splay to initiate a tumble. In both *S. enterica* and *E. coli*, the MS-ring, the rod, the P ring, and the C ring are conserved and highly similar.

There are the following additional differences between the flagellar motors in the two species. In spirochetes, the distal rod of external flagella is too short to penetrate the outer membrane resulting in the flagellar being wound around the protoplasmic body of the cell. In *Salmonella*, the filaments are exported and bundle to form filaments as in *E. coli*. Thus, there are specific differences in the mechanism of polymerization of FlgG. In *S. enterica*, MotA and MotB are anchored to the peptidoglycan layer through the periplasmic domain of MotB, however in spirochetes a periplasmic collar assembles around the MS-ring, and the stator interacts with this collar. *S. enterica* additionally has the L ring, acting as a bushing in the cell envelope, to aid rotation of the filament inside the outer membrane.

14.8 Conclusion

Motility is not essential to life, but the bacterial flagellar motor offers impressive motility to those bacterial species that do swim. This allows these species to migrate and move rapidly through a host. From an engineering perspective, the technology present in Nature's first wheel is remarkable: a self-assembling motor that can sense changes in its environment and rotate to propel bacteria to where life is better. Current man-made nanotechnology cannot achieve the complexity of the flagellar motor, and thus by further study of the motor informs the development of new bionanotechnology modeled on this motor. Additionally, with the rise of bacterial antibiotic resistance (Blair et al. 2014), a thorough understanding of essential bacterial systems will offer alternate approaches in the development of novel antibacterial agents.

References

Adler J (1966) Chemotaxis in bacteria. Science 153:708–716
Asai Y, Yakushi T, Kawagishi I, Homma M (2003) Ion-coupling determinants of Na+-driven and H+-driven flagellar motors. J Mol Biol 327:453–463
Auwaerter PG, Aucott J, Dumler JS (2004) Lyme borreliosis (Lyme disease): molecular and cellular pathobiology and prospects for prevention, diagnosis and treatment. ERM 6. doi:10.1017/S1462399404007276
Bai F et al (2010) Conformational spread as a mechanism for cooperativity in the bacterial flagellar switch. Science 327:685–689
Baker MAB, Berry RM (2009) An introduction to the physics of the bacterial flagellar motor. Contemp Phys 50:617
Beeby M, Ribardo DA, Brennan CA, Ruby EG, Jensen GJ, Hendrixson DR (2016) Diverse high-torque bacterial flagellar motors assemble wider stator rings using a conserved protein scaffold. PNAS. doi:10.1073/pnas.1518952113
Berg H (2003) The rotary motor of bacterial flagella. Ann Rev
Blair JMA, Webber MA, Baylay AJ, Ogbolu DO, Piddock LJV (2014) Molecular mechanisms of antibiotic resistance. Nat Rev Microbiol 13:42–51
Bray D (2002) Bacterial chemotaxis and the question of gain. PNAS 99:7–9
Chen S et al (2011) Structural diversity of bacterial flagellar motors. EMBO J 30:2972–2981
Chen X, Berg HC (2000) Torque-speed relationship of the flagellar rotary motor of Escherichia coli. Biophys J 78: 1036–1041
Cluzel P, Surette M, Leibler S (2000) An ultrasensitive bacterial motor revealed by monitoring signaling proteins in single cells. Science (New York, N.Y.) 287:1652–1655
Coburn B, Grassl GA, Finlay BB (2006) Salmonella, the host and disease: a brief review. Immunol Cell Biol 85:112–118
Delalez N, Armitage JP (2009) Parts exchange: tuning the flagellar motor to fit the conditions. Mol Microbiol 71:807–810
Delalez NJ et al (2010) Signal-dependent turnover of the bacterial flagellar switch protein FliM. PNAS 107:11347–11351
Delalez NJ, Berry RM, Armitage JP (2014) Stoichiometry and turnover of the bacterial flagellar switch protein FliN. MBio 5:e01216–14–e01216–14
DeRosier D (2006) Bacterial flagellum: visualizing the complete machine in situ. Curr Biol 16: R928–R930–R928–R930
Erhardt M, Namba K, Hughes KT (2010) Bacterial nanomachines: the flagellum and type III injectisome. Cold Spring Harb Perspect Biol 2:a000299
Fukuoka H, Sagawa T, Inoue Y, Takahashi H, Ishijima A (2014) Direct imaging of intracellular signaling components that regulate bacterial chemotaxis. Sci Signal 7:ra32
Lee LK, Ginsburg MA, Crovace C, Donohoe M, Stock D (2010) Structure of the torque ring of the flagellar motor and the molecular basis for rotational switching. Nature 466:996–1000
Lele PP, Branch RW, Nathan VSW, Berg HC (2012) Mechanism for adaptive remodeling of the bacterial flagellar switch
Lele PP, Hosu BG, Berg HC (2013) Dynamics of mechanosensing in the bacterial flagellar motor. PNAS 110:11839–11844
Loquet A et al. (2012) Atomic model of the type III secretion system needle. Nature 486:276–279

Olia AS, Prevelige PE, Johnson JE, Cingolani G (2011) Three-dimensional structure of a viral genome-delivery portal vertex. Nat Struct Mol Biol 18:597–603
Purcell E (1977) Life at low Reynolds number. Am. J, Phys
Sourjik V (2002) Binding of the Escherichia coli response regulator CheY to its target measured in vivo by fluorescence resonance energy transfer. Proc Natl Acad Sci USA 99:12669–12674
Sowa Y, Berry RM (2008) Bacterial flagellar motor. Q Rev Biophys 41:103–132
Steere AC, Hardin JA, Malawista SE (1977) Erythema chronicum migrans and Lyme arthritis: cryoimmunoglobulins and clinical activity of skin and joints. Science 196:1121–1122
Stock D, Leslie AGW, Walker JE (1999) Molecular architecture of the rotary motor in ATP synthase. Science 286:1700–1705
Tipping MJ, Delalez NJ, Lim R, Berry RM, Armitage JP (2013) Load-dependent assembly of the bacterial flagellar motor. MBio 4:e00551–13–e00551–13
Togashi F, Yamaguchi S, Kihara M, Aizawa SI, Macnab RM (1997) An extreme clockwise switch bias mutation in fliG of Salmonella typhimurium and its suppression by slow-motile mutations in motA and motB. J Bacteriol 179:2994–3003
Toker AS, Macnab RM (1997) Distinct regions of bacterial flagellar switch protein FliM interact with FliG, FliN and CheY1. J Mol Biol 273:623–634
Wang F, Yuan J, Berg HC (2014) Switching dynamics of the bacterial flagellar motor near zero load. PNAS 111:15752–15755
Xue R, Ma Q, Baker MAB, Bai F (2015) A delicate nanoscale motor made by nature-the bacterial flagellar motor. Adv Sci n/a–n/a
Yorimitsu T, Homma M (2001) Na+-driven flagellar motor of Vibrio. BBA-Bioenerg
Yuan J, Branch RW, Hosu BG, Berg HC (2012) Adaptation at the output of the chemotaxis signalling pathway. Nature 484:233–236
Yuan J, Fahrner KA, Berg HC (2009) Switching of the bacterial flagellar motor near zero load. J Mol Biol 390:394–400
Zhao X, Norris SJ, Liu J (2014) Molecular architecture of the bacterial flagellar motor in cells. Biochemistry 53:4323–4333

Chapter 15
Mechanics of Bacterial Cells and Initial Surface Colonisation

Sebastian Aguayo and Laurent Bozec

Abstract The mechanical properties of bacterial cells play an important role in crucial bacterial processes such as cell growth, colonisation and biofilm formation. Recent developments in the field of nanotechnology and atomic force microscopy (AFM) have made it possible to observe, characterise and understand the nanomechanic behaviour of live bacterial cells as never before. Unlike traditional techniques, AFM makes it possible to employ living bacteria in their physiological environment with minimal or no sample preparation. The technique of AFM nanoindentation opens new possibilities to study bacterial cell wall stiffness under different mechanical and buffer conditions. Also, by attaching bacterial cells to functionalised AFM cantilevers, single-cell force spectroscopy (SCFS) can be used to measure the adhesion of bacteria to biological and non-biological substrates at the nano-newton and pico-newton scale, and provide specific information on receptor–ligand interactions. By studying the biophysics of the bacterial–surface interaction with the abovementioned techniques, it has been possible to gain new insight on the early stages of bacterial colonisation and biofilm formation.

Keywords AFM · Biophysics · Microbiology · Nanotechnology · Force spectroscopy

15.1 Why Should We Consider the Mechanic Behaviour of Bacteria?

Bacteria are prokaryotic microorganisms that exist on virtually every environment on Earth. In many cases, they are not found individually, but attached to surfaces forming part of intricate microbial communities called *biofilms*. Biofilms are of particular interest to biomedicine as they can be found growing on medical devices,

S. Aguayo (✉) · L. Bozec
Department of Biomaterials and Tissue Engineering, UCL Eastman Dental Institute, University College London, London, UK
e-mail: sebastian.aguayo.13@ucl.ac.uk

© Springer International Publishing Switzerland 2016
M.C. Leake (ed.), *Biophysics of Infection*, Advances in Experimental Medicine and Biology 915, DOI 10.1007/978-3-319-32189-9_15

catheters, artificial implants (i.e. hip and dental implants) and valve replacements (Subramani et al. 2009; Abe et al. 2012; Schmidlin et al. 2013). Furthermore, the establishment of specific biofilms throughout the body contributes to a range of diseases that can be very difficult to control and treat (McConnell et al. 2010). Although biofilms have beneficial applications in some industries, such as water remediation and microbial leeching, many efforts are placed in searching for novel ways to control pathogenic biofilm formation, which poses a major issue to the medical field (McConnell et al. 2010; Gan et al. 2014).

The development of a biofilm consists on a series of sequential phases that enable the progressive incorporation of an increasing number of bacterial strains (Van Acker et al. 2014). This process, however, must be initiated by the *attachment of early bacterial colonisers to the surface* (Waar et al. 2005). It has been shown that particle geometry, surface roughness and elastic properties are important factors in the adhesion of biological particles and cells, and therefore key participants in bacteria–substrate interactions and biofilm formation (Oh et al. 2009). However, as individual bacterial cells behave in a very different manner than established/matured biofilms, it is important to consider them independently. This *bottom-up* approach to study bacterial and biofilm mechanics has proven extremely useful in recent years, and has given the field of microbiology new insights on the complex process of bacteria–surface coupling.

15.2 Atomic Force Microscopy: Probing into Nanoscale Mechanics

Atomic force microscopy (AFM) has been a valuable tool for the study of living bacterial samples. Introduced initially as an imaging tool (Binnig et al. 1986), the AFM has evolved into a highly specialised instrument to probe into both cellular and sub-cellular mechanics (Aguayo et al. 2015b). As the aim of this chapter is not to be an in-depth description of the workings of the instrument, only a brief overview of the AFM system will be covered in order to give the reader sufficient background to understand how mechanics data can be obtained from the sample and which techniques can be utilised to acquire data on both indentation and adhesion experiments.

The main components of an AFM setup are illustrated in Fig. 15.1a. Interestingly, although the AFM is a complex and highly specialised instrument, its basic working principle is rather simple and easy to comprehend. As observed, the AFM consists of a specialised *tip* attached to a flexible *cantilever*, which is scanned over a sample of interest. As the tip comes in contact (or close contact) with the sample it causes deflection of the cantilever, which is picked up by a *laser-photodiode system* (Nunez et al. 2005). Finally, the AFM *software* transduces the deflection readings into either an image or mechanics data depending on the experiment being carried out by the researcher. Currently, there are many different

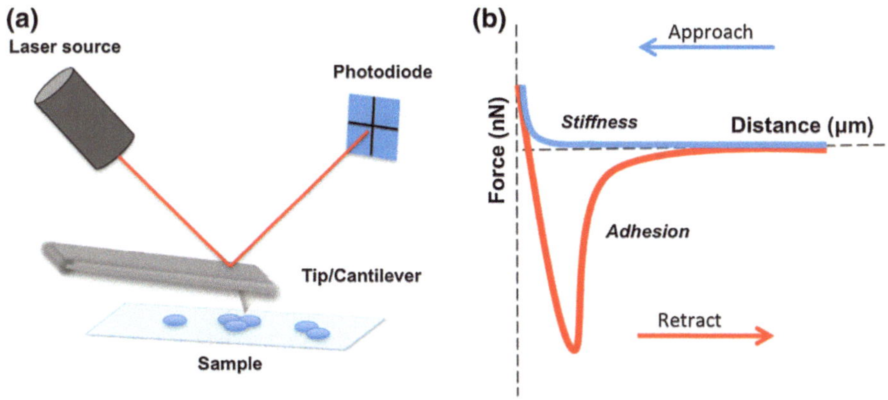

Fig. 15.1 **a** Basic working principle for an AFM. Deflection of the cantilever (produced by the tip–surface interaction) changes the incidence of the laser onto the photodiode detector. **b** Diagrammatic representation of a force–distance curve. Data on sample stiffness is usually obtained from the approach curve, while adhesion interactions are shown in the retraction portion of the curve (and are generally expressed as negative force values). Force curves for surfaces with different mechanical properties will show different approach and retraction patterns

AFM systems available on the covering a vast number of applications ranging from conventional imaging and nanomechanics to high-speed AFM systems (Ando et al. 2013; Jalili and Laxminarayana 2004).

This ability of the AFM to interact directly with substrates can be used to study the nanomechanical properties of bacterial cells, giving it an important advantage compared to traditional microscopy techniques such as electron microscopy (Pillet et al. 2014). Perhaps one of the most important advantages for microbiology is that the AFM allows measurements to be performed in liquid with little to no sample preparation, and therefore viable bacteria can be studied under physiological conditions (Helenius et al. 2008; Gaboriaud and Dufrene 2007). Employed buffers can be further modified by the addition of antibacterial solutions to study the impact of these substances on bacterial morphology, aggregation, cell wall mechanics or adhesion (Pillet et al. 2014). The applicability of AFM mechanics in microbiology goes beyond the use of bacterial cells, as the fabrication of single-molecule-coated AFM tips can be utilised to probe the mechanical behaviour of sub-cellular bacterial structures with pico-newton sensitivity (Hwang et al. 2015; Mendez-Vilas et al. 2007).

When employing the AFM to study the mechanical properties of samples, a *force curve* representing the relationship between force and distance is obtained (Fig. 15.1b). Initially, the tip/cantilever is situated at a distance from the sample, and is slowly approached towards the surface until a given loading force is reached (Fig. 15.1b, blue curve). After a user-determined contact time, the probe is retracted from the sample surface and brought back to the initial resting point (Taubenberger et al. 2013). The adhesion occurring between the tip and surface is observed as a dip towards negative force values during this retraction phase: therefore, adhesive

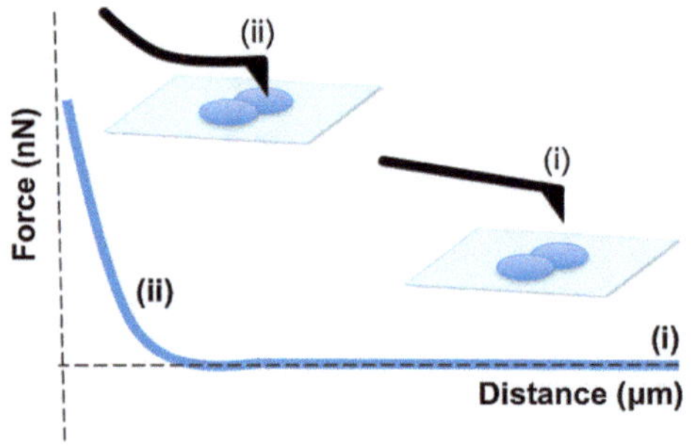

Fig. 15.2 Nanoindentation of bacterial cells attached to a substrate. Initially, the cantilever is away from the bacterium *i* and is slowly approached towards the surface with a determined loading force. Once the bacterium is deformed, an increase in force is observed from which the mechanical properties of the cell can be investigated *ii*. However, the introduction of more biologically significant models that consider adhesion between the bacterium and tip (such as the DMT model) employ the retraction curve for fitting

forces are usually expressed as negative force values (Fig. 15.1b, red curve). When dealing with bacteria, however, retraction curves tend to be more complex as they display a series of unbinding events, which can be further analysed to characterise the corresponding cell–surface interaction (Muller et al. 2009) .

How are force values obtained from the cantilever deflection? As the tip/cantilever behaves as a spring, force (F) values can be obtained according to *Hooke's Law*

$$\mathrm{F} = -k \times \mathrm{d} \tag{15.1}$$

where *k* is the cantilever *spring constant* and *d* is *vertical deflection* of the cantilever (Schaer-Zammaretti and Ubbink 2003; Bolshakova et al. 2004). Previous calibration of each cantilever before experimentation is essential in order to obtain correct spring constant values, as variations can be found even within cantilevers from the same batch (Sheng et al. 2007). Several different methodologies for calibration, such as thermal tuning and pushing the cantilever against a calibrated reference lever, have been developed in hopes of providing precise and consistent spring constant measurements. Most commercially available AFM systems incorporate reliable built-in tuning software, many of which are highly automated and user friendly (Atabek and Camesano 2007). Currently, some companies also provide researchers with pre-calibrated cantilevers that simplify the process even further.

These advantages that AFM brings to the field of microbiology allow researchers to design and perform a broad array of experiments to characterise the mechanic behaviour of single-bacterial cells (Wu and Zhou 2010). Although many approaches have been described throughout the literature, the present chapter will focus on three of the most utilised: *bacterial nanoindentation*, *single-cell force spectroscopy* and *single-molecule pulling*.

15.3 Probing the Elasticity of Bacterial Cells

First, the AFM can be used to understand the elastic properties of bacterial cells and how certain environmental and antibacterial factors can influence these parameters (Fig. 15.2) (Vadillo-Rodriguez et al. 2004). Upon indenting a sample with a determined loading force, it will respond by deforming its surface by a given amount. This stress/strain relationship can be utilised to determine *stiffness* of a sample, which is also applicable to bacterial indentation experiments. However, more useful to researchers is the *elastic modulus,* which can be obtained by determining values for *Young's modulus* according to several elastic models (Dokukin et al. 2013). Traditionally, the *Hertzian model for spherical indentation* has been applied to cell indentation data mainly because of simplicity, and can be expressed by the relationship between force (F) and indentation (δ)

$$F = \frac{4ER^{1/2}\delta^{3/2}}{3(1 - v^2)} \quad (15.2)$$

where E is the *Young's modulus*, R is the *radius of the indenter* and v is the *Poisson's ratio* of the indented sample (which for cells and biological samples is usually considered at 0.5). This approach, however, has several considerations that make it not appropriate for the interpretation of bacterial indentation experiments (Glaubitz et al. 2014). First, the Hertzian model assumes no adhesive interaction between the indenter and the surface, which is not the case for bacterial cells as their adhesive behaviour allows attachment to cantilevers (coated or un-coated) during indentation experiments. Furthermore, Hertzian modelling assumes that the indented surface is infinitely thick compared to the indenter, and therefore indentation should only occur on the surface of the sample and not affect the bulk material. As bacterial cells are compliant in nature, it is safe to assume that a load sufficient enough to deform the cell wall can also cause a certain degree of deformation of underlying cellular structures. These limitations have taken researchers to consider the application of more appropriate models such as the Derjaguin-Muller-Toporov (DMT) model for indentation, which also takes into account adhesion between the bacterial sample and indenter and is therefore more suitable for the real experimental situation (Loskill et al. 2014). Modern AFM setups are progressively incorporating these models into their analysis software to facilitate data analysis and interpretation in a user-friendly manner.

It is also important to consider that as bacteria are heterogeneous in composition, it remains difficult to apply models that completely reflect the overall deformation occurring in the bacterial cell. Different constituents such as the cytoplasm, as cell wall and extracellular matrix components, will have different deformation values when constant loads are being applied (Chen et al. 2012). Nevertheless, as values of bacterial elastic modulus vary according to many—i.e. cell wall structure and integrity, surrounding osmotic pressure—nanoindentation of bacteria is an efficient way of understanding how they respond to changes in their internal or external environments. As certain antibiotics can target cell wall synthesis, reduced Young's modulus values have been observed for bacteria after treatment with antibacterial drugs (Formosa et al. 2012a, b; Wu and Zhou 2009). Furthermore, this approach is not limited to individual bacterial cells as it can also be utilised to study the mechanical behaviour of biofilms at different stages of development (Abe et al. 2011) and before and after chemical treatment (Powell et al. 2013).

15.4 Forces Driving Bacteria–Substrate Adhesion

As mentioned previously, bacteria are usually found attached to surfaces or forming part of a biofilm. From here, some bacterial cells can detach and migrate to other places to colonise new surfaces (Winkelstroter et al. 2013). The ability of bacteria to bind to most biological and non-biological substrates is one of the most important factors for their high survival and endurance capacities in nature (Zhao et al. 2013). Therefore, many scientific fields are actively looking for new methods to either enhance or inhibit the adhesion of bacteria to surfaces, in hopes of improving industrial applications of biofilms or reducing the prevalence of bacterial infections in the population.

Before discussing some experimental results obtained by AFM adhesion experiments, it is important to understand the biophysics of bacterial attachment to surfaces. For simplification purposes, the adhesion of bacterial cells to substrates can be considered a two-stage process (Gordesli and Abu-Lail 2012). Initially, when a planktonic bacterium comes into close proximity with a substrate, there is a predominance of non-specific long-range interactions such as van der Waals and electrostatic forces (Harimawan et al. 2013). The *Derjaguin, Landau, Vervey and Overbeek (DLVO) theory*, mainly used in colloidal particle physics, is useful to understand the influence of both of these physicochemical interactions occurring in this initial 'docking' phase (Abu-Lail and Camesano 2003; Hermansson 1999). As the bacterium approaches a surface, the interplay of both attractive van der Waals and repulsive electrostatic forces determine the likelihood of the bacterial cell to attach to the substrate. If attractive forces overweigh repulsive ones, the bacterial cell will tend towards the surface of interest and allow the adhesion process to continue (Fig. 15.3). Subsequently, specific short-range interactions stabilise the bacterial–substrate interface, 'locking' the bacterium and surface together and giving place to a secondary phase of irreversible adhesion. At this point, bacteria can only be removed from the surface by mechanical or chemical treatment. Bacterial appendages such as pili and fimbriae, and surface proteins as *adhesins* and *microbial surface components recognising adhesive matrix molecules* (MSCRAMMs) play an important role in this stage (Dorobantu et al. 2009; Buck et al. 2010). Therefore, this secondary locking stage is mainly mediated by the interaction of specific adhesion molecules present on the surface of the bacterium with the substrate.

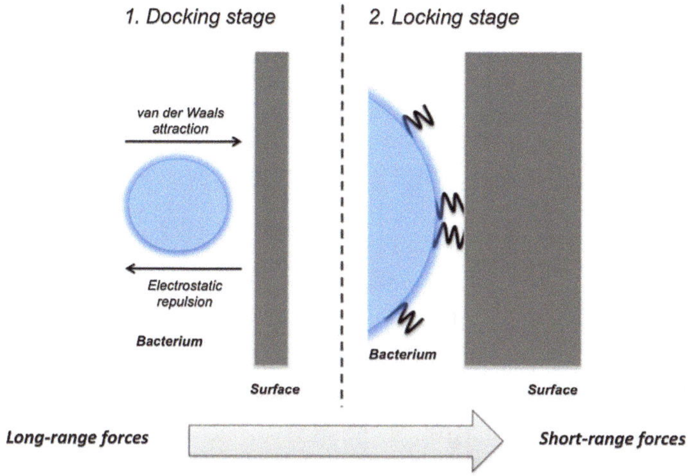

Fig. 15.3 Overview of the forces influencing bacterial attachment to surfaces

In vitro, it is possible to probe into the world of bacterial adhesion to surfaces by using AFM force spectroscopy. To do so, it is necessary to attach living bacterial cells to a cantilever tip by means of an immobilising agent (Beaussart et al. 2014). Immobilisation is the key element for effective cell-based AFM work, and must not only ensure correct attachment of the bacterium to the cantilever tip but also maintain cell viability throughout experimentation. Failing to do so may result in inconsistent datasets or detachment of bacteria during probing. Many different approaches have been reported to effectively attach bacteria to cantilevers, namely *gelatin*, *poly-L-lysine* and *poly-DOPA* among others, all of which confer a positive charge to the cantilever surface to which negatively charged bacteria can covalently bind (Louise Meyer et al. 2010; Beaussart et al. 2013a). This ensures stability throughout force spectroscopy measurements, as ineffective immobilisation may lead to incorrect measurements due to lateral displacement or complete bacterial detachment. Since current methodologies pursue the immobilisation of single or reduced number of bacteria on the cantilever tip, the technique has become known as *single-cell force spectroscopy* (SCFS) (Beaussart et al. 2013b). Bacteria can be attached directly to the cantilever tip, although several studies recommend the use of a tip-less cantilever mounted with a colloidal bead to ensure a more stable and reproducible contact area between bacterium and surface (Beaussart et al. 2013a, 2014; Aguayo et al. 2015a). Force curves resulting from SCFS experiments display characteristic patterns in the retraction segment, considerably different than the ones obtained on non-biological samples (Fig. 15.4).

In SCFS, a cell-modified cantilever is approached, placed into contact and retracted from a surface of interest to explore the adhesive behaviour between the bacteria and the substrate (Ubbink and Schar-Zammaretti 2005). Therefore, the most important information is obtained from the retraction segment of the resulting force curves. The approach segment of the curve usually displays a smoothened slope,

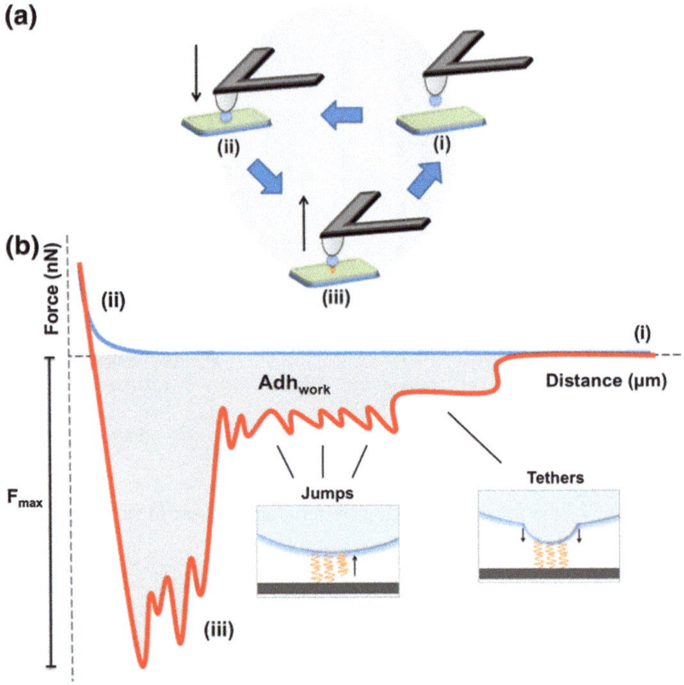

Fig. 15.4 Diagrammatic representation of a single-cell force spectroscopy (SCFS) experiment. **a** A bacterium is immobilised onto a glass-microsphere previously mounted onto an AFM cantilever. Initially, the bacterial probe is away from the surface *i* and is approached slowly towards the substrate of interest until contact *ii*. After a determined contact time, the probe is retracted from the surface *iii* and brought back to the initial position *i*. **b** Diagrammatic representation of a force–distance curve obtained for the unbinding of a cell probe. Probe positions *i*, *ii* and *iii* observed in 4**a** are shown. Insets depict the two most common situations observed during cell unbinding

consistent with the viscoelastic behaviour of the bacterium being gently pressed against the surface. The retraction segment, however, displays a great amount of information regarding the bacterium–surface interaction. Initially, a major unbinding peak representing the largest negative value in the vertical axis can be observed, from which the maximum adhesion force (F_{max}) can be obtained. F_{max} is usually the most reported parameter in SCFS experiments and is mainly representative of non-specific adhesion between the bacteria and substrate. Another relevant parameter known as adhesion work (Adh_{work}) can be extracted by integrating the retraction curve, obtaining an area under the curve indicative of the amount of energy needed to fully unbind the bacterium–surface interaction. Adh_{work} involves both the major peak and minor unbinding events, and therefore reflects the influence of both the non-specific and specific surface adhesion (Taubenberger et al. 2013).

Furthermore, the analysis of minor unbinding events observed in the retraction curve can give insight on the dynamic behaviour of bacterial adhesion molecules

(Evans and Calderwood 2007). Two distinct minor unbinding patterns can be detected throughout bacterial retraction curves. The most commonly observed are *jumps* that represent the unbinding of single-adhesive units and which sometimes are described in the literature as 'sawtooth-like' patterns (Yongsunthon and Lower 2006; Lower et al. 2005). *Tethers*, which represent the stretching of the cellular membrane before unbinding (Krieg et al. 2008), are not observed frequently in bacterial SCFS retraction curves as many bacteria are surrounded by structures (i.e. cell wall, capsule) that do not allow significant elongation of the cell before receptor unbinding.

An important parameter that can be controlled is the *dwelling time*, also known as surface delay, which represents the contact time between the bacterium and the surface before retracting the probe. As bacteria are dynamic cells, the influence of contact time on all previously mentioned adhesion parameters can be studied in great detail (Kashef and Franz 2015). Most importantly, control of the dwelling time allows researchers to simulate different stages of the physiological process of bacterial adhesion. Short dwelling times are representative of the early interaction between the bacterium and surface, while increased times can be used to simulate changes in adhesion forces once bacteria have settled on the surface. Many bacterial receptors have shown time-dependent bond strengthening at increasing contact times, effect which has been reported extensively in the literature (Mei et al. 2009; Vadillo-Rodriguez et al. 2004).

Finally, a mathematical approach known as *Poisson analysis* can also be applied to the minor unbinding events found in the retraction curve to deconstruct each minor peak into the corresponding *short-range* and *long-range* force components, as previously discussed (Fig. 15.3) (Chen et al. 2011). This approach assumes that short-range adhesion forces are constituted by the sum of a limited number of individual bonds, and therefore, the total bacterium–substrate adhesion force (F) can be expressed as:

$$F = (f_{SR} \times k) + F_{LR} \tag{15.3}$$

where f_{SR} is the value of a single-bond, k is the number of formed bonds and F_{LR} the value for long-range interactions. At very close distances (such as the ones observed during AFM force measurements), values for F_{LR} are minimal compared to F_{SR} and therefore it is safe to assume that variance in adhesion forces are mainly due to variations in the occurrence of bonds, k. As the distribution of unbinding forces between bacterial probes and surfaces usually follow a *Poisson distribution*, it is possible to plot a linear regression from which both F_{LR} and F_{SR} can be determined (Mei et al. 2009). F_{LR} is mostly representative of van der Waals and electrostatic interactions while F_{SR} reflects the value of specific bacterium–surface adhesion, and therefore decoupling bacterial adhesion into these two categories allows for deeper understanding of the specific mechanisms behind the adhesion of different bacterial strains. As an example, Abu-lail and Camesano have suggested that in many cases the formation of hydrogen bonds, with an individual bond value of ~0.13 nN, are key players of bacterial adhesion to hydrophilic surfaces (Abu-Lail and Camesano

2006). By determining the values for F_{SR} through Poisson analysis, it is possible to hypothesise the approximate number of hydrogen bonds between *Staphylococcus aureus* and a clinically-analogous titanium implant surface (Aguayo et al. 2015a). Lastly, it is important to consider that contrary to F_{SR} which is always positive, values for F_{LR} can be either positive or negative depending on if the overall force is repulsive or attractive, respectively (Fig. 15.3).

15.5 Pulling Single-Molecule Bacterial Complexes

The applicability of the AFM in microbiology is not only limited to the nanoindentation of microbial cells or to the probing of forces driving bacterial attachment to surfaces. The modelling of bacterium–surface unbinding events can yield important information on the length and nature of the specific molecules involved in adhesion to biological and non-biological surfaces. Several models have been reported to describe the stretching of elastic polymers during AFM experimentation (Storm and Nelson 2003; Francius et al. 2009).

For the case of protein unfolding, force-extension data can be modelled according to the *worm-like chain (WLC)* equation as

$$\mathrm{F(x)} = \frac{\mathrm{k_b}T}{\mathrm{l_p}}\left[\frac{1}{4}\left(1-\frac{x}{L}\right)^{-2}+\frac{x}{L}-\frac{1}{4}\right] \tag{15.4}$$

where F is force (N), k_b is the Boltzmann constant (J/kelvin), T is temperature (kelvin), x is molecular extension (m), l_p is the persistence length and L the total contour length of the molecule. The WLC model considers polypeptides as elastic molecules conformed by smaller indivisible units (persistence length, l_p), which add up to give the total length of the molecule (contour length, L) (Marszalek and Dufrene 2012). Instead of only being flexible between l_p segments, proteins are continuously flexible throughout the entire length of the molecule (Fig. 15.5). For fitting purposes, l_p is usually considered as the length of a single aminoacid ($\sim$0.36 nm) for polypeptide force-extension experiments (Herman et al. 2014).

Polysaccharides, on the other hand, are better fitted by the freely jointed chain (FJC) model, which can be used to probe the adhesion of lectins and other polysaccharide microbial molecules (Marszalek and Dufrene 2012; Francius et al. 2009). The FJC model considers the total length of a molecule as

$$L = n \times l \tag{15.5}$$

where L is the total unfolded length, n is the number of monomers and l is the length of a single monomer. Contrary to the WLC model, monomers are considered rigid units and thus molecules can only bend between segments (Fig. 15.5). In literature, both of these force-extension models have been employed to model the stretching of bacterial appendages and proteins, and more recently, to describe the

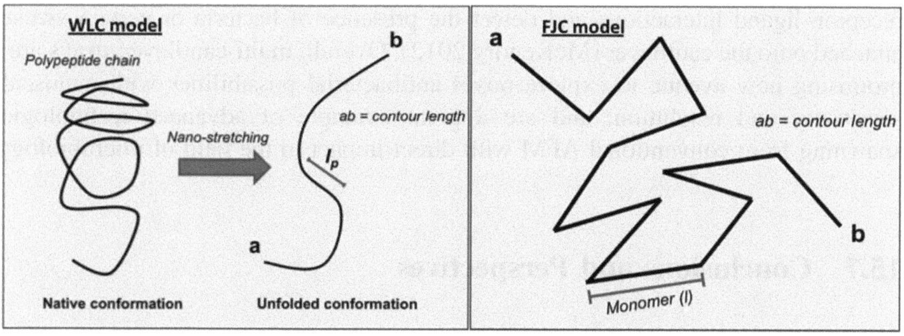

Fig. 15.5 Diagrammatic representation the worm-like chain (WLC) and the freely-jointed chain (FJC) models, which are commonly utilised to model the stretching polypeptide and polysaccharide chains, respectively

unbinding of bacterial cell probes from surfaces during SCFS experiments (Jacquot et al. 2014; Muller et al. 1999; Beaussart et al. 2014; Aguayo et al. 2015a; Rangel et al. 2013).

15.6 Multi-cantilever Arrays for Antibacterial Detection: Novel Applications of AFM in Microbiology

Recently, there has been increasing worries regarding the potential advent of widespread antimicrobial resistance. Nowadays, many researchers agree that the world is at the verge of a post-antibiotic era, with many drug-resistant strains and 'superbugs' already being reported (Hu et al. 2014; Ford et al. 2013). Among others, resistant strains of highly pathogenic bacteria such as methicillin-resistant *Staphylococcus aureus* (MRSA), *Escherichia coli* and *Mycobacterium tuberculosis* have been isolated. As it has been decades since the introduction of a novel class of antibacterial drugs (Cooper and Shlaes 2011), many efforts are currently being placed on the discovery and development of novel antibacterial therapies to aid in the fight against drug-resistant strains.

As a result, advanced applications for bacterial AFM work have been developed in hopes of optimising this process. By immobilising bacterial structural components—such as cell wall precursors—to cantilevers, researchers have managed to measure the interaction between a number of antibacterial substances and the molecules of interest (Ndieyira et al. 2013). As the antibiotic–target interaction generates mechanic stress that causes bending of the cantilever, it can be picked up by the laser-photodiode system and measured with nanoscale sensitivity. This allows the screening of vast amounts of antibacterial drugs in reduced periods of time. This approach has been already utilised to test and compare the effects of two powerful antibiotics, vancomycin and oritavancin, in solution (Ndieyira et al. 2014). These multi-cantilever systems can also be utilised to measure other receptor–ligand interactions and detect the presence of bacteria or certain viruses attached onto the cantilever (McKendry 2012). Overall, multi-cantilever arrays are a promising new avenue to explore novel antibacterial possibilities with nanoscale sensitivity and resolution, and are a great example of advanced technologies spawning from conventional AFM with direct impact in the field of microbiology.

15.7 Conclusions and Perspectives

As reviewed throughout this chapter, many recent advances have placed bacterial nanomechanics as a highly useful and reproducible technique to study bacterial cells and molecules at the nanoscale. Although great progress has been made in optimising the immobilisation of bacteria to substrates and cantilevers and improving viability throughout measurements, efforts must still be made to homologate data collection and analysis to allow for effective comparison between different research groups. In general, SCFS still remains a highly user-dependent technique and therefore it is still necessary to find ways to simplify setups to optimise instrument time and reduce complications.

As the field of nanoscale microbiology continues to expand, it is also necessary to explore new techniques such as pairing AFM with Raman or real-time fluorescence microscopy. Furthermore, the advent of new AFM systems with faster scanning rates and with the ability to obtain image, adhesion and stiffness maps simultaneously has opened the possibility to study the real-time effect of antibacterial agents or other environmental changes on bacterial cells. Although conventional bacterial nanomechanic techniques allow for the potential study of most bacteria, they are still confined mostly to aerobic strains, as AFM setups are yet to develop effective approaches for the study of anaerobic bacteria. Finally, with the increasing amounts of raw data generated with conventional fast-scan techniques, it is necessary to develop new and standardised batch data analysis tools to extract and process information in an effective manner. Nevertheless, nanoindentation and SCFS have proven to be important techniques to characterise the nanoscale behaviour of living bacteria and have contributed enormously to the understanding of single-bacterium microbiology.

References

Abe Y, Polyakov P, Skali-Lami S, Francius G (2011) Elasticity and physico-chemical properties during drinking water biofilm formation. Biofouling 27(7):739–750. doi:10.1080/08927014.2011.601300

Abe Y, Skali-Lami S, Block JC, Francius G (2012) Cohesiveness and hydrodynamic properties of young drinking water biofilms. Water Res 46(4):1155–1166

Abu-Lail NI, Camesano TA (2003) Role of ionic strength on the relationship of biopolymer conformation, DLVO contributions, and steric interactions to bioadhesion of Pseudomonas putida KT2442. Biomacromolecules 4(4):1000–1012

Abu-Lail NI, Camesano TA (2006) Specific and nonspecific interaction forces between Escherichia coli and silicon nitride, determined by poisson statistical analysis.[Erratum appears in Langmuir. 2008 Apr 15;24(8):4420]. Langmuir 22 (17):7296–7301

Aguayo S, Donos N, Spratt D, Bozec L (2015a) Nanoadhesion of Staphylococcus aureus onto titanium implant surfaces. J Dent Res. doi:10.1177/0022034515591485

Aguayo S, Donos N, Spratt D, Bozec L (2015b) Single-bacterium nanomechanics in biomedicine: unravelling the dynamics of bacterial cells. Nanotechnology 26(6):062001. doi:10.1088/0957-4484/26/6/062001

Ando T, Uchihashi T, Kodera N (2013) High-speed AFM and applications to biomolecular systems. Annu Rev Biophys 42:393–414. doi:10.1146/annurev-biophys-083012-130324

Atabek A, Camesano TA (2007) Atomic force microscopy study of the effect of lipopolysaccharides and extracellular polymers on adhesion of Pseudomonas aeruginosa. J Bacteriol 189 (23):8503–8509

Beaussart A, El-Kirat-Chatel S, Herman P, Alsteens D, Mahillon J, Hols P, Dufrene YF (2013a) Single-cell force spectroscopy of probiotic bacteria. Biophys J 104(9):1886–1892. doi:10.1016/j.bpj.2013.03.046 S0006-3495(13)00385-8[pii]

Beaussart A, Herman P, El-Kirat-Chatel S, Lipke PN, Kucharikova S, Van Dijck P, Dufrene YF (2013b) Single-cell force spectroscopy of the medically important Staphylococcus epidermidis-Candida albicans interaction. Nanoscale 5(22):10894–10900. doi:10.1039/c3nr03272h

Beaussart A, El-Kirat-Chatel S, Sullan RM, Alsteens D, Herman P, Derclaye S, Dufrene YF (2014) Quantifying the forces guiding microbial cell adhesion using single-cell force spectroscopy. Nat Protoc 9(5):1049–1055. doi:10.1038/nprot.2014.066

Binnig G, Quate CF, Gerber C (1986) Atomic force microscope. Phys Rev Lett 56(9):930–933

Bolshakova AV, Kiselyova OI, Yaminsky IV (2004) Microbial surfaces investigated using atomic force microscopy. Biotechnol Prog 20(6):1615–1622

Buck AW, Fowler VG Jr, Yongsunthon R, Liu J, DiBartola AC, Que YA, Moreillon P, Lower SK (2010) Bonds between fibronectin and fibronectin-binding proteins on Staphylococcus aureus and Lactococcus lactis. Langmuir 26(13):10764–10770. doi:10.1021/la100549u

Chen Y, Busscher HJ, van der Mei HC, Norde W (2011) Statistical analysis of long- and short-range forces involved in bacterial adhesion to substratum surfaces as measured using atomic force microscopy. Appl Environ Microbiol 77(15):5065–5070

Chen Y, Norde W, van der Mei HC, Busscher HJ (2012) Bacterial cell surface deformation under external loading. MBio 3(6). doi:10.1128/mBio.00378-12

Cooper MA, Shlaes D (2011) Fix the antibiotics pipeline. Nature 472(7341):32. doi:10.1038/472032a

Dokukin ME, Guz NV, Sokolov I (2013) Quantitative study of the elastic modulus of loosely attached cells in AFM indentation experiments. Biophys J 104(10):2123–2131. doi:10.1016/j.bpj.2013.04.019

Dorobantu LS, Bhattacharjee S, Foght JM, Gray MR (2009) Analysis of force interactions between AFM tips and hydrophobic bacteria using DLVO theory. Langmuir 25(12):6968–6976. doi:10.1021/la9001237

Evans EA, Calderwood DA (2007) Forces and bond dynamics in cell adhesion. Science 316 (5828):1148–1153. doi:10.1126/science.1137592

Ford CB, Shah RR, Maeda MK, Gagneux S, Murray MB, Cohen T, Johnston JC, Gardy J, Lipsitch M, Fortune SM (2013) Mycobacterium tuberculosis mutation rate estimates from different lineages predict substantial differences in the emergence of drug-resistant tuberculosis. Nat Genet 45(7):784–790. doi:10.1038/ng.2656

Formosa C, Grare M, Duval RE, Dague E (2012a) Nanoscale effects of antibiotics on P. aeruginosa. Nanomedicine 8(1):12–16. doi:10.1016/j.nano.2011.09.009

Formosa C, Grare M, Jauvert E, Coutable A, Regnouf-de-Vains JB, Mourer M, Duval RE, Dague E (2012b) Nanoscale analysis of the effects of antibiotics and CX1 on a Pseudomonas aeruginosa multidrug-resistant strain. Sci Rep 2:575. doi:10.1038/srep00575

Francius G, Alsteens D, Dupres V, Lebeer S, De Keersmaecker S, Vanderleyden J, Gruber HJ, Dufrene YF (2009) Stretching polysaccharides on live cells using single molecule force spectroscopy. Nat Protoc 4(6):939–946. doi:10.1038/nprot.2009.65

Gaboriaud F, Dufrene YF (2007) Atomic force microscopy of microbial cells: application to nanomechanical properties, surface forces and molecular recognition forces. Colloids Surf, B 54(1):10–19

Gan M, Zhou S, Li M, Zhu J, Liu X, Chai L (2014) Bioleaching of multiple heavy metals from contaminated sediment by mesophile consortium. Environ Sci Pollut Res Int. doi:10.1007/s11356-014-3759-x

Glaubitz M, Medvedev N, Pussak D, Hartmann L, Schmidt S, Helm CA, Delcea M (2014) A novel contact model for AFM indentation experiments on soft spherical cell-like particles. Soft Matter 10(35):6732–6741. doi:10.1039/c4sm00788c

Gordesli FP, Abu-Lail NI (2012) Combined Poisson and soft-particle DLVO analysis of the specific and nonspecific adhesion forces measured between L. monocytogenes grown at various temperatures and silicon nitride. Environ Sci Technol 46(18):10089–10098

Harimawan A, Zhong S, Lim CT, Ting YP (2013) Adhesion of B. subtilis spores and vegetative cells onto stainless steel—DLVO theories and AFM spectroscopy. J Colloid Interface Sci 405:233–241

Helenius J, Heisenberg CP, Gaub HE, Muller DJ (2008) Single-cell force spectroscopy. J Cell Sci 121(Pt 11):1785–1791. doi:10.1242/jcs.030999 121/11/1785[pii]

Herman P, El-Kirat-Chatel S, Beaussart A, Geoghegan JA, Foster TJ, Dufrene YF (2014) The binding force of the staphylococcal adhesin SdrG is remarkably strong. Mol Microbiol 93 (2):356–368. doi:10.1111/mmi.12663

Hermansson M (1999) The DLVO theory in microbial adhesion. Colloids Surf B 14(1–4): 105–119. doi:10.1016/S0927-7765(99)00029-6

Hu S, Li G, Li H, Liu X, Niu J, Quan S, Wang F, Wen H, Xu Y, Li Q (2014) Rapid detection of isoniazid resistance in mycobacterium tuberculosis using real-time PCR-based melting curve analysis. J Clin Microbiol. doi:10.1128/JCM.03395-13

Hwang G, Marsh G, Gao L, Waugh R, Koo H (2015) Binding force dynamics of Streptococcus mutans-glucosyltransferase B to Candida albicans. J Dent Res. doi:10.1177/0022034515592859

Jacquot A, Sakamoto C, Razafitianamaharavo A, Caillet C, Merlin J, Fahs A, Ghigo JM, Beloin C, Duval JF, Francius G (2014) Dynamic modulation of fimbrial extension and FimH-mannose binding force on live bacteria under pH changes: a molecular atomic force microscopy analysis. J Biomed Nanotechnol 10(11):3361–3372

Jalili N, Laxminarayana K (2004) A review of atomic force microscopy imaging systems: application to molecular metrology and biological sciences. Mechatronics 14(8):907–945. doi:10.1016/j.mechatronics.2004.04.005

Kashef J, Franz CM (2015) Quantitative methods for analyzing cell-cell adhesion in development. Dev Biol 401(1):165–174. doi:10.1016/j.ydbio.2014.11.002

Krieg M, Helenius J, Heisenberg CP, Muller DJ (2008) A bond for a lifetime: employing membrane nanotubes from living cells to determine receptor-ligand kinetics. Angew Chem Int Ed Engl 47(50):9775–9777. doi:10.1002/anie.200803552

Loskill P, Pereira PM, Jung P, Bischoff M, Herrmann M, Pinho MG, Jacobs K (2014) Reduction of the peptidoglycan crosslinking causes a decrease in stiffness of the Staphylococcus aureus cell envelope. Biophys J 107(5):1082–1089. doi:10.1016/j.bpj.2014.07.029

Louise Meyer R, Zhou X, Tang L, Arpanaei A, Kingshott P, Besenbacher F (2010) Immobilisation of living bacteria for AFM imaging under physiological conditions. Ultramicroscopy 110 (11):1349–1357. doi:10.1016/j.ultramic.2010.06.010 S0304-3991(10)00191-9 [pii]

Lower BH, Yongsunthon R, Vellano FP 3rd, Lower SK (2005) Simultaneous force and fluorescence measurements of a protein that forms a bond between a living bacterium and a solid surface. J Bacteriol 187(6):2127–2137. doi:10.1128/JB.187.6.2127-2137.2005

Marszalek PE, Dufrene YF (2012) Stretching single polysaccharides and proteins using atomic force microscopy. Chem Soc Rev 41(9):3523–3534. doi:10.1039/c2cs15329g

McConnell MD, Liu Y, Nowak AP, Pilch S, Masters JG, Composto RJ (2010) Bacterial plaque retention on oral hard materials: effect of surface roughness, surface composition, and physisorbed polycarboxylate. J Biomed Mater Res, Part A 92(4):1518–1527

McKendry RA (2012) Nanomechanics of superbugs and superdrugs: new frontiers in nanomedicine. Biochem Soc Trans 40(4):603–608. doi:10.1042/BST20120082

Mei L, Ren Y, Busscher HJ, Chen Y, van der Mei HC (2009) Poisson analysis of streptococcal bond-strengthening on saliva-coated enamel. J Dent Res 88(9):841–845. doi:10.1177/0022034509342523

Mendez-Vilas A, Gallardo-Moreno AM, Gonzalez-Martin ML (2007) Atomic force microscopy of mechanically trapped bacterial cells. Microsc Microanal 13(1):55–64

Muller DJ, Baumeister W, Engel A (1999) Controlled unzipping of a bacterial surface layer with atomic force microscopy. Proc Natl Acad Sci USA 96(23):13170–13174

Muller DJ, Helenius J, Alsteens D, Dufrene YF (2009) Force probing surfaces of living cells to molecular resolution. Nat Chem Biol 5(6):383–390. doi:10.1038/nchembio.181 nchembio.181 [pii]

Ndieyira JW, Watari M, McKendry RA (2013) Nanomechanics of drug-target interactions and antibacterial resistance detection. J Visualized Exp: JoVE 80:e50719. doi:10.3791/50719

Ndieyira JW, Kappeler N, Logan S, Cooper MA, Abell C, McKendry RA, Aeppli G (2014) Surface-stress sensors for rapid and ultrasensitive detection of active free drugs in human serum. Nat Nanotechnol 9(3):225–232. doi:10.1038/nnano.2014.33

Nunez ME, Martin MO, Chan PH, Duong LK, Sindhurakar AR, Spain EM (2005) Atomic force microscopy of bacterial communities. Methods Enzymol 397:256–268

Oh YJ, Lee NR, Jo W, Jung WK, Lim JS (2009) Effects of substrates on biofilm formation observed by atomic force microscopy. Ultramicroscopy 109(8):874–880. doi:10.1016/j.ultramic.2009.03.042

Pillet F, Chopinet L, Formosa C, Dague E (2014) Atomic Force Microscopy and pharmacology: from microbiology to cancerology. Biochim Biophys Acta 1840(3):1028–1050. doi:10.1016/j.bbagen.2013.11.019

Powell LC, Sowedan A, Khan S, Wright CJ, Hawkins K, Onsoyen E, Myrvold R, Hill KE, Thomas DW (2013) The effect of alginate oligosaccharides on the mechanical properties of Gram-negative biofilms. Biofouling 29(4):413–421. doi:10.1080/08927014.2013.777954

Rangel DE, Marin-Medina N, Castro JE, Gonzalez-Mancera A, Forero-Shelton M (2013) Observation of bacterial type I pili extension and contraction under fluid flow. PLoS ONE 8(6): e65563. doi:10.1371/journal.pone.0065563

Schaer-Zammaretti P, Ubbink J (2003) Imaging of lactic acid bacteria with AFM—elasticity and adhesion maps and their relationship to biological and structural data. Ultramicroscopy 97(1–4):199–208

Schmidlin PR, Muller P, Attin T, Wieland M, Hofer D, Guggenheim B (2013) Polyspecies biofilm formation on implant surfaces with different surface characteristics. J Appl Oral Sci 21(1): 48–55. doi:S1678-77572013000100048

Sheng X, Ting YP, Pehkonen SO (2007) Force measurements of bacterial adhesion on metals using a cell probe atomic force microscope. J Colloid Interface Sci 310(2):661–669

Storm C, Nelson PC (2003) Theory of high-force DNA stretching and overstretching. Phys Rev E: Stat, Nonlin, Soft Matter Phys 67(5 Pt 1):051906

Subramani K, Jung RE, Molenberg A, Hammerle CH (2009) Biofilm on dental implants: a review of the literature. Int J Oral Maxillofac Implants 24(4):616–626

Taubenberger AV, Hutmacher DW, Muller DJ (2013) Single-cell force spectroscopy, an emerging tool to quantify cell adhesion to biomaterials. Tissue Eng Part B Rev. doi:10.1089/ten.TEB.2013.0125

Ubbink J, Schar-Zammaretti P (2005) Probing bacterial interactions: integrated approaches combining atomic force microscopy, electron microscopy and biophysical techniques. Micron 36(4):293–320. doi:10.1016/j.micron.2004.11.005

Vadillo-Rodriguez V, Busscher HJ, Norde W, de Vries J, van der Mei HC (2004) Atomic force microscopic corroboration of bond aging for adhesion of Streptococcus thermophilus to solid substrata. J Colloid Interface Sci 278(1):251–254

Van Acker H, Van Dijck P, Coenye T (2014) Molecular mechanisms of antimicrobial tolerance and resistance in bacterial and fungal biofilms. Trends Microbiol. doi:10.1016/j.tim.2014.02.001

Waar K, van der Mei HC, Harmsen HJ, de Vries J, Atema-Smit J, Degener JE, Busscher HJ (2005) Atomic force microscopy study on specificity and non-specificity of interaction forces between Enterococcus faecalis cells with and without aggregation substance. Microbiology 151(Pt 7):2459–2464

Winkelstroter LK, Teixeira FB, Silva EP, Alves VF, De Martinis EC (2013) Unraveling microbial biofilms of importance for food microbiology. Microb Ecol. doi:10.1007/s00248-013-0347-4

Wu Y, Zhou A (2009) In situ, real-time tracking of cell wall topography and nanomechanics of antimycobacterial drugs treated Mycobacterium JLS using atomic force microscopy. Chem Commun (Camb) 45:7021–7023. doi:10.1039/b914605a

Wu Y, Zhou A (2010) Fluctuations in adhesion behavior of dividing/budding mycobacterium sp. strains JLS, KMS, MCS: an AFM evaluation. Micron 41(7):814–820

Yongsunthon R, Lower SK (2006) Force spectroscopy of bonds that form between a Staphylococcus bacterium and silica or polystyrene substrates. J Electron Spectrosc Relat Phenom 150(2–3):228–234. doi:10.1016/j.elspec.2005.06.012

Zhao G, Usui ML, Lippman SI, James GA, Stewart PS, Fleckman P, Olerud JE (2013) Biofilms and inflammation in chronic wounds. Adv Wound Care (New Rochelle) 2(7):389–399. doi:10.1089/wound.2012.0381

Chapter 16
Neutron Reflectivity as a Tool for Physics-Based Studies of Model Bacterial Membranes

Robert D. Barker, Laura E. McKinley and Simon Titmuss

Abstract The principles of neutron reflectivity and its application as a tool to provide structural information at the (sub-) molecular unit length scale from models for bacterial membranes are described. The model membranes can take the form of a monolayer for a single leaflet spread at the air/water interface, or bilayers of increasing complexity at the solid/liquid interface. Solid-supported bilayers constrain the bilayer to 2D but can be used to characterize interactions with antimicrobial peptides and benchmark high throughput lab-based techniques. Floating bilayers allow for membrane fluctuations, making the phase behaviour more representative of native membranes. Bilayers of varying levels of compositional accuracy can now be constructed, facilitating studies with aims that range from characterizing the fundamental physical interactions, through to the characterization of accurate mimetics for the inner and outer membranes of Gram-negative bacteria. Studies of the interactions of antimicrobial peptides with monolayer and bilayer models for the inner and outer membranes have revealed information about the molecular control of the outer membrane permeability, and the mode of interaction of antimicrobials with both inner and outer membranes.

Keywords Neutron reflectivity · Lipid membrane · Structure · Biomimetic · Antimicrobial · Lipopolysaccharide

---

R.D. Barker
Institut Laue-Langevin, 71 Avenue Des Martyrs,
CS 20156, 38042 Grenoble Cedex 9, France

L.E. McKinley · S. Titmuss (✉)
School of Physics and Astronomy, University of Edinburgh,
James Clerk Maxwell Building, The King's Buildings,
Peter Guthrie Tait Road, Edinburgh EH9 3FD, Scotland
e-mail: simon.titmuss@ed.ac.uk

R.D. Barker
School of Science and Engineering, University of Dundee,
James Clerk Maxwell Building, The King's Buildings,
Peter Guthrie Tait Road, Dundee DD1 4HN, UK

© Springer International Publishing Switzerland 2016

M.C. Leake (ed.), *Biophysics of Infection*, Advances in Experimental Medicine and Biology 915, DOI 10.1007/978-3-319-32189-9_16

16.1 Introduction

In this chapter we will explain, with illustrative examples, how neutron reflectivity can be used as a tool to understand the physics of bacterial membranes and their interactions with antimicrobial peptides. We will consider three case studies of relevance to bacterial infection: the interaction of antimicrobial peptides (AMP) with the inner membrane, the influence of lipopolysaccharide on the outer membrane of Gram-negative bacteria such as *E. coli* and its interaction with AMP. We emphasize at the outset that in the search for unifying features, model membrane systems *designed* to facilitate the greatest amount of physical insight, rather than aimed at reproducing a biochemically accurate biomimetic, will often be most appropriate. Recently, Clifton and co-workers have built on this approach to develop a realistic model for the bacterial periphery that is coupled to a solid substrate, and have studied its structure with neutron reflectivity (Clifton et al. 2015a). Such a biomimetic could form the basis for a sensor to be used in lab-based characterization of antimicrobials, and neutron reflectivity has an important role in validating the structure of this biomimetic.

16.1.1 Why Study Bacterial Membranes?

In 2013, the UK's Chief Medical Officer placed the development of antibiotic resistance by bacteria, and our collective failure to develop alternative control strategies, on the UK risk register. An unpalatable future in which even routine operations become hazardous, due to the risk posed by bacterial infection, will be the end result of this failure. For a physicist, bacterial membranes are interesting: although they are much more complex than the models we will employ, they are simpler than eukaryotic membranes and it is conceivable that their behaviour is largely controlled by purely physical variables. Furthermore, the bacterial inner membrane plays a key role in the synthesis of ATP, and the cell's conversion of energy. As use of energy is a key characteristic of life, understanding the bacterial membrane will get us closer to understanding how bacteria live, and hence how this might be controlled. The presence of a second (outer) membrane in Gram-negative bacteria, which has different biophysical properties, also poses some interesting physics questions.

16.1.2 Why Use a Physics-Based Approach?

Physicists seek to uncover universal behaviour. In the context of the biophysics of infection, we believe that finding features at the molecular-scale that are common to the membrane aspects of different systems of relevance to life and death of bacteria, will provide useful experimental input into the development of a framework that explains these features in terms of physical principles (membrane structure/thermodynamics/mechanics). We hope that such experimental data and physical framework might help in the development of *broad-spectrum* antimicrobial treatments. Why might this work? It can been said that, "Biology is all interfaces"—whilst this is a gross simplification, it is indisputable that all bodies interact with their environment through their interfacial regions. In the case of a bacteria, the periphery of the cell is defined by a cell membrane, and in the case of Gram-negative, two of them (inner and outer membranes). The bacterial cell membrane thus serves the purpose of the cell's first line of defence. Furthermore, it is also an integral part of the cell's metabolism. This suggests that interfering with the physical properties (mechanical and electrical integrity) might cause the cell problems.

16.1.3 What Do We Want to Learn?

The methodology we outline is based on the premise that if we can understand the physical basis for the manner in which peptides interact with the bacterial cell membranes, it may be possible to tune/manipulate these interactions in a way that will enable the control of bacterial growth or the development of improved strategies to kill bacteria. To achieve this, we want to learn where the peptides sit in the membrane and what effect they have on the structural integrity of the membrane. This requires a coarse-grained structural technique, which will provide structural information on the length scale of the lipids and peptides. In real cells, the membrane is a dynamic environment, with fluctuations playing a significant role in biological function, so this information needs to be obtained from membranes that are free to exhibit such fluctuations. We will demonstrate that neutron reflectivity meets these criteria, with the one caveat that the studies have to be conducted on model membranes rather than native membranes in bacteria. We will introduce the different model membranes that can be employed, explaining the rationale, strengths and weaknesses for each one. The aim and scope of this chapter is somewhat different from exhaustive reviews of the field (Pabst et al. 2010; Wacklin 2010; Junghans et al. 2015).

16.1.4 Neutron Reflectivity as an Ideal Probe of Model Bacterial Membranes

16.1.4.1 Physics of Neutron Reflectivity

The physics of a neutron reflectivity experiment is the same as observing interference colours in the reflection of light from an oily film on a puddle. The same optical principles that hold for light, hold for neutrons, except that by using thermal (where the thermal bath is that of a moderator at $\sim$20 K) neutrons, the radiation has a wavelength λ in the range 0.2–3 nm i.e. $\sim$1000× smaller than that of visible light. As such the films, which will give rise to strong interference effects, are 1000× thinner than the oily film on the puddle. Specifically, strong interference effects will be observed from films that are 3–5 nm thick, precisely the length scale of a bacterial membrane. The other optical criterion for observing interference effects is that there is a change in the refractive index across the film. Refractive indices (n) for neutrons in media are generally very close to unity, and with a few exceptions, are slightly smaller than 1,[1]

$$n = 1 - \frac{\lambda^2 \rho}{2\pi}. \tag{16.1}$$

This means that neutrons will undergo total external reflection from most materials (if incident from air): below the critical angle, the reflection is total; above it the reflectivity is determined by the variation in the scattering length density ρ perpendicular to the interface. The scattering length density of a layer (l) is determined by the chemical composition of the components ($\mathcal{N}_k(l)$ gives the number of each atom k, which has scattering length b_k in a layer of volume $V(l)$) of the film and the density profile ($\mathcal{N}_k(l)/V(l)$),

$$\rho(l) = \frac{\Sigma_k \mathcal{N}_k(l) b_k}{V(l)}. \tag{16.2}$$

Specular neutron reflectivity is a relatively low-resolution structural technique, with the limit on the resolution being imposed by the relatively low maximum value of the scattering vector Q_z perpendicular to the surface, for which a reflectivity $R(Q_z)$ can be measured before the incoherent background obscures the reflection; the specular condition is illustrated by the left panel of Fig. 16.1.[2] The resolution limit means that neutron reflectivity can provide structural information on the length scale of sub-units of molecules. In the case of lipids, then resolving the density profile of the lipid head groups, the methylene chain of the lipid tails and the methyl end-groups of the lipid tails might be an appropriate level of coarse-graining. We assert that the presence of significant thermal fluctuations in these systems makes this the appropriate length scale on which to be aiming for structural information, as distinct from crystallography, which is typically performed on crystals at cryogenic temperatures and has a different role to play in understanding biological function. Applying this coarse-graining to Eq. (16.2) leads to Eq. (16.3).

[1]For this reason, it is convenient (and usual) to characterize a material's neutron optical properties by a scattering length density, ρ.

[2]The incoherent background is largely due to the presence of hydrogen, which has a very high incoherent cross-section, in the sample/sub-phase, which is unavoidable in biologically relevant samples; in solid/liquid experiments this can be minimized to some extent by using a low sub-phase volume, such as in the laminar flow cell illustrated in the right panel of Fig. 1.

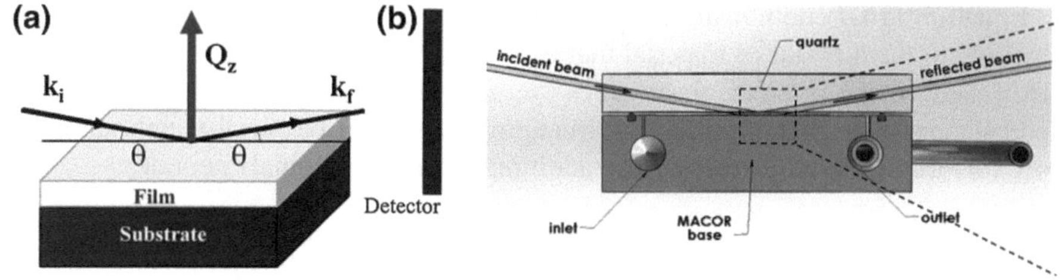

Fig. 16.1 **a** The geometrical optics of specular reflection from a thin film between two bulk media. The scattering vector $Q_z = \frac{4\pi}{\lambda}\sin\theta$, where θ is the grazing angle of incidence and λ the wavelength of the neutrons, is related to the change in the momentum of the neutron perpendicular to the interface. **b** Schematic of a solid/liquid reflectivity experiment in which the neutrons are incident through a single-crystal superphase, which forms one half of a laminar flow cell, which can be used to exchange the water sub-phase. Adapted from Junghans et al. (2015)

$$\rho(l) = \Sigma_j \rho_j(l) = \frac{\Sigma_j \mathcal{N}_j(l) b_j}{V(l)} = \Sigma_j \phi_j(l) \rho_j(l) \tag{16.3}$$

Thus, measuring the reflectivity gives a means to determine the average density profile $\phi_j(l)$ perpendicular to the surface, which is the observable we are interested in.[3]

To scratch below the surface of the geometrical optics analogy of the interference colours from an oily film observed with light, a kinematic approach can be taken to obtain the following expression for the reflectivity of a thin film of material characterized by some density distribution which has a second moment σ:

$$R(Q) = \frac{16\pi^2}{Q^2} m^2 \exp(-Q^2\sigma^2) - \frac{32\pi^2}{Q^2} \Delta\rho \langle z \rangle m \exp\left(\frac{-Q^2\langle z^3 \rangle}{6\langle z \rangle}\right) + \frac{16\pi^2}{Q^4} \Delta\rho^2 \tag{16.4}$$

where m is proportional to the adsorbed amount at the interface, $\langle z^n \rangle$ is the nth power of distance from the interface averaged over the adsorbate distribution and $\sigma = (\langle z^2 \rangle - \langle z \rangle^2)^{1/2}$ is the second moment of the adsorbate distribution ($\sigma^2 \approx \langle z^3 \rangle / (6\langle z \rangle)$). At the air/contrast-matched water interface, only the first term contributes to the reflectivity, and a plot of $\ln(R(Q)Q^2)$ as a function of Q^2 has a slope determined by the second moment of the adsorbate distribution $-\sigma^2$.

[3]At a fundamental level, the average (pseudo)potential experienced by the neutron is $\bar{V} = \frac{2\pi\hbar^2}{m}\rho$. Since the momentum change in specular reflection is solely perpendicular to the interface, it is the force on the neutron perpendicular to the interface that is important in determining the reflectivity. Since the force is given by the gradient of the potential, it is clear why it is the gradient of the scattering length density profile that is important in determining the reflectivity. The same conclusion can be drawn in an explicit mathematical form by application of the Born approximation $R(Q) = \frac{16\pi^2}{Q^4}|\tilde{\rho}(Q)|^2$, where $\tilde{\rho}(Q) = \int_{-\infty}^{\infty} \frac{d\rho(z)}{dz} \exp(-iQz)dz$ is the Fourier transform of the derivative of the scattering length density profile perpendicular to the interface.

Equation (16.4) demonstrates in mathematical form that neutron reflectivity is sensitive to the amount of material that is at an interface and to how that material is distributed: these are the observables we require to develop our physics-based understanding of what controls the structural integrity of bacterial membranes, and how this might be perturbed by the addition of (anti-microbial) peptides.

16.1.4.2 Neutron Reflectivity as a Bridge of the Reality Gap

Neutron reflection experiments cannot be done on individual bacteria, so what place does it have in an issue dedicated to the *Biophysics of Infection*? We start from the premise that the physical (as compared to the biochemical) properties have a role to play in the life and death of bacteria. The physical properties (mechanical and thermodynamic) will be a function of the membrane composition but also of any applied fields (electrical, stress, chemical potential, thermal, pressure). Bacterial cells are robust, and across the diverse range of bacteria examples can be found to thrive over wide ranges of temperature, pH and salt; however a given bacteria will actually thrive over a relatively limited range of each of these potentials. This makes doing sufficiently systematic investigations of the effect of these potentials on real living bacteria unfeasible. This means that a model (or biomimetic) system must be used to facilitate such studies. One approach is to perform experiments in silico using molecular dynamics (MD) simulations at various levels of coarse-graining (Marrink et al. 2009; Illya and Deserno 2008; Woo and Wallqvist 2011; Chen et al. 2012). Such in silico experiments are attractive as it is possible to try things that are not necessarily feasible experimentally. Typically, a MD simulation is based around a supercell geometry: a finite patch of lipid bilayer (or monolayer) is sandwiched between suitable water layers, and then periodic boundary conditions are applied in the plane of the bilayer to generate an infinite 2D-bilayer; this supercell is then repeated in the direction perpendicular to the bilayer, to allow the computational simplification that a structure periodic in 3D affords.

The size of the patch that is repeated in the plane is limited by the available computational power, as the computational cost increases with the number of atoms in the simulation. There are two ways in which this number can be kept manageable: either the patch size is limited (a bilayer of $14 \times 14 \times 2$ is not unusual) or by coarse-graining the structure from individual atoms into beads (i.e. by changing the constituent beads from atoms to molecular sub-units). The limited patch size that is possible for fully atomistic simulations presents a serious problem when trying to understand how the membrane of a bacterium responds collectively to some externally applied field (chemical, electrical, pressure etc.), where the elastic response of the membrane as a whole may play as important a role as the local interaction energy.

The importance of such elastic energy terms was demonstrated by a coarse-grained dissipative particle dynamics simulation of the interaction of the AMP maganin 2 with a bilayer composed of 4:1 DPPC/POPG by Woo and

Wallqvist (2011). By using a large patch size (0.1 μm), they were able to suggest that the peptide induced disordered toroidal pores and that when the peptides are allowed to interact from only one side of the bilayer, they induce a buckling that produces a quasi-spherical bud connected by a narrow neck.

In the case of fluctuating soft systems, coarse-graining does not really remove any meaningful structural information, but it could affect the accuracy of the force fields and so the thermodynamic properties of the system constructed from those coarse-grained beads, and hence indirectly, the structure adopted at the minimum of the free energy.

In some respects, a neutron reflectivity experiment represents an experimental in vitro realization of the in silico MD simulation. Whilst the planar geometry and coarse-graining are inherent in the measurement, there are two subtle but important advantages in the reflectivity experiment: the coarse-graining is only applied to *interpret* the data, our bilayers and monolayers are built from real lipids, and so by definition have the correct force-field; and our bilayers and monolayers span areas of 10 cm^2, so they are free to fluctuate, bend and buckle. For this reason, we view neutron reflectivity as providing a bridge across the reality gap between microbial growth studies, such as those described by Jepson and Poon (2014), Jepson et al. (2016) and molecular dynamics simulations such as described by Carr (2015).

To consider the bridge between reflectivity experiments and microbial AMP assays, a simple calculation is useful. In our bilayer experiments, we use a ~ 25 cm^2 floating bilayer with a typical area per lipid molecule of 50 Å^2, so there would be 10^{-9} moles of lipids present. The floating bilayer is in contact with a sub-phase of volume ~ 2.5 mL, which can be exchanged by laminar flow. We would typically flow a 10 mL volume of peptide solution (at $\sim$ μM concentration) through the flow cell, so the bilayer would be exposed to 10^{-9} moles of peptide; hence overall there would be a peptide to lipid ratio of $P/L \sim 1$. How that compares in detail to microbial assays, depends on how that assay is performed (Jepson 2014; Jepson et al. 2016), but it is not unreasonable (Melo and Castanaho 2012).

We will illustrate the utility of the neutron reflectivity approach with examples that are relevant to the inner membrane (interaction with antimicrobial peptide sequences) and the outer membrane (influence of lipopolysaccharide on membrane structure and hence mechanical integrity).

16.2 Illustrative Examples of the Application of Neutron Reflectivity to the Physics of Bacterial Infection

We will first consider examples pertaining to the interaction of antimicrobial peptides with the inner membrane, and then examples related to the lipopolysaccharide-bearing outer membrane. In both cases, we will describe the findings resulting from reflectivities measured from both monolayers (at the air/water interface) and bilayers (either supported or floating on a solid substrate).

Both of these approaches have a role to play when used appropriately, and we will highlight the strengths of each approach.

16.2.1 The Interaction of Antimicrobial Peptides with Monolayers at the Air/Water Interface

Much of the pioneering use of reflectivity techniques to the study of peptides interacting with models for bacterial membranes was carried out by the group of K. Y. C. Lee, using monolayers spread on a Langmuir trough. Although such films are clearly not bilayers, and so represent a step away from the native microbial membrane across the reality gap, used appropriately, such studies can greatly augment our understanding of the way in which peptides interact with lipid membranes. If the peptide only interacts with one leaflet of the membrane, then the monolayer is already a good model. This is the case for the 11-residue membrane targeting sequence (*mts*) of the cell division protein MinD which is expressed inside the cell with the function of helping to locate the mid-point of an intact cell, such that the MinD-*mts* can only interact with the inner leaflet, which is the target of our own ongoing investigations (McKinley 2015). Figure 16.2 illustrates the scheme of a typical neutron reflection experiment to investigate the interaction of peptides with a lipid monolayer.

Gidalevitz et al. studied the interaction of protegrin-1 (PG-1), an 18-amino acid amidated peptide, which is part of the porcine immune system and appears to be functionally analogous to human defensins, with monolayers comprising of DPPG or POPG/POPC or lipid A as models for the membranes of the bacteria that PG-1 is known to be toxic to, which includes *E. coli* (Gidalevitz et al. 2003). Making measurements on a Langmuir trough using a combination of epifluorescence and X-ray reflectivity/grazing incidence X-ray diffraction (GIXRD), they show that insertion of PG-1 into the monolayer models for bacterial membranes has a disordering effect on the lipid layer. The X-ray reflectivity data requires an additional sub-phase layer to fit the data compared to the two-layer model which is sufficient

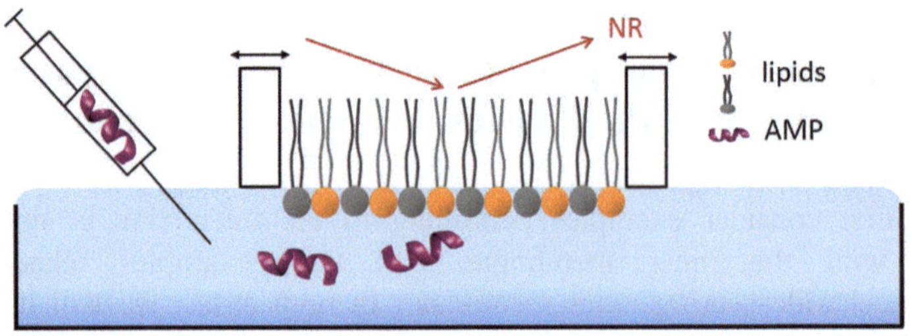

Fig. 16.2 The scheme of a typical reflectivity experiment to study the interaction of a peptide with a lipid monolayer

for the pure lipid layers. In this three-layer model, the air-side layer has an electron density corresponding to the lipid tails, the next layer corresponds to the lipid head groups, whilst the final layer, located on the water side of the head groups, has an electron density consistent with the peptide and a thickness of 27 Å, which equals the longest dimension of the peptide. By making their measurements at constant surface pressure of $\pi = 20$ and 30 mN/m, they were able to monitor the change in the area of the frame after peptide was injected beneath the barrier to a sub-phase concentration of 0.025 mg/mL (11 μM), chosen to match the lytic concentration in microbial studies. At 30 mN/m, they observe a much more pronounced increase in the change in area for POPG ($\Delta A/A = 33\,\%$) compared to POPC ($\Delta A/A = 7\,\%$). They describe this as insertion, even though their structural studies indicate that the peptide is located beneath the headgroups in the proximal sub-phase region. From the GIXRD, they are able to observe a loss of the in-plane ordering in the packing of the lipid molecules, although their analysis of the X-ray reflectivity data does not attempt to correlate the electron densities determined in the upper two layers with a lipid area per molecule.

By contrast, Clifton et al. combine neutron reflectivity with X-ray reflectivity to explicitly determine both the lipid area per molecule and the protein coverage in their study of the interaction of the plant defence proteins α1- and α2-purothionin with DPPG monolayers (Clifton et al. 2012). This study not only nicely illustrates some of the strengths of neutron reflectivity for the study of biological materials, but also clearly demonstrates a role for monolayer studies.

A key strength of neutron reflectivity is afforded by the very different scattering lengths of hydrogen and deuterium. This means that by using deuterated lipids it is possible to construct a lipid monolayer which will reflect neutrons differently whilst retaining very similar biochemical properties. This technique, known as *contrast variation*, enables the volume fractions of the different components to be evaluated explicitly. Constructing separate monolayers from h-lipids and d-lipids results in a scattering length density profile for the layers that can be written:

$$\rho_{\mathrm{h}} = \rho_{\mathrm{h-lipid}}\phi_{\mathrm{lipid}} + \rho_{\mathrm{protein}}\phi_{\mathrm{protein}} + \rho_{\mathrm{water}}\phi_{\mathrm{water}} \tag{16.5}$$

$$\rho_{\mathrm{d}} = \rho_{\mathrm{d-lipid}}\phi_{\mathrm{lipid}} + \rho_{\mathrm{protein}}\phi_{\mathrm{protein}} + \rho_{\mathrm{water}}\phi_{\mathrm{water}} \tag{16.6}$$

These represent a pair of simultaneous equations that can be solved for the volume fraction of lipid in the layer:

$$\rho_{\mathrm{h}} - \rho_{\mathrm{d}} = (\rho_{\mathrm{h-lipid}} - \rho_{\mathrm{d-lipid}})\phi_{\mathrm{lipid}} \tag{16.7}$$

which can then be substituted back into Eq. (16.5) or (16.6) to solve for the volume fraction of protein; it is then a trivial matter to convert these volume fractions into the corresponding area per molecule A_m using

$$A_m = \frac{v_m}{t\phi} \tag{16.8}$$

where v_m is the lipid (or protein) molecular volume and t the thickness of the layer, which is also determined from fitting the reflectivity data. Using this approach, Clifton et al. were able to quantify that the lipid area per molecule following the injection of α2-Pth is 12 % greater than that following the injection of α1-Pth; this level of molecular information allowed them to correlate the relative activity of defence proteins with their hydrophobicity.

This pair of studies nicely illustrates the advantages that monolayer studies can offer over bilayer studies:

(i) it is possible to measure *and* vary the surface pressure and area per molecule on a Langmuir trough; adsorption experiments can be carried out at either constant pressure or constant area. This means that it is possible to access, in a thermodynamically controlled fashion, regions of the phase diagram that may vary from the canonically accepted surface pressure regime of a bacterial mimetic bilayer.

(ii) It is possible to use Brewster angle microscopy to provide direct images of the phase behaviour.

16.2.1.1 Brewster Angle Microscopy (BAM): Revealing Phase Behaviour

In Brewster angle microscopy, contrast is provided by the sensitivity to small changes in the optical thickness of a film, when the light is close to the Brewster angle (53.1° in the case of monolayers at the air/water interface). This means that changes in the packing density of lipid monolayers will produce contrast in BAM, hence providing a direct visualization of phase behaviour on the 10 μm scale. Furthermore, since specular reflectivity provides information about the scattering length density projected onto the surface normal, if there is lateral structure it is also important to know about it, and its relative length scale, such that the specular reflectivity can be modelled appropriately. It is incorrect to state that for a meaningful specular reflectivity analysis, the layer must be laterally homogenous; it is possible to analyze laterally inhomogeneous layers, providing that the appropriate averaging scheme over the different domains is employed. This requires an image of the lateral structure, which BAM can not only provide but the contrast mechanism for which is closely related to the molecular packing parameters, which also affects the specular neutron reflectivity.

A note of caution should be added as it is possible that the reason the PG-1 is observed to interact beneath the head group region is that the single monolayer leaflet is simply not thick enough to fully incorporate the 18-residue peptide. The obvious way to remove this concern is to instead work with bilayer models.

16.2.2 Bilayer Models for Inner Bacterial Membranes and Their Interactions with Antimicrobial Peptides

There are two approaches which have been followed for the construction of lipid bilayers as models for membranes: supported lipid bilayers which are simpler (Wacklin 2010) and floating lipid bilayers which are more realistic (Fragneto et al. 2012; Hughes et al. 2008, 2014; Barker 2011).

16.2.2.1 Supported Bilayers

Fernandez et al. have used supported bilayers formed by liposome deposition as a platform to study the interaction with the antimicrobial peptides aurein 1.2 (Fernandez et al. 2012b) and maculatin 1.1 (Fernandez et al. 2012a). The supported lipid bilayer approach is convenient as it allows for in situ self-assembly of the lipid bilayer. This approach lends itself to flow-based biophysical screening techniques such as quartz crystal microbalance with dissipation (QCM-D) and dual polarization interferometry (DPI). These laboratory-based techniques, that essentially measure added mass, have a higher throughput than is possible with central facilities-based neutron reflectivity measurements, but provide only indirect access to structural information. In the case of QCM-D, determining unique structural information would require the implementation of hydrodynamic modelling. The prevalence of fingerprinting-type studies *in lieu* of this suggests that this modelling is more difficult than that involved in the calculation of neutron reflectivity. Neutron reflectivity measurements have an important role in benchmarking these techniques and augmenting the structural information they provide.

In the liposome deposition method, solutions of the liposomes are prepared by the rehydration of dried lipid films assembled from a chloroform-based solution that has the same composition as is desired for the supported lipid bilayer. These liposomes can then be injected into the flow cell, whether it be for DPI, QCM-D or neutron reflectometry measurements. On contact with the solid substrate (silicon oxynitride in the case of DPI, quartz in the case of QCM-D, and typically a silicon oxide layer on silicon in the case of neutron reflectivity) the liposomes rupture, forming a hopefully complete lipid bilayer of the same composition as the original chloroform-based solution used to prepare the dried lipid film from which the liposomes were rehydrated (Fernandez et al. 2012a). Fernandez and co-workers highlight the importance of preparing consistent, homogenous and *defect-free* bilayers if they are to be used for peptide-binding studies (Fernandez et al. 2012a). The area per lipid molecule and corresponding bilayer thickness they determine by DPI for a 4:1 DMPC/DMPG model for a bacterial membrane is (50 ± 1.8) Å^2 and (44.9 ± 1.7) Å, respectively. Whilst the value for the bilayer thickness they determine from their neutron reflectivity studies (performed on a different substrate) is consistent at (43 ± 3) Å, the value for the area per lipid molecule at (75 ± 6) Å^2 is not. They explain this discrepancy in terms of an incomplete (83 ± 7 %) bilayer coverage in the neutron reflectivity experiments. This highlights a significant drawback with supported bilayer methodology: the nature of the bilayer is determined by the interaction with the solid substrate and the specific surface chemistry that this interface presents. The incomplete bilayer coverage also presents a complication in the quantitative interpretation of the neutron reflectivity data. Either the bilayer is homogenous and sparser than might normally be expected, as the typical area per molecule for a lipid bilayer at the canonical leaflet surface pressure of 30 mN/m is more like that measured in the DPI experiments, or the layer is patchy. In the former case, one might anticipate that it is easier for any peptides to insert, since there is more free volume in the bilayer than there might be expected to be in a realistic bacterial membrane, which would call into question the significance of any conclusions drawn about peptide insertion into such a bilayer. In the second case, the reflectivity should properly be evaluated as an incoherent superposition of the reflectivity from a bare substrate (17 % of substrate, in the case considered here) and the reflectivity from a bilayer covered silicon substrate (83 % of substrate, in the case considered here). Either of these effects will compromise the ability to reliably locate peptide molecules within the bilayer. As the peptide molecule will be highly hydrated, it will bring a significant amount of water with it, such that an inserted highly hydrated peptide is difficult to uniquely distinguish from a patch of bare surface, or a sparse lipid monolayer in which the excess free volume will be occupied by water. At a more fundamental level, the demonstrable impact that the interaction with the substrate has on the bilayer means that the phase behaviour of the bilayer cannot be regarded to be purely a function of the bilayer composition and thermodynamic variables (temperature, pressure, applied electric field).

To study the interaction of a supported bilayer with the 18-residue peptide $p - Ant\,p_{43-58}$, Fragneto et al. instead used Langmuir–Blodgett (to deposit the inner leaflet) and Langmuir–Schaeffer (to deposit the outer leaflet) techniques to prepare the gel-phase-supported DPPC and 9:1 DPPC/DPPS bilayers; the interaction with peptide was investigated using the same deposition techniques, but from lipid monolayers spread on sub-phases containing the peptide at 0.7 μM (Fragneto et al. 2000). They find that in the case of the DPPC bilayer, the peptide is uniformly distributed through the bilayer; but in the case of the mixed bilayer, in which the head group region has a negative charge due to the phosphatidylserine lipids, the peptide, which is rich in cationic residues, is localized in the headgroup region. Although the peptide initially interacts with the monolayer (i.e. the individual leaflets, from which the bilayer was deposited in the gel phase), this study served as a proof-of-principle for the feasibility of using specular neutron reflectivity to locate 10–20 residue peptides in lipid bilayers.

16.2.2.2 Floating Bilayers

A better approach is provided by the floating bilayer method that was developed by Hughes et al. (2008), from the approach of Fragneto (Charitat et al. 1999; Fragneto et al. 2012), and refined by Barker (2011). In the floating bilayer approach, the lipid bilayer that is to be the subject of the investigation sits above a water layer of 2–5 nm, which sits between the bilayer and a functionalized (typically) silicon substrate. As conceived by Charitat et al., the functionalization of the silicon substrate is provided by a lipid bilayer that is deposited using the Langmuir–Blodgett technique. Such a bilayer might be expected to experience comparable interactions with the underlying solid substrate as in the case of supported bilayers prepared by liposome deposition considered above. To produce a fluid bilayer, they extended this approach to deposit a second bilayer: in this case, the third monolayer was deposited using Langmuir–Blodgett with the fourth monolayer, forming the outer leaflet of the outer bilayer, being deposited using Langmuir–Schaeffer (see Fig. 16.1 in Fragneto et al. (2012)).

The structure of DPPC bilayers assembled in this fashion was then determined using specular neutron reflectivity from four different combinations of lipid bilayer deuteration level and sub-phase H_2O/D_2O ratio, to provide different sensitivities to different molecular sub-groups via the contrast variation method described previously. They find that this outer bilayer is: more homogenous and reproducible than the supported bilayers formed by vesicle fusion and more flexible, as the $\sim$3 nm water cushion means that the Helfrich fluctuations are not suppressed (Helfrich 1978)—these fluctuations are evident from the Yoneda peak in the off-specular scattering, and fits to the reflectivity require a roughness that is 6–7 Å greater than that of the underlying substrate.

Unfortunately, this methodology is not universally applicable to all lipids, being limited to those that are in the gel phase above room temperature (e.g. saturated phosphocholines longer than C16). Hughes and co-workers introduced a series of refinements to the approach with the aim of extending the range of lipids that can be assembled into floating bilayers (Hughes et al. 2002, 2008, 2014). The key refinement is to replace the inner solid-supported bilayer with a covalently grafted self-assembled monolayer (SAM) that presents a lipid headgroup functionality: ω-thiolipids on gold or 1-Palmitoyl-2-[16-(acryloyloxy)hexadecanoyl]-*sn*-glycero-3-phosphorylcholine (al-PC) coupled via a reactive acryl silane SAM on a silicon oxide terminated silicon substrate. The covalent coupling to the substrate and, in the case of al-PC, the possibility of cross-linking, results in increasingly robust SAMs. Careful optimisation of each stage in the process has allowed a platform to be developed whereby it is possible to reproducibly form high coverage layers which can be precharacterized and are stable for multiple uses, in contrast to the more complex approach pioneered by Fragneto, which requires the building up of a supporting layer prior to deposition of the floating bilayer for each new sample. However, even with these refinements the deposition of the floating bilayer, which is depicted schematically in Fig. 16.3, can be sensitive to the precise orientation of the substrate as it is pushed through the monolayer in the Langmuir–Schaeffer step. To address this, the new Level'O'Matic (or LOM), was developed—a laser scanning mechanism removing the human alignment of the sample in the Langmuir–Schaeffer step, during which most deposited membranes are lost (Barker 2011). With these improvements, the fundamental difficulties in building these

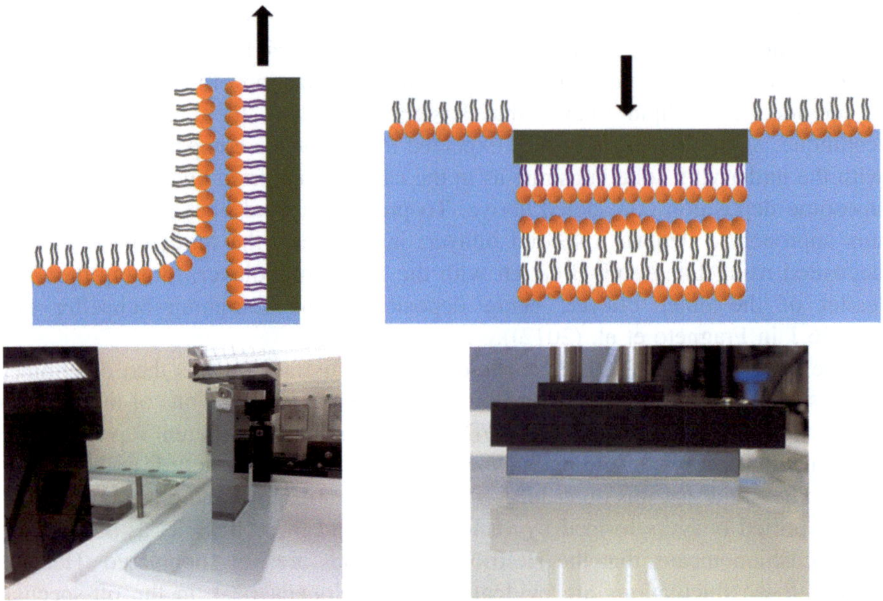

Fig. 16.3 The assembly of floating bilayers by Langmuir–Blodgett (*left*) and Langmuir–Schaeffer (*right*) deposition

model systems have been significantly reduced, broadening their usage beyond the specialist and enabling more complex biological problems to be addressed.

Significantly for the assembly of realistic mimetics of bacterial membranes, Hughes et al. have now described the extension of the approach to unsaturated lipids (Hughes et al. 2014). This study also indicates another key feature of the neutron for extracting structural information: it is a spin-$\frac{1}{2}$ particle, such that polarized neutron beams can be created. The scattering length density of a magnetic layer, such as permalloy in a magnetic field, is different for the spin up/down states; opening up the possibility of obtaining two *magnetic* contrast measurements simultaneously. This technique was conceived at NIST, as providing a reference that would enable the phase problem, inherent in a reflectivity measurement, to be circumvented by facilitating a direct inversion of the data (Majkrzak et al. 1998). Although Hughes et al. still transform the reflectivity to a structure by means of a fitting procedure, rather than direct inversion, it is clear that the simultaneous measurement of the two magnetic contrasts serves as a strong constraint on the fitting procedure, increasing the robustness of the extraction of structural details. Recently, one of us (Barker), has proposed the exploitation of this constraint to fit spin asymmetry rather than pure reflectivity. This approach appears to have the potential to increase the sensitivity to small perturbations of the bilayer structure caused by the interaction with small peptides (Jagalski et al. 2015).

16.2.2.3 The Challenge of Studying the Interaction of Antimicrobial Peptides with Floating Bilayer Models of the Inner Bacterial Membrane

That 15 years have elapsed since Fragneto's proof of principle for peptide location in a gel-phase-supported lipid bilayer, without there being an equivalent study for a floating bilayer model of the inner bacterial membrane, is a reflection of the challenge that it presents. The nature of the challenge is two-fold:

(i) bilayer models for bacterial membranes should contain 25 % anionic lipids and ideally contain unsaturated lipids—such bilayers are less stable and harder to prepare than DPPC floating bilayers;
(ii) the presence of the negative charge in the head group region and/or the unsaturated tails mean that such bilayers undergo greater fluctuations—this makes the fitting of the reflectivity profiles using the standard parameterized box models (extensions of the approach described in 2.1 to bilayers) more difficult.

It is a challenge that we have been working on recently. Specifically, we have constructed floating bilayers that are 3:1 DPPC/DPPG and 3:1 POPC/POPG; in the former case, although we deposit in the gel-phase, we then anneal through the gel/fluid phase transition and study the interaction with peptide in the fluid phase. We have investigated the interactions of these floating bilayers with the antimicrobial peptides pexiganan, which has been the subject of microbial killing assays in the School of Physics and Astronomy at the University of Edinburgh (Jepson 2014; Jepson et al. 2016). Although the detailed analysis of this data is ongoing (Titmuss 2013), we find that the floating bilayer undergoes a large-scale structural rearrangement, depicted schematically in Fig. 16.4, at a sub-phase concentration of peptide that is comparable to the minimum inhibitory concentration (MIC) determined for this peptide in Edinburgh (Jepson 2014; Jepson et al. 2016). The challenge to the data analysis is that the fluctuations mean that the apparent Gaussian roughness of the layers corresponding to the molecular sub-units (lipid head groups and lipid tails), are greater than the thickness of the lipid head group region. We are currently applying the recently developed continuous distribution method of compositional space modelling to the analysis of the data (Schekhar et al. 2011; Heinrich and Losche 2014).

16.2.3 Models for the Outer Membrane of Gram-Negative Bacteria and Their Interactions with Antimicrobial Peptides

The outer leaflet of the outer membrane is decorated by lipopolysaccharide (LPS), as illustrated schematically in Fig. 16.5. The lipopolysaccharide layer resembles a

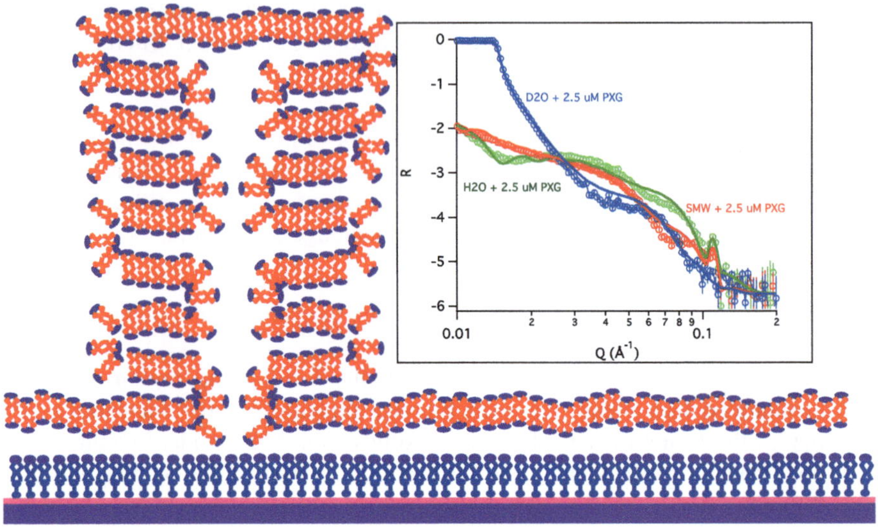

Fig. 16.4 A cartoon to illustrate the nature of the reconstruction of a 3:1 DPPC/DPPG floating bilayer that has been exposed to a sub-phase concentration of the AMP pexiganan. The presence of a Bragg feature at $Q_z = 0.11\ \text{Å}^{-1}$ on the *inset* reflectivity profiles allows us to determine the lamellar repeat (from its position Q_z^*) and the number of repeat units (from it's width δQ_z^*). The form of the reflectivity at low Q_z is only consistent with an intact floating bilayer, and the height of the Bragg peak gives the fraction of the surface that is covered by the lamellar repeats. That only a fraction of the surface is covered by the lamellar structure necessitates the implementation of an incoherent superposition of the reflectivity from an intact bilayer, and the reflectivity from an intact bilayer decorated by a lamellar structure. That the bilayer must remain intact allows us to propose the structure depicted in the cartoon. Only 8 lamellar repeat units are shown, but from the width of the Bragg peak we expect there to be 16 (Titmuss 2013)

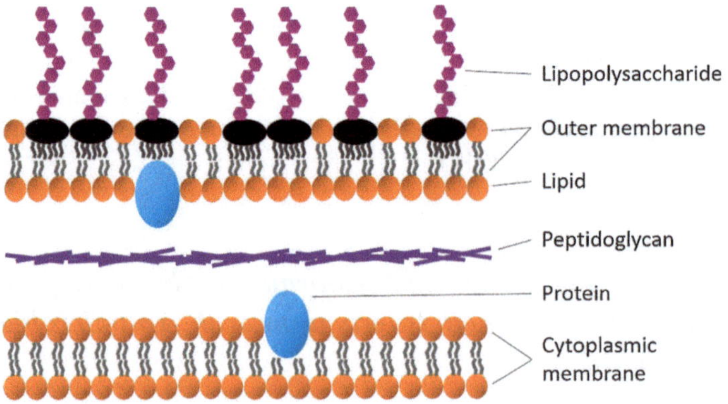

Fig. 16.5 The cell envelope of Gram-negative bacteria such as *E. coli* is defined by an inner cytoplasmic membrane and an outer membrane, the outer leaflet of which is decorated by lipopolysaccharide (LPS)

polymer brush, and its biological functions are thought to include acting as a first line of defence and controlling the outer membrane permeability. The LPS is endotoxic, making this outer layer highly relevant to studies of the physics of infection.

As the name suggests, lipopolysaccharides incorporate both lipid and sugar functional groups. LPS can be considered to comprise of three parts; a lipid part (lipid A) which is covalently linked to a core polysaccharide region, followed by a variable O-antigen chain of disaccharides. LPS that comprises all three parts is termed smooth, whilst mutants lacking the O-antigen chain are termed rough; rough LPS can be designated from Re to Ra, depending on where the core region terminates.

Clifton and co-workers have been systematically building up the complexity of the models for LPS layers that can be studied with reflectivity techniques (Brun et al. 2013; Clifton et al. 2013, 2015a, b). The starting point for this series of investigations was to study a rough mutant of LPS (RcLPS), which comprises the lipid A moiety and the first seven sugar residues of the core region. In a further illustration of the contrast variation technique, they use both unlabelled and deuterated RcLPS, with the latter being produced by bacteria cultured in media with increasing deuterium content. Using the same approach as described earlier for their studies of plant defence proteins, they combine neutron reflectivity with X-ray reflectivity, GIXRD and BAM. They find that RcLPS adopts an oblique hexagonal packing at all surface pressures, whereas lipid A adopts a hexagonal packing or (at pressures above 20 mN/m) distorted hexagonal packing, indicating that the sugar residues affect the lateral interactions between the molecules.

Recently, one of us (Barker) has used Langmuir monolayers as a mimetic for the outer leaflet of the outer membrane, in order to investigate how the structure of rough lipopolysaccharide affects the molecular interactions with the mammalian antimicrobial peptides LL37 and lactoferricin (Bello et al. 2015).

The key findings of this study are that shorter rough LPS oligosaccharides induce an ordering effect on outer membrane mimetics, whilst longer rough LPS oligosaccharides exert a slight steric barrier against AMP penetration; that excess peptides localize into non-interacting layers adjacent to outer membrane mimetics and that LL37 penetrates deeper into LPS-containing outer membrane mimetics than lactoferricin.

Using the same rough mutant LPS, Schneck et al. showed that divalent ions displace monovalent ions from the core region (Schneck et al. 2010). In a set of experiments which combined neutron reflectivity and GIXRD at the air/water interface, with specular neutron reflectivity at the solid/liquid interface, Clifton and co-workers were able to demonstrate that divalent calcium ions bind to the core region of the rough mutant LPS (RaLPS) films, producing more ordered structures in comparison to divalent cation free monolayers (Clifton et al. 2015b). Removal of the calcium from the sub-phase of an asymmetric solid-supported model for a Gram-negative bacterial outer membrane, which initially comprises of an inner DPPC leaflet (deposited directly onto the oxide-terminated interface of a silicon single crystal) and a RaLPS outer leaflet, lead to a flipping of the lipids between leaflets to produce a more symmetrical distribution of DPPC and RaLPS across the two leaflets. The authors state that in the absence of divalent ions, there are repulsive electrostatic interactions between the core regions of the RaLPS, which feature multiple phosphate groups. They claim that the driving force for the net flipping of RaLPS into the inner leaflet is the minimization of this repulsive interaction. They suggest that this provides a molecular explanation for the observation that divalent ions stabilize the outer membranes against penetration by antimicrobials. We would suggest that this could be a consequence of the lower area per molecule that is possible at a given surface pressure in the presence of divalent ions, as a consequence of the reduction in the effective area per lipid due to the lower osmotic contribution to the surface pressure caused by counterion release; the lower area per molecule will make the outer membrane less permeable.

A better approach is provided by the floating bilayer method described in Sect. 2.2.2. In an elegantly conceived experiment that builds on many of the techniques and methodologies we have discussed, Clifton et al. have recently reported an accurate in vitro model of the *E. coli* envelope (Clifton et al. 2015a). Using an ω-thiolipid SAM functionalized gold surface deposited onto a 137 Å thick permalloy layer on top of a silicon oxide-terminated silicon substrate, they first form a deuterated DPPC monolayer as the inner leaflet of their asymmetric bilayer by Langmuir–Blodgett, before using Langmuir–Schaeffer deposition of the RaLPS. The deuterium labelling of the initially inner DPPC layer enables them to use neutron reflectivity to characterize the extent to which DPPC transfers into the outer (initially) rough LPS layer. They characterize this as an asymmetry (LPS/PC) and find a range of asymmetries for the two leaflets. In what they term the best bilayers, the outer layer is 79 % LPS and the inner layer is 8 % LPS.

Using polarized neutron reflectivity combined with 3 different sub-phase contrasts (H_2O, D_2O and $75\,\%\,D_2O/25\,\%\,H_2O$) provides six different contrasts. This greatly constrains the fitting procedure, providing good sensitivity to subtle structural features such as the leaflet asymmetry described above.

To characterize the usefulness of this construct as a tool to assist in the future development of antibiotics, they tested its response to removal of the divalent cations and to antimicrobial proteins. Removal of divalent cations by EDTA sequestration reduces the asymmetry of both leaflets by 20 %. Treatment of the bilayer by the human antimicrobial protein lactoferrin reduced the bilayer coverage by 12 % and the asymmetry by 30 %, and lead to an increase in the thickness of the lipid A core region by 90 Å—which they attribute to the protein binding with its major axis parallel to the membrane normal. Treatment with lysozyme, also part of the human innate immune system, resulted in an increase in the thickness of the lipid A headgroup region by 20 Å; although there was no loss in bilayer coverage, the roughness did increase from 9 to 13 Å. That the model asymmetric bilayers exhibit a biological response that is consistent with the in vivo response of bacterial membranes to these antimicrobial treatments is encouraging for the use of bilayers as a tool to assist the development of future antimicrobial treatments.

To further increase the realism of these outer membrane mimetics, the rough LPS should be replaced by smooth LPS, as most Gram-negative bacteria contain smooth LPS. The higher water solubility of the smooth LPS compared to the rough LPS conferred by the longer polysaccharide chain makes this more challenging. Schneck et al. have deposited a monolayer of smooth LPS onto a silane-hydrophobized silicon substrate, such that the O-antigen chain extends into the sub-phase, to address the conformation of the O-antigen chain in the absence and presence of calcium in the sub-phase (Schneck et al. 2009). They find that the O-antigen chain forms a shorter, denser layer in the presence of divalent calcium.

16.3 Summary and Outlook

We conclude by summarizing the key points from this chapter and by providing a brief outlook to the future application of neutron reflectivity as a tool to tackle the membrane aspects of the biophysics of infection.

16.3.1 Summary

- Neutron reflectivity provides structural information from monolayer and bilayer models for bacterial membranes at the level of (sub-) molecular units. This allows the area per lipid to be characterized, which can be related to membrane permeability; and the thickness of the membrane and the location of any interacting peptides to be determined, which can provide insight into the mode of action of AMP.
- Model membranes can be constructed with compositions that range from highly idealized, to facilitate the systematic investigation of the fundamental physical principles that govern membrane behaviour, through to realistic membrane mimetics.
- Monolayers spread at the air/water interface can be thought to represent one of the two leaflets making up inner and outer membranes. In such monolayer films, the leaflets are strongly confined to 2D. Such experiments allow for a direct investigation of the interactions of AMP with the headgroup region of one leaflet, and allow for experiments in which the surface pressure and area per molecule can be varied, making systematics investigations of the influence of these parameters possible. However, the interaction behaviour of larger peptide fragments may not be representative of that which would be observed with native membranes comprising of two leaflets.
- Supported bilayers, in which the lipids are deposited direct onto a solid substrate, enable these interaction studies to be extended to two leaflets, but in this case both leaflets are essentially constrained to 2D. This is useful for determining if interactions of peptides with the headgroup region are fully representative. Neutron reflectivity measurements on such model bilayers are also useful for the quantitative benchmarking of higher throughput techniques such

as DPI and QCM-D, which both use rather similar solid-supported bilayers as substrates.
- In floating bilayers, the model membrane sits on a 2–5 nm cushion of water, which allows it to fluctuate as it might in a bacterial periphery. This means that the phase behaviour should solely be a reflection of the composition and any imposed thermodynamic potentials (temperature, transmembrane voltage, peptide solution).
- The process of forming these floating bilayers has been continuously refined, such that it is now possible to construct model membranes that have compositions that accurately represent both the inner and outer bacterial membranes. In addition to the fundamental physical insights that such accurate model membranes can provide, as they also exhibit realistic biological behaviour, they can be used as a platform to help the development of new antimicrobials.

16.3.2 Outlook

The ensemble nature of the model membranes used in neutron reflectivity, and the controlled sample environments, means that as a technique it is well suited to making studies on systems at thermodynamic equilibrium. When acting as agents for infection, bacterial cells are living and so by definition not at thermodynamic equilibrium. From a physicist's perspective, the way in which the bacterial membrane is coupled into the cell's life cycle and use of energy, is through the transmembrane potential that always exists across the inner membrane. Floating bilayer samples assembled onto gold layers are well suited to the application of a transmembrane potential, as the gold layer can serve as a working electrode, and a counter-electrode can be incorporated into the base of the laminar flow cell (i.e. into the MACOR part of Fig. 16.1). In our recent preliminary experiments (McKinley 2015), that use floating bilayers assembled on a gold layer deposited on a permalloy layer on silicon, we have exploited magnetic contrast and polarized reflectivity to observe transmembrane potential dependent effects on the bilayer structure and on the binding of the MinD-*mts* peptide, at biologically relevant transmembrane potentials. We believe that the combination of magnetic contrast to resolve the small structural changes induced by the adsorption of small peptides, and floating bilayers assembled on gold layers to facilitate the application of biologically relevant transmembrane potentials, will provide a powerful tool to study the way in which the efficacy of antimicrobials depends on membrane potential and hence the bacterial metabolic state.

Acknowledgments We acknowledge the STFC for the award of beamtime at ISIS (RB1220350) and the Institut Laue-Langevin (8-02-637/8-02-682/8-02-704), and the EPSRC for a studentship to Laura McKinley (EP/L504956/1).

References

Barker RD (2011) al-PC SAM approach to supported bilayers: Developing a novel model membrane system. PhD thesis, School of Chemistry, University of Bath

Bello G, Bodin A, Lawrence MJ, Barlow D, Mason AJ, Barker RD, Harvey RD (2015) The influence of rough lipopolysaccharide structure on molecular interactions with mammalian antimicrobial peptides. BBA: Biomembranes 1858(2):197–209. doi:10.1016/j.bbamem.2015.11.007

Carr M (2015) Advancing our understanding of lipid bilayer interaction: a molecular dynamics study. PhD thesis, School of Physics and Astronomy, University of Edinburgh

Charitat T, Bellet-Amalric E, Fragneto G, Graner F (1999) Adsorbed and free lipid bilayers at the solid-liquid interface. Eur Phys J B 8:583–593

Chen L, Gao L, Fang W, Golubovic L (2012) How the antimicrobial peptides kill bacteria: computational physics insights. Commun Comput Phys 11(3):709–725

Clifton LA, Sanders M, Kinane C, Arnold T, Edler KJ, Neylon C, Green RJ, Frazier RA (2012) The role of protein hydrophobicity in thionin-phospholipid interactions: a comparison of $\alpha 1$ and $\alpha 2$-purothionin adsorbed anionic phospholipid monolayers. PCCP 14:13569–13579

Clifton LA, Skoda MWA, Daulton EL, Hughes AV, Le Brun AP, Lakey JH, Holt SA (2013) Asymmetric phospholipid: lipopolysaccharide bilayers; a gram-negative bacterial outer membrane mimic. J R Soc Interface 10(89):20130810

Clifton LA, Holt SA, Hughes AV, Daulton EL, Arunmanee W, Frank H, Khalid S, Jefferies D, Charlton TR, Webster JRP, Kinane CJ, Lakey JH (2015a) An accurate in vitro model of the *E. coli* envelope. Angew Chem Int Ed 54(41):11952–11955

Clifton LA, Skoda MWA, Le Brun AP, Ciesielski F, Kuzmenko I, Holt SA, Lakey JH (2015b) Effect of divalent cation removal on the structure of gram-negative bacterial outer membrane models. Langmuir 31(1):404–412. doi:10.1021/la504407v

Fernandez DL, Le Brun AP, Lee TH, Bansal P, Aguilar MI, James M, Separovic F (2012a) Structural effects of the antimicrobial peptide maculatin 1.1 on supported lipid bilayers. Eur Biophys J

Fernandez DL, Le Brun AP, Whitwell TC, Sani MA, James M, Separovic F (2012b) The antimicrobial peptide aurein 1.2 disrupts model membranes via the carpet mechanism. Phys Chem Chem Phys 14:15739–15751

Fragneto G, Graner F, Charitat T, Dubos P, Bellet-Amalric E (2000) Interaction of the third helix of antennapedia homeodomain with a deposited phospholipid bilayer: a neutron reflectivity structural study. Langmuir 16(10):4581–4588

Fragneto G, Charitat T, Daillant J (2012) Floating lipid bilayers: models for physics and biology. Eur Biophys J 41:863–874

Gidalevitz D, Ishitsuka Y, Muresan AS, Konvalov O, Waring AJ, Lehrer RI, Lee KYC (2003) Interaction of antimicrobial peptide protegrin with biomembranes. PNAS 100(11):6302–6307

Heinrich F, Lösche M (2014) Zooming in on disordered systems: neutron reflection studies of proteins associated with fluid membranes. Biochim Biophys Acta 1838:2341–2349

Helfrich W (1978) Steric interaction of fluid membranes in multilayer systems. Z Naturforsch 33a:305–315

Hughes AV, Goldar A, Gerstenberg MC, Roser SJ, Bradshaw JP (2002) A hybrid sam phospholipid approach to fabricating a free supported lipid bilayer. PCCP 4:2371–2378

Hughes AV, Howse JR, Dabkowska A, Jones RAL, Lawrence MJ, Roser SJ (2008) Floating lipid bilayers deposited in chemically grafted phosphatidylcholine surfaces. Langmuir 24:1989–1999

Hughes AV, Holt SA, Daulton W, Soliakov A, Charlton TR, Roser SJ, Lakey JH (2014) High coverage fluid-phase floating bilayers supported by ω-thiolipid self-assembled monolayers. J R Soc Interface 11(20140):447

Illya G, Deserno M (2008) Coarse-grained simulation studies of peptide-induced pore formation. Biophys J 95:4163–4173

Jagalski V, Barker RD, Thygesen MB, Gotfryd K, Krüger MB, Shi L, Maric S, Bovet N, Moulin M, Haertlin M, Pomorski TG, Loland CJ, Cárdenas M (2015) Grafted biomembranes containing membrane proteins—the case of the leucine transporter. Soft Matter 11:7707–7711

Jepson AK (2014) *E. coli* motility and growth: a biophysical study. PhD thesis, School of Physics, University of Edinburgh

Jepson AK, Schwarz-Linek J, Ryan L, Ryadnov MG, Poon WCK (2016) What is the 'minimum inhibitory concentration' (MIC) of pexiganan acting on Escherichia coli? In: Leake MC (ed.), Biophysics of infection. A cautionary case study. Springer. doi: 10.1007/978-3-319-32189-9_4

Junghans A, Watkins EB, Barker RD, Singh S, Smith HL, Pocivavsek L, Majewski J (2015) Analysis of biosurfaces by neutron reflectometry: from simple to complex interfaces. Biointerphases 10(019):014

Le Brun AP, Clifton LA, Halbert CE, Lin B, Meron M, Holden PJ, Lakey JH, Holt SA (2013) Structural characterization of a model gram-negative bacterial surface using lipopolysaccharides from rough strains of escherichia coli. Biomacromolecules 14(6):2014–2022. doi:10.1021/bm400356m, URL http://dx.doi.org/10.1021/bm400356m, pMID: 23617615

Majkrzak CF, Berk NF, Dura J, Satija SK, Karim A, Pedulla J, Deslattes RD (1998) Phse determination and inversion in specular neutron reflectometry. Phys B 248:338–342

Marrink SJ, de Vries AH, Tieleman DP (2009) Lipids on the move: simulations of membranes pores, domains, stalks and curves. Biochim Biophys Acta 1788:149–168

McKinley LE (2015) Reports on 8-02-704 and 8-02-709. Technical report, Institut Laue Langevin

Melo MN, Castanaho MARB (2012) The mechanism of action of antimicrobial peptides: lipid vesicles vs. bacteria. Front Immunol 3:1–4

Pabst G, Kucerka N, Nieh MP, Rheinstädter MC, Katsara J (2010) Applications of neutron and x-ray scattering to the study of biological model membranes. Chem Phys Lipids 163:460–479

Schekhar P, Nanda H, Lösche M, Heinrich F (2011) Continuous distribution model for the investigation of complex molecular architectures near interfaces with scattering techniques. J Appl Phys 110(102):216

Schneck E, Papp-Szabo E, Quinn BE, Konovalov OV, Beveridge TJ, Pink DA, Tanaka M (2009) Calcium ions induce collapse of charded o-side chains of lipopolysaccharides from pseudomonas aeruginosa. J R Soc Interface 6:S671–S678

Schneck E, Schubert T, Konovalov BE Quinn OV, Gutsmann T, Brandenburg K, Olivera RG, Pink DA, Tanaka M (2010) Quantitative determination of ion distributions in bacterial lipopolysaccharide membranes by grazing-incidence x-ray fluorescence. PNAS 107:9147–9151

Titmuss S (2013) Report on 8-02-684. Technical report, Institut Laue Langevi

Wacklin HP (2010) Neutron reflection from supported lipid membranes. Curr Opin Colloid Interface Sci 15:445–454

Woo HJ, Wallqvist A (2011) Spontaneous buckling of lipid bilayer and vesicle budding induced by antimicrobial peptide maganin 2: a coarse-grained simulation study. J Phys Chem B 115:8122–8129

Chapter 17
Mechanisms of *Salmonella* Typhi Host Restriction

Stefania Spanò

Abstract *Salmonella enterica serovar* Typhi (*S.* Typhi) is the cause of typhoid fever, a life-threatening bacterial infection that is very common in the developing world. Recent spread of antimicrobial resistant isolates of *S.* Typhi makes typhoid fever, a global public health risk. Despite being a common disease, still very little is known about the molecular mechanisms underlying typhoid fever and *S.* Typhi pathogenesis. In contrast to other *Salmonellae*, *S.* Typhi can only infect humans. The molecular bases of this human restriction are mostly unknown. Recent studies identified a novel pathway that contributes to *S.* Typhi human restriction and is required for killing *S.* Typhi in macrophages of nonsusceptible species. The small Rab GTPase Rab32 and its guanine nucleotide exchange factor BLOC-3 are the critical components of this pathway. These proteins were already well known as important regulators of intracellular membrane transport. In particular, they are central for the transport of enzymes that synthetize melanin in pigment cells. The recent findings that Rab32 and BLOC-3 are required for *S.* Typhi host restriction point out to a novel mechanism restricting the growth of bacterial pathogen, dependent on the transport of still unknown molecule(s) to the *S.* Typhi vacuole. The identification of this novel antimicrobial pathway constitutes a critical starting point to study molecular mechanisms killing bacterial pathogens and possibly identify novel antimicrobial molecules.

Keywords *Salmonella* Typhi · Typhoid fever · Rab GTPases · Hermansky–Pudlak syndrome · Lysosome-related organelles · Host–pathogen interaction

S. Spanò (✉)
Institute of Medical Sciences, College of Life Sciences and Medicine,
University of Aberdeen, Foresterhill, Aberdeen AB25 2ZD, UK
e-mail: stefania.spano@abdn.ac.uk

© Springer International Publishing Switzerland 2016

M.C. Leake (ed.), *Biophysics of Infection*, Advances in Experimental Medicine and Biology 915, DOI 10.1007/978-3-319-32189-9_17

17.1 Typhoid Fever and *S.* Typhi

Typhoid fever is a serious systemic infection caused by the intracellular bacterium *S.* Typhi (Parry et al. 2002). Typhoid fever is an ancient plague that affected the human history through devastating outbreaks, such as the plague of Athens (Papagrigorakis et al. 2006). Nowadays, typhoid fever is rare in developed countries, mostly associated to travellers returning from affected areas of the world (Crump and Mintz 2010). However, typhoid fever is still a very common infection in poorer areas of the world, especially Southeast Asia and sub-Saharan Africa (Buckle et al. 2012; Mogasale et al. 2014). Overall, it is estimated to affect more than 27 millions individuals and kill around 200,000 people worldwide. Like for other bacterial infections, the treatment of typhoid fever is becoming a challenge due to the emergence of multidrug-resistant strains. Typhoid infections are traditionally treated with ampicillin, chloramphenicol, and trimethoprim-sulfamethoxazole. However, the emergence of antibiotic resistance in *S.* Typhi isolates renders many cases of typhoid resistant to these treatments (Crump et al. 2015). The majority of *S.* Typhi strains isolated recently belongs to a multidrug-resistant clade, known as H58, which often (more than 60 % of isolates) simultaneously carry genes conferring resistance to ampicillin, trimethoprim, sulfonamides, chloramphenicol, and streptomycin (Wong et al. 2015). The rapid spread of antibiotic resistant *S.* Typhi isolates renders typhoid infection a global public health risk.

17.2 *Salmonella* Interaction with Host Cells

Despite being an ancient disease, the molecular mechanisms underlying typhoid fever and *S.* Typhi pathogenesis had not been investigated until recently. Most of the experimental works to understand general mechanisms of pathogenesis has been done using the related pathogen, *Salmonella enterica* serovar Typhimurium (*S.* Typhimurium). Researchers have focused on *S.* Typhimurium, because (a) *S.* Typhi can only infect humans and, consequently, no convenient animal model is available to study *S.* Typhi infections and (b) *S.* Typhimurium causes a systemic disease in mice that has features in common with typhoid fever, although not manifesting with fever and other symptoms typical of typhoid fever. It is important to emphasize, however, that *S.* Typhimurium causes in humans an acute, relatively mild gastroenteritis, called salmonellosis, which is a disease very different from typhoid fever.

S. Typhi and *S.* Typhimurium are both facultative intracellular pathogen (House et al. 2001; Dougan and Baker 2014). They are able to actively invade cells of the intestinal epithelium and replicate within these cells (Jones and Falkow 1996; Galan 2001). In the lymphatic tissue of the intestinal mucosa, these pathogens encounter a variety of other host cells and are internalized by phagocytic cells, such as macrophages. Within macrophages, *S.* Typhi and *S.* Typhimurium are often able to survive and replicate (Hurley et al. 2014; Behnsen et al. 2015). In contrast to other bacteria that are rapidly killed once internalized by phagocytic cells, *Salmonella enterica* (*Salmonella*) is able to establish a specialized membrane compartment derived from the endolysosomal system, called *Salmonella*-containing vacuole. In spite of having features in common with a lysosome, which includes the presence of lysosomal markers and the vATPase responsible for its acidification, this vacuole is an ideal niche that supports the survival and the replication of this bacterium (Bakowski et al. 2008; Malik-Kale et al. 2011). The *Salmonella*-containing vacuole is in a close and dynamic relationship with the endolysosomal system (Drecktrah et al. 2008). Recent super-resolution microscopy revealed interesting morphological features of the membrane system that connect this vacuole with other intracellular compartments (Krieger et al. 2014). Essential to the ability to invade host cells and establish a *Salmonella*-containing vacuole are two type III secretion systems. These are specialized secretion systems that deliver bacterial proteins, called bacterial effectors directly to the cytosol of the host cells to manipulate their functions for their own benefit (Galan and Wolf-Watz 2006; Chatterjee et al. 2013). Around 40 *Salmonella* type III secretion effectors have been identified to date (LaRock et al. 2015). The functions of some of these effectors have been clarified in the past 20 years. Some of these bacterial effectors (i.e., SopE, SopB) are able to induce *Salmonella* invasion of epithelial cells through the stimulation of actin rearrangements and formation of a macropinocytic cup (Hardt et al. 1998; Terebiznik et al. 2002; Hernandez et al. 2004; Zhou et al. 2001). Another effector (i.e., SifA) is responsible for inducing long membranous protrusions, called *Salmonella*-induced filaments (Ohlson et al. 2008; Brumell et al. 2001, 2002; Beuzon et al. 2000). The role of these filaments in *Salmonella* pathogenesis is still unclear. However, it was recently proposed that these filaments have a role in acquisition of nutrients (Popp et al. 2015).

17.3 Genetic Differences Distinguish *S.* Typhi and *S.* Typhimurium

Despite sharing critical virulence factors and mechanisms of invasion of host cells, *S.* Typhi and *S.* Typhimurium are different pathogens, which cause different diseases in humans and show different host specificities. The complete sequencing of the first *S.* Typhi and *S.* Typhimurium strains in 2001 opened an era of comparative genome analysis (Parkhill et al. 2001; McClelland et al. 2001; Sabbagh et al. 2010). Around 600 genes are unique to *S.* Typhi and around 480 are unique to *S.* Typhimurium, suggesting that these genes may play critical role in the unique pathogenic traits of these bacteria. A unique feature of *S.* Typhi is its human restriction, i.e., the ability to infect only humans. *S.* Typhi is never been isolated from any other animal species in the wild. This is in contrast to the vast majority of the other *Salmonella*, which can infect a broad range of vertebrate species. The inability to infect animal species other than humans implies that *S.* Typhi has lost the ability to counteract the immune systems of other animal species. The sequencing of *S.* Typhi genome revealed that *S.* Typhi genome has gone through degradation, with around 200 genes that have become pseudogenes, and some genes that have never been acquired by *S.* Typhi (Parkhill et al. 2001; McClelland et al. 2001). This is not surprising, since genome degradation and elimination of genetic information is commonly associated with host adaptation (Toft and Andersson 2010). However, only recently it was shown that the absence of one of these genes is causally connected to host restriction in *Salmonella* (Spano and Galan 2012).

17.4 Mechanisms Underlying *S.* Typhi Unique Pathogenesis and Human Restriction

In the past decade, a combination of genetic, biochemical, and imaging analyses at the interface between *S.* Typhi and its host provided significant advancement on the understanding of the mechanisms underpinning *S.* Typhi unique pathogenesis and human restriction.

17.4.1 Insights from a Unique Bacterial Toxin into the Function of Host Rab GTPases

One of the genetic features typical of *S.* Typhi is typhoid toxin, a unique A_2B_5 toxin (Song et al. 2013; Spano et al. 2008; Spano and Galan 2008). In contrast to all the other AB toxins, typhoid toxin is not expressed when the bacterium is in the extracellular environment (Haghjoo and Galan 2004). It is only expressed a few hours after bacterial internalization into host cells, once the bacterium has established its intracellular vacuole (Haghjoo and Galan 2004). Typhoid toxin is then delivered from the *S.* Typhi vacuole to the extracellular space through an unprecedented transport pathway (Spano et al. 2008) (Fig. 17.1). Once in the extracellular environment, it binds a surface receptor and it is internalized and transported into target cells, which can be different from the infected cells where the toxin is initially expressed (Spano et al. 2008; Song et al. 2013). Therefore, *S.* Typhi is expected to target, through the typhoid toxin paracrine delivery, cells different from the ones that are infected. This unique delivery pathway has likely evolved to prevent immune system recognition. The identification of this paracrine pathway raised very intriguing questions about how the toxin, expressed when the bacterium resides inside an intracellular vacuole, is transported to extracellular space. Searching for host factors required for the transport of typhoid toxin out of the vacuole, the host GTPase Rab29 (also known as Rab7L1) was identified and shown to be required for efficient transport of typhoid toxin from the *S.* Typhi vacuole (Spano et al. 2011). Rab29 is membrane traffic regulator that localizes on the Golgi complex in mammalian cells and is recruited to the *S.* Typhi-containing vacuole, when these cells are infected with *S.* Typhi (Spano et al. 2011). Rab29 is a member of a small subfamily of Rab GTPases, which also includes Rab32 and Rab38. Interestingly, similarly to Rab29, Rab32 and Rab38 localize to the *S.* Typhi-containing vacuole (Spano and Galan 2012). Rab32 and Rab38 are Rab GTPases associated with lysosome-related organelles and required for the transport of critical enzymes for the maturation of these organelles, such as melanocytic enzymes in pigment cells (Wasmeier et al. 2006; Bultema et al. 2012).

17.4.2 GtgE, a Bacterial Effector Targeting Rab GTPases

The initial observation that Rab29 was present on the *S.* Typhi-containing vacuole, but not on the *S.* Typhimurium-containing vacuole, stimulated a series of genetic and biochemical experiments to clarify the molecular basis of such a difference. The results showed that a bacterial effector, called GtgE, delivered by *S.* Typhimurium, but absent from *S.* Typhi, targets Rab29 and the two related Rab GTPases Rab32 and Rab38 (Spano et al. 2011; Spano and Galan 2012). GtgE is a specific cysteine protease targeting this group of highly related GTPases (Spano and Galan 2012; Spano et al. 2011). This effector is delivered by the type III secretion system and encoded in a prophage region of *S.* Typhimurium, which is common to a large number of broad-host *Salmonella* serovars (Spano 2014). Delivery of GtgE by broad-host *Salmonella* serovars results in cleavage of Rab29, Rab32, and Rab38 and their removal from the surface of the vacuoles containing broad-host *Salmonella* serovars (Spano and Galan 2012; Spano et al. 2011).

17.4.3 GtgE, a Molecular Tool to Identify a Novel Antimicrobial Pathway

The finding that GtgE is a specific protease targeting three Rab GTPases that localize to the *S.* Typhi-containing vacuole, combined with the observation that it is exclusively expressed by broad-host Salmonella, suggested that this bacterial effector could regulate the biology of infected cells and affect *Salmonella* host specificity. To test this hypothesis experiments were performed in mouse primary macrophages, which, in contrast to human macrophages, clear *S.* Typhi infection (Vladoianu et al. 1990; Schwan et al. 2000). Experiments in mouse bone marrow-derived macrophages showed that exogenous expression of GtgE in *S.* Typhi confers this bacterium the ability to survive in these cells. A *S.* Typhi strain expressing GtgE also shows ability to colonize mice. Further experiments linked the ability of GtgE to enable *S.* Typhi survival in mouse primary macrophages to the removal of Rab32. The results showed

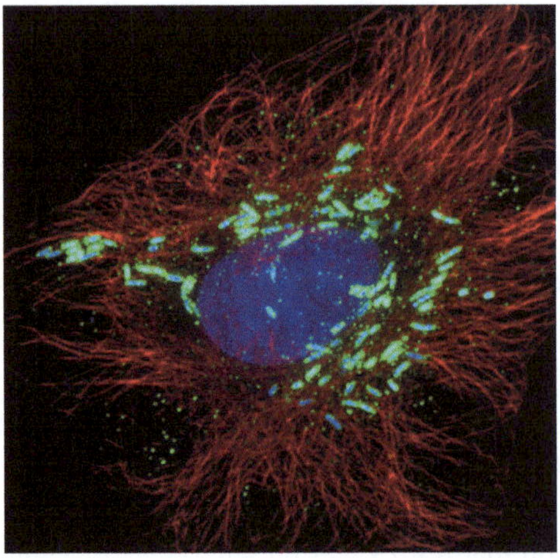

Fig. 17.1 Typhoid toxin in *S.* Typhi-infected epithelial cells. Henle-407 intestinal epithelial cells infected with *S.* Typhi are visualized by immunofluorescence 24 hours post-infection. Typhoid toxin (in *green*) appears associated to the bacteria as well as present in small punctate structures that radiate from the bacteria toward the cell periphery, representing typhoid toxin en-route to the extracellular environment. Infected cell microtubules are labeled in *red* and host and bacterial DNAs are labelled in *blue*

the Rab32 is critical to control the clearance of *S.* Typhi in mouse primary macrophages. Indeed, *S.* Typhi is able to survive in macrophages depleted of Rab32, even if the bacterium does not express GtgE (Spano and Galan 2012). Rab32 and its homologue Rab38 regulates transport of melanocytic enzymes to melanosomes, specialized organelles present in melanocytes and other pigments cells (Wasmeier et al. 2006). A Rab38 spontaneous mutation is present in mice (*chocolate* mouse) and is associated with a light coat color (Loftus et al. 2002). Other mutations are associated in mouse with hypopigmentation and in humans with the Hermansky–Pudalk syndrome, which manifests with hypopigmentation and bleeding defects. Many of these mutations involve genes encoding for the components of three protein complexes, called Biogenesis of Lysosome-related Organelles Complex (BLOC)-1, 2 and 3 (Dell'Angelica 2004; Di Pietro and Dell'Angelica 2005). Genetic associations as well biochemical and imaging analysis had clearly implicated Rab32, Rab38, and the BLOCs in post-Golgi trafficking of melanocytic enzymes to maturing melanosomes (Marks et al. 2013; Wasmeier et al. 2006). Several observations indicated a general role for these molecules in trafficking to other lysosome-related organelles, such as platelet dense granules and lytic granules (Wei and Li 2013). Interestingly and similarly to Rab32, BLOC-3 is required to control *S.* Typhi clearance in mouse primary macrophages (Spano and Galan 2012). Remarkably, BLOC-3 has been recently reported to be a guanine nucleotide exchange factor for Rab32 and Rab38, i.e., the molecule catalyzing the exchange of GDP with GTP during the GTPase cycle of these Rabs (Gerondopoulos et al. 2012), indicating that GTPase cycling is required for clearing *S.* Typhi from mouse macrophages. These results strongly suggested an

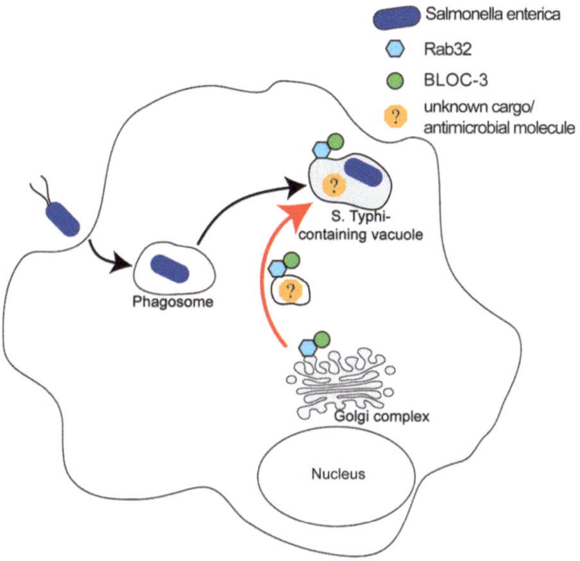

Fig. 17.2 A Rab32 and BLOC-3-dependent pathway restrict the growth of *S.* Typhi in mouse macrophages. The model illustrates that a Rab32 and BLOC-3-dependent trafficking event (*red arrow*) is required to deliver an unknown antimicrobial cargo to the *S.* Typhi-containing vacuole

intriguing model in which the Rab32/BLOC-3-dependent trafficking pathway transports an antimicrobial factor to the *S.* Typhi vacuole, preventing *S.* Typhi survival (Spano 2014) (Fig. 17.2). The results also suggested that the *S.* Typhi-containing vacuole should be classified as a lysosome-related organelle of macrophages (Spano and Galan 2013).

17.5 Pathogen Killing Mechanisms in Macrophages

How is *S.* Typhi killed in mouse macrophages? Macrophages are the main cell type devoted to bacterial killing in mammals. They possess an arsenal of molecular weapons to recognize, internalize, and kill bacteria. Disparate signaling pathways are involved in recognizing bacteria and activating antimicrobial responses (Kawai and Akira 2010; Newton and Dixit 2012). Some of the effector mechanisms activated by these pathways and directly responsible for bacterial death are also well known. Production of molecular radicals is one of the best-characterized mechanisms for the killing of bacterial pathogens. Reactive oxygen and nitrogen species, which are produced through the activity of the NADPH oxidase and the inducible nitric oxide synthase (iNOS), respectively, are highly toxic to bacteria (Fang 2004, 2011; Slauch 2011). The multisubunit NADPH oxidase assembles on the phagolysosome membrane and creates superoxide anion in the phagosome by reducing oxygen. Superoxide is enzymatically or spontaneously reduced to hydrogen peroxide, which is further reduced by free iron in the so-called Fenton reaction to create hydroxyl radical, the most toxic of the oxygen radicals. The cytosolic iNOS, instead, generates the membrane diffusible nitric oxide. Nitric oxide generated in one cell can likely kill extracellular bacteria or bacteria within neighboring cells (Burton et al. 2014). The hydroxyl radical and nitric oxide can also react to form peroxynitrite, the most damaging of all the radicals (Brunelli et al. 1995). All these species are cytotoxic to bacteria, as they target multiple molecular functions. However, they can also induce a bacterial stress response that neutralizes the damage at different levels, either by removing these toxic molecules or limiting and repairing the damage produced (Imlay 2013). Bacterial pathogens, including *Salmonella*, are well equipped to respond to the radical attack of macrophages (Fang 2011; Slauch 2011). A battery of detoxifying enzymes, mostly reductases, ensures that the concentrations of molecular radicals in maintained below the lethal concentrations.

Despite the fact that *Salmonella* is endowed with molecular means to neutralize ROS, experimental and clinical observations showed that NADPH oxidase is critical to control *Salmonella* infections in mouse as in humans. Mice deficient for NADPH oxidase are more susceptible to *S.* Typhimurium infection (Mastroeni et al. 2000). Mutations of the NADPH oxidase gene in humans are associated to a condition called chronic granulomatous disease (CGD) that causes patients to be more susceptible to bacterial infections, including *Salmonella* infections. This has been related to the inability of macrophages to kill bacteria. Both NADPH oxidase and iNOS are required to kill *S.* Typhimurium in ex vivo peritoneal macrophages (Vazquez-Torres et al. 2000a, b), however iNOS does not seem to be essential to control early phases of *S.* Typhimurium infections in mice (Mastroeni et al. 2000).

Other bacterial killing mechanisms are reported, including toxic accumulation of heavy metals, such as copper. Copper is antimicrobial and there is increasing evidence that copper is used by macrophages to restrict pathogens (Darwin 2015; Neyrolles et al. 2015). A complex system of copper homeostasis is present in macrophages, which includes the transporter ATP7A that pumps copper into the phagolysosome (White et al. 2009). Copper has also been shown to be redistributed in macrophages in response to *S.* Typhimurium infection, suggesting that it is part of a specific macrophage strategy to clear infection (Achard et al. 2012). As for other antimicrobial strategies used by macrophages, *Salmonella* is equipped to neutralize it, through an efflux system made by the two P-type ATPases CopA and GolT, SodCII and a periplasmic copper chaperone, CueP (Osman et al. 2013; Nies and Herzberg 2013). These observations suggest that, like for other killing mechanisms, very subtle concentration changes can make a difference between bacterial life and death.

Another strategy used by the host to eliminate bacteria is the production of cationic antimicrobial peptides, such as defensins and cathelicidins. These are amphiphatic pore-forming peptides that kill bacteria by permeabilizing the bacterial membrane (LaRock and Nizet 2015). The only antimicrobial peptide produced by macrophages is cathelicidin. The cathelicidin-related antimicrobial peptide appears to be involved in controlling a virulence defective mutant of *S.* Typhimurium infection in mice (Rosenberger et al. 2004). However, cathelicidin is not required for controlling *Salmonella* infections in humans (Strandberg et al. 2012). That makes cathelicidin-related antimicrobial peptide, a good candidate for being transported to the vacuole by the Rab32/BLOC-3-dependent pathway. At present, it is not known about where antimicrobial peptides are expressed and processed and how they are transported to the lumen of the *Salmonella*-containing vacuoles.

17.6 Conclusions

In the future, it will be critical to understand if any of these macrophage strategies to clear bacterial pathogens is related to the Rab32/BLOC-3 restriction of *S.* Typhi in mouse macrophages. In particular, it will be interesting to analyze if this transport pathway deliver components of the enzymatic activities required to synthetize specific antimicrobial molecules (e.g., NADPH oxidase) or metal transporters. Based on the recent findings, the *S.* Typhi vacuole resembles more a lysosome-related organelle rather than a phagolysosome. High-resolution proteomics and metabolomics can contribute to identify the molecules specifically delivered to this vacuole and help define the composition of this complex organelle. Advanced microscopy, including super-resolution microscopy and single-molecule imaging, can contribute to define the structure and composition of this vacuole. Overall, identifying novel molecules used by the mouse host to kill *S.* Typhi can shed light on novel antimicrobial processes and help identifying new opportunities to treat bacterial diseases.

References

Achard ME, Stafford SL, Bokil NJ et al (2012) Copper redistribution in murine macrophages in response to Salmonella infection. Biochem J 444(1):51–57. doi:10.1042/BJ20112180

Bakowski MA, Braun V, Brumell JH (2008) Salmonella-containing vacuoles: directing traffic and nesting to grow. Traffic 9(12):2022–2031. doi: 10.1111/j.1600-0854.2008.00827.x

Behnsen J, Perez-Lopez A, Nuccio SP et al (2015) Exploiting host immunity: the Salmonella paradigm. Trends Immunol 36(2):112–120. doi:10.1016/j.it.2014.12.003

Beuzon CR, Meresse S, Unsworth KE et al (2000) Salmonella maintains the integrity of its intracellular vacuole through the action of SifA. EMBO J 19(13):3235–3249. doi:10.1093/emboj/19.13.3235

Brumell JH, Goosney DL, Finlay BB (2002) SifA, a type III secreted effector of Salmonella typhimurium, directs Salmonella-induced filament (Sif) formation along microtubules. Traffic 3 (6):407–415. doi: tra030604 [pii]

Brumell JH, Rosenberger CM, Gotto GT et al (2001) SifA permits survival and replication of Salmonella typhimurium in murine macrophages. Cell Microbiol 3(2):75–84. doi: cmi87 [pii]

Brunelli L, Crow JP, Beckman JS (1995) The comparative toxicity of nitric oxide and peroxynitrite to Escherichia coli. Arch Biochem Biophys 316(1):327–334. doi: S0003-9861 (85)71044-2 [pii]

Buckle GC, Walker CL, Black RE (2012) Typhoid fever and paratyphoid fever: systematic review to estimate global morbidity and mortality for 2010. J Glob Health 2(1):010401. doi: 10.7189/jogh.02.010401

Bultema JJ, Ambrosio AL, Burek CL et al (2012) BLOC-2, AP-3, and AP-1 proteins function in concert with Rab38 and Rab32 proteins to mediate protein trafficking to lysosome-related organelles. J Biol Chem 287(23):19550–19563. doi:10.1074/jbc.M112.351908

Burton NA, Schurmann N, Casse O et al (2014) Disparate impact of oxidative host defenses determines the fate of Salmonella during systemic infection in mice. Cell Host Microbe 15 (1):72–83. doi:10.1016/j.chom.2013.12.006

Chatterjee S, Chaudhury S, McShan AC et al (2013) Structure and biophysics of type III secretion in bacteria. Biochemistry 52(15):2508–2517. doi:10.1021/bi400160a

Crump JA, Mintz ED (2010) Global trends in typhoid and paratyphoid Fever. Clin Infect Dis: Off Publ Infect Dis Soc Am 50(2):241–246. doi:10.1086/649541

Crump JA, Sjolund-Karlsson M, Gordon MA et al (2015) Epidemiology, clinical presentation, laboratory diagnosis, antimicrobial resistance, and antimicrobial management of invasive salmonella infections. Clin Microbiol Rev 28(4):901–937. doi:10.1128/CMR.00002-15

Darwin KH (2015) Mycobacterium tuberculosis and Copper: a newly appreciated defense against an old foe? J Biol Chem 290(31):18962–18966. doi:10.1074/jbc.R115.640193

Dell'Angelica EC (2004) The building BLOC(k)s of lysosomes and related organelles. Curr Opin Cell Biol 16(4):458–464. doi:10.1016/j.ceb.2004.05.001

Di Pietro SM, Dell'Angelica EC (2005) The cell biology of Hermansky-Pudlak syndrome: recent advances. Traffic 6(7):525–533. doi:10.1111/j.1600-0854.2005.00299.x

Dougan G, Baker S (2014) Salmonella enterica Serovar Typhi and the pathogenesis of typhoid fever. Annu Rev Microbiol 68:317–336. doi:10.1146/annurev-micro-091313-103739

Drecktrah D, Levine-Wilkinson S, Dam T et al (2008) Dynamic behavior of Salmonella-induced membrane tubules in epithelial cells. Traffic 9(12):2117–2129. doi:10.1111/j.1600-0854.2008.00830.x

Fang FC (2004) Antimicrobial reactive oxygen and nitrogen species: concepts and controversies. Nat Rev Microbiol 2(10):820–832. doi:10.1038/nrmicro1004

Fang FC (2011) Antimicrobial actions of reactive oxygen species. MBio 2(5). Print 2011. doi: 10.1128/mBio.00141-11

Galan JE (2001) Salmonella interactions with host cells: type III secretion at work. Annu Rev Cell Dev Biol 17:53–86

Galan JE, Wolf-Watz H (2006) Protein delivery into eukaryotic cells by type III secretion machines. Nature 444(7119):567–573. doi: nature05272 [pii]

Gerondopoulos A, Langemeyer L, Liang JR et al (2012) BLOC-3 mutated in Hermansky-Pudlak syndrome is a Rab32/38 guanine nucleotide exchange factor. Curre Biol: CB 22(22):2135–2139. doi:10.1016/j.cub.2012.09.020

Haghjoo E, Galan JE (2004) Salmonella typhi encodes a functional cytolethal distending toxin that is delivered into host cells by a bacterial-internalization pathway. Proc Natl Acad Sci USA 101 (13):4614–4619

Hardt WD, Chen LM, Schuebel KE et al (1998) S. typhimurium encodes an activator of Rho GTPases that induces membrane ruffling and nuclear responses in host cells. Cell 93(5):815–826. doi: S0092-8674(00)81442-7 [pii]

Hernandez LD, Hueffer K, Wenk MR et al (2004) Salmonella modulates vesicular traffic by altering phosphoinositide metabolism. Science 304(5678):1805–1807. doi:10.1126/science.1098188

House D, Bishop A, Parry C et al (2001) Typhoid fever: pathogenesis and disease. Curr Opin Infect Dis 14(5):573–578

Hurley D, McCusker MP, Fanning S et al (2014) Salmonella-host interactions: modulation of the host innate immune system. Front Immunol 5:481. doi:10.3389/fimmu.2014.00481

Imlay JA (2013) The molecular mechanisms and physiological consequences of oxidative stress: lessons from a model bacterium. Nat Rev Microbiol 11(7):443–454. doi:10.1038/nrmicro3032

Jones BD, Falkow S (1996) Salmonellosis: host immune responses and bacterial virulence determinants. Annu Rev Immunol 14:533–561. doi:10.1146/annurev.immunol.14.1.533

Kawai T, Akira S (2010) The role of pattern-recognition receptors in innate immunity: update on Toll-like receptors. Nat Immunol 11(5):373–384. doi:10.1038/ni.1863

Krieger V, Liebl D, Zhang Y et al (2014) Reorganization of the endosomal system in Salmonella-infected cells: the ultrastructure of Salmonella-induced tubular compartments. PLoS Pathog 10(9):e1004374. doi:10.1371/journal.ppat.1004374

LaRock CN, Nizet V (2015) Cationic antimicrobial peptide resistance mechanisms of streptococcal pathogens. Biochim Biophys Acta 1848(11 Pt B):3047–3054. doi: 10.1016/j.bbamem.2015.02.010

LaRock DL, Chaudhary A, Miller SI (2015) Salmonellae interactions with host processes. Nat Rev Microbiol 13(4):191–205. doi:10.1038/nrmicro3420

Loftus SK, Larson DM, Baxter LL et al (2002) Mutation of melanosome protein RAB38 in chocolate mice. Proc Natl Acad Sci USA 99(7):4471–4476. doi:10.1073/pnas.072087599

Malik-Kale P, Jolly CE, Lathrop S et al (2011) Salmonella: at home in the host cell. Front Microbiol 2:125. doi:10.3389/fmicb.2011.00125

Marks MS, Heijnen HF, Raposo G (2013) Lysosome-related organelles: unusual compartments become mainstream. Curr Opin Cell Biol 25(4):495–505. doi: 10.1016/j.ceb.2013.04.008

Mastroeni P, Vazquez-Torres A, Fang FC et al (2000) Antimicrobial actions of the NADPH phagocyte oxidase and inducible nitric oxide synthase in experimental salmonellosis. II. Effects on microbial proliferation and host survival in vivo. J Exp Med 192(2):237–248

McClelland M, Sanderson KE, Spieth J et al (2001) Complete genome sequence of Salmonella enterica serovar Typhimurium LT2. Nature 413(6858):852–856. doi:10.1038/35101614

Mogasale V, Maskery B, Ochiai RL et al (2014) Burden of typhoid fever in low-income and middle-income countries: a systematic, literature-based update with risk-factor adjustment. Lancet Glob Health 2(10):e570–e580. doi:10.1016/S2214-109X(14)70301-8

Newton K, Dixit VM (2012) Signaling in innate immunity and inflammation. Cold Spring Harb Perspect Biol 4(3). doi: 10.1101/cshperspect.a006049

Neyrolles O, Wolschendorf F, Mitra A et al (2015) Mycobacteria, metals, and the macrophage. Immunol Rev 264(1):249–263. doi:10.1111/imr.12265

Nies DH, Herzberg M (2013) A fresh view of the cell biology of copper in enterobacteria. Mol Microbiol 87(3):447–454. doi:10.1111/mmi.12123

Ohlson MB, Huang Z, Alto NM et al (2008) Structure and function of Salmonella SifA indicate that its interactions with SKIP, SseJ, and RhoA family GTPases induce endosomal tubulation. Cell Host Microbe 4(5):434–446. doi:10.1016/j.chom.2008.08.012

Osman D, Patterson CJ, Bailey K et al (2013) The copper supply pathway to a Salmonella Cu, Zn-superoxide dismutase (SodCII) involves P(1B)-type ATPase copper efflux and periplasmic CueP. Mol Microbiol 87(3):466–477. doi:10.1111/mmi.12107

Papagrigorakis MJ, Yapijakis C, Synodinos PN et al (2006) DNA examination of ancient dental pulp incriminates typhoid fever as a probable cause of the Plague of Athens. Int J Infect Dis 10 (3):206–214. doi: S1201-9712(05)00178-5 [pii]

Parkhill J, Dougan G, James KD et al (2001) Complete genome sequence of a multiple drug resistant Salmonella enterica serovar Typhi CT18. Nature 413(6858):848–852

Parry CM, Hien TT, Dougan G et al (2002) Typhoid fever. N Engl J Med 347(22):1770–1782

Popp J, Noster J, Busch K et al (2015) Role of host cell-derived amino acids in nutrition of intracellular Salmonella enterica. Infect Immun. doi: IAI.00624-15 [pii]

Rosenberger CM, Gallo RL, Finlay BB (2004) Interplay between antibacterial effectors: a macrophage antimicrobial peptide impairs intracellular Salmonella replication. Proc Natl Acad Sci USA 101(8):2422–2427. doi: 101/8/2422 [pii]

Sabbagh SC, Forest CG, Lepage C et al (2010) So similar, yet so different: uncovering distinctive features in the genomes of Salmonella enterica serovars Typhimurium and Typhi. FEMS Microbiol Lett 305(1):1–13. doi:10.1111/j.1574-6968.2010.01904.x

Schwan WR, Huang XZ, Hu L et al (2000) Differential bacterial survival, replication, and apoptosis-inducing ability of Salmonella serovars within human and murine macrophages. Infect Immun 68(3):1005–1013

Slauch JM (2011) How does the oxidative burst of macrophages kill bacteria? Still an open question. Mol Microbiol 80(3):580–583. doi: 10.1111/j.1365-2958.2011.07612.x

Song J, Gao X, Galan JE (2013) Structure and function of the Salmonella Typhi chimaeric A(2)B (5) typhoid toxin. Nature 499(7458):350–354. doi:10.1038/nature12377

Spano S (2014) Host restriction in Salmonella: insights from Rab GTPases. Cell Microbiol 16 (9):1321–1328. doi:10.1111/cmi.12327

Spano S, Galan JE (2008) A novel pathway for exotoxin delivery by an intracellular pathogen. Curr Opin Microbiol 11(1):15–20. doi:10.1016/j.mib.2007.12.002

Spano S, Galan JE (2012) A Rab32-dependent pathway contributes to Salmonella typhi host restriction. Science 338(6109):960–963. doi:10.1126/science.1229224

Spano S, Galan JE (2013) A novel anti-microbial function for a familiar Rab GTPase. Small GTPases 4(4)

Spano S, Liu X, Galan JE (2011) Proteolytic targeting of Rab29 by an effector protein distinguishes the intracellular compartments of human-adapted and broad-host Salmonella. Proc Natl Acad Sci USA 108(45):18418–18423. doi:10.1073/pnas.1111959108

Spano S, Ugalde JE, Galan JE (2008) Delivery of a Salmonella Typhi exotoxin from a host intracellular compartment. Cell Host Microbe 3(1):30–38

Strandberg KL, Richards SM, Gunn JS (2012) Cathelicidin antimicrobial peptide expression is not induced or required for bacterial clearance during salmonella enterica infection of human monocyte-derived macrophages. Infect Immun 80(11):3930–3938. doi:10.1128/IAI.00672-12

Terebiznik MR, Vieira OV, Marcus SL et al (2002) Elimination of host cell PtdIns(4,5)P(2) by bacterial SigD promotes membrane fission during invasion by Salmonella. Nat Cell Biol 4 (10):766–773. doi:10.1038/ncb854

Toft C, Andersson SG (2010) Evolutionary microbial genomics: insights into bacterial host adaptation. Nat Rev Genet 11(7):465–475. doi:10.1038/nrg2798

Vazquez-Torres A, Jones-Carson J, Mastroeni P et al (2000a) Antimicrobial actions of the NADPH phagocyte oxidase and inducible nitric oxide synthase in experimental salmonellosis. I. Effects on microbial killing by activated peritoneal macrophages in vitro. J Exp Med 192 (2):227–236

Vazquez-Torres A, Xu Y, Jones-Carson J et al (2000b) Salmonella pathogenicity island 2-dependent evasion of the phagocyte NADPH oxidase. Science 287(5458):1655–1658. doi: 8324 [pii]

Vladoianu IR, Chang HR, Pechere JC (1990) Expression of host resistance to Salmonella typhi and Salmonella typhimurium: bacterial survival within macrophages of murine and human origin. Microb Pathog 8(2):83–90

Wasmeier C, Romao M, Plowright L et al (2006) Rab38 and Rab32 control post-Golgi trafficking of melanogenic enzymes. J Cell Biol 175(2):271–281. doi:10.1083/jcb.200606050

Wei AH, Li W (2013) Hermansky-Pudlak syndrome: pigmentary and non-pigmentary defects and their pathogenesis. Pigment Cell Melanoma Res 26(2):176–192. doi:10.1111/pcmr.12051

White C, Lee J, Kambe T et al (2009) A role for the ATP7A copper-transporting ATPase in macrophage bactericidal activity. J Biol Chem 284(49):33949–33956. doi: 10.1074/jbc.M109.070201

Wong VK, Baker S, Pickard DJ et al (2015) Phylogeographical analysis of the dominant multidrug-resistant H58 clade of Salmonella Typhi identifies inter- and intra-continental transmission events. Nat Genet 47(6):632–639. doi:10.1038/ng.3281

Zhou D, Chen LM, Hernandez L et al (2001) A Salmonella inositol polyphosphatase acts in conjunction with other bacterial effectors to promote host cell actin cytoskeleton rearrangements and bacterial internalization. Mol Microbiol 39(2):248–259. doi: mmi2230 [pii]

Chapter 18
Insights into Biological Complexity from Simple Foundations

L. Albergante, D. Liu, S. Palmer and T.J. Newman

Abstract We discuss an overtly "simple approach" to complex biological systems borrowing selectively from theoretical physics. The approach is framed by three maxims, and we show examples of its success in two different applications: investigating cellular robustness at the level of gene regulatory networks and quantifying rare events of DNA replication errors.

Keywords Simplicity · Theoretical physics · Probability theory · Gene regulatory networks · DNA replication · Big data

18.1 Introduction

The cell is often seen as the building block of life, and as such holds within it many of the secrets of biology. As much as one can speak of "the cell" as an entity, given the enormous variety of cell types, the cell has successfully retained its enigmatic qualities for many decades, despite huge research investment and the attention of many of the world's most able minds. The sustained difficulties in forming an overarching understanding of the cell, and the tissues composed thereof, and the continual discovery of new layers of process and regulation lead us inevitably to speak of "biological complexity". The challenge of biological complexity most often invokes a reductionist response, viz. in order to understand the complexity, we must understand in detail the molecular constituents of the cell and their interactions. This has led to mini revolutions badged under genomics, proteomics, transcriptomics, etc. and more generally systems biology. While the authors in no way dispute the importance of reductionist approaches, in this chapter we would like to remind the reader that there are parallel strategies through which to achieve

---

L. Albergante · D. Liu · S. Palmer · T.J. Newman (✉)
School of Life Sciences, University of Dundee, Dundee DD1 5EH, UK
e-mail: T.Newman@dundee.ac.uk

© Springer International Publishing Switzerland 2016

M.C. Leake (ed.), *Biophysics of Infection*, Advances in Experimental Medicine and Biology 915, DOI 10.1007/978-3-319-32189-9_18

an understanding of complexity. One of these is to seek simpler higher-level mechanisms that in and of themselves provide means for selective pressure, and hence are of fundamental biological importance. We have, in our research group over the past five years, found an approach based almost entirely on "simplicity" to be remarkably effective in providing new insights into the workings of cells and other biological systems. The purpose of this chapter is to provide some rationale for simplicity as an approach, using, with care, some insights from physics and to give two examples of its power in applications to gene regulation (Albergante et al. 2014) and DNA replication (Newman et al. 2013). We provide very brief sketches of these applications, with the aim of focusing on the approach itself, as the biological details and results can be found in published work.

18.2 Three "Equations"

In a recent essay (Newman 2015), three maxims, which have the ring of equations, were introduced to help frame the rationale for bringing some concepts from physics into biology, particularly in regard to complementing reductionism with an approach based on simple (but not necessarily elegant) higher-level theories. We provide here, a brief summary of the point of each equation, and refer the reader to Newman 2015 for a more extended discussion.

18.2.1 Physics Does not Equal Biology

In bringing concepts from physics, it is crucial to recognise at the outset that one does not in any sense have in mind "pasting" physics onto biology. Non-living and living systems have profound differences, the most notable being that living systems have arisen from a process of evolution spanning billions of years, and hence contain within them countless layers of contingency. This is one of the major drivers for biological complexity and most likely forbids a typical physics approach seeking theories based on "elegance".

18.2.2 Small Does not Imply Fundamental and Vice Versa

This is the central idea, that understanding a system does not necessarily require knowledge of the components of that system. The properties of the system may arise from the enormous *number* of components and their interactions without being very sensitive to the nature of the components themselves. Hence "fundamental" properties do not map on, necessarily, to properties of constituents. Examples of this abound in physical systems, e.g. our best theory of the dynamical properties of fluids, namely, fluid mechanics, refers in no way to the molecular properties of fluids, and our best theory of electronic properties of solids (insulators, semiconductors, etc.) relies on emergent concepts that are only weakly dependent on atomic properties (cf. solid state theory). The application of this type of theory to biology has so far been rather limited, though there is most certainly a recognition of the importance of related concepts such as emergence; examples are Gibson et al. (2006) and Krieger and Simons (2015).

18.2.3 Simple Does not Imply Obvious

Higher-scale theories of life will be, by their very nature, simpler than detailed molecular theories that attempt to capture the multitude of interactions. For this reason, one might think that biological complexity cannot arise from theoretical simplicity. This third "equation" reminds us that simple theories often provide predictions for arbitrarily complex (or non-obvious) phenomena. Again, examples abound in physics, from theories which are simple in their form, such as relativity, quantum mechanics and electromagnetism, and yet which provide an astonishingly rich variety of predictions.

18.3 Big Data

We reflect briefly on the advent of "big data" which presents huge opportunity to biology, and yet which also must be treated with some caution. We have found that big data provides an invaluable lens through which to view (and test) predictions from our simple theories.

The availability of high-throughput experimental techniques and the establishment of large efforts such as the "1000 Genome Project" (The 1000 Genomes Project Consortium 2015) and the "Human Microbiome Project" (The NIH HMP Working Group et al. 2009) have led to an unprecedented amount of data on the genotype and phenotype of cells and species. Despite the availability of such large amounts of data, many of the mechanisms and selective pressures that control and shape biological entities remain unknown. The difficulty in extracting these from big data are connected both to the complexity of biological information and to the analytical approaches currently available.

An *exploratory approach* is often employed to study large biological datasets: statistical techniques are used with the aim of finding patterns that distinguish different conditions (e.g. cancer vs. non-cancer cells). This approach has been able to identify differences that would have been impossible to discern by the human eye and has been frequently used as an initial step to allow further, more mechanistic explorations (Tinti et al. 2014). Nonetheless, despite the advancement of our knowledge of statistics and algorithms, strong limitations remain. For example, computational considerations force researchers to restrict the classes of differences that can be assessed. Moreover, statistical techniques generally rely on some kind of null model and it is often unclear whether the null models used are appropriate for subsequent genetic exploration.

18.4 Two Examples of the "Simple Approach"

The following sections describe two examples, selected from our recent work, which highlight the power of simple models in providing important insights into the features of very different biological processes. These examples share important features: the focus on the fundamental features of the problem under consideration, the use of a very limited number of parameters (even zero), the derivation of biologically relevant predictions, and detailed challenge and testing using experimentally derived datasets.

18.4.1 Robustness of Cell States and BQS

Biological systems are remarkably robust to both internal and external disturbances (Silva-Rocha and de Lorenzo 2010). For example, cells are able to function properly despite the presence of large fluctuations in gene expression, in the number of proteins, and in the concentration of extracellular molecules. Cells are very complex entities formed by many sub-systems that are responsible for different functionalities. Many of these sub-systems are ultimately controlled by gene regulation.

Gene regulation is complex: at the single gene level, protein complexes are responsible for the initiation and control of transcription and translation. At the genomic level, tens to hundreds of genes—usually called transcription factors (TFs)—directly regulate the expression of thousands of downstream genes including TFs themselves. Experimental studies have elucidated many of the specific molecular details of transcriptional control. However, assessing the role of robustness at the genomic level requires a more theoretical approach.

Networks provide an excellent way to model systems that rely on widespread interactions amongst a large number of components (Newman 2010). Therefore, it is not surprising that they have been extensively employed to study gene interactions. Gene regulatory networks (GRNs) have been used to characterise genome-wide transcriptional control (Alon 2007). In this type of network, each node represents a gene and an edge is inserted between two nodes if one is a TF that regulates the expression of the other (Fig. 18.1a). Reliable experimental techniques exist to obtain GRNs in a relatively straightforward way (e.g. ChIP-Seq) and therefore GRNs associated with different organisms are available in the literature (Salgado et al. 2013; Harbison et al. 2004; Gerstein et al. 2012).

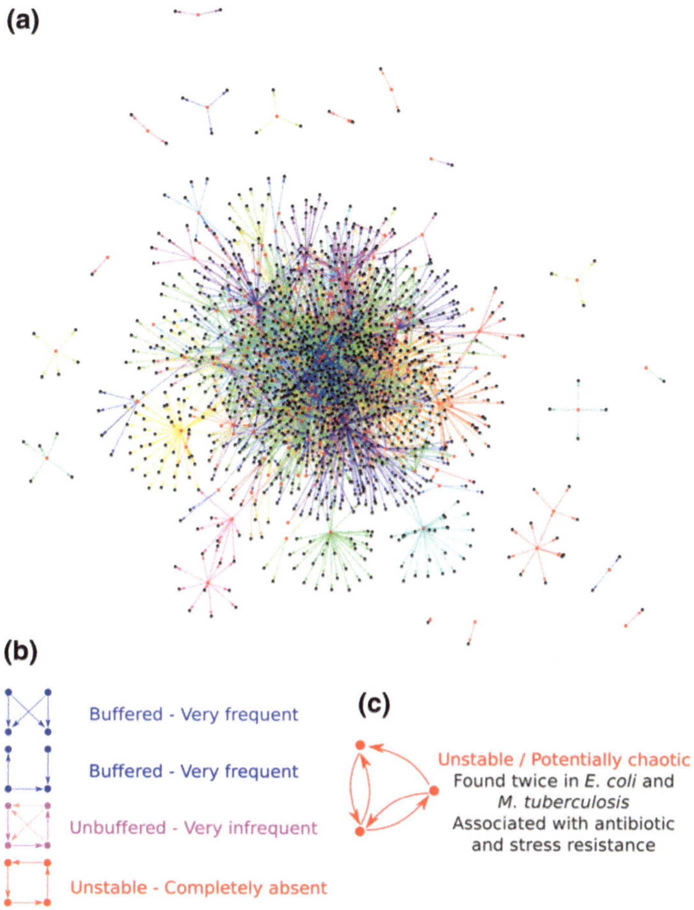

Fig. 18.1 Transcriptional networks and their local stability properties. **a** The transcriptional network of *E. coli* has been derived from Salgado et al. (2013) as described in Albergante et al. (2014). Each *node* represents a gene. *Red nodes* are TF genes, while *black nodes* are non-TF genes. *Arrows* are colour coded depending on the source TF. **b** Different four gene motifs are reported with their stability class and frequency in GRNs. In the *violet* unbuffered motif, *light red arrows* indicate potential arrows that are not part of the motif itself. **c** The chaotic motif reported is very unlikely to emerge by chance and yet is the only type of long feedback loop observed in both *E. coli* and *M. tuberculosis*

The theory of Buffered Qualitative Stability (BQS) aims to understand to what extent robustness and stability shape the topology of GRNs (Albergante et al. 2014). BQS is based on two key assumptions: (1) the robustness of GRNs is independent of the precise values of the thousands of parameters that characterise network-wide gene interactions, and (2) the robustness of GRNs must be preserved over evolutionary timescales, and hence the GRN must remain robust even if genetic innovations, in the form of new links in the network, are introduced.

Therefore, the GRN must be “qualitatively stable” and this stability must be “buffered”.

BQS is based on the theory of Qualitative Stability (QS), introduced in 1965 by economists Quirk and Ruppert (1965). QS provides a set of topological rules that, when verified, guarantee a network to be stable. These rules are very stringent, and a network is extremely unlikely to verify them by chance. The most relevant rule of QS in the context of GRNs is the absence of any feedback loops formed by three or more genes (“long feedback loops”). Using this theory as a starting point, we derived the central postulate of BQS: “to be stable over evolutionary times, GRNs must be configured to minimise the number of actual and potential feedback loops”.

The extremely simple nature of this postulate allowed us to derive six strong predictions on the structure of GRNs. Each of these predictions was either to be right or wrong when challenged with data as the theory contains no parameters.

We found that all of the predictions from BQS were verified by analysis of experimentally derived GRNs of *E. coli*, *M. tuberculosis*, *P. aeruginosa*, yeast, an immune-derived mouse cell line, and a non-cancer human cell line (GM12878). In particular, we were able to show that GRNs are indeed buffered against the creation of long feedback loops both at the local and at the global scale. Our analysis shed new light on the selective pressures that control the emergence of commonly encountered network motifs, such as the feed-forward loop and the bi-fan, and highlighted previously unnoticed common features of these GRNs (Fig. 18.1b).

Interestingly, a very limited deviation from the predictions of BQS was observed in network structures associated with antibiotic and stress response in both *E. coli* and *M. tuberculosis* (Fig. 18.1c). This suggests that genetic regulation may be able to dynamically control the stability of GRNs depending on the specific environmental conditions encountered by cells. For example, the unstable and potentially chaotic response that can be provided by long feedback loops can lead to a large diversity in gene expression levels and therefore to phenotypic heterogeneity, which will promote the emergence of bacteria better suited to survive harsh conditions.

Additional insights into the role of robustness came from our analysis of the GRN of a leukaemia cell line (K562). Despite having a comparable number of nodes and edges to GM12878, the GRN of K562 is very different. A very large number of long feedback loops is present and many of the features associated with BQS are absent. This suggests commonalities between the behaviour of cancer cells and bacteria in having mechanisms built into GRNs that can rapidly generate phenotypic heterogeneity. In particular, we found that the instability of the GRN of K562 could be controlled by targeting specific pairs of genes. Essentially, BQS could be reinstated in K562 through elimination of two or three nodes of the network. This finding may have significance for targeted gene therapy in cancer.

The theory of BQS is an important step in understanding how robustness influences the topology of networks and, due to the generality of our approach, our theory could be easily extended to study networks in a range of complex systems in biology and beyond.

18.4.2 Error Rates in DNA Replication

During the cell cycle, how does a cell ensure the replication of all of its DNA exactly once—i.e. not missing any DNA, and not replicating any DNA more than once? This problem has been intensively studied by experimental biologists for many years, and answered through the discovery of "DNA licensing" (Blow and Dutta 2005). As shown schematically in Fig. 18.2a, replication forks are nucleated at special sites on the DNA known as replication origins (ROs). These ROs are "licensed" for replication prior to S-phase in the cell cycle, and are eligible as

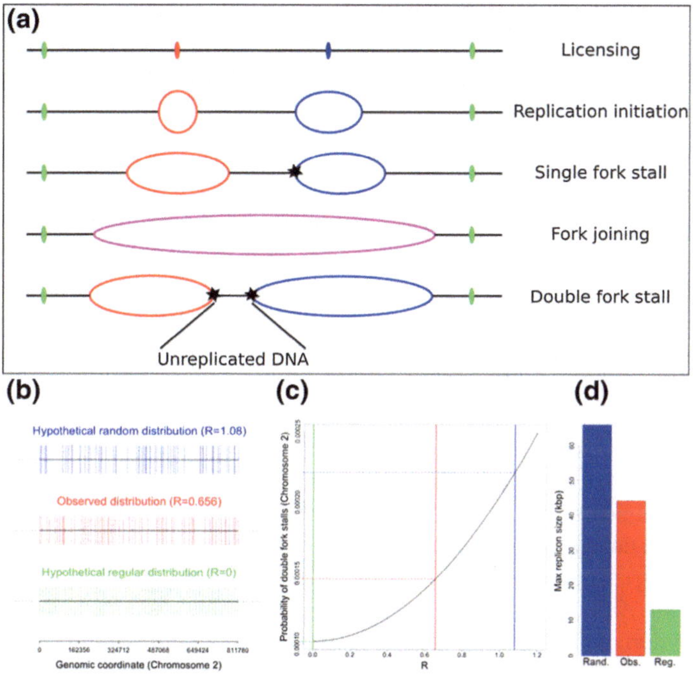

Fig. 18.2 Replication origins and their connection with DNA double fork stalls. **a** A cartoon of replication dynamics is reported. Licensed ROs provide starting points for DNA replication forks. When a single replication fork stalls, a nearby active fork can make sure that the DNA is completely replicated. When two forks stall with no licensed ROs between them, a double fork stall occurs and part of the DNA remains unreplicated. **b** The observed distribution of replication origins in chromosome 2 of yeast (*red*), obtained from Siow et al. (2012), is compared with a hypothetical random distribution (*blue*) and to a perfectly regular distribution (*green*). *R* indicates the ratio of the standard deviation over the mean of the distribution of the inter-RO distances. The smaller the value of *R*, the more regular the distribution. Note how the observed distribution is more regular than expected by chance. **c** The probability of a double fork stall in chromosome 2 of yeast is plotted as a function of the value *R*, with the same colour conventions of panel b. Note how increasing regularity lowers the probability. **d** The size of the largest replicon is reported for the different distributions of panel a. Note how the size of the largest replicon in the observed distribution is smaller than expected by chance

candidates for the nucleation of replication machinery. Once S-phase is initiated, no other ROs can be licensed. This mechanism prevents the re-replication of already replicated DNA. The only hitch occurs if a replication fork stalls—this is quite rare, since the efficiency of replication machinery is exceedingly high. In most cases, a stalling event can be rescued by a fork approaching from the opposite direction. If that fork also stalls, there is a potentially catastrophic situation of unreplicated DNA lying between the two stalled forks. This in turn can be avoided by having an overabundance of ROs, so that the intervening DNA between two stalled forks will likely contain a licensed but unused RO, so that eventually this will nucleate replication machinery and complete replication of the intervening gap is possible. The system is beautifully constructed, but not failsafe. There is still a tiny chance that two replication forks approaching each other in a small region of DNA containing no ROs will actually both stall (which we call a double fork stall, or DFS), in which case an unrecoverable unreplicated region of DNA will result. Questions of biological interest are: exactly how rare is this rare event?, what can the cell do in terms of overabundance of ROs and placement of ROs on the DNA to minimise these rare events?, and how much more difficult does it become to avoid this catastrophe as the size of the genome increases?

In attempting to answer these questions, one could start with a detailed description of the beautiful internal workings of the replication machinery, with a zoo of rate constants and Hill coefficients, calculating the stochastic dynamics of replication forks as they compete for a reservoir of nucleotides. Instead, we took the "simple approach", ignoring all molecular details and even ignoring all dynamics as well. In Newman et al. (2013), we developed the simplest possible probabilistic model of this process, seeking only to calculate the probability of a DFS error once all ROs had been used (or else removed by traversing forks). The model was based on a set of biologically plausible assumptions, and it contained only one a priori unknown parameter: the median distance a replication fork would travel along the DNA before stalling (which is inversely proportional to the minute per nucleotide rate of stalling). Experimental data (Maya-Mendoza et al. 2007) suggests this has a value around 10 megabases.

We first derived an equation for the probability of a DFS within a generic region of DNA between two adjacent ROs. This was then used to derive an expression for the probability of DFS errors across the entire genome. Minimising this latter expression allowed us to derive a number of predictions: that ROs should be as evenly spaced as possible, that endmost ROs should be placed very near to chromosome ends, and that large gaps between ROs should be avoided (summarised in Fig. 18.2b–d, where R is a measure of positional regularity of ROs). These predictions were tested against high-quality data on RO positions in the yeast genome (which has a size of approx. 12 megabases) obtained from OriDB (Siow et al. 2012) and were found to be correct. Strikingly, we were also able to use the data to estimate the median stalling distance from our theory, and obtained the value 12.7 megabases, remarkably close to the experimental value. We could also estimate the probability of a genome-wide DFS error, and found this to be roughly 0.001, which is of the same order of magnitude as chromosome loss errors in yeast, which makes biological sense: there would be no gain in reducing DFS errors below the rate of a different catastrophic event. Finally, we could show that once genomes get significantly larger, particularly to the size of the human genome (of diploid size ~6 Gigabases), DFS errors become of order unity, no matter how carefully spaced are the ROs, implying that new repair mechanisms are required for longer genomes. A detailed analysis of the human case, using our simple theory, is in preparation and contains many new results which illustrate how these repair mechanisms have had to be tailored to rates of DFSs in longer genomes.

18.5 Conclusions

Our main message in this brief chapter is that in biological research, reductionism, while extremely useful, is certainly not the only approach. We have described an alternative approach, namely, looking at each biological problem more holistically, and keeping concepts as simple as possible. Within this approach, we advocate to keep the number of postulates and parameters to an absolute minimum, and to use big data as a lens through which to rigorously test predictions.

There is nothing new to this style of theory; indeed, this is the approach that has been used for centuries in physics. Crucially, though, we are not pasting physics onto biology. We make no use of the fundamental laws of physics, nor seek elegance or "beauty" in our theories. It is the simplicity from theoretical physics that we use, along with a guiding principle that biological systems are "emergent structures" and therefore robust and explainable at a phenomenological, or coarse-grained, level.

Our group has been continually surprised at how effective this approach appears to be. For example, we have used the "simple approach" to derive rare event statistics of highly complex processes such as cancer metastasis (Cisneros and Newman 2014), finding both interesting mechanisms for metastatic growth and surprisingly good overlap with physiological measurements. We also have forthcoming papers in which this approach has shed light on modularity in GRNs, topologies of embryonic tissue and a role of immunosenescence on cancer incidence. In all cases, we test our predictions successfully against detailed biological data.

In an age of systems biology, in which it has become possible to instruct computers to solve very large systems of differential equations with hundreds of parameters, we find it refreshing to march boldly in the opposite direction. We ask similar questions, seeking to understand complex biological systems such as intracellular networks and tissue morphologies and dynamics, but we draw the line at models with two parameters, preferring in fact one, or better still zero.

It is perhaps fitting, therefore, to close by quoting the last author's favourite composer:

Before you play two notes learn how to play one note – and don't play one note unless you've got a reason to play it. Mark Hollis (1998)

Acknowledgments The authors would like to thank Professor Julian Blow for innumerable enlightening and inspiring discussions on the research projects described in this chapter.

References

Albergante L, Blow JJ, Newman TJ (2014) Buffered qualitative stability explains the robustness and evolvability of transcriptional networks. eLife 3. http://doi.org/10.7554/eLife.02863

Alon U (2007) An introduction to systems biology : design principles of biological circuits. Chapman and Hall/CRC

Blow JJ, Dutta A (2005) Preventing re-replication of chromosomal DNA. Nat Rev Mol Cell Biol 6 (6):476–486. http://doi.org/10.1038/nrm1663

Cisneros LH, Newman TJ (2014) Quantifying metastatic inefficiency: rare genotypes versus rare dynamics. Phys Biol 11(4):046003

Gerstein MB, Kundaje A, Hariharan M, Landt SG, Yan K-K, Cheng C, Snyder M (2012) Architecture of the human regulatory network derived from ENCODE data. Nature 489 (7414):91–100. http://doi.org/10.1038/nature11245

Gibson MC, Patel AB, Nagpal R, Perrimon N (2006) The emergence of geometric order in proliferating metazoan epithelia. Nature 442(7106):1038–1041. http://doi.org/10.1038/nature05014

Harbison CT, Gordon DB, Lee TI, Rinaldi NJ., Macisaac KD., Danford TW, Young RA (2004) Transcriptional regulatory code of a eukaryotic genome. Nature 431(7004):99–104. http://doi.org/10.1038/nature02800

Krieger T, Simons BD (2015) Dynamic stem cell heterogeneity. Development 142(8):1396–1406. http://doi.org/10.1242/dev.101063

Mark Hollis (1998) TV Interview, Okay Tone, DR2, Danish National Television, 22 Feb 1998

Maya-Mendoza A, Petermann E, Gillespie DA, Caldecott KW, Jackson DA (2007) Chk1 regulates the density of active replication origins during the vertebrate S phase. EMBO J 26(11):2719–2731. http://doi.org/10.1038/sj.emboj.7601714

Newman MEJ (2010) Networks: an introduction. Oxford University Press, Oxford; New York

Newman T (2015) Biology is simple. Phys Biol 12(6):063002

Newman TJ, Mamun MA, Nieduszynski CA, Blow JJ (2013) Replisome stall events have shaped the distribution of replication origins in the genomes of yeasts. Nucleic Acids Res 41 (21):9705–9718. http://doi.org/10.1093/nar/gkt728

Quirk J, Ruppert R (1965) Qualitative economics and the stability of equilibrium. Rev Econ Stud 32(4):311–326. http://doi.org/10.2307/2295838

Salgado H, Peralta-Gil M, Gama-Castro S, Santos-Zavaleta A, Muñiz-Rascado L, García-Sotelo JS, Collado-Vides J (2013) RegulonDB v8.0: omics data sets, evolutionary conservation, regulatory phrases, cross-validated gold standards and more. Nucleic Acids Res 41(D1):D203–D213. http://doi.org/10.1093/nar/gks1201

Silva-Rocha R, de Lorenzo V (2010) Noise and robustness in prokaryotic regulatory networks. Ann Rev Microbiol 64(1):257–275. http://doi.org/10.1146/annurev.micro.091208.073229

Siow CC, Nieduszynska SR, Müller CA, Nieduszynski CA (2012) OriDB, the DNA replication origin database updated and extended. Nucleic Acids Res 40(D1):D682–D686. http://doi.org/10.1093/nar/gkr1091

The 1000 Genomes Project Consortium (2015) A global reference for human genetic variation. Nature 526(7571):68–74

The NIH HMP Working Group, Peterson J, Garges S, Giovanni M, McInnes P, Wang L, Guyer M (2009) The NIH human microbiome project. Genome Res 19(12):2317–2323. http://doi.org/10.1101/gr.096651.109

Tinti M, Dissanayake K, Synowsky S, Albergante L, MacKintosh C (2014) Identification of 2R-ohnologue gene families displaying the same mutation-load skew in multiple cancers. Open Biol 4(5). http://doi.org/10.1098/rsob.140029

Chapter 19
Force Spectroscopy in Studying Infection

Zhaokun Zhou and Mark C. Leake

Abstract Biophysical force spectroscopy tools—for example, optical tweezers, magnetic tweezers, atomic force microscopy—have been used to study elastic, mechanical, conformational and dynamic properties of single biological specimens from single proteins to whole cells to reveal information not accessible by ensemble average methods such as X-ray crystallography, mass spectroscopy, gel electrophoresis and so on. Here, we review the application of these tools on a range of infection-related questions from antibody-inhibited protein processivity to virus-cell adhesion. In each case, we focus on how the instrumental design tailored to the biological system in question translates into the functionality suitable for that particular study. The unique insights that force spectroscopy has gained to complement knowledge learned through population averaging techniques in interrogating biomolecular details prove to be instrumental in therapeutic innovations such as those in structure-based drug design.

Keywords Optical tweezers · Magnetic tweezers · Atomic force microscopy

19.1 Introduction

Force spectroscopy refers to a series of experimental techniques capable of linear and rotational force (the latter also referred to as torque) transduction onto and the measurement of microscopic length scale objects and smaller, biological samples in particular. The forces can arise from the transfer of photon momenta (optical tweezers), magnetic interaction between a generated $\boldsymbol{B}$ field and a magnetic

Z. Zhou (✉) · M.C. Leake
Departments of Physics and Biology, Biological Physical Sciences Institute,
University of York, York YO10 5DD, UK
e-mail: zz840@york.ac.uk

© Springer International Publishing Switzerland 2016
M.C. Leake (ed.), *Biophysics of Infection*, Advances in Experimental Medicine
and Biology 915, DOI 10.1007/978-3-319-32189-9_19

micro-object (magnetic tweezers), mechanical manipulation (atomic force microscopy, micropipette and microneedle) and liquid flow pressure. The samples are usually imaged or tracked with optical microscopy—brightfield, fluorescence imaging and so on. But the configuration of force spectroscopy instruments have uniquely allowed other measurement methods, such as back focal plane interferometry. These methods are explained below. With its capability to transduce force to perturb the conformation of the biological samples, force spectroscopy has found applications in studying elasto-mechanical properties of biological polymers such as the over-stretching (Smith et al. 1996), supercoiling (Strick et al. 1996), braiding (Strick et al. 1998), buckling (Strick et al. 1996) and denaturing (King et al. 2013) of nucleic acids and torsional compliance of bacteria flagella (Block et al. 1989), and probing muscle proteins for shock-absorbing properties (Leake et al. 2003, 2004; Linke and Leake 2004; Leake et al. 2006; Bullard et al. 2006), all being techniques which use physics to probe biology one molecule at a time (Leake 2013). Another advantage afforded by these techniques stems from the fact that they study one copy of the sample at a time, be it a biological macromolecule, a cell or a virus, which, in contrast to ensemble methods that have studied biological molecules and organisms prior to the advent of force spectroscopy, does not need to temporally and spatially average over population measurements. As a result, force spectroscopy techniques can interrogate the time-resolved series of molecular actions that comprise a complete catalytic or translocation cycle. Among the achievements are the characterisation of topoisomerase actions in uncoiling double-stranded nuclei acids (Charvin et al. 2005) and the walking patterns of myosin (Finer et al. 1994) on actin fibres and kinesin (Block et al. 1990)/dynein (Gennerich et al. 2007) on cytoskeletal microtubules. Also worth mentioning is the increasing spatial-temporal resolution achieved over the years since the inception of force spectroscopy. Super-resolution fluorescence microscopy and interferometry position tracking with split-photodiode detection afford sensitive detection whilst continual improvements in the instrumental design—acoustic and mechanical vibration isolation, atmospheric air purification by replacement with helium gas, active feedback sample drift cancelling and so on—minimise instrumental noise to fundamental sources such as Brownian noise and shot noise. Angstrom resolution has already been achieved, which is necessary to resolve movement on, for example, DNA double strands with base pair repeating distances of 3.4 Å. This opens the door to the observation of nucleic acid motor proteins: topoisomerases, helicases (Fili et al. 2010), gyrases (Gore et al. 2006) and RNA polymerases (Abbondanzieri et al. 2005), etc.

Questions linked to infection have been studied with force spectroscopy experiments to advance our understanding in infection so we can more effectively prevent, diagnose and treat infectious diseases. These experiments have mostly focused on molecular and cellular levels but they involve many aspects of infection including the metabolism and behaviour of infectious microorganisms, the effect of drugs on infectious microorganisms, the properties of our immune cells and so on.

Heterogeneity among individual cells can arise as a response to uneven substrate distributions, as a result of detecting different signal molecules, or simply from stochastic variations. Population averaging measurements are unable to distinguish between statistical distributions, such as between normal and bimodal distributions, both common among cell expressions. As Lidstrom et al. (2010) pointed out, normally distributed phenotypes imply a cooperative binding mechanism while a bimodal distribution imply an on/off switch mechanism determined by a threshold level with some stochastic elements. In contrast, information pertaining to the state of every cell allows the determination of the distribution and thus the biological mechanism. Similar necessity for single entity measurements is seen in cell responses, as indicated by the level of some measurable parameters, to perturbations. Immediate response from part of the population to a perturbation has the same ensemble appearance as a slow or delayed response from the entire population, which again calls for the measurement of single individuals.

Structure-based drug design is a directed and systematic approach of discovering new drugs that does not resort to traditional trial-and-error testing of myriads of candidate chemicals. Rather, the structural features that enable the drug to block, inhibit or activate a target protein are designed according to the structural features on the target protein. So it is crucial to possess detailed and accurate knowledge of the target proteins. Traditional ensemble methods to investigate structural properties of proteins—biochemical studies, NMR, mass spectroscopy, X-ray crystallography—as well as the traditional single-molecule imaging method electron microscopy, have had great successes in discovering molecular structures which has led to breakthroughs in drug development. However, during a catalytic cycle or force generation action, the conformation of or charge distribution on a protein may change throughout the process taking a few distinctive forms. The aforementioned techniques are limited by the necessity of sample preparations that isolate and fix proteins so the rich conformational states will not be sufficiently present during detection. In contrast, force spectroscopy tools allow proteins to complete a set of actions as they do in living, physiological settings so the change of states at each phase of the process can be characterised with the high spatial-temporal resolution afforded by force spectroscopy tools. Apart from the ability to characterise molecular structures during molecular activity, force spectroscopy can also characterise the molecules' hindered actions when they are inhibited either by naturally existing chemicals or manmade drugs. Naturally existing inhibitory chemicals may be secreted by a nearby organism to suppress competitors. The molecular details of the inhibitory mechanism will shed light on the response of molecules to the binding of the inhibitor and inspire the design of new drugs. Manmade drugs need to be assessed for the effectiveness and the current main approach is in silico simulation, which is sometimes prohibited by heavy computational loads. Direct empirical observations will fill in the gaps where simulations are infeasible.

This chapter will summarise the designs and measurement mechanisms of three examples of force spectroscopy instruments and their applications in studying biological agents which, among other achievements, tremendously advance our ability at understanding infectious diseases.

19.2 Introduction to Force Spectroscopy Tools

Optical tweezers, magnetic tweezers and atomic force microscopy, among others, are three types of microscopic force transduction instruments capable of applying controlled forces and torques onto single molecules and cells as a means of perturbing the biological samples to reveal their mechanical and/or dynamical properties. Some of them also serve as a contact-mode imaging technique with molecular precision. Their mechanisms and recent advances have been described in detail in the literature (Neuman and Nagy 2008; Moffitt et al. 2008; De Vlaminck 2012).

19.2.1 Optical Tweezers (OT)

Photons carry momenta and when they enter a medium of different refractive index, they deflect and impart part of their momenta to the medium; this serves as the underlying principle of OT in force application to microscopic length scale objects. A collimated laser beam is typically focused via a microscope objective lens to form a diffraction-limited spot so that light rays converge strongly in all three dimensions towards the centre of the spot (Fig. 19.1a, b, e). Any microscopic object of higher refractive index than the surrounding medium will experience gradient forces that push it to the centre of the trap (Fig. 19.1c, d)—a ray trace diagram can show this (Ashkin 1992). Since the gradient force scales linearly with light intensity, the maximum force is limited by the sample heating at high laser powers. But this is greater than many biological forces at molecular and cellular levels. OT routinely achieves sub-nanometre and sub-millisecond resolutions since the optical system of OT is compatible with quadrant photodiode detection that images the interference pattern due to the bead in the beam waist formed at the back focal plane of the condenser lens.

19.2.2 Magnetic Tweezers (MT)

A magnetic dipole $\boldsymbol{m}$ in a magnetic field $\boldsymbol{B}$ experiences a magnetic force $\boldsymbol{f}$ (Fig. 19.2c) and torque $\boldsymbol{\tau}$ (Fig. 19.2d): the force acting on the dipole is proportional to the field gradient: $\boldsymbol{f} = (\boldsymbol{m} \cdot \nabla)\boldsymbol{B}$ (Shevkoplyas et al. 2007) and points along the direction of the gradient; the torque scales with the field itself: $\boldsymbol{\tau} = \boldsymbol{B} \times \boldsymbol{m}$ (Mosconi 2011) and points in the direction of the field. MT uses this principle to trap and rotate magnetic particles: a $\boldsymbol{B}$ field is generated either with permanent magnets (Fig. 19.2a) or electromagnets (Fig. 19.2b) and a microscopic magnetic object (called the 'handle') is placed inside the field. Thus, all translational and rotational degrees of freedom of the handle can be controlled. The biological sample

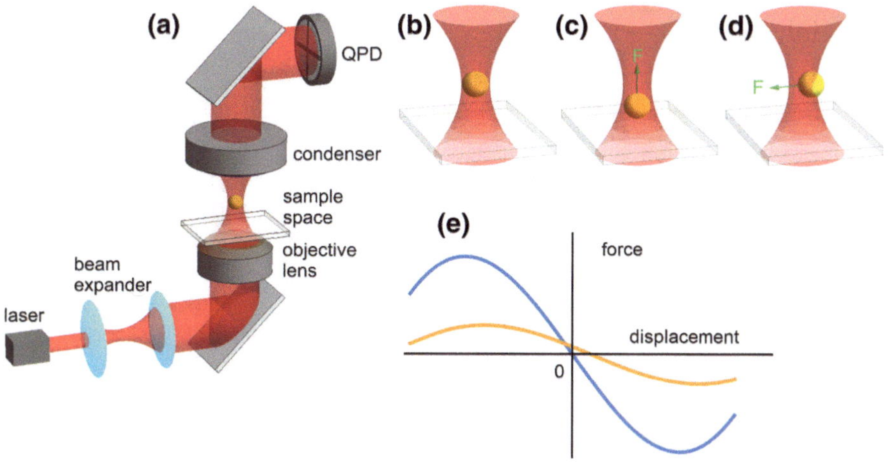

Fig. 19.1 The force transduction mechanism of optical tweezers (OT). **a** An overview of a basic OT. The laser emits a collimated infrared beam that is expanded with a telescope beam expander to overfill the objective lens. The objective lens focuses the laser beam to a diffraction-limited spot at the beam waist (the narrowest place of the beam). The trapped bead scatters the beam and creates an interference pattern at the back focal plane of the condenser, which is imaged via a positive lens (not shown) at a quadrant photo diode to determine the bead position with sub-nanometre resolution. **b** Shows the bead in equilibrium position—along the beam axis and slightly past the point of highest intensity. Apart from trapping force, there is also a scattering force due to the absorption of some photons and thus the absorption of their forward moving momenta, so the bead is actually slightly past the beam waist. In the main text we have ignored this effect for simplicity. When the bead is displaced downward as shown in (**c**) and sideways as shown in (**d**), there appears a restoring force pushing the bead towards the equilibrium position, represented by the *green arrow* and the *letter F*. **e** Plots force-displacement curves for axial (*orange*) and radial (*blue*) bead displacements from the centre of the trap. The relationship is approximately linear in the vicinity of the equilibrium position, allowing calculation of trapping force from measured displacement

is usually covalently attached to the handle and any translation and rotation of the bead are passed on to the sample. The magnetic force can easily reach biologically relevant values of tens of pN: titin unfolds at applied force of 20–30 pN (Kellermayer et al. 1997) and double-stranded DNA overstretches at 65 pN (Smith et al. 1996). The minimum of applicable force is just as important since it determines whether the smallest biological force can be measured. In the case of MT, this value can be as low as thermal forces. The ease with which MT applies torque and twists lends MT its popularity. Compared to optical tweezers, the creation of a stable and low-noise rotating magnetic field is economical and technically straightforward. Biological values of torque are in the range of a few to a few tens of pN nm: the DNA double strands separate when a 9 pN nm torque and >0.5 pN tension are applied against the double helix (Strick et al. 1999) and the ATP synthase F_1 motor generates a torque $\sim$40 pN nm MT.

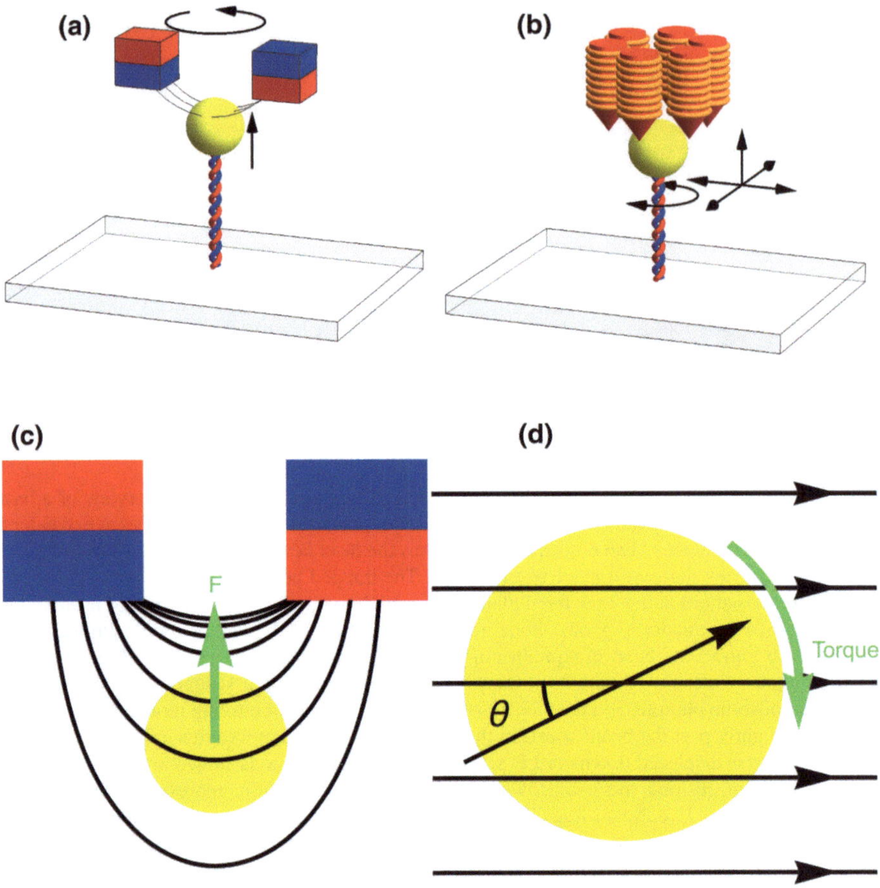

Fig. 19.2 The force and torque transduction mechanism of magnetic tweezers (MT). The B field is created either by permanent magnets (**a**) or electromagnets (**b**). **a** Shows an experimental assay where a double-stranded DNA is tethered to the cover slip at the *bottom* and a bead at the *top*. The bead can be pulled and twisted by moving the two magnets: moving the magnets closer to the bead increases the field gradient at the bead so increases the magnetic force; rotating the magnets will rotate the bead. **b** Is the electromagnet equivalent that features six coils (Gosse and Croquette 2002) and has the added capability of applying horizontal forces compared to the configuration in (**a**). A magnetic bead in this ***B*** field is subject to a magnetic force proportional to the ***B*** field gradient as shown in (**c**). The force is indicated by the *green arrow*. Like OT in which a bead has the tendency of moving towards laser intensity maximum, the magnetic bead tends to move towards the ***B*** field maximum. **d** Illustrates the mechanism that gives rise to magnetic torques: misalignment of the magnetisation of the magnetic bead from the background ***B*** field, shown as the angle θ. The torque is indicated by the *curved green arrow*

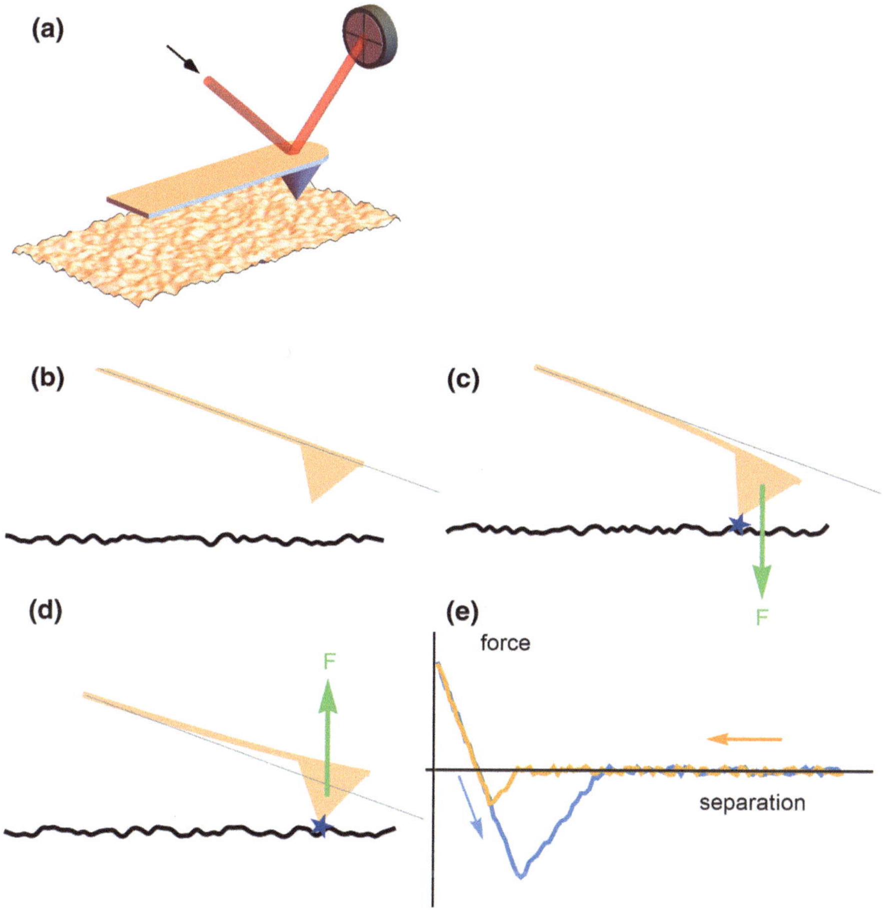

Fig. 19.3 Atomic force microscope (AFM) and its imaging mechanisms. **a** Schematics of AFM imaging a surface profile. The cantilever has a stylus attached to one end. The stylus is brought into physical contact with the sample surface. When the cantilever scans the surface, the local height changes mechanically move the stylus vertically thus deflecting the cantilever. A laser beam reflects off the upper face of the cantilever and a split-photodiode (SPD) catches the reflected light, which allows the measurement of the deflection. **b–d** Show the force levels that the surface transduces to the cantilever and the bending of the cantilever. The green arrows represent forces. In **b** the separation between the cantilever and the substrate is large and there is no interaction. In **c** the stylus is close enough to the substrate so that a Van der Waals attraction bends the stylus downward. **d** As the cantilever further approaches the substrate, a Coulomb repulsion develops that overcomes the Van der Waals attraction and the cantilever feels a push upward. **e** Shows an idealised representation of the force changes as the stylus approaches and leaves the substrate, as indicated by the *orange* and *blue curves* respectively. The dents in the *curves* are due to Van der Waals attraction

19.2.3 Magnetic Tweezers (MT)

AFM is composed of a microscopic cantilever with a stylus attached to one end, which reaches the surface of a biological sample (Fig. 19.3a). The stylus interacts with the sample through the combination of Van der Waals attraction and Coulomb repulsion (Fig. 19.3b–e), resulting in deflection of the cantilever (Binnig et al. 1986). A laser beam is directed to the cantilever, reflects off the top surface and is imaged on to a split-photodiode that reads out the beam deflection. This way angstrom resolution is readily achievable. AFM has two distinct applications: imaging and force measurement. In imaging mode, the cantilever performs a 2D raster scan over the sample surface and reads out the height profile. The high resolution stems from the fact that the imaging is not light based but rather physical contact mediated, so far-field diffraction no longer applies. AFM does have the limit of low temporal resolution as the scanning tip has to physically reach each 'pixel' on the sample surface. In force measurement mode, the AFM tip no longer scans the sample surface but only moves vertically via a piezo electric actuator. The stylus is tethered to the biological molecule at one end and the chamber floor attaches to the other end. The cantilever can then be pulled to reveal a force-displacement curve, which reveals the conformational change of the biological sample.

19.3 Application of Force Spectroscopy Tools in Studying Infection

In studying infections, force spectroscopy has been applied to a wide range of biological subjects: pathogenic proteins, nucleic acids and pathogenic/immune cell behaviours. Below we summarise a number of selected case studies to illustrate how the design of force spectroscopy tools allow them to achieve high resolution measurements of biological samples in physiological settings and how the results help progress our understanding in infection.

19.3.1 Resolving Pathogenic Protein Actions

Nucleic acid proteins are vital to the health and functioning of all organisms. For instance, nucleic acids activities are facilitated with a large number of proteins or protein complexes. During DNA replication, topoisomerases (Champoux 2001) and gyrases (Gore et al. 2006) introduce both positive and negative twists into DNA to relieve extra turns built up from helicase unwinding and DNA local rotational constrains. Helicases (Patel and Donmez 2006) unwind and separate DNA double strands so DNA polymerases have access to the single strands for adding new nucleotides to form daughter double strands. During transcription, RNA polymerases create complementary RNA strands from single DNA strands (Murakami and Darst 2003). The list goes on. All of these activities and the mechanical properties of nucleic acids themselves are subjects of study for molecular tools.

RNA polymerases (RNAP) in bacteria are essential for nucleic acid transcription and can be inhibited resulting in bacterial death. In nature, bacteriophages have targeted the sigma factor in RNAP to modulate RNAP behaviours for the phages' benefit: *Xanthomonas oryzae* bacteriophage Xp10 encodes transcription regulator P7, an anti-sigma factor that displaces sigma factor during engagement of the RNAP with the promoter DNA. Sigma factor is an RNAP subunit that allows promoters to locate starting sites on DNA for transcription, and phages modulate sigma so RNAPs transcribe their DNA rather than bacterial DNA (Liu et al. 2014). Understanding the mechanisms of RNAP actions inspires the development of antibacterial drugs. With optical tweezers, *Escherichia coli* RNA polymerase has been measured to move along DNA in discreet steps of 3.7 ± 0.6Å (Fig. 19.4c, d) (Abbondanzieri et al. 2005), about the size of one DNA base pair, so they are shown to move one bp at a time. The sub-nm measurement resolution requires supreme beam stability from the optical tweezers used to make the measurement.

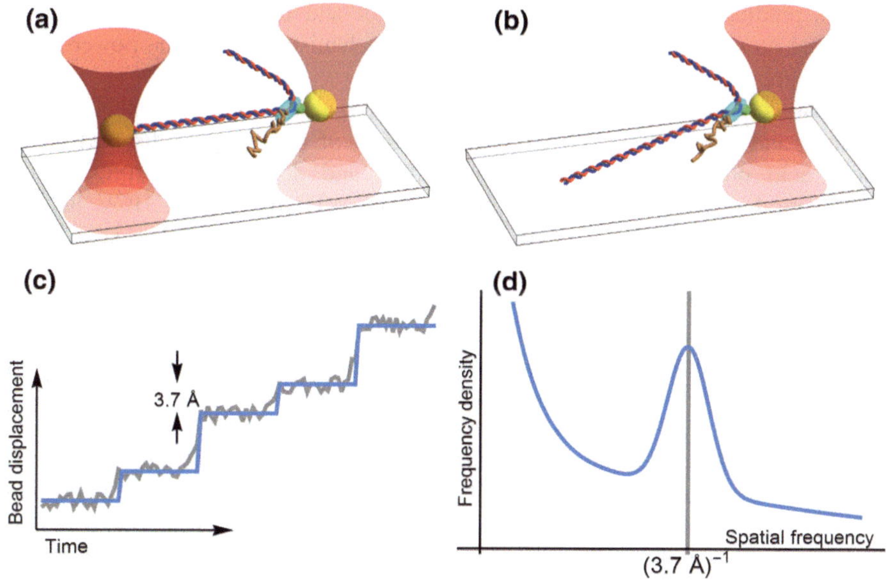

Fig. 19.4 Optical trap set-up to study RNAP activity and schematics of results. **a** Two laser traps are created at the sample space to trap two beads simultaneously. The advantage of this configuration over an alternative single trap design in which one end of the DNA is fixed to the surface of the coverslip as shown in (**b**) is the avoidance of any mechanical noise or drift from the coverslip from transducing to the biological sample itself thus better resolution can be achieved. **c** A schematic showing the displacement over time measurement that illustrates the identification of 3.7 Å step size. **d** A spatial frequency plot of the data in (**c**) identifying a peak at $(3.7\ \text{Å})^{-1}$

The DNA being transcribed is tethered to a bead on one end. The RNAP is tethered to another bead while it moves along the DNA. The two beads are trapped separately with two optical traps—the 'dumbbell' configuration (Fig. 19.4a), which removed the need of tethering one end of the DNA to the reaction chamber (an alternative set-up, Fig. 19.4b), decoupling the system from the microscope stage thus eliminating stage drift. Also, as the two traps arise from the same laser source by splitting the beam into two beams of orthogonal polarisation so laser pointing fluctuations do not result in relative shift in the two traps. Another measure to minimise the noise level is by sheltering the instrumentations in a helium-filled box to reduce interaction between air density fluctuations with the laser beam. This reduces low frequency noises to 1/100. The single molecule OT provides the unique capability to controllably apply a force to the RNAP to either assist its advancement along the DNA or to hinder it. This allowed the authors to plot RNAP movement velocity as a function of applied force, at a few different NTP concentrations. These relationships let the researchers deduce the kinetic model for RNA translocation. Existing competing models include Brownian ratchet mechanism (Schafer et al. 1991) in which the RNAP is free to diffuse both forward and backward along the DNA but incoming NTP acts as a ratchet to favour the forward motion and power stroke mechanism (Jiang and Sheetz 1994) in which the incorporation of NTP releases pyrophosphate that provides energy to RNAP translocation. The two models predict distinguished velocity-force plot and the assay verifies that the measured plot matches the Brownian ratchet mechanism and not the power stroke mechanism.

Hepatitis C virus (HCV) helicase as a drug target has captured much attention (Steimer and Klostermeier 2012) as deactivation of this enzyme disables all the nuclei acid processes that require unwinding and double-strand splitting thus halting viral replication. With the energy from ATP hydrolysis, HCV helicase moves along RNA or DNA to remove proteins bound to the nucleic acid and to unzip any complementary strands. A detailed understanding of the mechanism of its operation will facilitate the development of effective HCV helicase inhibitors. Despite much research attention, the mechanism of HCV helicase translocation is not fully understood due to the lack of spatial and temporal resolution in experimental measurements until single molecular measurements. X-ray crystallography reveals detailed HCV helicase structural information but it does not provide any information on the mechanistic cycles in which HCV helicase movement and ATP hydrolysis are coupled. Using optical tweezers, the series of actions have been resolved (Dumont 2006). An RNA hairpin was attached to two beads on both ends each via a DNA linkage (Fig. 19.5a). One bead is tethered to an optical trap to allow measurement of force and the other bead to a micropipette for mechanical sub-nanometre position control. With active position feedback, the instrument can adjust the separation between the two beads to maintain either constant-force levels or changing force levels. Thus, the separation of duplex RNA can be monitored at various pulling forces (Fig. 19.5b), with the presence or absence of HCV helicase and different ATP concentrations. This assay allowed the measurement of movement in steps of multiples of 11 ± 3 bp and unwinding steps of multiples of $3.6 \pm$

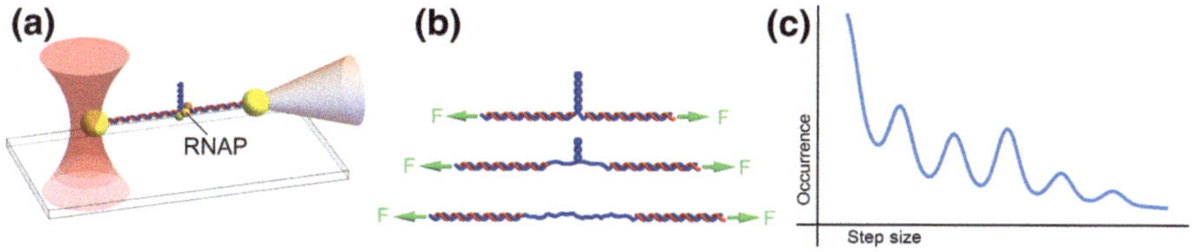

Fig. 19.5 OT assay to probe the mechanical cycles of HCV NS3 helicase. **a** the OT configuration featuring an RNA hairpin (*blue*) tethered on both ends to beads via ssDNA handles (*red*). One end is trapped in OT with active feedback allowing constant-force mode. The other end is fixed onto a pipette tip with sub-nanometre mechanical positioning. Remarkably, the trap is created with two counter propagating beams so the forward-going scattering forces due to the two beams are cancelled, resulting in stiffer trapping along the axial direction. This way lower numerical aperture objective lenses can be used and all the light leaving the trap can be measured and analysed to determine the rate of momentum change (Smith et al. 2003). **b** From *top* to *bottom*: the RNA hairpin denatures at catalysis of NS3 Helicase. The application of constant weak forces on both ends of the structure pulls the structure taut. Time traces of bead displacements such as the one shown in Fig. 19.4c can be determined from this assay, which allows the characterisation of step sizes shown in (**c**). **c** Discrete step sizes of various magnitudes are shown through the peaks in the step size versus occurrence plot

1.3 bp (Fig. 19.5c). The authors could conclude from their measurements the action model in which the translocator site on the helicase contacts the DNA 11 bp ahead of the helix opener site. The helicase advances in an inchworm fashion while the trailing helix opener separates the double strands.

As part of virus replication, DNA packaging motors push DNA into pre-assembled virus capsid against the tremendous internal forces due to the near-crystalline DNA density. Using optical tweezers, Smith et al. could measure the speed, pauses, slips and forces of packaging and establish the speed dependence on loadings (Liphardt et al. 2001). In their set-up, a bacteriophage $\varphi 29$ portal motor packages double-stranded DNA into the phage capsid (Fig. 19.6a). The capsid is tethered onto a bead whose position is fixed with a micropipette. The loose end of the DNA attaches to another bead trapped in the OT. The measurements are carried out in two modes: (1) constant-force feedback mode where the tension in the DNA is kept at 5 pN by adjusting the bead separation (Fig. 19.6b, c) and (2) zero feedback mode where the bead separation stays stationary and tension in DNA increases as the motor protein packages DNA (Fig. 19.6d). In constant feedback mode, the packaging rate initially stayed constant but decreased quickly as the capsid filled up, indicating that the packaging rate is sensitive to internal pressure. 5 μm long DNA strand packaging have been observed indicating high processivity. In zero feedback mode, the force builds up to about 55 pN before the motor stalls, the maximum motor force output. This also indicates the level of internal force due to DNA packaging. Small parts of packaged DNA also occasionally slips out of the capsid, shown as abrupt drops in force that lasts about 0.014 s; this also provides evidence that there is little resistance to DNA movement and thus little energy dissipation. Finally for DNA ejection, the high internal force at dense packing supports passive ejection for early stages but motor assisted ejection for latter stages.

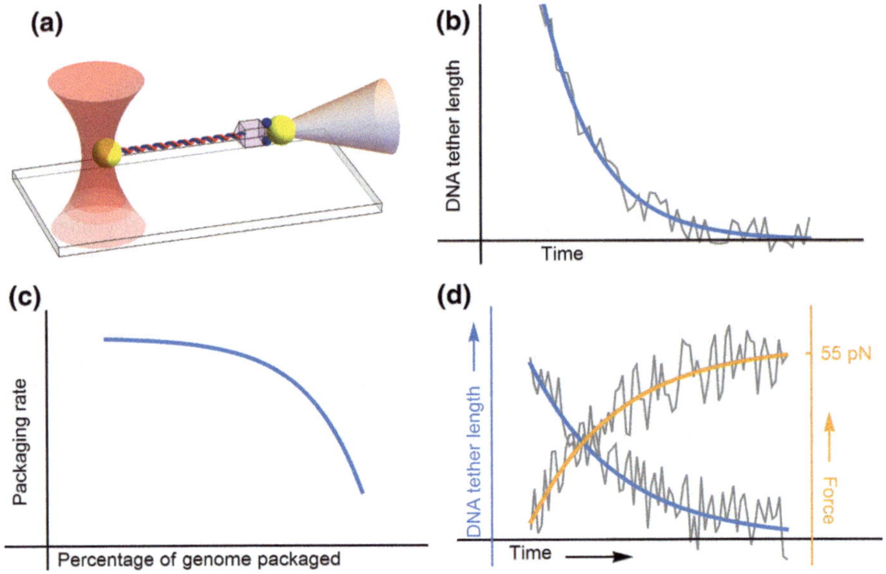

Fig. 19.6 OT assay to probe motor proteins that package DNA into virus capsid. **a** Schematics of the assay: the capsid of the virus is fixed to a pipette tip while the other end of the DNA is trapped in the OT. **b** The constant-force mode. DNA outside the capsid is shortened over time as motor proteins package the genome into the capsid. **c** Derives from data in (**b**) and shows the drop of packaging rate as the packaging progresses. **d** The assay operates in no-feedback mode so the separation between the trapping beam and the pipette tip stays stationary. The tension in DNA builds up as the motor protein pulls the DNA and packages it into the capsid. The protein halts at approximately 55 pN, suggesting the maximum pulling strength of this protein. This also hints on the internal pressure that can be built up via packaging. The authors argued that this pressure may be used for DNA ejection inside host bodies

19.3.2 Resolving Inhibited Protein Actions

OT has also been applied to the characterisation of the actions of inhibited RNAP. Antibacterial peptide Microcin J25 (MccJ25) is produced by strains of *Escherichia coli* that bind to the secondary channel of RNAP such that nucleotide substrates cannot enter the enzyme active site. Adelman et al. first used biochemical assays and phosphor imaging (an ensemble biochemical method that quantifies and localises radioactively labelled nucleic acids in sample preparations such as gels and blots) to study the effect of MccJ25 on RNAP transcription and concluded that MccJ25 significantly inhibits transcription elongation even at saturating NTP concentrations and increases the appearance of transcriptional pauses. However, their assays could not quantify the reduction of elongation rate, nor could they distinguish whether MccJ25 caused additional pauses or extended the duration of existing pauses. Also, the complexity of the abortive initiation process made it difficult to extract mechanical details of the inhibition. The authors therefore resorted to optical tweezers that probe a single elongation complex with MccJ25 binding at a time to reveal the aforementioned parameters. In their OT set-up, one end of a DNA substrate is tethered to a trapped micro-bead via streptavidin-biotin link. The transcription site on the DNA binds to an RNAP which in turn is tethered to the chamber surface via (anti-HA)-(HA-epitope) link (Fig. 19.7a). The OT pulls taught the DNA section between the bead and the RNAP so that any advancement of RNAP along the DNA manifests as the change in DNA length between the bead and RNAP. With feedback control, the tension in DNA is kept constant. Thus RNAP movement can be measured with interferometry bead displacement tracking in real time. Apart from consistent conclusions with biochemical studies over the decrease of RNAP processivity with increasing concentration of MccJ25, the optical tweezers assay also reveals through the elongation-time plot (Fig. 19.7b) of individual transcription events that MccJ25 inhibition leads to increases in both frequency and duration of transcriptional pauses but the rate of 12.5 nt/s was not affected (Fig. 19.7c). This suggests that MccJ25 stops rather than slows transcription: when it blocks secondary channels, it blocks completely, since partially blocked channels will manifest as slower transcription rates. Also, since the timing of pauses is determined as well, the authors could find out that the stopping events are randomly distributed (Adelman et al. 2004). These findings are not accessible via ensemble detection methods.

The inhibition of DNA repairing protein RecA in *Mycobacterium tuberculosis* has been studied with magnetic tweezers. When double-stranded DNA is broken, one of the repair mechanisms, homologous recombination, pairs one of the damaged strands with the corresponding section on the complementary strand of a similar or identical DNA ('the recipient DNA'), such as a sister chromatin, in order

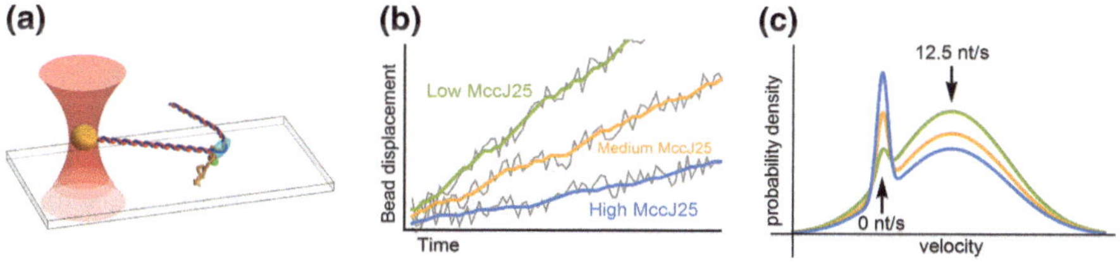

Fig. 19.7 OT assay to probe the inhibition of antibacterial peptide Microcin J25 (MccJ25) on the activity of bacterial RNA polymerase. **a** The assay set-up differs from that in Fig. 19.4a in that only one trap is used here. The RNAP end of the DNA is tethered to the camber surface, subject to the transmission of mechanical noise to the biological sample. This configuration is simpler to implement while providing enough resolution over the time scale of ~100 s to characterise the transcription rate. **b** Shows time traces of RNAP advancements along DNA at three concentrations of the inhibitor MccJ25. Individual moving and pausing events can be identified as tilting and horizontal parts in the *curve* to allowing the authors to plot transcription rate distribution in (**c**): it turns out that MccJ25 does not alter transcription rate but rather increases the frequency of pauses

for the synthesis of a new strand to replace the damaged strand. In this process, RecA polymerises to form filament that binds to the single strand to assist the process. The regulatory protein RecX inhibits the polymerisation and promotes the depolymerisation of RecA filaments. Meanwhile, mechanical force arises from the stretching of the RecA-ssDNA complex as the complex translocates to reach the recipient DNA. This force reduces the inhibition of RecA filament due to RecX. Using MT, Le et al. (2014) characterised the responses of the inhibitory effects to different force levels and various concentrations of RecX. A DNA is tethered vertically between a magnetic bead and the coverslip surface. Two magnets with opposite poles are held above the magnetic bead to provide the upward pulling force (Figs. 19.2a, 19.8a). With video imaging and defocusing diffraction pattern analysis, the vertical position of the bead can be determined with nanometre resolution. Force-extension curves are plotted for naked ssDNA and full-length RecA-bound ssDNA (Fig. 19.8b). The latter shows much higher rigidity: naked DNA is initially coiled and extends rapidly as tension increases but RecA-bound DNA is extended initially and does not stretch much further as force increases. This establishes a method to determine the extent of RecA filament polymerisation and depolymerisation: it varies the vertical bead position. At low forces, RecX inhibits RecA filament formation and encourages depolymerisation (Fig. 19.8c, d) but this effect is reduced at higher applied tension in the filament to the extent that partially disassembled filament can re-grow (Fig. 19.8e). This suggests a mechanosensitive regulation mechanism: during homologous recombination, the tension in DNA-RecA complex increases. To keep both ends of the broken DNA together, the DNA-RecA complex stabilises as a response to the tension.

19.3.3 Investigating Immune Cells

Ashkin et al. showed OT capability of trapping bacteria, viruses (Ashkin and Dziedzic 1987) and human cells (Ashkin and Dziedzic 1987) as early as 1987. In contrast, due to the lack of ferro-/para-magnetic properties, biological samples cannot be directly manipulated by MT, but a magnetic microparticle can be inserted into a cell and the magnetic translation of this particle perturbs the structure and dynamics of the cell to reveal cell properties—magnetic microrheology. Alternatively, a trapped magnetic particle can be brought into contact of cells to probe adhesion properties.

Macrophages are immune cells that engulf and/or destroy any objects in the body that are not recognised as healthy body cells for not having the right type of surface signalling proteins, potentially because they are invaders, pathogens in particular. The understanding of these cells is crucial to understanding the immune system. The viscoelasticity of the macrophage cytoplasm has direct implications on cell motion and deformation, intracellular transportation and phagocytosis, all of which have direct impacts on the macrophages' immune response. Further,

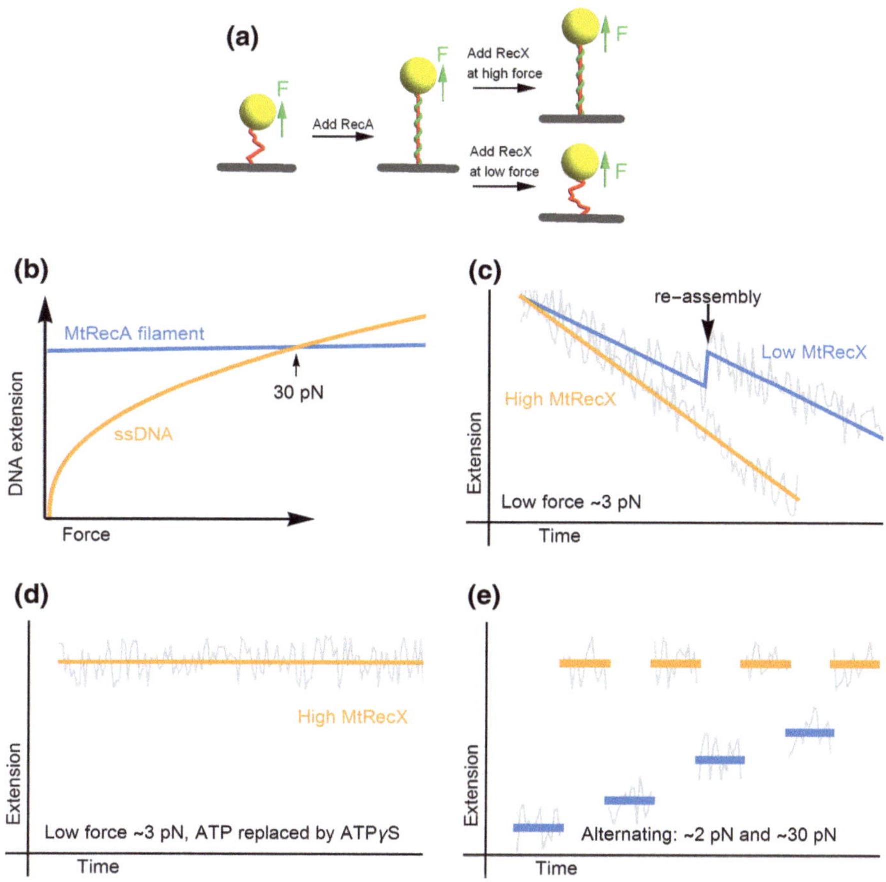

Fig. 19.8 Magnetic tweezers assay to probe RecX inhibition of RecA filament formation and decomposition. **a** A graphical representation of the main findings. The ssDNA (*red curve*) curls up at low force (*left*) but addition of RecA forms a rigid filament along the DNA, straightening the DNA thus pushing the bead away from the chamber surface (*middle*). Further addition of RecX depolymerises the RecA filament at low forces (*right*, *bottom*) but the inhibition effect is less prominent at high forces (*right*, *top*). **b** DNA extension versus force curves with and without RecA to show the difference in the rigidity of the two scenarios. This establishes a method to measure the amount of filament formation. **c** Time traces of extension showing RecA formation at low and high RecX respectively. **d** When ATP is replaced by non-hydrolysable homologue, ATPγS, no RecA filament decomposition is detected, pointing to ATP dependence of RecX inhibition. **e** Alternating low and high forces showing the roles that forces play on the effectiveness of RecX inhibition on RecA filament formation

viscoelastic parameters allow quantitative characterization of the structure of the cytoplasm. Cytoplasmic viscoelastic moduli are determined from viscoelastic creep curve: a constant local stress applied on the material for a period of time deforms the sample (the deformation known as strain) and the stress-strain versus time is plotted. The curve consists of three qualitatively distinctive parts: a fast elastic response followed by a slower response to the applied stress and finally relaxation when the stress is removed. From the curve, viscosity, relaxation time and shear constant can be calculated. Magnetic tweezers are used to apply the stress: micron-sized magnetic beads are phagocytized by macrophages. The magnetic force to probe the cytoplasm needs to reach nanonewtons, higher than usual MT designs are capable of. But this can be done with electromagnetic tweezers by having many coil turns, large electric current and high magnetic core permeability whilst bringing the core close to the sample. Bausch et al. (1998) devised a field generation unit composing of a single coil with 1200 turns wound round a soft iron core that supports up to 4 A current. The step force is in one direction only, so one coil suffices. This allows the tip of the core piece to reach as close as 10 μm to the sample. With CCD camera video microscopy and single particle tracking algorithm, the spatial resolution reached 10 nm and temporal resolution 0.04 s (Bausch et al. 1999). The authors measured a shear modulus ranging from 20 to 735 Pa, viscosity 210 ± 143 Pa s. Also non-magnetic beads are tracked to characterise the displacement field and the authors concluded that the cytoplasm is composed of densely packed filaments with soft gaps. They also could measure local active forces within the cytoplasm potentially due to motor proteins or local flow of cytoplasm. Prior methods to measure forces have resorted to averaging during long periods of time and have been hampered by the high level of heterogeneity in the cytoplasm of macrophages.

19.3.4 Characterising Virus Attachment to Host Cells

Plasmodium, commonly known as malaria parasites, is a genus of unicellular eukaryotic organisms that have more than 100 species, many of which can infect humans and animals to cause malaria. The parasites found in mosquito saliva are at a reproductive stage when they are called *sporozoites* and are capable of infecting new hosts. Following a mosquito bite, the sporozoites are deposited under the skin of the recipient and they migrate to enter the host's circulation system and finally the liver where the parasites move on to the next stage of life cycle. Cellular mobility is essential to plasmodium reproduction and infection of their hosts. Sporozoites have the shape of a crescent (Fig. 19.9). The cell surface features adhesion-mediating proteins for sites that adhere to cellular substrates in physiological environments but also to carbon and gold surfaces in experimental settings. When sporozoites navigate along a surface, new adhesion sites form in the leading edge and existing sites disassemble at the trailing edge. The complex interplay between surface proteins and signalling events involved in this gliding motility is made accessible via optical tweezers manipulation of the parasites. Hegge et al. (2012) devised an OT assay capable of trapping and manipulating a part of the crescent parasite and characterising the forces that the laser trap applies to the cell. To avoid blocking adhesion sites on the surface of the sporozoites, the sporozoites are trapped directly rather than via an attached bead. The major challenge lies in the

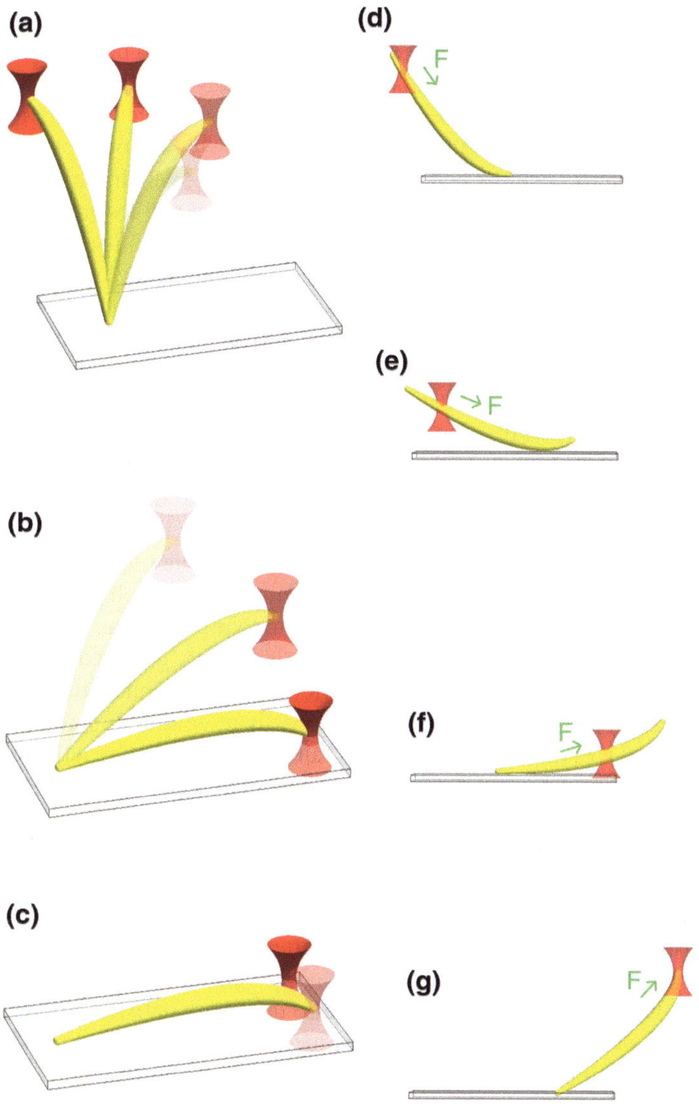

Fig. 19.9 Optical tweezers (OT) manipulate a malaria parasite. A few modes of rotating and moving the parasite are shown to illustrate the capability of OT in interacting with the parasite. **a** One end of the parasite adheres to the coverslip via its surface adhesion protein. The rest of the parasite is free from the glass so the parasite is not rotationally constrained. The other end of the parasite is trapped in the OT focus. The relative position of the trap with respect to the coverslip rotates around the adhesion point and parasite rotates with the trap. **b** The trap position relative to the coverslip is closed down to bring the free end of the parasite to the coverslip. **c** Once the free end touches the coverslip, adhesion sites may form and if the 100 mW trap is moved away now, the parasite will not follow the trap, indicating strong adhesion. **d–g** When only one adhesion site is formed between the parasite and the coverslip and when the trap moves along the long axis, it is possible to push the parasite forward. The *green arrows* indicate the direction of pushing and pulling. During the movement, the adhesion site migrates along the parasite surface

non-spherical shape of the cells, as well as the heat damage to the cells. The long axis of the sporozoites exceeds the extent of the trap focus so only a part of the cell is trapped and manipulated. The cells are soft and thus bend under the force so trap movements of a few microns do not necessarily move the cells. This means standard thermal noise characterisation of the trap stiffness does not apply. The authors use flow pressure to push the cell out of the trap during which the escape velocity and thus trap stiffness can be determined. The attaching and locomotion behaviour of parasites that are genetically modified to lack specific proteins are manipulated to uncover the distinct roles that each protein plays as well as the interaction between species of proteins. For example, it was established that both the surface proteins TRAP and S6 are required for correct initial adhesion, that S6 determines the strength of adhesion for resistance against externally applied axial force (Fig. 19.9d–g) and that TRAP determines resistance against rotation (Fig. 19.9c).

Mammen et al. devised the 'OPTCOL' (optically controlled collision) assay to measure the probability of adhesion of viruses to host cells upon collision. OPTCOL is a pair of independently controlled traps (the 'dumbbell' configuration) in which one trap manipulates the host cell and the other manipulates a 5 μm bead whose surface is coated with viruses (Mammen 1996). The traps first move towards each other to bring the viruses into contact with the host cell and then the traps move away so adherence/separation between the two bodies can be determined by observing if the two bodies attach or separate. Statistics gathered are used to calculate the probability of attachment at various inhibitor concentrations. Previous ensemble methods suspend viruses and host cells at certain concentrations in the solution. Due to size differences, host cells fall to the bottom of reaction chamber by gravity while viruses do not. This creates collision events between the two bodies. Attachments between multiple hosts and viruses eventually form a macroscopic aggregate. The limitation of this method is that inhibitor concentration has a minimum measurable value below which different types of inhibitors appear to be equally effective. The significance of OT compared to ensemble methods, apart from measuring inhibitor effectiveness at extremely low concentrations, includes the ability to vary collision velocity which mimics different situations inside human bodies and which has potent influence over attachment statistics, the capability to extend agents of interest to a wide range of mesoscopic objects and provide detailed adhesion distributions over a collection of individual collisions.

19.4 Conclusion

Introduced in the 1980s, force spectroscopy is still in its youth. In recent years, advancements in components such as lasers and CMOS camera that make up force spectroscopy instruments have allowed the increase of resolution in force spectroscopy instruments. In addition, combinations of spectroscopy tools with each other allow unique manipulation and measurement capabilities so a wider range of biological systems can be probed. For example, magneto-optical tweezers

(Crut 2007; Zhou et al. 2015) are a chimera of OT and MT that can manipulate particles in all translational and rotational degrees of freedom without the technical challenges in implementing rotation-enabled OT (La Porta and Wang 2004) as MT easily applies twists or in implementing high resolution MT (Dulin et al. 2015) as OT intrinsically has high resolution. Combination of force spectroscopy such as super-resolution microscopy adds direct optical visualisation to molecular activities (Miller et al. 2015). Another area that will likely see significant development is device multiplicity in the sense that multiple copies of biological samples are simultaneously experimented on high throughput devices. Due to heterogeneity among biological samples, force spectroscopy investigations need to be statistical in nature. Holographic OT (Padgett and Di Leonardo 2011) and DNA curtains (Greene et al. 2010) are some examples of configuration and device designs to allow multiplicity. Many biological systems relevant to infection remain to be investigated and, with the knowledge already obtained with ensemble methods, experimental results from force spectroscopy will prove to be useful.

References

Abbondanzieri EA, Greenleaf WJ, Shaevitz JW, Landick R, Block SM (2005) Direct observation of base-pair stepping by RNA polymerase. Nature 438:460–465

Adelman K, Yuzenkova J, La Porta A, Zenkin N, Lee J, Lis JT, Borukhov S, Wang MD, Severinov K (2004) Molecular mechanism of transcription inhibition by peptide antibiotic Microcin J25. Mol Cell 14(6):753–762. doi:10.1016/j.molcel.2004.05.017

Ashkin A (1992) Forces of a single-beam gradient laser trap on a dielectric sphere in the ray optics regime. Biophys J 61(2):569–582. doi:10.1016/S0006-3495(92)81860-X

Ashkin A, Dziedzic JM (1987) Optical trapping and manipulation of viruses and bacteria. Science 235(4795):1517–1520

Bausch AR, Moller W, Sackmann E (1999) Measurement of local viscoelasticity and forces in living cells by magnetic tweezers. Biophys J 76:573–579

Bausch AR, Ziemann F, Boulbitch AA, Jacobson K, Sackmann E (1998) Local measurements of viscoelastic parameters of adherent cell surfaces by magnetic bead microrheometry. Biophys J 75:2038–2049

Binnig G, Quate CF, Gerber C (1986) Atomic force microscope. Phys Rev Lett 56 (9):930–933. doi:10.1103/PhysRevLett.56.930

Block SM, Blair DF, Berg HC (1989) Compliance of bacterial flagella measured with optical tweezers. Nature 338(6215):514–518. doi:10.1038/338514a0

Block SM, Goldstein LS, Schnapp BJ (1990) Bead movement by single kinesin molecules studied with optical tweezers. Nature 348:348–352

Bullard B, Garcia T, Benes V, Leake MC, Linke WA, Oberhauser AF (2006) The molecular elasticity of the insect flight muscle proteins projectin and kettin. Proc Natl Acad Sci USA 103 (12):4451–4456

Champoux JJ (2001) DNA topoisomerases: structure, function, and mechanism. Annu Rev Biochem 70:369–413. doi:10.1146/annurev.biochem.70.1.369

Charvin G, Strick TR, Bensimon D, Croquette V (2005) Tracking topoisomerase activity at the single-molecule level. Annu Rev Biophys Biomol Struct 34:201–219

Crut A, Koster DA, Seidel R, Wiggins CH, Dekker NH (2007) Fast dynamics of supercoiled DNA revealed by single-molecule experiments. Proc Natl Acad Sci USA 104(29):11957–11962. doi:10.1073/pnas.0700333104

De Vlaminck ID C (2012) Recent advances in magnetic tweezers. Annu Rev Biophys 41:453–472. doi:10.1146/annurev-biophys-122311-100544

Dulin D, Cui TJ, Cnossen J, Docter MW, Lipfert J, Dekker NH (2015) High spatiotemporal-resolution magnetic tweezers: calibration and applications for dna dynamics. Biophys J 109(10):2113–2125. doi:10.1016/j.bpj.2015.10.018

Dumont S (2006) RNA translocation and unwinding mechanism of HCV NS3 helicase and its coordination by ATP. Nature 439:105–108

Fili N, Mashanov GI, Toseland CP, Batters C, Wallace MI, Yeeles JTP, Dillingham MS, Webb MR, Molloy JE (2010) Visualizing helicases unwinding DNA at the single molecule level. Nucleic Acids Res 38(13):4448–4457. doi:10.1093/nar/gkq173

Finer JT, Simmons RM, Spudich JA (1994) Single myosin molecule mechanics: piconewton forces and nanometre steps. Nature 368(6467):113–119. doi:10.1038/368113a0

Gennerich A, Carter AP, Reck-Peterson SL, Vale RD (2007) Force-induced bidirectional stepping of cytoplasmic dynein. Cell 131(5):952–965. doi:10.1016/j.cell.2007.10.016

Gore J, Bryant Z, Stone MD, Nöllmann M, Cozzarelli NR, Bustamante C (2006) Mechanochemical analysis of DNA gyrase using rotor bead tracking. Nature 439(7072):100–104

Gosse C, Croquette V (2002) Magnetic tweezers: micromanipulation and force measurement at the molecular level. Biophys J 82:3314–3329

Greene EC, Wind S, Fazio T, Gorman J, Visnapuu ML (2010) DNA curtains for high-throughput single-molecule optical imaging. Methods Enzymol 472:293–315. doi:10.1016/S0076-6879(10)72006-1

Hegge S, Uhrig K, Streichfuss M, Kynast-Wolf G, Matuschewski K, Spatz JP, Frischknecht F (2012) Direct manipulation of malaria parasites with optical tweezers reveals distinct functions of Plasmodium surface proteins. ACS Nano 6(6):4648–4662. doi:10.1021/nn203616u

Jiang MY, Sheetz MP (1994) Mechanics of myosin motor: force and step size. BioEssays 16(8):531–532. doi:10.1002/bies.950160803

Kellermayer MS, Smith SB, Granzier HL, Bustamante C (1997) Folding-unfolding transitions in single titin molecules characterized with laser tweezers. Science 276(5315):1112–1116

King GA, Gross P, Bockelmann U, Modesti M, Wuite GJL, Peterman EJG (2013) Revealing the competition between peeled ssDNA, melting bubbles, and S-DNA during DNA overstretching using fluorescence microscopy. Proc Natl Acad Sci USA 110(10):3859–3864. doi:10.1073/pnas.1213676110

La Porta A, Wang MD (2004) Optical torque wrench: angular trapping, rotation, and torque detection of quartz microparticles. Phys Rev Lett 92(19):190801

Le S, Chen H, Zhang X, Chen J, Patil KN, Muniyappa K, Yan J (2014) Mechanical force antagonizes the inhibitory effects of RecX on RecA filament formation in Mycobacterium tuberculosis. Nucleic Acids Res 42(19):11992–11999. doi:10.1093/nar/gku899

Leake MC (2013) The physics of life: one molecule at a time. Philos Trans Royal Soc B: Biol Sci 368(1611):20120248

Leake MC, Grützner A, Krüger M, Linke WA (2006) Mechanical properties of cardiac titin's N2B-region by single-molecule atomic force spectroscopy. J Struct Biol 155(2):263–272

Leake MC, Wilson D, Bullard B, Simmons RM (2003) The elasticity of single kettin molecules using a two-bead laser-tweezers assay. Febs Lett 535 (1–3):55–60. doi:10.1016/S0014-5793(02)03857-7

Leake MC, Wilson D, Gautel M, Simmons RM (2004) The elasticity of single titin molecules using a two-bead optical tweezers assay. Biophys J 87(2):1112–1135. doi:10.1529/biophysj.103.033571

Lidstrom ME, Konopka MC (2010) The role of physiological heterogeneity in microbial population behavior. Nat Chem Biol 6(10):705–712. doi:10.1038/nchembio.436

Linke WA, Leake MC (2004) Multiple sources of passive stress relaxation in muscle fibres. Phys Med Biol 49(16):3613

Liphardt J, Onoa B, Smith SB, Tinoco I, Bustamante C (2001) Reversible unfolding of single RNA molecules by mechanical force. Science 292:733–737

Liu B, Shadrin A, Sheppard C, Mekler V, Xu Y, Severinov K, Matthews S, Wigneshweraraj S (2014) A bacteriophage transcription regulator inhibits bacterial transcription initiation by sigma-factor displacement. Nucleic Acids Res 42(7):4294–4305. doi:10.1093/nar/gku080

Mammen M (1996) Optically controlled collisions of biological objects to evaluate potent polyvalent inhibitors of virus-cell adhesion. Chem Biol 3:757–763

Miller H, Zhou Z, Wollman AJ, Leake MC (2015) Superresolution imaging of single DNA molecules using stochastic photoblinking of minor groove and intercalating dyes. Methods. doi:10.1016/j.ymeth.2015.01.010

Moffitt JR, Chemla YR, Smith SB, Bustamante C (2008) Recent advances in optical tweezers. Annu Rev Biochem 77:205–228. doi:10.1146/annurev.biochem.77.043007.090225

Mosconi F, Alleamnd JF, Croquette V (2011) Soft magnetic tweezers: a proof of principle. Rev Sci Instr 82(3):034302. doi:10.1063/1.3531959

Murakami KS, Darst SA (2003) Bacterial RNA polymerases: the whole story. Curr Opin Struct Biol 13(1):31–39

Neuman KC, Nagy A (2008) Single-molecule force spectroscopy: optical tweezers, magnetic tweezers and atomic force microscopy. Nat Methods 5(6):491–505. doi:10.1038/nmeth.1218

Padgett M, Di Leonardo R (2011) Holographic optical tweezers and their relevance to lab on chip devices. Lab Chip 11(7):1196–1205. doi:10.1039/c0l00526f

Patel SS, Donmez I (2006) Mechanisms of helicases. J Biol Chem 281(27):18265–18268. doi:10.1074/jbc.R600008200

Schafer DA, Gelles J, Sheetz MP, Landick R (1991) Transcription by single molecules of RNA polymerase observed by light microscopy. Nature 352(6334):444–448. doi:10.1038/352444a0

Shevkoplyas SS, Siegel AC, Westervelt RM, Prentiss MG, Whitesides GM (2007) The force acting on a superparamagnetic bead due to an applied magnetic field. Lab Chip 7(10):1294–1302. doi:10.1039/b705045c

Smith SB, Cui Y, Bustamante C (1996) Overstretching B-DNA: the elastic response of individual double-stranded and single-stranded DNA molecules. Science 271(5250):795–799

Smith SB, Cui YJ, Bustamante C (2003) Optical-trap force transducer that operates by direct measurement of light momentum. Method Enzymol 361:134–162

Steimer L, Klostermeier D (2012) RNA helicases in infection and disease. RNA Biol 9(6):751–771. doi:10.4161/rna.20090

Strick TR, Allemand JF, Bensimon D, Bensimon A, Croquette V (1996) The elasticity of a single supercoiled DNA molecule. Science 271:1835–1837

Strick TR, Allemand JF, Bensimon D, Croquette V (1998) Behavior of supercoiled DNA. Biophys J 74:2016–2028

Strick TR, Bensimon D, Croquette V (1999) Micro-mechanical measurement of the torsional modulus of DNA. Genetica 106(1–2):57–62

Zhou Z, Miller H, Wollman A, Leake M (2015) Developis. Photonics 2(3):758

Chapter 20
Imaging Immunity in Lymph Nodes: Past, Present and Future

James Butler, Amy Sawtell, Simon Jarrett, Jason Cosgrove, Roger Leigh, Jon Timmis and Mark Coles

Abstract Immune responses occur as a result of stochastic interactions between a plethora of different cell types and molecules that regulate the migration and function of innate and adaptive immune cells to drive protection from pathogen infection. The trafficking of immune cells into peripheral tissues during inflammation and then subsequent migration to draining lymphoid tissues has been quantitated using radiolabelled immune cells over 40 years ago. However, how these processes lead to efficient immune responses was unclear. Advances in physics (multi-photon), chemistry (probes) and biology (animal models) have provided immunologists with specialized tools to quantify the molecular and cellular mechanisms driving immune function in lymphoid tissues through directly visualising cellular behaviours in 3-dimensions over time. Through the temporal and spatial resolution of multi-photon confocal microscopy immunologists have developed new insights into normal immune homeostasis, host responses to pathogens, anti-tumour immune responses and processes driving development of autoimmune pathologies, by the quantification of the interactions and cellular migration involved in adaptive immune responses. Advances in deep tissue imaging, including new fluorescent proteins, increased resolution, speed of image acquisition, sensitivity, number of signals and improved data analysis techniques have provided unprecedented capacity to quantify immune responses at the single cell level. This quantitative information has facilitated development of high-fidelity mathematical and computational models of immune function. Together this approach is providing new mechanistic understanding of immune responses and new insights into how immune modulators work. Advances in biophysics have therefore revolutionised our understanding of immune function, directly impacting on the development of next generation immunotherapies and vaccines, and is

J. Butler · A. Sawtell · S. Jarrett · J. Cosgrove · R. Leigh · M. Coles (✉)
Centre for Immunology and Infection, Department of Biology, York YO10 5DD, UK
e-mail: mark.coles@york.ac.uk

J. Butler · S. Jarrett · J. Cosgrove · J. Timmis · M. Coles
York Computational Immunology Laboratory, York YO10 5DD, UK

J. Butler · S. Jarrett · J. Cosgrove · J. Timmis
Department of Electronics, University of York, York YO10 5DD, UK

© Springer International Publishing Switzerland 2016
M.C. Leake (ed.), *Biophysics of Infection*, Advances in Experimental Medicine and Biology 915, DOI 10.1007/978-3-319-32189-9_20

providing the quantitative basis for emerging technology of simulation-guided experimentation and immunotherapeutic design.

Keywords Multi-photon · Immunity · Cellular interactions · Imaging · Migration · Modelling · Lymph nodes

20.1 Background: Overview of the Immunological Puzzle

Over the last 500 million years vertebrates have evolved ever-complex immune systems to provide protective immune responses to pathogens while maintaining tolerance to self and living in symbiosis with their microbiota. This process is driven through *activating* and *regulatory* networks of immune cells, cytokines and receptors that together provide for efficient initiation and timely resolution of immune responses. Pathogens trigger both innate (rapid) and adaptive (delayed) immune responses that can lead to clearance and long term immunity. Understanding how this process is controlled is key to the development of new vaccines and immunotherapies for infectious disease, and understanding the puzzle of how the immune response limits tissue damage while effectively clearing pathogens, a requirement that has driven the evolution of effective mechanisms to turn on and then rapidly turn off immune responses, including immune activating and inhibitory receptors, gene regulatory networks and through a balance of effector and regulatory cells.

Immune responses are initiated by a complex series of events, initiated by pathogen opsonisation and activation of immediate innate effectors through damage and pathogen recognition receptors, followed rapidly by the orchestrated cellular interactions that underpin all aspects of immunity. These are controlled by immune cell migration into tissues guided by localised gradients of chemotactic molecules, cell–cell interactions antigen and pathogen recognition receptors, and localised cytokine signalling. For effective induction of long-lived immunity to infection and potent effector function, these cellular events are coordinated within the specialised, highly organised microenvironments of lymphoid tissues that contain distinct areas adapted to provide different functions. In mammals, primary immune responses occur in tissue-draining lymph nodes, specialised secondary immune tissues that co-evolved with the advanced immune systems of mammals (Fig. 20.1). The complex architecture of the lymph node is underpinned by a dynamic network of stromal cells. Through the provision of survival, activation and migratory factors, stromal cells create and maintain a niche to ensure that the right cells interact at the right time and place. Understanding how the dynamic structural organisation of the lymph node (LN) regulates efficient immune responses is key to understanding the mechanisms driving generation of high affinity adaptive immune responses. Thus the key challenges in immunology are to understand how different exogenous and endogenous signals coordinate immune responses to pathogens within the complex physiology of lymphoid tissues, and how these processes are subverted by pathogens to evade the immune response. Solving this puzzle requires quantitative

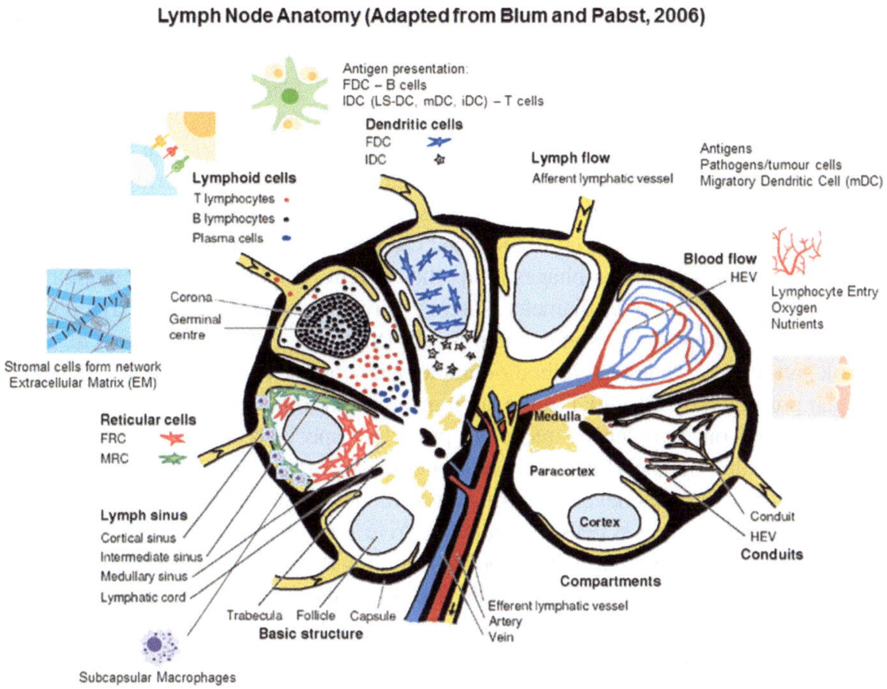

Fig. 20.1 Immune responses triggered by vaccines, pathogens and tumours in peripheral tissues occur in lymph nodes, highly structured tissues with organised zones that drive adaptive immune responses. Key events in immune system homeostasis, and the process of initiation and resolution of adaptive immune responses have been quantified using multiphoton microscopy. The basic organizational unit of the human lymph node is the lobule, bounded by trabeculae which are continuous with the surrounding capsule. Each lobe is separated into cortical, paracortical and medullary region which possess different cellular composition and organisation. Lymphocytes enter the blood through High Endothelial Vessels (HEV), while migratory dendritic cells enter through the lymphatic sinus. Antigen can enter the lymph node either by direct trafficking by dendritic cells, uptake by subcapsular macrophages (>70 Kd) or through the conduit network (<70 Kd). Lymphocytes egress through the medullary sinus to re-enter into circulation (Blum and Pabst 2006)

measurement of immune responses in real-time in tissues using an array of imaging technologies including multi-photon confocal microscopy.

20.2 Why Deep Tissue Imaging Revolutionised Understanding of Immune Response Initiation in Lymph Nodes

Pathogen triggered immune responses are initiated in peripheral tissues through the activation of localised innate antigen presenting cells (APC) (e.g. dermal dendritic cells) inducing up-regulation of the chemokine receptor CCR7. Expression of this molecule permits entry into the lymph node cortex through responsiveness to CCL21 chemokine gradients surrounding the lymphatic capillaries leading to entry into the draining lymphatic vessels. CCR7 also regulates migration in response to localised CCL19 gradients within the lymph node cortex, where the activated APCs interact with T cells and prime adaptive immune responses (Weber et al. 2013; Russo et al. 2016). These dendritic cells accumulate in a specialised structure called the cortical ridge where they are scanned by the naïve and memory T cells. In contrast, B-cell mediated immunity is triggered through either antigen (a molecule capable of inducing an immune response) uptake by subcapsular macrophages lining the efferent lymphatics within lymph nodes, or if less than 70 KDa in diameter, directly through conduits into the lymph node parenchyma and uptake by non-cognate B cells (Sixt et al. 2005; Rantakari et al. 2015). Whole proteins or antigen-antibody-complement complexes (in contrast to peptide antigen presented by DCs to T cells) are transferred to follicular dendritic cells (FDCs) and presented on complement receptors providing specialised structures to competitively stimulate cognate B-cell activation and drive formation of the germinal centre reaction (a structure that develops during an antibody response where selective pressure is placed on responding B cells to efficiently drive protective antibody responses). The initiation of high-affinity antibody responses requires B-cells to acquire cognate antigen and to subsequently interact with T cells (De Silva and Klein 2015). This occurs as a result of chemokine regulated interactions between effector T helper cells and B cells driven by co-stimulatory signals for B and T cells (CD40-CD40L and B7-CD28). This provides a check and balance in high-affinity antibody responses in the germinal centre reaction leading to the selection of the optimal B cell repertoire. This highly dynamic process requires orchestration of lymphocyte and dendritic cell migration, dynamic tissue remodelling in the B-cell follicle and efficient cell-cell interactions so that the right cells (individual antigen specific B and T cell clones) are in the right place (contact with an FDC (B cells), DC (T cells) that expresses specific cognate antigen) at the right time (interactions have to occur sequentially and are time-limited) for an effective antigen specific immune responses (De Silva and Klein 2015). This process is entirely stochastic with the response to antigen being dependent on lymphocyte and dendritic cell motility, chemotaxis and antigen specific receptor expression (Wei et al. 2003).

In summary, an intricate set of cellular and molecular interactions governs the induction of immune responses in complex and dynamic secondary lymphoid tissues such as the lymph node. Understanding pathogen mediated immune responses requires quantification of the three-dimensional (3D) organisation of tissues and real-time imaging of cellular interactions in the context of tissue geometry; something that can only be achieved by a combination of imaging technologies capable of resolving immune cell movement through complex 3D tissues over short timescales.

20.3 Why Understanding Immune Cell Dynamics and Function Has Created a Revolution in Treatment of Human Disease

Understanding the mechanisms that control the turning on (inflammation) and turning off (regulation) of immune responses has led to breakthroughs in the treatment of autoimmune disease through targeting cytokines (e.g. anti-TNF (infliximab), anti-IL17 (ixekizumab), sIL1R (anakinra)), cytokine receptors or effector immune populations (e.g. B cells (rituximab)). More recently improved understanding of immune responses to tumours has led to the development of novel treatments for tumours through potentiating immune responses by inhibiting regulatory mechanisms (e.g. anti-CTLA4 (ipilimumab), anti-PD1 (nivolumab) (Beck et al. 2010; Weiner et al. 2010). These approaches have transformed the treatment of autoimmune diseases and are driving the development of ground-breaking immunotherapies that are transforming the treatment of advanced cancers that were previously untreatable. The key to clinical efficacy has been that key checks and balances in the immune response are controlled in tissue draining lymph nodes through regulating the cycles of antigen priming by dendritic cells, effector-cell activation and localisation of cytokine producing cells, which have all been quantified and understood through multi-photon imaging. Translating this approach to pathogen infection is more challenging due to the capacity of pathogens to actively evade immune responses, often hijacking the immune system to the benefit of the pathogen through the production of decoy receptors, hiding within immune cells and manipulating inflammation to 'metastasise' through the body (e.g. Tuberculosis). Biophysical technologies are however providing unprecedented capacity to image host-pathogen interactions, providing new insights into the dynamics of infection and immune cells and helping to identify molecular and cellular mechanisms that may boost immune responses leading to pathogen clearance. Therapeutic antibodies will likely have a more limited role in treatment for infectious disease due their inherent cost, particularly in the developing world, however novel treatments that can push the balance in favour of pathogen clearance are likely to transform therapies for infectious disease and have the potential to generate anti-microbial strategies that do not rely on the development of ever more complex antibiotics thus avoiding problems associated with development of antibiotic resistance. This is an ambitious goal however and first requires precise quantitative technologies to further understand immune function.

20.4 Methodologies to Quantifying Immune Responses: Role of Multi-dimensional Imaging to Understand and Quantify Immune Responses at the Single Cell Level

To quantify the complexity of immune responses, immunologists have developed and used a combination of technologies to quantitate molecular and cellular mechanisms during immune responses. These include flow cytometry, immunohistochemistry, western blotting, quantitative polymerase chain reaction, ELISA and in vitro functional assays, however these technologies are in themselves inherently limited in spatial, temporal or cellular resolution and insufficient alone to understand the underlying physics and biology of immune responses. The development of a plethora of biologically-compatible chemical dyes and fluorescent proteins, in combination with advances in imaging technologies, have provided the potential to image immune function at the single-cell level in four-dimensions (4D = 3D space + time). By using a range of imaging technology it has been possible to quantify immune responses from single molecular diffusion in lipid membranes, to analysis of immune cell interactions in tissues (Fig. 20.2). Application of total internal reflection fluorescence microscopy (TIRF) and super-resolution imaging techniques has permitted real time imaging of single molecular diffusion in the immunological synapse between effector lymphocytes and antigen presenting cells revolutionizing our understanding of molecular dynamics and organization of cell membrane proteins and intracellular signaling adapters during initiation of immune responses (Angus and Griffiths 2013; Rossy et al. 2013). Confocal imaging has permitted analysis of immune cell migration and interactions, whole tissue light sheet has permitted the analysis of the 3D organization of tissues, and luminescent imaging, PET/SPECT and MRI has permitted analysis of immune responses at the whole organism level. This combination of imaging technologies has provided the capability to quantify immune responses from single molecule through to organism level immunity. This has provided key mechanistic insights into the molecular mechanisms that regulate immune responses. Key recurrent themes from quantitative imaging have been that the 3D organization is the driving force in immune function, whether it be spatial organization of kinases, signaling intermediates and phosphatases within cells, receptors, integrins and ligands on cell membranes, organization of cells in secondary lymphoid tissues or higher level tissue organisation. This is not been possible using histology sections or cell suspensions. Thus there is a need to apply a range of imaging technologies to capture the key dynamics of immune function in lymph nodes.

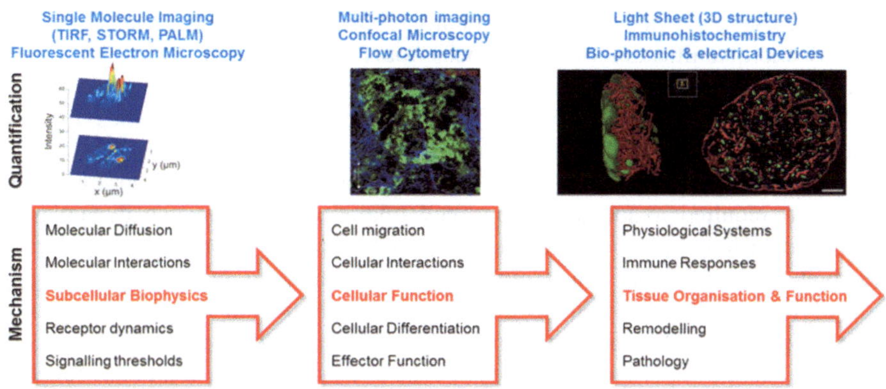

Fig. 20.2 Understanding immune function will require a spectrum of different technologies from single molecule imaging through to whole tissue imaging. This will provide mechanistic understanding at multiple levels from molecular dynamics of receptors and signalling molecules through to physiological function of whole lymph nodes

20.5 Past: The Multi-photon Revolution in Immunology: A Unique Technology that Permitted Single Cell Quantification of Immune Responses in 4D (Space and Time)

From the 1980s onwards, the development of multi-colour flow cytometry and higher resolution thin sectioning of immune tissues combined with an ever-expanding number of antibodies and fluorochromes drove rapid progress in immunology, due to the capacity of the technology to quantify gene expression and classify at a single cell level live immune populations and their effector function. This technology led to new models of the key events in antigen initiated immune responses (Catron et al. 2004). Although flow cytometry and immunohistochemistry have been fundamental to the modern understanding of immune responses as a technology it cannot address the role of space, time and organisation which is critical to immune responses. It was the development of multi-photon confocal microscopy by Watt Web that revolutionised our understanding of cell dynamics in immune responses by permitting real time imaging of immune responses deeper into living tissues (Zipfel et al. 2003; Dong et al. 2000). Multi-photon confocal microscopy has been used to extensively image cells in tissues in a wide range of biological tissues using a combination of fluorescent proteins, dyes and quantum dots (Germain et al. 2006; Michalet et al. 2005). Two photon imaging has many advantages over single photon microscopy including much greater depth penetration, lowered phototoxicity and generation of second harmonics, which highlight the collagen fibres and the extracellular matrix that form the conduit network, the capsule and structural support in lymph nodes (Wei et al. 2003). The development of better analytical software and mathematical analysis techniques has permitted a highly quantitative analysis of different parameters in cell motility and quantification of cellular behaviours (Beltman et al. 2009).

Immune responses in lymph nodes occur within the cortex and the paracortex of the lymph node which requires imaging through the capsule of the LN, the lymphatic vasculature and the marginal reticulum, and thus conventional single photon confocal microscopy lacked the depth of laser penetration (limited to maximum ~60 μm) required to image immune cell migration, therefore the dynamics of immune cells in lymph nodes were not previously appreciated or have been wrongly interpreted. Since 2002, when pioneering work utilizing in vivo cell imaging quantifying live T-cell dynamics and behavior in lymphoid tissues, multiphoton microscopy has revolutionized and enhanced our understanding of how immune cells function, and the importance of tissue structure and organization in immune function (Cahalan and Parker 2008). This has been extensively reviewed (Robey and Bousso 2003; Bousso and Robey 2004; Tarakhovsky 2005; von Andrian 2003; Mempel et al. 2004a; Wei et al. 2003; Huang et al. 2004; Sumen et al. 2004), and key findings have permitted quantification of T—DC interaction, T-cell proliferation, T effector function, regulatory T-cell function, stromal—lymphocyte interactions, lymphocyte entry and exit from lymph nodes, B-T interactions, B cell—FDC interactions, dynamics of B cells in germinal center interactions and antigen up take, transport and processing in lymph nodes (summarized in Fig. 20.1). This has required the use of specific immune or stromal cell populations expressing fluorescent proteins (FP) that are expressed on specific subsets of immune cells. This can occur through the use of either direct expression of FPs through tissue specific promoters or use of cre recombinase crossed to Rosa26 or ubiquitin reporter strains that contain one or more (e.g. brainbow or Ubow) FPs (Ghigo et al. 2013). An alternative method is through the adoptive transfer of labelled host cells using cell labelling agents into recipient mice (e.g. CFSE or CMTMR), application of bone marrow chimeras to permit imaging of ubiquitous FPs to label stromal and endothelial cells, and fluorescent labelled pathogens (Coombes and Robey 2010). Imaging has been performed in either live animals under anesthesia (most commonly on the popliteal LN) or by explanting the LN into a dish with physiologically balanced medium. This approach has certain advantages but lacks key physiological cues. One of the challenges in maintaining normal physiology, lymphocytes are characterized by their highly energetic migration, a process that is absolutely dependent in lymph nodes on oxygen tension and temperature, additionally localized inflammation resulting from surgery can significantly modify outcomes thus imaging can lead to artifacts as a result of exposing and imaging the tissue (Germain et al. 2006).

During an immune response the first step is the acquisition of antigen so it can be efficiently presented by specialized dendritic cells to T cells and follicular dendritic cells to B cells. Utilizing a combination of imaging technology it was shown that a special population of resident dendritic cells are found integrated into the lymphatic sinus, actively scanning lymph fluid, and then once activated, rapidly stimulating CD4+ and CD8 T cells (Mempel et al. 2004b; Gerner et al. 2015). Analysis of T-cell migration and interaction with DCs during the different stages of priming, proliferation and development of effector cell function has provided new insights into the key mechanisms determining the efficacy of immune responses and has provided improved understanding of how chemotaxis, and intrinsic cell motility, modulates immune cell migration and increases the efficacy of the immune response (Harris et al. 2012; Ulvmar et al. 2014; Ackerknecht et al. 2015). More recently it has been shown that efficient immune cell migration occurs in the context of their localized microenvironments including secreted homeostatic chemokines, cytokines and extracellular matrix which are secreted by specialized stromal cells, fibroblastic reticular cells in the T cell zone, marginal reticular cells (MRC) in the LN proximal to the lymphatic sinus, follicular dendritic cells (FDCs) supporting B-cell follicles and specialized stromal cells in the medullary region and dark zone of the B-cell follicle. The role of these networks in immune cell migration and survival has been quantified (Bajénoff et al. 2006). Analysis of these stromal networks can rapidly modulate during immune responses undergoing dynamic remodeling including stretching during immune responses (Acton et al. 2014) thus controlling the dynamics of immune responses (Fletcher 2015).

The activation of B cells is involves migration and serial interactions between different cell types that provide both antigen, survival and activation signals and stimulate antigen specific B cell proliferation and affinity maturation. All of these processes that have been extensively quantified and tracked using MP imaging. The activation process needs to be exquisitely orchestrated to drive high affinity antibody responses while maintaining both tolerance, and not undergoing somatic transformation into hematological malignancy, which occurs when cells fail to migrate correctly (Muppidi et al. 2014). The process of initiating B cell responses is initiated through antigen movement from the lymphatic sinus to the FDC occurring through both macrophage dependent (large antigens) and conduit dependent processes (small antigens) (Batista and Harwood 2009). B cells that are activated through binding to antigen bound by CR1/2 on FDCs enter into a carefully orchestrated set of events involving T-cell mediated stimulation and subsequent entry into the dark zone of the germinal center. The germinal center is a transient niche, providing a key checkpoint in the selection and expansion of antigen specific B cells (Shulman et al. 2014). This movement and migration processes is regulated through G protein receptor signaling involving gradients of CXCL13 and CXCL12 chemokines, sphingosphine-1-phosphate and EBI2 ligand (Green and Cyster 2012; Arnon et al. 2013). The expression of key molecules that regulate lymphocyte interactions (Cannons et al. 2010) and activation are required to provide for the efficient stimulation of protective antibody responses (Corcoran and Tarlinton 2016). Through combining deep tissue imaging with genomics, flow cytometry and analysis of protective antibody responses a picture of how the germinal center reaction is orchestrated and regulated has emerged.

Understanding of immune responses has been further extended to understand both innate immune cells and lymphoid cell migration in peripheral tissues including skin, joints, intestines and liver providing new insights into the process of immune-surveillance and mechanisms driving innate and adaptive cell migration in peripheral tissues (Weninger et al. 2014). Through imaging of tumors and inflammatory pathologies the immune cell dynamics have been quantified and new insights into molecular cues (e.g. leukotrienes, chemokines) regulating cellular behaviors have been determined. In the skin, leukotriene B4 has been shown to have an essential role in regulating neutrophil migration and swarming (Lämmermann et al. 2013).

20.6 Present: The Current State of Multi-photon Imaging in Understanding LN Biology

Advances in multi-photon imaging technologies have been advanced through the development of more sensitive PMTs (gallium arsenide), better electronics creating lower signal to noise ratios, improved optics including high numerical aperture water dipping objectives, filters and optical layout, longer wavelength 2-photon lasers, and better tracking algorithms have created incremental but important advances allowing the development of highly sensitive easy to use systems from a range of microscope manufacturers and optical systems providers (Germain et al. 2006). Fundamentally MP imaging is reaching a ceiling where imaging improvements are becoming incremental and unlikely to significantly improve imaging due to the increasingly difficulty to significantly increase the efficiency of detectors and penetration of light through the LN. Advances in chemical dyes including increasing 2-photon absorption characteristics, application of quantum dots and next generation of fluorescent proteins have permitted new insights into biological systems allowing simultaneous detection of four or more cell populations, increased signal to noise ratios and lowering overall photo-bleaching and toxicity permitting longer imaging sessions. One of the fundamental problems with tissue and/or temporal specific fluorescent proteins is that they are active once expressed thus there is no physical control of fluorescent labelling. The development of photoactivatable proteins has permitted micro-anatomical cell labelling in LNs, this was very elegantly used by Victoria and Nussenzweig to quantify cellular dynamics in the GC reactions by specifically photo-converting a photoactivatable green fluorescent protein (PA-GFP) expressed in a B cell within either the light zone or dark zone of the B cell follicle (Victoria et al. 2010). This permitted analysis of cell labelled in time and space by both multi-photon confocal microscopy and flow cytometry. Using this technological breakthrough, the key limiting factor in the GC reaction was identified to be competition for T cell help rather than antigen availability on the FDCs.

Much of the early work quantifying immune responses has focused on the popliteal LN due to its easy accessibility, convenient route for administering substances and small size permitting imaging of entire lymph nodes. However tissues like the lung and associated lymphoid tissues are key sites of immune responses, Max Krummel and colleagues have developed technology that permitted optimal imaging through the combination of a fast resonate scanner multiphoton confocal microscope (30 fps) with a ventilator that is synchronised to data acquisition (120 breaths per minute) which yielded imaging at 20 fps following respiratory cycle removal, and the software to adjust for breathing cycle movement (Looney et al. 2011). This permitted direct analysis of antigen uptake of alveolar dendritic cells and subsequent interactions with T cells, furthering understanding of mechanisms driving airway hyper-reactivity and tumour metastasis to the lung (Thorton et al. 2012; Headley et al. 2016).

20.7 Present: Understanding Cell Signalling Pathways Using Multiphoton Imaging

Cell migration, interactions and function required integration of signals from outside the cell to intracellular signalling networks that control gene transcription, molecular motors that induce cell migration and control the metabolic state of the cell. Traditionally cell signalling networks have been quantified using western blotting and other proteomic based technologies. Multi-photon confocal microscopy has provided a key tool when combined with cell tracker dyes and tissue specific florescent proteins to determine the role of different cell signalling pathways have in cell migration and effector cell function. One key approach has been to generate mice genetically deficient for a specific protein (e.g. receptor tyrosine kinase) in specific cell types to analyse signalling pathways in immune cell function. However, this approach is limited to proteins that do not effect either cell development or early stages in their activation/migratory processes. An alternative is through temporal addition of small molecule inhibitors or blocking antibodies into tissue permitting quantifying of the effects of specific receptors/ligands and associated signalling pathways on immune cell function. To do this we developed a LN tissue explant model where substances could be added into explanted lung draining LNs at fixed time points post infection with *Streptococcus pneumoniae* into the lung (Fig. 20.3). This permitted direct quantification of effects of specific inhibitors on neutrophil migration in the medullary region of the LN. Using this approach, we imaged GFP expressing neutrophils in the lung draining LN, and then continued to image the migration and behaviours of neutrophils post treatment with small molecular inhibitors or blocking antibodies. This technique permitted us to characterise the similarities and differences between neutrophil migration in different tissues and how this compared to other innate immune cell types. We believe this approach can provide deeper insight into how drugs work in tissues and identify the key molecular pathways that can effectively modulate immune cellular behaviours.

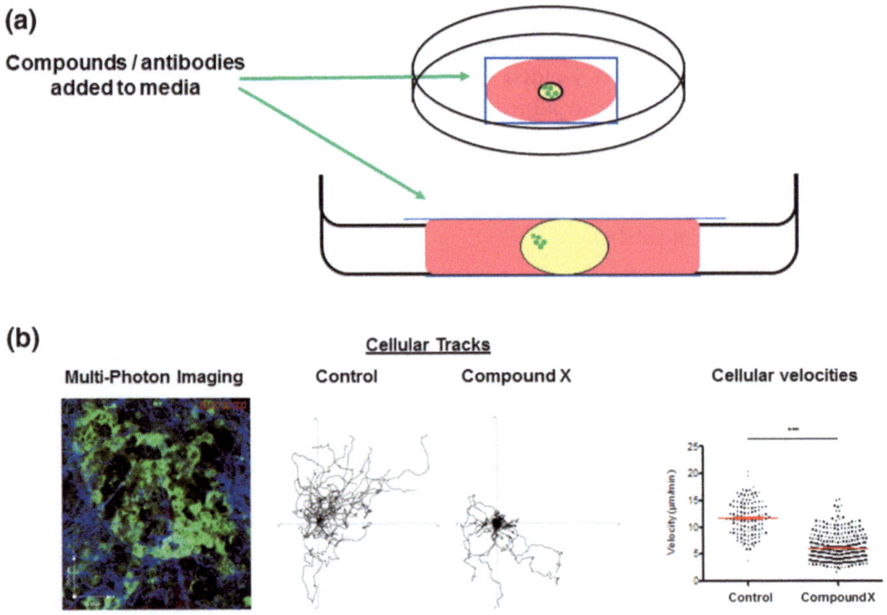

Fig. 20.3 Application of multi-photon microscopy to analyse signalling pathways regulating cell migration. Explanted lymph nodes were cultured in tissue culture medium in a glass bottom dish and cellular migration analysed using multi-photon microscopy. Compounds or antibodies (and appropriate controls) were perfused into the tissue culture medium after 30 min of imaging (**a**). Cells were then tracked and velocities and meandering indexed calculated (**b**)

20.8 Future: Creation of a Lymph Node 3D Atlas

Lymph nodes are highly complex 3D structures supported through vascular networks that control cellular input and exit from lymph nodes, provide nutrients and oxygen to the parenchyma of the LN, mesenchymal stroma. One of the key challenges is to develop a highly accurate 3D atlas of cellular localisation and gene expression within a lymph node. Such a tool would permit new insights through provision of a reference map to a lymph node (Kislitsyn et al. 2015). The emergence of high resolution multichannel light sheet microscopy (Ahrens et al. 2013) and novel clearing technologies (Tainaka et al. 2014) has provided the technologies to undertake a single cell map of LN geography. These technologies have inherent limits on the number of dyes and fluorescent proteins that can be simultaneously analysed, the ultimate goal would of be to generate a complete genomic analysis of a lymph node where transcriptional profiles can be linked to individual cells within a lymph node. We propose that this challenge of a transcriptional map of a LN is possible through using photoactivatable proteins and dyes that permit isolation and transcriptional profiling of specific cells in a LN where the geographical location is known prior to isolation (Fig. 20.4).

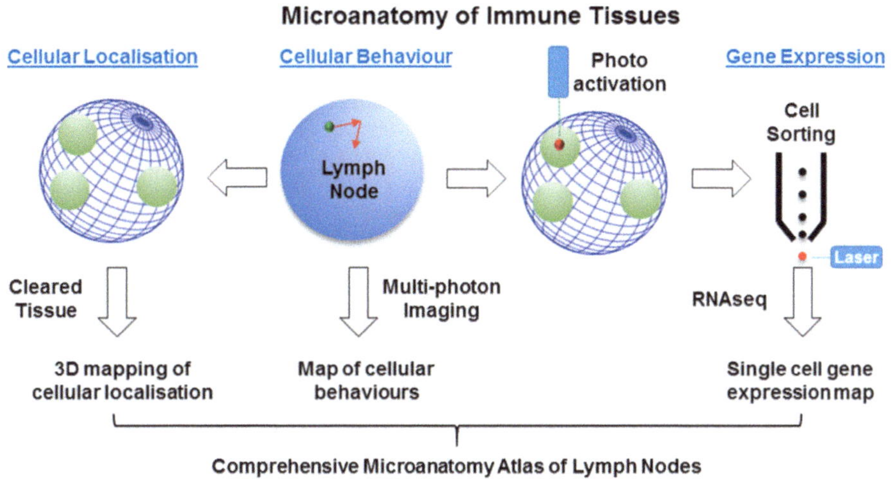

Fig. 20.4 Generation of a lymph node anatomical atlas of gene expression and cellular behaviours using a combination multi-photon confocal microscopy to generation maps of cellular behaviours, photo-activatable fluorescent proteins/dyes to specifically label cells using the multiphoton laser and subsequent capture and analysis by RNAseq or other genomic technique. Using a combination of fluorescent conjugated antibodies and fluorescent proteins and clearing technologies the cellular localisation of all cells within the lymph node can be mapped. Together this data has the potential to yield a comprehensive micro-anatomical atlas of cells in a lymph node an essential resource to quantify immune responses

20.9 Future: Combining Simulation with Imaging to Create a Synergistic Cycle of Modelling and Experimentation

Multiphoton confocal imaging has provided a unique tool to quantitate immune responses providing highly numerical data on cell migration and interactions that are key to parameterising mathematical and computational models of immune responses (Fig. 20.5) (Witt et al. 2005). Models have been used to understand the biophysical mechanisms of cell migration and interactions through analysis of data sets and resolved how 3D cellular topologies dictate immune cell signalling (Harris et al. 2012; Beltman et al. 2007; Angermann et al. 2012). To consolidate current biological understanding across time and length scales, we and others have developed multiscale models. (Ludewig et al. 2012; Cilfone et al. 2014; Kirschner and Linderman 2009; Alden et al. 2012; Meyer-Hermann et al. 2012), which incorporate ordinary differential equations and other modelling techniques into agent-based models permitting 3D spatial modelling over time of tens or hundreds of thousands of individual cells in a model replicating key micro-anatomical details (Kislitsyn et al. 2015; Cosgrove et al. 2015) and resulting in the emergence of complex phenomena observable experimentally. Building on the work of Bocharov and colleagues we have developed 3D stromal cell network structures on which

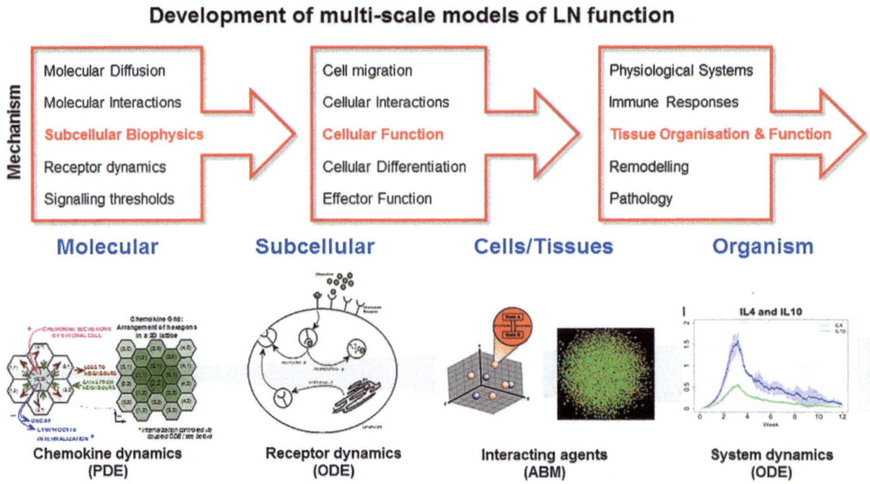

Fig. 20.5 A number of different modelling techniques can be hybridised together to generate a multi-scale model of LN function parameterised using quantitative imaging technologies (Fig. 20.1). These include ordinary and partial differential equations (ODE/PDE), interacting rule based agents, cellular automata and Monte Carlo simulations

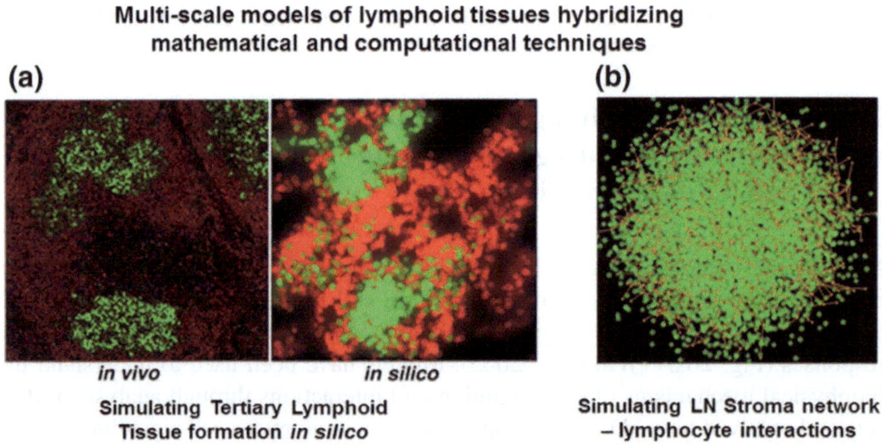

Fig. 20.6 Application of multi-scale computational modelling to understand lymphoid tissue function. Models can replicate key features of lymphoid tissues found in vivo (**a**) and the complex 3D network of stromal networks (*gold*) upon which lymphocytes migrate (*blue*) (**b**)

lymphocytes can migrate and interact (Fig. 20.6). This approach provides a unique platform to study mechanisms of immune function and simulate therapeutic interventions (Cosgrove et al. 2015). Advances in computing power and more advanced tools for the development and analysis of computational models are providing a platform for the development of predictive models (Tas et al. 2016;

Meyer-Hermann et al. 2009, 2012; Alden et al. 2013, 2014). Multiphoton and single molecule imaging approaches provide the tools to develop a synergistic cycle of modelling and experimentation, helping to identify the key parameters in a system and thus driving the development of new experimental questions in an iterative cycle. Outcomes from experiments, in turn, advance the theoretical models and thus our quantitative understanding of key immunological processes. This technological advance has the capacity to change how immunology is performed, where imaging and modelling take a central place in approaches developed to understand the molecular mechanisms driving immune responses guiding the development of the next generation of therapeutics.

References

Ackerknecht M, Hauser MA, Legler DF, Stein JV (2015) In vivo TCR signaling in CD4(+) T cells imprints a cell-intrinsic, transient low-motility pattern independent of chemokine receptor expression levels, or microtubular network, integrin, and protein kinase C activity. Front Immunol 8(6):297. doi:10.3389/fimmu.2015.00297.eCollection 2015

Acton SE, Farrugia AJ, Astarita JL, Mourão-Sá D, Jenkins RP, Nye E, Hooper S, van Blijswijk J, Rogers NC, Snelgrove KJ, Rosewell I, Moita LF, Stamp G, Turley SJ, Sahai E, Reis e Sousa C (2014) Dendritic cells control fibroblastic reticular network tension and lymph node expansion. Nature 514(7523):498–502

Alden K, Timmis J, Andrews PS, Veiga-Fernandes H, Coles MC (2012) Pairing experimentation and computational modeling to understand the role of tissue inducer cells in the development of lymphoid organs. Front Immunol 3:172. doi:10.3389/fimmu.2012.00172

Alden K, Read M, Timmis J, Andrews P, Veiga-Fernades H, Coles M (2013) Spartan: a comprehensive tool for understanding uncertainty in simulations of biological systems. PLOS Comput Biol 9(2):e1002916. doi:10.1371/journal.pcbi.1002916

Alden K, Andrews PS, Veiga-Fernandes H, Timmis J, Coles MC (2014) Utilising a simulation platform to understand the effect of domain model assumptions. Nat Comput. doi:10.1007/s11047-014-9428-7

Ahrens MB, Orger MB, Robson DN, Li JM, Keller PJ (2013) Whole-brain functional imaging at cellular resolution using light-sheet microscopy. Nat Methods 10:413–420

Angermann BR, Klauschen F, Garcia AD, Prustel T, Zhang F, Germain RN, Meier-Schellersheim M (2012) Computational modeling of cellular signaling processes embedded into dynamic spatial contexts. Nat Methods 9(3):283–289

Angus KL, Griffiths G (2013) Cell polarisation and the immunological synapse. Curr Opin Cell Biol 24:85–89

Arnon TI, Horton RM, Grigorova IL, Cyster JG (2013) Visualization of splenic marginal zone B-cell shuttling and follicular B-cell egress. Nature 493(7434):684–688

Bajénoff M, Egen JG, LKoo LY, Laugier JP, Brau F, Glaichenhaus N, Germain RN (2006) Stromal cell networks regulate lymphocyte entry, migration, and territoriality in lymph nodes. Immunity 25(6):989–1001

Batista FD, Harwood NE (2009) The who, how and where of antigen presentation to B cells. Nat Rev Immunol 9(1):15–27

Beck A, Wurch T, Bailly C, Corvaia N (2010) Strategies and challenges for the next generation of therapeutic antibodies. Nat Rev Immunol 10:345–352

Beltman JB, Maree AFM, Lynch JN, Miller MJ, de Boer RJ (2007) Lymph node topology dictates T cell migration behaviour. J Exp Med 204(4):771–780

Beltman JN, Maree AFM, de Boer RJ (2009) Analysing immune cell migration 9:789–798

Blum KS, Pabst R (2006) Keystones in lymph node development. J Anat 209(5):585–595
Bousso P, Robey EA (2004) Dynamic behaviour of T cells and thymocytes in lymphoid organs as revealed by two-photon microscopy. Immunity 21:349–355
Cahalan MD, Parker I (2008) Choreography of cell motility and interaction dynamics imaged by two-photon microscopy in lymphoid organs. Annu Rev Immunol 26:585–626
Cannons JL1, Qi H, Lu KT, Dutta M, Gomez-Rodriguez J, Cheng J, Wakeland EK, Germain RN, Schwartzberg PL (2010) Optimal germinal center responses require a multistage T cell: B cell adhesion process involving integrins, SLAM-associated protein, and CD84. Immunity 26;32 (2):253–65
Catron DM, Itano AA, Pape KA, Mueller DL, Jenkins MK (2004) Visualising the first 50 hr of the primary immune response to a soluble antigen. Immunity 21:341–347
Cilfone N, Kirschner D, Linderman JJ (2014) Strategies for Efficient numerical implementation of hybrid multi-scale agent-based models to describe biological systems. Cell Mol Bioeng. doi:10.1007/s12195-014-0363-6
Coombes JL, Robey EA (2010) Dynamic imaging of host-pathogen interactions in vivo. Nat Rev Immunol 10:353–364
Corcoran LM, Tarlinton DM (2016) Regulation of germinal center responses, memory B cells and plasma cell formation-an update. Curr Opin Immunol 39:59–67
Cosgrove J, Butler J, Alden K, Read M, Kumar V, Cucurull-Sanchez L, Timmis J, Coles M (2015) Agent-based modeling in systems pharmacology. CPT: Pharmacom Syst Pharmacol 4 (11):615–629
De Silva NS, Klein U (2015) Dynamics of B cells in germinal centres. Nat Rev Immunol. 15 (3):137–148
Dong CY, Kim KH, Buehler C, Hsu L, Kim H (2000) So PTC, probing deep-tissue structures by two-photon fluorescence microscopy, emerging tools for single cell analysis: advances in optical measurement technologies. Wiley-Liss, pp 221–237
Fletcher AL, Acton SE, Knoblich K (2015) Lymph node fibroblastic reticular cells in health and disease. Nat Rev Immunol 15(6):350–361
Germain RN, Miller MJ, Dustin M, Nussenzweig MC (2006) Dynamics imaging of the immune system: progress, pitfalls and promise. Nat Rev Immunol 6:497–507
Gerner MY, Torabi-Parizi P, Germain RN (2015) Strategically localized dendritic cells promote rapid T cell responses to lymph-borne particulate antigens. Immunity 42:172–185
Green JA, Cyster JG (2012) S1PR2 links germinal center confinement and growth regulation. Immunol Rev 247(1):36–51
Gong C1, Mattila JT, Miller M, Flynn JL, Linderman JJ, Kirschner D (2013) Predicting lymph node output efficiency using systems biology. J Theor Biol. 335:169–184
Ghigo C, Mondor I, Jorquera A, Nowak J, Wienert S, Zahner SP, Clausen BE, Luche H, Malissen B, Klauschen F, Bajénoff M (2013) Multicolor fate mapping of Langerhans cell homeostasis. J Exp Med 210(9):1657–1664
Harris TH, Banigan EJ, Christian DA, Konradt C, Woino EDT, Norose K, Wilson EH, John B, Weninger W, Luster AD, Liu AJ, Hunter CA (2012) Generalized Lévy walks and the role of chemokines in migration of effector CD8+T cells. Nature 486:545–548
Headley MB, Bins A, Nip A, Roberts EW, Looney MR, Gerard A, Krummel MF (2016) Visualization of immediate immune responses to pioneer metastatic cells in the lung. Nature 24;531(7595):513–7
Huang AY, Qi H, Germain RN (2004) Illuminating the landscape of in vivo immunity: insights from dynamic in situ imaging of secondary lymphoid tissues. Immunity 21(3):331–339
Kirschner DE, Linderman JL (2009) Mathematical and computational approaches can complement experimental studies of host pathogen interactions. Cell Microb 11(4). doi:10.1111/j.1462-5822.2009.01281
Kislitsyn A, Savinkov R, Novkovic M, Onder L, Bocharov G (2015) Computational approach to 3D modeling of the lymph node geometry. Computation 3(2):222–234. doi:10.3390/computation3020222

Lämmermann T, Afonso PV, Angermann BR, Wang JM, Kastenmüller W, Parent CA, Germain RN (2013) Neutrophil swarms require LTB4 and integrins at sites of cell death in vivo. Nature 498(7454):371–375

Looney MR, Thorton EE, Sen D, Lamm WJ, Glenny RW, Krummel MF (2011) Stabilized imaging of immune surveillance in the mouse lung. Nat Methods 8(1):91–96

Ludewig B, Stein JV, Sharpe J, Cervantes-Barragan L, Thiel V, Bocharov G (2012) Eur J Immunol 42:3116–3125

Mempel TR, Scimone ML, Mora JR, von Andrian UH (2004a) In vivo imaging of leukocyte trafficking in blood vessels and tissues. Curr Opin Immunol 16(4):406–417

Mempel TR, Henrickson SE, Von Andrian UH (2004b) T-cell priming by dendritic cells in lymph nodes occurs in three distinct phases. Nature 427(6970):154–159

Meyer-Hermann M, Figge MT, Toellner KM (2009) Germinal centres seen through the mathematical eye: B-cell models on the catwalk. Trends Immunol 30(4):157–164

Meyer-Hermann M, Mohr E, Pelletier N, Zhang Y, Victora GD, Toellner KM (2012) A theory of germinal center B cell selection, division, and exit. Cell Reports 2(1):162–74

Michalet X, Pinaud FF, Bentolila LA, Tsya JM, Doose S, Li JJ, Sundaresan G, Wu AM, Gambhir SS, Weiss S (2005) Quantum dots for live cells, in vivo imaging and diagnostics. Science 307:538–544

Muppidi JR, Schmitz R, Green JA, Xiao W, Larsen AB, Braun SE, An J, Xu Y, Rosenwald A, Ott G, Gascoyne RD, Rimsza LM, Campo E, Jaffe ES, Delabie J, Smeland EB, Braziel RM, Tubbs RR, Cook JR, Weisenburger DD, Chan WC, Vaidehi N, Staudt LM, Cyster JG (2014) Loss of signalling via Gα13 in germinal centre B-cell-derived lymphoma. Nature 516 (7530):254–258

Phan TG, Grigorova I, Okada T, Cyster JG (2007) Subcapsular encounter and complement-dependent transport of immune complexes by lymph node B cells. Nat Immunol 8(9):992–1000

Rantakari P, Auvinen K, Jäppinen N, Kapraali M, Valtonen J, Karikoski M, Gerke H, Iftakhar-E-Khuda I, Keuschnigg J, Umemoto E, Tohya K, Miyasaka M, Elima K, Jalkanen S, Salmi M (2015) The endothelial protein PLVAP in lymphatics controls the entry of lymphocytes and antigens into lymph nodes. Nat Immunol. 16(4):386–396

Robey EA, Bousso P (2003) Visualizing thymocyte motility using 2-photon microscopy. Immunol Rev 195:51–57

Rossy J, Pageon SV, Davis DM, Gaus K (2013) Super-resolution microscopy of the immunological synapse. Curr Opin Immunol 25(3):307–312

Russo E, Teijeira A, Vaahtomeri K, Willrodt AH, Bloch JS, Nitschké M, Santambrogio L, Kerjaschki D, Sixt M, Halin C (2016) Intralymphatic CCL21 promotes tissue egress of dendritic cells through afferent lymphatic vessels. Cell Reports. pii:S2211-1247(16)30026-2. doi:10.1016/j.celrep.2016.01.048

Shulman Z, Gitlin AD, Weinstein JS, Lainez B, Esplugues E, Flavell RA, Craft JE, Nussenzweig MC (2014) Dynamic signaling by T follicular helper cells during germinal center B cell selection. Science 345(6200):1058–1062

Sixt M, Kanazawa N, Selg M, Samson T, Roos G, Reinhardt DP, Pabst R, Lutz MB, Sorokin L (2005) The conduit system transports soluble antigens from the afferent lymph to resident dendritic cells in the T cell area of the lymph node. Immunity 22(1):19–29

Sumen C, Mempel T, Mazo IB, von Andrian UH (2004) Intravital microscopy: visualizing immunity in context. Immunity 21:315–329

Tainaka K, Kubota SI, Suyama TQ, Susaki EA, Perrin D, Ukai-Tadenuma M, Ukai H, Ueda HR (2014) Whole-body imaging with single-cell resolution by tissue decolorization. Cell 159 (4):911–24

Tarakhovsky A (2005) On imaging, speed, and the future of lymphocyte signaling. J Exp Med 201:505–508

Tas JM, Mesin L, Pasqual G, Targ S, Jacobsen JT, Mano YM, Chen CS, Weill JC, Reynaud CA, Browne EP, Meyer-Hermann M, Victora GD (2016) Visualizing antibody affinity maturation in germinal centers. Science 351(6277):1048–1054

Thorton EE, Looney MR, Bose O, Sen D, Sheppard D, Locksley R, Huang X, Krummel MF (2012) Spatiotemporal separated antigen uptake by aveolar dendritic cells and airway presentation to T cells in the lung. J Exp Med 209(6):1183–1199

Ulvmar MH, Werth K, Braun A, Kelay P, Hub E, Eller K, Chan L, Lucas B, Novitzky-Basso I, Nakamura K, Rülicke T, Nibbs RJ, Worbs T, Förster R, Rot A (2014) The atypical chemokine receptor CCRL1 shapes functional CCL21 gradients in lymph nodes. Nat Immunol 15(7):623–630

Victoria GD, Schwixker TA, Fooksman DR, Kamphorst AO, Meyer-Hermann M, Dustin ML, Nussenzweig MC (2010) Germinal centre dynamics revealed by multiphoton microscopy with a photoactivatable fluorescent reporter. Cell 143:592–605

von Andrian UH, Mempel TR (2003) Homing and cellular traffic in lymph nodes. Nat Rev Immunol 3(11):867–878

Weber M, Hauschild R, Schwarz J, Moussion C, de Vries I, Legler DF, Luther SA, Bollenbach T, Sixt M (2013) Interstitial dendritic cell guidance by haptotactic chemokine gradients. Science 339(6117):328–332. doi:10.1126/science.1228456

Wei SH, Parker I, Miller MJ, Cahalan MD (2003) A stochastic view of lymphocyte motility and trafficking within the lymph node. Immunol Rev 195:136–159

Weiner LM, Rishi Surana R, Wang S (2010) Monoclonal antibodies: versatile platforms for cancer immunotherapy. Nat Rev Immunol 10:317–327

Weninger W, Biro M, Jain R (2014) Leukocyte migration in the interstitial space of non-lymphoid organs. Nat Rev Immunol 14(4):232–246

Witt C, Raychaudhuri S, Chakraborty AK (2005) Movies, measurement, and modeling: the three Ms of mechanistic immunology. J Exp Med 201(4):501–504

Zipfel WR, Williams RM, Webb WW (2003) Nonlinear magic: multiphoton microscopy in the biosciences. Nat Biotechnol 21(11):1369–1377

Chapter 21
Novel Approaches to Manipulating Bacterial Pathogen Biofilms: Whole-Systems Design Philosophy and Steering Microbial Evolution

Alexandra S. Penn

Abstract Understanding and manipulating bacterial biofilms is crucial in medicine, ecology and agriculture and has potential applications in bioproduction, bioremediation and bioenergy. Biofilms often resist standard therapies and the need to develop new means of intervention provides an opportunity to fundamentally rethink our strategies. Conventional approaches to working with biological systems are, for the most part, "brute force", attempting to effect control in an input and effort intensive manner and are often insufficient when dealing with the inherent non-linearity and complexity of living systems. Biological systems, by their very nature, are dynamic, adaptive and resilient and require management tools that interact with dynamic processes rather than inert artefacts. I present an overview of a novel engineering philosophy which aims to exploit rather than fight those properties, and hence provide a more efficient and robust alternative. Based on a combination of evolutionary theory and whole-systems design, its essence is what I will call *systems aikido*; the basic principle of aikido being to interact with the momentum of an attacker and redirect it with minimal energy expenditure, using the opponent's energy rather than one's own. In more conventional terms, this translates to a philosophy of equilibrium engineering, manipulating systems' own self-organisation and evolution so that the evolutionarily or dynamically stable state corresponds to a function which we require. I illustrate these ideas with a description of a proposed manipulation of environmental conditions to alter the stability of co-operation in the context of *Pseudomonas aeruginosa* biofilm infection of the cystic fibrosis lung.

Keywords Steering complex adaptive systems • Systems aikido • Permaculture design • Bacterial biofilms • Evolution of co-operation • *Pseudomonas aeruginosa*

A.S. Penn (✉)
Evolution and Resilience of Industrial Ecosystems Project (ERIE), Department of Sociology/Centre for Environmental Strategy, University of Surrey, Guildford, Surrey GU2 7XH, UK
e-mail: a.penn@surrey.ac.uk

© Springer International Publishing Switzerland 2016
M.C. Leake (ed.), *Biophysics of Infection*, Advances in Experimental Medicine and Biology 915, DOI 10.1007/978-3-319-32189-9_21

21.1 Engineering Philosophy for Complex Adaptive Systems

21.1.1 Managing Living Systems: A Need for New Approaches

Many of our most pressing societal problems involve understanding and managing *complex adaptive systems* (or CAS); systems of many interacting, evolving parts that give rise to unexpected system-level behaviours. Moreover, many important complex systems are biological, from multispecies communities adapting to climatic change to evolving microbial communities implicated in chronic infection. It is becoming increasingly clear that complex adaptive systems need specific approaches to their management (Frei and Marzo Serugendo 2012; Bedau et al. 2010; Kay et al. 1999; May et al. 2008). By dint of the often non-linear interactions between numerous interconnected components, CAS commonly display so-called emergent effects, which are difficult to predict from knowledge of individual components' properties. Such complex interconnected systems also tend to contain numerous feedbacks which may not be immediately apparent. When we interact with them, we thus often produce unexpected indirect effects. Cascade effects in ecology are well-known examples (e.g. Fortin et al. 2005; Brett and Goldman 1996). Perhaps, most crucially, biological complex systems evolve. They change in response to our intervention. Perhaps the most obvious and currently critical example is the evolution of antimicrobial resistance in bacterial populations. Despite the indisputable success and importance of the use of antibiotics, it is clear that in providing such intense selection pressures on rapidly evolving bacterial populations, we have forged a rod for our own backs by generating drug-resistant strains (Aminov 2010). Effectively, we are playing to the bacteria's strengths with regards to their capacity for rapid adaptation. Many of our technologies or approaches to living systems have this same Sisyphusian character: pushing metaphorical rocks up hills, only to be forced to begin again when they inevitably tumble down.

To improve this situation when attempting to manipulate or manage biological systems, it is crucial to fully take into account the fact that we are *not* engineering or interacting with static artefacts. Rather than attempting to suppress biological change, we need approaches to engineering and manipulating adaptive, dynamic, living *processes*. Biological systems have their own powerful dynamics and when we work against them we tend to produce energetically inefficient systems. Given most living systems' substantial momentum and inherent adaptability, it makes sense to think about methods to exploit rather than fight their dynamics. From this perspective, we may be able to use their strengths productively rather than unsuccessfully attempt to suppress them.

21.1.2 Insights from Whole-Systems Design

I propose that whole-systems design approaches and philosophies coupled with insights from evolutionary theory provide a useful approach to these problems. In particular, I have taken inspiration from permaculture design (Mollison and Holmgren 1978; Mollison 1988; Holmgren 2002; Whitefield 2004). A system for creating and managing human ecosystems developed by forest ecologists Bill Mollison and David Holmgren in the 1970s. Permaculture (from "permanent agriculture" or "permanent culture") combines a continuous design cycle of observation, analysis, design and implementation with heuristics and philosophies drawn from the observation of pre-climax woodland ecosystems, with the aim of creating resilient, self-supporting human ecosystems that function with low work inputs.

Mollison and Holmgren's essential insight was that the constant input of energy and effort required to maintain agricultural productivity was not required to maintain the diversity and productivity of natural woodland ecosystems. In particular, in arable agriculture and horticulture, energetic and material inputs are constantly required in the form of ploughing and other mechanical soil interventions, and the application of herbicides, pesticides and fertilisers. However, left to itself, the field would return to scrub then woodland, going through secondary ecological succession. The process of agriculture requires constant work to hold the ecosystem in the early stages of succession, a continual input of energy to resist the system's own dynamics, or in some sense to keep it out of equilibrium. The continual resetting of the successional process, either via ploughing or by slash and burn is used to create the correct conditions for our crops:-fast-growing annuals bred from early succession species. This system is productive, but requires an enormous input of energy in continually resetting the ecological dynamics. It also inadvertently creates the ideal conditions for weeds, themselves the fast-growing, early succession species from which many of our crops were bred. The continued use of such a system also damages soil structure and microbial ecosystems, thus increasing the need for fertiliser inputs and repeated mechanical intervention. Effectively, our agricultural practice, like our use of antibiotics, creates a need for constant work to maintain the status quo. The central idea of permaculture is that if we consider working with, rather than against, the ecological dynamics we could lead the system through to a pre-climax agroforestry system via thoughtful succession planting. Eventually reaching a productive, slower-changing, successional state where the ratio of energy in: product out is much more favourable. Mollison described this approach as performing "aikido on the landscape" (Mollison and Holmgren 1978), understanding and working with ecosystem dynamics. In the martial art of aikido, we use our opponent's momentum and energy against them. When they run at us, we intelligently chose the point to intervene, to interact with them and cause them to fall. Currently, our interaction with biological systems is more akin to karate: we expend a great deal of energy using our force and strength against our opponent, creating a need for constant work.

For the most part, permaculture thinking and methodology has been used in agroforestry, horticultural or other land-based contexts; however it explicitly attempts to incorporate social, political and economic factors and contexts and is increasingly applied across a wide range of more abstract "human ecosystems" such as bioregions, cities and industrial networks (Bell 2005; Hopkins 2008). I suggest that its core ideas can be usefully applied or inspire new approaches across a broad range of living or life-like systems, including the manipulation of microbial communities. Although the universal validity of some of permaculture's ecological heuristics is debatable, its core philosophy of exploiting the natural dynamics of systems to make effective and efficient design interventions is a sound basis for a design approach to complex adaptive systems in general. By entering into thoughtful interaction with our "opponent" system and espousing a philosophy of targeted effective interaction, steering biological systems into states which we desire by understanding and exploiting their dynamics, we may be able to perform a more general *systems aikido* (Penn et al. 2010).

21.2 Steering Complex Microbial Communities

21.2.1 Manipulating Co-operation in Bacterial Biofilms

As a specific potential application of a systems aikido approach, I will discuss ideas for the manipulation of microbial biofilm communities, focussing in particular on bacterial co-operation. The vast majority of bacteria live in complex multispecies communities. Often within *biofilms*; large 3-dimensional assemblies of cells held together by an extracellular polysaccharide matrix (O'Toole and Ghannoum 2004). Biofilm control is vitally important, from both prevention and enhancement perspectives. Biofilms are ecologically important as drivers of climatic and ecological cycles, play vital symbiotic roles within, for example, gut microbiota and have immense potential utility for bioproduction, bioremediation and bioenergy (Hall-Stoodley et al. 2004). However, they have been estimated to be implicated in more than 60 % of human infections and display enhanced antimicrobial resistance when compared to non-aggregated cells (Parsek and Singh 2003; Costerton et al. 1999; Stewart and Costerton 2001). Bacteria living within biofilms display many behaviours different from single cells, including the co-operative production of public goods: substances required for survival which are produced by individuals but freely available for use by others. Examples include extracellular matrix, iron-chelating siderophores, quorum sensing molecules and surfactants (Crespi 2001; O'Toole and Ghannoum 2004; West et al. 2007). Public good production, and other biofilm-specific co-operative behaviours, may enhance virulence in infections, resistance to antibiotics or enhance collective properties which we might wish to exploit and is thus important in biofilm functions which matter to us. If public goods are costly to produce, then the possibility of *cheating*, using but not making them exists. This renders co-operation vulnerable to the evolution or proliferation of cheats or free-loading individuals. Indeed, in an unstructured population in which both cheats and co-operators require a costly public good, produced only by co-operators, a 'tragedy of the commons' would be expected (Hardin 1968; Rankin et al. 2007). That is, cheats would out-compete co-operators leading ultimately to the disappearance of the resource on which both depend and to a population crash.

However, although non-producer cheats arise commonly in vivo, including within biofilms associated with chronic infection (De Vos et al. 2001; Schaber et al. 2004), co-operation is remarkably stable within these systems. This stability, which must be actively maintained in some way, although many possible mechanisms have been postulated, remains to be fully explained (Buckling et al. 2007; Griffin et al. 2004; Kreft 2004; Rainey and Rainey 2003).

In order to apply a systems aikido approach in the context of a microbial community, it is necessary to identify both system dynamics which we can exploit and effective points of intervention which we can manipulate in order to steer the system. In effect, we must identify the system's *momentum* and find its *levers*. We know that co-operation strongly affects biofilm behaviour in ways that are important to us, for example, increasing virulence and antibiotic resistance, and that it is *dynamically maintained* in vivo. Also that it is *costly* to maintain. This suggests sources of system dynamics or momentum and an area to search for effective levers which we can use. Given the importance of public good production in biofilm function and its potential instability, disrupting or encouraging bacterial co-operation could be a potential intervention point with which to influence the behaviour of biofilms in a variety of contexts. We can try to use an aikido approach to manipulating bacteria using what we know about the circumstances under which co-operation does or does not evolve. By altering the environmental conditions in which bacterial colonies are developing, we can change the context of evolution, changing the relative pay-offs to cheats and co-operators. We may therefore be able to change the proportion of cheats, thus affecting the structure and stability of colonies. Our aim would be to indirectly manipulate systems' population dynamics, so that the ultimate community composition in a given scenario corresponds to a system behaviour that suits our purposes. Specifically, changing the context in which co-operator/cheat interactions unfold by manipulating parameters such as resource availability or durability of public goods which may alter the outcome of biofilm development in certain ways. Our aim is to change the rules of the game so that, what the bacteria want to do (their evolutionarily stable strategy) is what we want them to do. Effectively we try to create a win-win scenario (See Fig. 21.1).

A large amount of established evolutionary theory exists about under what conditions co-operation can evolve and we can use this to suggest potential specific system levers. Given the requirement for a public good, anything which preferentially apportions the benefits of co-operation to co-operators can support the evolution of co-operation. For example, spatial localization and repeated interactions, due to factors such as population viscosity, group structure or diffusibility of

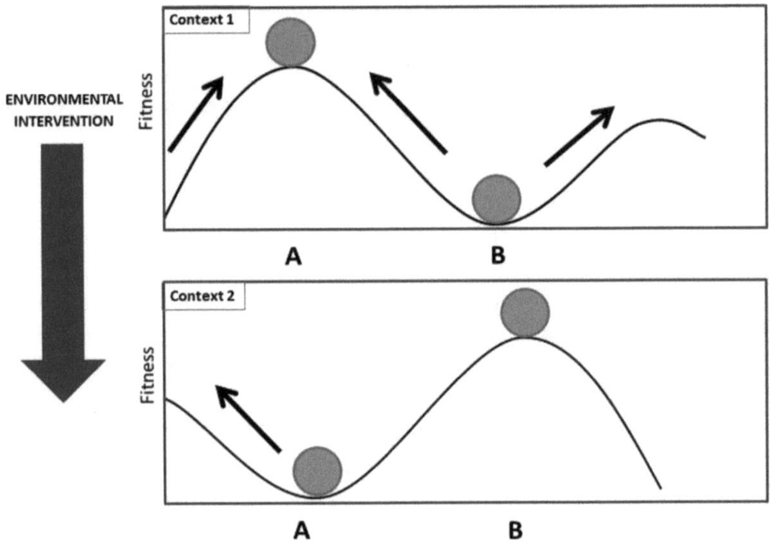

Fig. 21.1 Conceptual sketch of impact of a well-chosen environmental manipulation on fitness landscape experienced by targeted bacteria. By changing the environmental state from context 1 to 2, we aim to change the selective context or fitness landscape such that A, the selectively favoured community composition in 1, becomes unstable and such that community composition B is now favoured

public goods, usually promote evolution of co-operation (Levin and Kilmer 1974; Wilson 1975; Michod and Sanderson 1985; Nowak and May 1992; Mitteldorf and Wilson 2000; Lehmann and Keller 2006; El Mouden et al. 2010; Wilson 1987) (although see (Hauert and Doebeli 2004) for a counter example). Imposed or self-organised group structure and appropriate dispersal mechanisms can also stabilise co-operation and synergy. In particular, repeated random group formation, dispersal and re-formation have been shown in models to provide a potential mechanism for suppression of cheats (Wilson 1975, 1980, 1987). Under these conditions, a 'Simpson's paradox'(Simpson 1951) may arise, in which the global proportion of co-operators can increase even whilst it decreases within any one group. With the outcome of selection dependent on parameters of the group aggregation and dispersal process, such as initial group size, time in groups and total group growth with respect to population composition. This effect has been proposed as a possible mechanism by which co-operation could be sustained in biofilms and has been shown to be effective in vitro in synthetic planktonic systems of *E. coli* with population structure provided by plate wells (Chuang et al. 2009). However, we showed that the specific conditions for it to be effective did not occur in *P. aeruginosa* biofilms in continuous culture (Penn et al. 2012).

Each of the proposed mechanisms which may promote co-operation in a given system has particular parameters which alter the ultimate balance between the success of co-operators and cheats. Those parameters which can be directly or indirectly influenced by the experimentalist or human designer have the potential to act as system levers. By combining theory and mathematical modelling, it should be possible to suggest specific parameters of interest. These of course must then be verified empirically and explored with regards to their practicality. We must be able to manipulate usable parameters with real, practical interventions and with a reasonably consistent response. Factors such as speed, magnitude and stability of system response and indirect effects on aspects of system (macro) state other than the target must all be considered.

21.2.2 A Test Case in *Pseudomonas aeruginosa* Biofilms: Engineering a Tragedy of the Commons

To give a practical context, I will discuss the potential example of manipulating evolutionary dynamics in *Pseudomonas aeruginosa* biofilms. *P. aeruginosa* in biofilm form is frequently implicated in chronic infection in immunocompromised patients, in particular, in the cystic fibrosis lung. As such, new methods of managing these communities could have significant clinical relevance. Given that co-operation is costly and difficult to evolve and maintain, but exists and is functionally important in *P. aeruginosa* biofilms, it could be an important target for systems aikido-style interventions in this system. Within the context of such chronic infection biofilms, an initial approach could be to promote cheating. Could we manipulate the evolutionary context such that we could engineer a tragedy of the commons (Hardin 1968)? Could we thus break-up the biofilm or reduce its virulence? The fundamental idea would be to change the conditions such that cheats are favoured in the short-term to the long-term detriment of the whole biofilm. That is to render co-operation selectively unfavourable by changing the environment. We propose to test these ideas in vitro in an experimental co-operator cheat system which models the cystic fibrosis lung colonised by pathogenic *P. aeruginosa* biofilms. Although experiments are in early stages, this method is as yet untested and the example should currently be considered as a thought experiment illustrating the approach.

For this example, the vast majority of the system understanding required to identify potential levers is drawn from the vast pre-existing literature on evolutionary dynamics and on *P. aeruginosa*; well-studied as both a model organism and an opportunistic pathogen. Wild-type *P. aeruginosa* bacteria produce iron-scavenging molecules known as siderophores which they release to the environment. These constitute a public good, as they are individually costly to produce, may be used by any cell in the vicinity including non-producers, but provide fitness benefits when iron is limiting (Chincholkar et al. 2007). Mutant *P. aeruginosa* cheats do not produce the siderophores and can exploit co-operative individuals which produce siderophore molecules that diffuse into the surrounding medium. Within the lung, iron is withdrawn by the host organism in order to restrict bacterial growth and is hence limiting. Host organisms release apotransferrins which bind free iron, siderophore molecules produced by individual bacteria compete with the host apotransferrins to chelate iron and render it absorbable by bacterial cells (Guerinot 1994; Ratledge and Dover 2000). Cheats which freeload rather than producing siderophores have been observed *in vivo* accumulating over time (De Vos et al. 2001), however co-operation appears to be stable. It is known that the proportion of cheats in a bacterial population can affect clinical outcomes via a reduction in public good production. For example, the presence of quorum sensing-defective mutants in clinical isolates can reduce the production of virulence factors and lead to the formation of weak biofilms (Schaber et al. 2004). Availability of iron is important for biofilm formation (Singh et al. 2002) and the presence of non-siderophore producing mutants in *P. aeruginosa* biofilms has been shown to decrease total biofilm mass in vitro (Harrison and Buckling 2009). Given the need for a sufficient siderophore concentration to support the bacterial population, we expect that if cheats came to predominate in the population, the virulence of an infection might be reduced. By manipulating environmental conditions to promote such a population composition we hope to open up new avenues for therapeutic intervention.

Based on our knowledge of the *Pseudomonas aeruginosa* siderophore system, the most obvious possibility would seem to be to remove the need for the public good and thus favour the evolution of cheats which do not pay the metabolic cost of siderophore production. After a short-term intervention to increase the proportion of cheats in the biofilm sufficiently to disrupt its function, we would revert to the original conditions, in which co-operation is required, producing a sudden decrease in overall biofilm fitness and viability and potentially providing a window for other therapies.

The basic outline of our suggested procedure would be to flood the infected site with the public good, in this case siderophores (or available iron), removing need for its production. We would consequently expect cheats to increase in frequency with respect to co-operators within the biofilm population, as they bear no metabolic cost of siderophore production and iron is no longer limiting. Once the community structure has changed to a sufficient degree to weaken the biofilm, iron could be removed rapidly, e.g. using an apotransferrin, renewing the need for a co-operative population. Collapse of the bacterial population could then be possible if adaptation or change in population structure is slow with respect to the negative impact on cheats of lack of public good. If such an intervention is successful in manipulating community composition and disrupting biofilm, it might be sufficient therapy in its own right. Or, perhaps more likely, could provide a window for a second stage of therapy which exploited the biofilm's weakened state with further antimicrobial treatments.

In summary, our suggestion is to change the selective environment so that a, normally evolutionarily unstable, community composition and genetic state we desire is artificially stabilised and hence evolves. We then remove the scaffolding which has allowed the changed population to persist, leaving a maladapted and weakened biofilm vulnerable to further intervention. As our goal in this situation is removal of biofilm, a true win-win scenario is not possible. In other contexts, such as the construction of useful microbial communities or the manipulation of symbiotic human microbiota communities, this two stage process may not be required. Rather we may be able to find an attractor which allows bacteria to stably exist with a community population and genetic composition which suits our requirements.

21.3 Broader Context and Further Work

21.3.1 Bacterial Systems Aikido in a Steering Complex Systems Design Cycle

I have previously discussed moving from design philosophy to a practical methodology for steering complex adaptive systems in socio-economic systems (Penn et al. 2016; Penn Forthcoming). The general framework which I have proposed combines adaptive management methodology (Waltner-Toews and Kay 2005) with participatory modelling and network controllability analysis (Liu et al. 2011) in an ongoing cycle of system intervention and management adaptation as the system adapts in response. A variety of approaches are used throughout this design and adaptive management cycle, including: developing system understanding through modelling and participation; choosing system goals; identification of appropriate points of intervention in specific systems; design of interventions and designing-in of tools and monitoring for adaptive management given unexpected indirect effects and continuous adaptation of components.

The management of living systems and technologies ultimately requires a similar cyclic approach as systems will inevitably adapt in response to interventions. However, in this initial proposal I have focused on one core component of the cycle, identifying appropriate system *dynamics* and *levers* which could be exploited for system manipulation. Rather than searching for key lower-level variables or components of the system to cause cascading change however, I have suggested manipulating particular mechanisms or processes of the bacterial population by changing their evolutionary context. Changing the environment allows indirect system-wide interventions which may be easier to implement than controlling lower-level variables. As we are attempting to co-opt the rapid evolutionary and ecological dynamics of the bacteria for our own ends, this approach also has the potential advantage of being more effective with less effort. Using the system's own capacity to do work, working with rather than against its dynamics, we are employing a systems aikido approach.

Unlike the complex socio-ecological-economic systems to which these ideas have previously been applied, our biofilm systems have the distinct advantages of having a single goal imposed by us, being relatively well-understood and lacking tricky ethical issues if we wish to experiment further to refine our knowledge of their behaviour. Even with these advantages, however, in any real context, a cystic fibrosis lung or a building surface coated with a functional biofilm, the bacteria in interaction with their changing environment have innumerable degrees of freedom and will adapt and change. Constructing and maintaining a dynamic selection pressure via the environment which can maintain our desired biofilm behaviour over time will still require us to enter into interaction with these systems over the long term, engaging in a constant process of system interaction and steering.

21.3.1.1 Developing a Portfolio of Interventions

I have presented a first suggestion for how we might manipulate a particular sort of evolutionary dynamic in a particular organism. However, I consider this to be an example of a broad possible portfolio of tools and methods for steering adaptive systems by altering either the fitness landscape which they experience or the attractor structure of their dynamics. Potential classes of evolutionary interventions include: filling fitness valleys to escape local optima; expansion of basins of attraction of difficult to access, but favourable attractors and manipulating the effective level of selection within the system. All these possibilities remain to be explored theoretically and operationalized for empirical testing in specific systems.

Within *P. aeruginosa* biofilms or other structured microbial populations, other interventions could be designed to indirectly manipulate group structure such that the benefits of co-operation are not predominantly delivered to co-operators. This will require a more developed understanding of precisely how co-operation is maintained in any given system. If, for example, an aggregation and dispersal process were important, we might be able to manipulate key parameters such as time spent in groups, size of group propagules, diffusibility of the public good or relative growth rates of groups with different population structures by changing environmental parameters such as resource levels, medium flow rates or viscosity (Penn et al. 2012).

Different sorts of manipulations will be required dependent on whether we wish to remove problematic biofilm or microbial communities, or to produce functionally useful or less harmful community compositions. Different applications may also be amenable to either active or passive interventions. The former in which we take time-limited or varying actions to steer the system, the latter in which we are able to build contexts into the system to provide a continuous selective pressure that manipulates self-organisation or evolution to a state which we desire. The development of understanding of systems sufficient to allow the discovery of effective levers and the design of selective contexts will require extensive work combining, theory, modelling and empirical testing. Additionally of course, the exploration of the potential challenges and benefits of applying these methods in real world contexts from both practical and ethical perspectives will be crucial in taking the approach forward.

21.4 Summary

The ultimate goal of this programme of work is to incorporate design philosophy and practice into the management of living systems and technologies. The bedrock of this approach is the idea that by understanding systems and intervening thoughtfully with them or their environment, we can co-opt their dynamics for our own purposes. By doing so, we transform the character of our management from protracted input of work and energy to maintain the system in an energetically or evolutionary unstable state, to targeted interactions which change the stability landscape of the system such that our goals correspond with energetically favourable outcomes.

I have presented preliminary ideas on using this so-called *systems aikido* approach in the management of bacterial pathogen biofilms. In particular, I have suggested that given that bacterial co-operation is costly and difficult to evolve and maintain, but exists and is functionally important in biofilms, it could be an important target for systems aikido-style interventions. The maintenance of co-operation is a dynamic process, providing us with a source of system momentum which we can attempt to redirect. If we can determine effective system levers or points of intervention, we should be able to change the context of the evolutionary process such that cheating rather than co-operation is favoured and hence evolves. By indirectly manipulating the relative pay-offs of cheating and co-operation via changing the environment we would hope to be able to change the evolutionary stable state of the system to a population composition dominated by cheats.

To illustrate this idea, I have made a specific suggestion of manipulating a siderophore co-operator-cheat system in pathogenic *Pseudomonas aeruginosa* biofilms by temporarily providing the bacteria with an oversupply of the public good. It is suggested that by thus removing the requirement to produce the siderophore, cheats, which do not pay the metabolic cost of public good production, should come to dominate the biofilm population. If the external supply of the siderophore can then be rapidly removed, this predominance of cheats should weaken and reduce the virulence of the biofilm. Although the selection pressure to co-operate would then be reinstated, this intervention could provide a window for a secondary antimicrobial therapy to act on a weakened biofilm.

This proposed intervention has yet to be tested empirically and clearly, even if successful in vitro, significant further work on the practicality and effectiveness of the approach in a clinical setting would be required. Although these ideas are at early stages, it is hoped that the concept can provide refreshing new ideas and perspectives and inspire work on both theory and practical examples. Ultimately, application of systems design methodology has the potential to provide novel, efficient and effective tools and methods for managing living and life-like systems which embrace rather than fight their complexity.

Acknowledgments Alexandra Penn is funded by the UK Engineering and Physical Sciences Research Council for programme grant EP/H021779/1 (Evolution and Resilience of Industrial Ecosystems (ERIE)). The support of MILES: Mathematics and Modeling in Life and Social Sciences (University of Surrey) for the Novel Approaches to Engineering Microbial Communities project is gratefully acknowledged. Thanks to Claudio Avignone Rossa, Kaylee Herbert, Doug Hodgson, Unal Sumer, Zoe Pounce, Jaime Gonzalez Gutierrez De La Conch, Jose Jimenez, Rebecca Hoyle, Ellen Dowell, Paddi Clark, Richard Watson, Tim Conibear, Jeremy Webb, and Alex Kraaijeveld. Particular thanks go to Bryn Thomas for his inspirational teaching and work in permaculture design.

References

Aminov RI (2010) A brief history of the antibiotic era: lessons learned and challenges for the future. Front Microbiol 1

Bedau MA, McCaskill JS, Packard NH, Rasmussen S (2010) Living technology: exploiting life's principles in technology. Artif Life 16(1):89–97

Bell G (2005) The Permaculture Way. Permanent Publications

Brett MT, Goldman CR (1996) A meta-analysis of the freshwater trophic cascade. Proc Natl Acad Sci 93(15):7723–7726

Buckling A, Harrison F, Vos M, Brockhurst MA, Gardner A, West SA, Griffin A (2007) Siderophore-mediated cooperation and virulence in Pseudomonas aeruginosa. FEMS Microbiol Ecol 62(2):135–141

Chincholkar S, Chaudhari B, Rane M (2007) Microbial siderophores in human and plant health-care. In: Microbial siderophores. Springer, pp 205–217

Chuang JS, Rivoire O, Leibler S (2009) Simpson's paradox in a synthetic microbial system. Science 323(5911):272–275

Costerton JW, Stewart PS, Greenberg E (1999) Bacterial biofilms: a common cause of persistent infections. Science 284(5418):1318–1322

Crespi BJ (2001) The evolution of social behavior in microorganisms. Trends Ecol Evol 16 (4):178–183

De Vos D, De Chial M, Cochez C, Jansen S, Tümmler B, Meyer J-M, Cornelis P (2001) Study of pyoverdine type and production by Pseudomonas aeruginosa isolated from cystic fibrosis patients: prevalence of type II pyoverdine isolates and accumulation of pyoverdine-negative mutations. Arch Microbiol 175(5):384–388

El Mouden C, West SA, Gardner A (2010) The enforcement of cooperation by policing. Evolution 64(7):2139–2152

Fortin D, Beyer HL, Boyce MS, Smith DW, Duchesne T, Mao JS (2005) Wolves influence elk movements: behavior shapes a trophic cascade in Yellowstone National Park. Ecology 86 (5):1320–1330

Frei R, Marzo Serugendo G (2012) The future of complexity engineering. Open Eng 2(2):164–188

Griffin AS, West SA, Buckling A (2004) Cooperation and competition in pathogenic bacteria. Nature 430(7003):1024–1027

Guerinot ML (1994) Microbial iron transport. Annu Rev Microbiol 48(1):743–772

Hall-Stoodley L, Costerton JW, Stoodley P (2004) Bacterial biofilms: from the natural environment to infectious diseases. Nat Rev Microbiol 2(2):95–108

Hardin G (1968) The tragedy of the commons. Science 162(3859):1243–1248

Harrison F, Buckling A (2009) Siderophore production and biofilm formation as linked social traits. ISME J 3(5):632–634

Hauert C, Doebeli M (2004) Spatial structure often inhibits the evolution of cooperation in the snowdrift game. Nature 428(6983):643–646

Holmgren D (2002) Principles and pathways beyond sustanability. Holmgren Design Services, Hepburn

Hopkins R (2008) The transition handbook: from oil dependency to local resilience. Green Books

Kay JJ, Regier HA, Boyle M, Francis G (1999) An ecosystem approach for sustainability: addressing the challenge of complexity. Futures 31(7):721–742

Kreft J-U (2004) Biofilms promote altruism. Microbiology 150(8):2751–2760

Lehmann L, Keller L (2006) The evolution of cooperation and altruism—a general framework and a classification of models. J Evol Biol 19(5):1365–1376

Levin BR, Kilmer WL (1974) Interdemic selection and the evolution of altruism: a computer simulation study. Evolution 527–545

Liu Y-Y, Slotine J-J, Barabási A-L (2011) Controllability of complex networks. Nature 473 (7346):167–173

May RM, Levin SA, Sugihara G (2008) Complex systems: ecology for bankers. Nature 451 (7181):893–895

Michod R, Sanderson M (1985) Behavioral structure and the evolution of cooperation. Evolution: essays in honor of John Maynard Smith, pp 95–104

Mitteldorf J, Wilson DS (2000) Population viscosity and the evolution of altruism. J Theor Biol 204(4):481–496

Mollison B (1988) Permaculture: a designer's manual

Mollison B, Holmgren D (1978) Permaculture one. Transworld Publications, Morebank, NSW Australia

Nowak MA, May RM (1992) Evolutionary games and spatial chaos. Nature 359(6398):826–829

O'Toole GA, Ghannoum MA (2004) Introduction to biofilms: conceptual themes

Parsek MR, Singh PK (2003) Bacterial biofilms: an emerging link to disease pathogenesis. Annu Rev Microbiol 57(1):677–701

Penn AS (Forthcoming) Steering a complex adaptive system: a complexity science design methodology applied to an industrial ecosystem in the Humber region, UK. In: Garcia-Diaz C, Olaya C (eds) Social systems engineering: the design of complexity series in computational and quantitative social science. Wiley

Penn AS, Conibear TC, Watson RA, Kraaijeveld AR, Webb JS (2012) Can Simpson's paradox explain co-operation in Pseudomonas aeruginosa biofilms? FEMS Immunol Med Microbiol 65 (2):226–235

Penn AS, Knight CJK, Chalkias G, Velenturf APM, Lloyd DJB (2016) Extending participatory fuzzy cognitive mapping with a control nodes methodology: a case study of the development of a bio-based economy in the Humber region. In: Gray S, Gray S, Jordan R, Pallisimio M (eds) Environmental modeling with stakeholders. Springer International Publishing AG

Penn AS, Watson RA, Kraaijeveld A, Webb J (2010) Systems aikido—a novel approach to managing natural systems. In: ALIFE, 2010. pp 577–580

Rainey PB, Rainey K (2003) Evolution of cooperation and conflict in experimental bacterial populations. Nature 425(6953):72–74

Rankin DJ, Bargum K, Kokko H (2007) The tragedy of the commons in evolutionary biology. Trends Ecol Evol 22(12):643–651

Ratledge C, Dover LG (2000) Iron metabolism in pathogenic bacteria. Annu Rev Microbiol 54 (1):881–941

Schaber JA, Carty NL, McDonald NA, Graham ED, Cheluvappa R, Griswold JA, Hamood AN (2004) Analysis of quorum sensing-deficient clinical isolates of Pseudomonas aeruginosa. J Med Microbiol 53(9):841–853

Simpson EH (1951) The interpretation of interaction in contingency tables. J R Stat Soc Ser B (Methodol) 238–241

Singh PK, Parsek MR, Greenberg EP, Welsh MJ (2002) A component of innate immunity prevents bacterial biofilm development. Nature 417(6888):552–555

Stewart PS, Costerton JW (2001) Antibiotic resistance of bacteria in biofilms. Lancet 358 (9276):135–138

Waltner-Toews D, Kay J (2005) The evolution of an ecosystem approach: the diamond schematic and an adaptive methodology for ecosystem sustainability and health. Ecol Soc 10(1):38

West SA, Diggle SP, Buckling A, Gardner A, Griffin AS (2007) The social lives of microbes. Annu Rev Ecol Evol Syst 53–77

Whitefield P (2004) Earth care manual: a permaculture handbook for Britain and other temperate climates. Permanent Publications

Wilson DS (1975) A theory of group selection. Proc Natl Acad Sci 72(1):143–146

Wilson DS (1980) The natural selection of populations and communities. Benjamin/Cummings Publishing Company

Wilson DS (1987) Altruism in Mendelian populations derived from sibling groups: the haystack model revisited. Evolution 1059–1070

Erratum to: The Type I Restriction Enzymes as Barriers to Horizontal Gene Transfer: Determination of the DNA Target Sequences Recognised by Livestock-Associated Methicillin-Resistant *Staphylococcus aureus* Clonal Complexes 133/ST771 and 398

Kai Chen, Augoustinos S. Stephanou, Gareth A. Roberts, John H. White, Laurie P. Cooper, Patrick J. Houston, Jodi A. Lindsay and David T.F. Dryden

Erratum to:
Chapter 7 in: M.C. Leake (ed.), *Biophysics of Infection*, Advances in Experimental Medicine and Biology 915, DOI 10.1007/978-3-319-32189-9_7

The original version of this chapter was inadvertently published without the mention of adenine methylation sites for CC1 S. aureus isolates were already determined by the laboratory of Dr. Monk as shown in the MBbio paper in Figure 1. The erratum book and the chapter have been updated with the change.

The updated original online version for this chapter can be found at
DOI 10.1007/978-3-319-32189-9_7

K. Chen · A.S. Stephanou · G.A. Roberts · J.H. White · L.P. Cooper · D.T.F. Dryden (✉)
EaStCHEM School of Chemistry, University of Edinburgh the King's Buildings,
Edinburgh EH9 3JJ, UK
e-mail: david.dryden@ed.ac.uk

P.J. Houston · J.A. Lindsay (✉)
Institute of Infection and Immunity, St George's, University of London,
Cranmer Terrace, London SW17 0RE, UK
e-mail: jlindsay@sgul.ac.uk

Present Address:
K. Chen
Shenyang Research Institute of Chemical Industry, 8 Shenliao Dong Road,
Shenyang Liaoning, People's Republic of China

Present Address:
P.J. Houston
The Pirbright Institute, Ash Road, Pirbright, Woking GU24 0NF, UK

© Springer International Publishing Switzerland 2016

M.C. Leake (ed.), *Biophysics of Infection*, Advances in Experimental Medicine and Biology 915, DOI 10.1007/978-3-319-32189-9_22

The authors apologise for this oversight. The correct information is given below:

7.3.4 Identification of Adenines Targeted for Methylation When There Is Ambiguity in the Target Site

Although the orientation of the site and the respective adenine bases that are subject to methylation is clearly determined for CC133/771-1 and CC398-1, this is not always the case using the outlined approach. For example, in our previous paper we assigned the orientation of CC5-1 (SauN315I) as $ATC(N)_5CCT$ (Roberts et al. 2013), but subsequent work using Type I RM systems with S subunits sharing sequence identity to one of the TRDs in CC5-1 clarified that the recognition site of CC5-1 (SauN315I) should be defined in the opposite orientation as $AGG(N)_5GAT$ (unpublished results). Moreover, in our previous work on CC1-1 and CC1-2 we were unable to assign the adenine base in one half of the site that is subject to methylation because there were two possible locations within each target recognition sequence. We have since determined the methylation site using an ATPase assay. Specifically, hemimethylated DNA duplexes containing the target recognition sites for either CC1-1 or CC1-2 in which one or other of the possible methylation sites was methylated were designed and generated. A DNA duplex of a similar length lacking the target recognition site was used as a negative control. An ATP-coupled assay was then employed to assess which of the alternative DNA duplexes promote ATPase activity after addition of the reconstituted nuclease for CC1-1 or CC1-2, Fig. 7.4. A duplex containing the unmethylated recognition site will promote the ATPase activity of the complete nuclease. Using this approach we were able to verify that the methylation sites for CC1-1 are $CC\underline{A}Y(N)_5T\underline{T}AA$ and those for CC1-2 are $CC\underline{A}Y(N)_6\underline{T}GT$ (underlined bases indicate methylation site either on the A shown or on the A on the complementary strand). These methylation sites have also been confirmed recently using single molecule real-time sequencing methods (Monk et al. 2015).

Abstract The Type I DNA restriction–modification (RM) systems of *Staphylococcus aureus* are known to act as a significant barrier to horizontal gene transfer between *S. aureus* strains belonging to different clonal complexes. The livestock-associated clonal complexes CC133/771 and CC398 contain Type I RM systems not found in human MRSA strains as yet but at some point transfer will occur. When this does take place, horizontal gene transfer of resistance will happen more easily between these strains. The reservoir of antibiotic resistance, virulence and host-adaptation genes present in livestock-associated MRSA will then potentially contribute to the development of newly evolving MRSA clones. The target sites recognised by the Type I RM systems of CC133/771 and CC398 were identified as $C\underline{A}G(N)_5R\underline{T}GA$ and $\underline{A}CC(N)_5R\underline{T}GA$, respectively. Assuming that these enzymes recognise the methylation state of adenine, the underlined A and T bases indicate the unique positions of methylation. Target methylation points for enzymes from CC1 were also identified. The methylation points for CC1-1 are $CC\underline{A}Y(N)_5T\underline{T}AA$ and those for CC1-2 are $CC\underline{A}Y(N)_6\underline{T}GT$ with the underline indicating the adenine methylation site as confirmed recently by Monk et al. (2015) (MBio 6:e00308–15, 2015), which finally clears up the ambiguity noted previously (Roberts et al. 2013, Nucleic Acids Res 41:7472–7484) for the half sites containing two adenine bases.

The original article was corrected.

Index

© Springer International Publishing Switzerland 2016

M.C. Leake (ed.), *Biophysics of Infection*, Advances in Experimental Medicine and Biology 915, DOI 10.1007/978-3-319-32189-9

MIX
Papier aus verantwortungsvollen Quellen
Paper from responsible sources
FSC® C105338

If you have any concerns about our products,
you can contact us on
ProductSafety@springernature.com

In case Publisher is established outside the EU,
the EU authorized representative is:
Springer Nature Customer Service Center GmbH
Europaplatz 3, 69115 Heidelberg, Germany

Printed by Libri Plureos GmbH
in Hamburg, Germany